VOLUME SEVEN HUNDRED AND THIRTY FIVE

METHODS IN ENZYMOLOGY

Thiamine and Thiamine-Pyrophosphate (TPP) Riboswitches

METHODS IN ENZYMOLOGY

VOLUME SEVEN HUNDRED AND THIRTY FIVE

METHODS IN ENZYMOLOGY

Thiamine and Thiamine-Pyrophosphate (TPP) Riboswitches

Edited by

FRANCK CHARMANTRAY
Researcher in Biocatalysis, CNRS, France

BASTIEN DOUMÈCHE
Lyon 1 University Claude Bernard, France

ELSEVIER

ACADEMIC PRESS
An imprint of Elsevier

Academic Press is an imprint of Elsevier
50 Hampshire Street, 5th Floor, Cambridge, MA 02139, United States
125 London Wall, London, EC2Y 5AS, United Kingdom

First edition 2026

ISBN: 978-0-443-47328-9
ISSN: 0076-6879

For information on all Academic Press publications
visit our website at https://www.elsevier.com/books-and-journals

Publisher: Zoe Kruze
Editorial Project Manager: Saloni Vohra
Production Project Manager: James Selvam
Cover Designer: Muthukumaran Thangaraj

Typeset by MPS Limited, India

Contents

Contributors

Jean Alric
CEA, CNRS, BIAM, Photosynthesis & Environment, Aix Marseille University, Saint Paul-Lez-Durance, France

Esther R. Angert
Department of Microbiology, Cornell University, Ithaca, NY, United States

Ulschan Bathe
Biotechnology of Horticultural Crops, School of Life Sciences, Technical University of Munich, Freising, Germany

Lucien Bettendorff
Laboratory of Neurophysiology, GIGA Institute, University of Liège, Liège, Belgium

Gerard Carrera i Cardona
Lab for Nanobiology, Department of Chemistry, KU Leuven, Leuven, Belgium

Adrien Chauvier
Department of Biology, Faculty of Science, RNA Group, Université de Sherbrooke, Sherbrooke, QC, Canada

Keonny Cosek
Department of Botany and Plant Pathology, Oregon State University, Corvallis, OR, United States

Nicholas Cox
Department of Microbiology and Cell Science, Institute of Food and Agricultural Science, University of Florida, Gainesville, FL, United States

Stefan Crielaard
Institute for Molecules and Materials, Radboud University, Nijmegen, The Netherlands

Edmar Ramos de Oliveira Filho
Department of Horticultural Sciences, University of Florida, Gainesville, United States

Naiara Almeida de Toledo
Departamento de Botânica, Instituto de Biociências, Universidade de São Paulo, Sao Paulo, SP, Brazil

Henrique Moura Dias
Department of Agronomy, Horticulture and Plant Science, South Dakota State University, Brookings, SD, United States

Jan P. Dobert
BIOS Lab on a Chip Group, MESA+ Institute for Nanotechnology, Technical Medical Centre, University of Twente, Enschede, The Netherlands

Katie A. Edwards
Department of Pharmaceutical Sciences, Binghamton University, Binghamton; Department of Microbiology, Cornell University, Ithaca, NY, United States

Teresa B. Fitzpatrick
Vitamins & Environmental Stress Responses in Plants, Department of Plant Sciences, University of Geneva, Geneva, Switzerland

Sarah A. Flowers
Department of Neuroscience, University of Virginia, Charlottesville, VA, United States

Elizabeth L. Frank
Department of Pathology, University of Utah Health, Salt Lake City, UT, United States

J. Donato García-García
Biotechnology of Bioengineering, Tecnológico de Monterrey, Guadalajara, Mexico

Anuran K. Gayen
Department of Horticultural Sciences, University of Florida, Gainesville, United States

Gary E. Gibson
Brain and Mind Research Institute, Weill Cornell Medicine, New York, NY, United States

Aymeric Goyer
Department of Botany and Plant Pathology, Oregon State University, Corvallis, OR, United States

Daniel B. Haack
Department of Biochemistry and Molecular Biophysics, University of California, San Diego, La Jolla, CA, United States

David Handy
Department of Botany and Plant Pathology, Oregon State University, Corvallis, OR, United States

Mirette Hashem
Department of Biology, Faculty of Science, RNA Group, Université de Sherbrooke, Sherbrooke, QC, Canada

Anna Hübenthal
Department of Biotechnology, Institute for Technical Microbiology, Technische Hochschule Mannheim, Mannheim, Germany

Kerry S. Jones
Nutritional Biomarker Laboratory, Institute of Metabolic Science – Epidemiology, University of Cambridge, Cambridge, United Kingdom

Dimitrios Kaloudas
Laboratory of Synthetic Biology and Bioinformatics, Faculty of Biology, Sofia University "St. Kliment Ohridski", Sofia, Bulgaria

Albert Koulman
Nutritional Biomarker Laboratory, Institute of Metabolic Science - Epidemiology, University of Cambridge, Cambridge, United Kingdom

Clifford E. Kraft
Department of Natural Resources and the Environment, Cornell University, Ithaca, NY, United States

Daniel A. Lafontaine
Department of Biology, Faculty of Science, RNA Group, Université de Sherbrooke, Sherbrooke, QC, Canada

Florian L.R. Lucas
BIOS Lab on a Chip Group, MESA+ Institute for Nanotechnology, Technical Medical Centre, University of Twente, Enschede, The Netherlands

Matthias Mack
Department of Biotechnology, Institute for Technical Microbiology, Technische Hochschule Mannheim, Mannheim, Germany

Madison M. Maille
Department of Chemistry, University of California, San Diego, La Jolla, CA, United States

Julie A. Maupin-Furlow
Department of Microbiology and Cell Science, Institute of Food and Agricultural Science; Genetics Institute, University of Florida, Gainesville, FL, United States

Damon A. Parkington
Nutritional Biomarker Laboratory, Institute of Metabolic Science - Epidemiology, University of Cambridge, Cambridge, United Kingdom

Nikolet Pavlova
Laboratory of Synthetic Biology and Bioinformatics, Faculty of Biology, Sofia University "St. Kliment Ohridski", Sofia, Bulgaria

Robert Penchovsky
Laboratory of Synthetic Biology and Bioinformatics, Faculty of Biology, Sofia University "St. Kliment Ohridski", Sofia, Bulgaria

J. Carlos Penedo
Laboratory for Biophysics and Biomolecular Dynamics, SUPA School of Physics and Astronomy; Laboratory for Biophysics and Biomolecular Dynamics, Biomedical Sciences Research Complex, School of Biology, Centre of Biophotonics, University of St. Andrews, St Andrews, United Kingdom

Eileen A. Randall
Department of Natural Resources and the Environment, Cornell University, Ithaca, NY, United States

Sofia B. Relvas
BIOS Lab on a Chip Group, MESA+ Institute for Nanotechnology, Technical Medical Centre, University of Twente, Enschede, The Netherlands

Boris Rudolfs
Department of Biochemistry and Molecular Biophysics, University of California, San Diego, La Jolla, CA, United States

Patrick St-Pierre
Department of Biology, Faculty of Science, RNA Group, Université de Sherbrooke, Sherbrooke, QC, Canada

Navtej Toor
Department of Biochemistry and Molecular Biophysics, University of California, San Diego, La Jolla, CA, United States

Martina Traykovska
Laboratory of Synthetic Biology and Bioinformatics, Faculty of Biology, Sofia University "St. Kliment Ohridski", Sofia, Bulgaria

Daphne A.L. van den Homberg
Institute for Molecules and Materials, Radboud University, Nijmegen, The Netherlands

Kristen Van Gelder
Agriculture and Agri-Food Canada, Saskatoon, SK, Canada

Marie-Anne Van Sluys
Departamento de Botânica, Instituto de Biociências, Universidade de São Paulo, Sao Paulo, SP, Brazil

Willem A. Velema
Institute for Molecules and Materials, Radboud University, Nijmegen, The Netherlands

Kai Wang
Faculty of Synthetic Biology & Institute of Emerging Agricultural Technology, Shenzhen University of Advanced Technology, Shenzhen, P.R. China; Vitamins & Environmental Stress Responses in Plants, Department of Plant Sciences, University of Geneva, Geneva, Switzerland

Katherine R. Weber
Department of Microbiology and Cell Science, Institute of Food and Agricultural Science, University of Florida, Gainesville, FL, United States

Kyly C. Whitfield
Department of Applied Human Nutrition, Mount Saint Vincent University, Halifax, Nova Scotia, Canada

Carsten Wloka
BIOS, Lab on a Chip group, MESA+ Institute for Nanotechnology, Technical Medical Centre, University of Twente, Enschede, The Netherlands

Patricia C. Wolfe
Department of Pharmaceutical Sciences, Binghamton University, Binghamton, NY, United States

Preface

Thiamine (vitamin B_1) was first recognized through the seminal work of T. Kanehiro and C. Eijkman, as the factor responsible for preventing beriberi, a neurological disease caused by malnutrition.[1] Thiamine deficiency occurs in poorly diversified diets, particularly those rich in refined carbohydrates (e.g., polished rice or refined flour). It is also degraded by an enzyme (thiaminase) in raw fish, explaining why beriberi was more prevalent in Asiatic countries at the end of the 19th and the beginning of the 20th century. Following an intense competition involving several research groups at the beginning of the last century, its structure was first determined by R.R. Williams, R.E. Waterman, and J.C. Keresztesy in 1933 and its chemical synthesis was achieved in 1936 by R.R. Williams and J.K. Cline.[2] The name thiamine was coined by this group from *thio-*, (reflecting the sulfur-containing thiazole ring), and *vitamine*. The history of thiamin discovery and synthesis was elegantly reviewed by Heinz M. Wuest as early as 1962[3] and is now the topic of numerous books and online open-access resources (e.g., Wikipedia). Later, thiamine deficiency was found to be involved in Wernicke–Korsakoff syndrome, optic neuropathy or Leigh's disease, among others.[4]

Since then, the field of thiamine research has expanded considerably, encompassing not only medical studies but also major advances in biochemistry, including the elucidation of its biosynthesis in bacteria[5] and plants,[6] the identification of its phosphorylated derivatives,[7] the most

[1] C. Eijkman, C. Eine Beri Beri-ähnliche Krankheit der Hühner. *Archiv f. pathol. Anat.* **1897**, 148, 523–532. https://doi.org/10.1007/bf01937576.

[2] R.R. Williams, J.K. Cline. Synthesis of vitamin B_1. *J. Am. Chem. Soc.* **1936**, 58, 8, 1504–1505. https://doi.org/10.1021/ja01299a505.

[3] H.M. Wuest. The history of thiamine. Ann. N. Y. Acad. Sci. **1962**, 98, 385-400. https://doi.org/10.1111/j.1749-6632.1962.tb30561.x.

[4] a) N.J.M. Arts, S.J.W. Walvoort, R.P.C. Kessels. Korsakoff's syndrome: a critical review. *Neuropsychiatr. Dis. Treat.* **2017**, 13, 2875–2890. https://doi.org/10.2147/ndt.s130078; b) S.M. Gratton, B.L. Lam. Visual loss and optic nerve head swelling in thiamine deficiency without prolonged dietary deficiency. *Clin. Ophthalmol.* **2014**, 8, 1021–1024 (Case report). https://doi.org/10.2147/opth.S64228; c) E.C. Dooling, E.P. Richardson. Ophthalmoplegia and ondines curse. *Arch. Ophthalmol.* **1977**, 95, 1790–1793.

[5] T.P. Begley, A. Chatterjee, J.W. Hanes, A. Hazra, S.E. Ealick. Cofactor biosynthesis—still yielding fascinating new biological chemistry. Curr. Opin. Chem. Biol. **2008**, 12, 118–125. https://doi.org/10.1016/j.cbpa.2008.02.006.

[6] A. Goyer. Thiamine in plants: Aspects of its metabolism and functions. Phytochem. **2010**, 71, 1615–1624. https://doi.org/10.1016/j.phytochem.2010.06.022.

[7] L. Bettendorff. Update on thiamine triphosphorylated Derivatives and Metabolizing Enzymatic Complexes. *Biomolecules*. 2021, 11, 1645. https://doi.org/10.3390/biom11111645.

important of which is thiamine pyrophosphate (TPP), and, more recently, the discovery of TPP riboswitches in bacteria.[8]

It has now been nearly thirty years since a *Methods in Enzymology* volume was devoted to thiamine. In 1997, 9 chapters focusing on thiamine among other cofactors were published in volume 279 of *Methods in Enzymology: Vitamins and Coenzymes part I*, edited by Donald B. McCormick, John W. Suttie, Conrad Wagner.[9] Given the central role of thiamine in metabolism, its function as the precursor of TPP, and the remarkable advances made in molecular biology, bioinformatics, plant biotechnology, analytical chemistry, enzymatic assays, nanopore technologies, cryo-EM, and genetics, the time has come to bring thiamine and its analogues back into the spotlight.

When we began serving as editors for this volume dedicated to thiamine, we attempted to contact Dr. Andrew Hanson from the University of Florida, because of his major contributions to our understanding of thiamine biosynthesis in plants.[10] We were deeply saddened to learn that Dr. Hanson had passed away only a few months earlier. We would like to dedicate this volume to him, beginning with Chapter 1, contributed by Kristen Van Gelder and Dr. Hanson's former collaborators. This chapter describes the molecular approaches for evolving thiazole synthase in plants. Kristen Van Gelder also kindly agreed to share a few words with us on her former supervisor:

"Dr. Andrew Hanson was a pioneer in the study of B vitamins, specifically folate, and later on, thiamin. He was a strong proponent of using comparative genomics to parse out genes in plant biochemical pathways, which led to the identification of plant thiamin transporters, kinases, hydrolases, and phosphatases. His ability to tackle these large genomic databases and identify sequences of importance,

[8] a) W. Winkler, A. Nahvi, R.R. Breaker. Thiamine derivatives bind messenger RNAs directly to regulate bacterial gene expression. Nature. **2002**, 419, 952–956. https://doi.org/10.1038/nature01145; b) A.S. Mironov, I. Gusarov, R. Rafikov, L.E. Lopez, K. Shatalin, R.A. Kreneva, D.A. Perumov, E. Nudler. Sensing small molecules by nascent RNA: a mechanism to control transcription in bacteria. Cell. **2002**, 111, 747–756. https://doi.org/10.1016/S0092-8674(02)01134-0.

[9] Donald B. McCormick, John W. Suttie, Conrad Wagner. **1997**. *Methods in Enzymology: Vitamins and Coenzymes part I*. Academic Press (Cambridge). ISBN: 978-0-12-182180-7.

[10] a) S. Gerdes, C. Lerma-Ortiz, O. Frelin, S.M.D. Seaver, C.S. Henry, V. Crécy-Lagard, A.D. Hanson. Plant B vitamin pathways and their compartmentation: a guide for the perplexed. *J. Exp. Bot.* **2012**, 63, 2, 695–709. https://doi.org/10.1093/jxb/ers208; b) A.D. Hanson, J.S. Amthor, J. Sun, T.D. Niehaus, J.F. Gregory, S.D. Bruner, Y. Ding. Redesigning thiamin synthesis: Prospects and potential payoffs. Plant Sci. **2018**, 273, 92–99. https://doi.org/10.1016/j.plantsci.2018.01.019; c) A.D. Hanson, G.A. Beaudoin, D.R. McCarty, J.F. Gregory II. Does abiotic stress cause functional B vitamin deficiency in plants? Plant Physiol. **2016**, 172, 2082–2097, https://doi.org/10.1104/pp.16.01371.

resulted in the discovery of catalytic thiazole synthases (THI4s) in cereal crops, wherein historically, only suicidal THI4s had been identified. Dr. Hanson was also a very early advocate of plant synthetic biology and an early adapter of directed evolution, using the yeast continuous directed evolution platform, OrthoRep, to evolve bacterial THI4s to function under plant-like conditions."

In Chapter 2, Marie-Anne Van Sluys, Henrique Moura Dias et coll. present a detailed modular bioinformatics workflow for the comparative analysis of thiamine biosynthesis genes across Bacteria, Archaea, and Eukarya. The next two chapters are dedicated to thiamin in plants. Goyer and coll. provide methods to increase the levels and stability of thiamin in white rice using molecular biology approaches (Chapter 3) while Wang et coll. focus on thiamine triphosphate and other thiamine derivatives in chloroplasts (Chapter 4). This first section on thiamine metabolism concludes with the Chapter 5 by Maupin-Furlow et coll. who study the underexplored thiamine biosynthesis in Archaea using a panel of molecular methods.

As previously mentioned, thiamine and its phosphorylated derivatives must be quantified in biological samples. Therefore, the second section of this volume describes assays and sample preparation protocols to accurately quantify these compounds. Whitfield and coll. describe the different biomarkers and methods of assessment of thiamine in human blood samples (Chapter 6). Jones and coll. (Chapter 7) focuses on the detection of thiamine pyrophosphate in human brain samples as a post-mortem marker of neurological symptoms and Bettendorff describes an HPLC method to detect thiamine pyrophosphokinase activity (Chapter 8). Edwards and coll. describe the measurement of Thiaminase I activity, an enzyme associated with thiamine deficiency in ecological and veterinary contexts (Chapter 9). The final two chapters of this section describe a completely different approach to quantify thiamine using nanopores. Wloka, Lucas, and coll. first describe the assembly of the nanopores sensing system with data acquisition (Chapter 10), then they explain in detail how these data should be analyzed (Chapter 11).

The last section of this *Methods in Enzymology* volume contains chapters devoted to thiamine riboswitches, regulatory RNA elements that sense thiamine metabolites and control gene expression. The structure elucidation of TPP-riboswitches using high resolution cryo-EM by Toor and coll. (Chapter 12) opens this section, providing structural insights into these highly dynamic macromolecules. Methodologies to study riboswitches are proposed, first by Penchovsky and coll. who demonstrate the effectiveness of chimeric antisense oligonucleotides conjugated with cell-penetrating peptide (Chapter 13), then by Mack and Hübenthal using a specialized dual-luciferase reporter gene assay (Chapter 14). Velema and coll. propose the use of photoaffinity probes based on thiamine monophosphate to identify new thiamine (or TPP) riboswitches. Lafontaine, Penedo and coll. conclude this volume by explaining how single-molecule fluorescence resonance energy transfer can be used with TPP riboswitch to monitor RNA folding within transcription complexes.

It was nearly two years ago that David Christianson contacted us to start this collaborative publishing project with volume 722 of *Methods in Enzymology*, now extended with volumes 735 & 736. This long but pleasant work is now over, and it has been a unique opportunity to work with all the contributors of this volume, from such different research fields. As co-editors of this volume 735 dedicated to thiamine, we would like to thank them all without whom this volume could not have been published. We cannot thank them enough for the time and energy they put into this collaborative project. What we did not initially anticipate is the remarkable importance of thiamine in all these research fields beyond our own areas of expertise. It appears that editing this volume was also a challenge as it is virtually impossible to cover all topics and related methods to study thiamine-dependent systems. Therefore, we encourage you to also consult two other volumes of *Methods in Enzymology* covering enzymes that use thiamine pyrophosphate as a cofactor: Volume 722: Transketolases and Volume 736: TPP-dependent enzymes.

We also express our gratitude to Dr. David Christianson and Dr. Karen Allen, the Editors-in-Chief of *Methods in Enzymology* for these 2 years of collaboration. We are grateful for the confidence you placed in us for these new volumes of the series. This work could simply not be brought to completion without the outstanding coordination provided by Zoe Kruze and the unfailing support of Saloni Vohra. We sincerely thank you for your kindness and availability.

We now invite readers to explore the chapters that follow and discover the wide range of methodologies presented by the authors for investigating the chemistry, biology, and metabolism of vitamin B_1.

SECTION 1

Thiamine metabolism and assays

CHAPTER ONE

Evolving thiazole synthases for expression in plants

Edmar Ramos de Oliveira Filho[a], Anuran K. Gayen[a], J. Donato García-García[b], Ulschan Bathe[c], and Kristen Van Gelder[d,*]
[a]Department of Horticultural Sciences, University of Florida, Gainesville, United States
[b]Biotechnology of Bioengineering, Tecnológico de Monterrey, Guadalajara, Mexico
[c]Biotechnology of Horticultural Crops, School of Life Sciences, Technical University of Munich, Freising, Germany
[d]Agriculture and Agri-Food Canada, Saskatoon, SK, Canada
*Corresponding author. e-mail address: kristen.vangelder@agr.gc.ca

Contents

Abstract

Thiazole synthases (THI4s) catalyze the synthesis of the thiazole moiety of thiamin (vitamin B_1) from glycine, NAD, and a sulfur atom. The sulfur atom can originate from an active site cysteine residue, as is the case for cysteine-dependent THI4s (cys-THI4s), or from an external sulfide source, as it is for sulfide-dependent THI4s (noncys-THI4s). Cys-THI4s are also known as suicidal enzymes, as catalysis leads to enzyme inactivation, followed by protein degradation. These enzymes can be considered energy intensive and catalytically inefficient enzymes. Replacement of plant cys-THI4s with noncys-THI4s which are not self-inactivating though still slow catalysts, could lead to

Methods in Enzymology, Volume 735
ISSN 0076-6879, https://doi.org/10.1016/bs.mie.2026.06.013

energy conservation through decreased protein turnover. However, noncys-THI4s are anaerobic or microoxic enzymes and thus would be poor replacements for plant cys-THI4s that are aerobic enzymes. Engineering a plant cys-THI4 to be a true catalyst or a noncys-THI4 to function in plant-like conditions would be necessary to achieve safe respiratory energy. In this chapter, we will provide methods outlining: (1) how to use comparative genomics to select the best candidate THI4s to be engineered via rational design or continuous directed evolution (CDE); (2) how to use rational design to engineer THI4s; and (3) how to use OrthoRep, a yeast-based CDE tool, to engineering noncys-THI4s to function under plant-like conditions.

1. Introduction

In fungi, plants, and some archaea and bacteria, the thiazole moiety of thiamin (vitamin B_1) is catalyzed by an aerobic cysteine-dependent thiazole synthase (cys-THI4), a self-inactivating, single-turnover enzyme, by combining NAD and glycine alongside a sulfur atom from an active site cysteine residue (Dorrestein et al., 2004; Chatterjee et al., 2011; Hwang et al., 2014; Zhang et al., 2016; Joshi et al., 2020). In some archaea and bacteria as well as in the grain of cereals, thiazole is synthesized from an anaerobic or microoxic sulfide-dependent thiazole synthase (noncys-THI4), that acquires the sulfur atom from an external sulfide source, thus avoiding self-inactivation. Noncys-THI4s can be considered truly catalytic enzymes but generally complete only a few reaction cycles before the enzyme is degraded (Eser et al., 2016; Joshi et al., 2020; Joshi et al., 2021). From a bioenergetics perspective, THI4s are costly enzymes due to their high turnover, yet essential, as thiamin is a key cofactor in several reactions (e.g., pyruvate decarboxylase, transketolase, pyruvate oxidase) across all kingdoms of life (Muller et al., 1993; Hanson et al., 2018). To free up approximately 2–4 % of the plant energy budget, plant cys-THI4s could be redesigned to be non-suicidal like noncys-THI4s in cereal grain, while still functioning under aerobic conditions (Hanson et al., 2018; Joshi et al., 2020). Alternatively, noncys-THI4s could be evolved to function under aerobic conditions, replacing the plant cys-THI4. This increased effectiveness in the plant energy budget could be put towards other metabolic processes like increased biomass and/or yield, or supporting industrially relevant, potentially new-to-nature metabolic pathways.

There are three tools we propose that can be used in combination to determine (1) what genes to redesign, (2) what key active site residues may be substituted, and (3) what residues beyond the active site may be changed. As the first step, we recommend using comparative genomics across a diverse range of gene homologs to identify unique homologs that may be well placed for engineering. Second, we recommend using rational design which uses known properties of the protein structure and catalytic mechanism to help determine residues that may be best suited for substitution to either increase catalytic activity, prevent inhibition, or alter other enzymatic properties (e.g., thermotolerance, oxygen-tolerance, etc.). As a third step, we recommend directed evolution (DE). DE uses random mutagenesis across the whole gene or in small distinct sections of the gene to generate a mutant library. Mutants can then be selected for in vivo under a desired selection pressure or in vitro, generally using a qualitative high-throughput assay. Improved variants can be achieved after multiple rounds of mutagenesis and selection (Kuchner & Arnold, 1997). Several DE systems exist and have been described and reviewed elsewhere (e.g., Packer & Liu, 2015; Wang et al., 2021). Continuous directed evolution (CDE) advances DE a step further, enabling both hyper mutagenesis and selection to occur in vivo, and should conceivably allow the gene of interest (GOI) to be evolved indefinitely, providing time and space for multi-mutational pathways to be transversed and increasing the likelihood at which superior mutations are obtained and fixed (Zhong & Liu, 2019). There are several CDE systems to choose from including OrthoRep, BacORep, PACE, EvolvR, and eMutaT7, among others (Esvelt et al., 2011; Ravikumar et al., 2014; Halperin et al., 2018; Park & Kim, 2021; Tian et al., 2023).

In the method below, we show how comparative genomics, rational design, and the CDE system OrthoRep, can be used to choose what *THI4* genes to evolve, and which amino acid residues can be substituted in bacterial THI4s for functioning under aerobic conditions – a requirement of their expression in plants.

2. Comparative genomics

2.1 Comparing *THI4* genes

2.1.1 Before you begin: Protein sequences

Recover the protein sequence of interest from curated databases (e.g., NCBI) and prepare the amino acid sequence in 1-letter format for downstream analyses. When looking for microbial *THI4s,* the query sequence used was the THI4 protein sequence from Deltaproteobacteria bacterium RBG 16_54_11 (GenBank OGP80537).

2.1.2 Sequence analysis pipeline in IMG/M (JGI)

The steps below and Fig. 1, describe how to identify and obtain information about homologous genes and examine gene-neighborhood context for a protein of interest across different genome and metagenome datasets using the Integrated Microbial Genomes & Microbiomes (IMG/M) from the Joint Genome Institute (JGI).

1. Navigate to the IMG/M webpage: https://img.jgi.doe.gov/cgi-bin/m/main.cgi
2. From the top menu, hover over "Search" and click on "All Isolate Genomes" under the BLAST section.
3. Paste the 1-Letter Amino Acid protein sequence of interest into the sequence field. At this step, the E-value parameter may be set to 1e−20 to restrict results to more reliable hits.

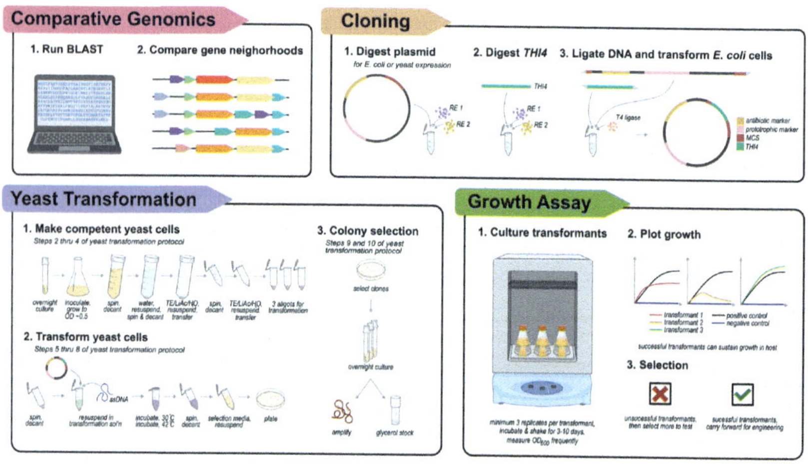

Fig. 1 Schematic of protocols to compare genes and genomes. Relevant sequences are identified through a BLAST search using a known sequence and gene neighborhoods across species can be used to identify unknown/non-annotated genes, hypothesize gene function based on nearby genes, or to identify genes belonging to the same biosynthetic pathway. Bacterial open reading frames often contain genes all belonging to one biosynthetic pathway. Candidate genes can be cloned into a desired plasmid, and transformed into *E. coli* or *S. cerevisiae*. Growth assays can be completed to determine complementarity in the host when the native host gene is knocked out. Growth assays can be run over three to ten days (depending on host) and OD_{600} can be measured frequently (every 8–24 h) and plotted. Complementing genes can be carried forward for engineering. If no genes completement, then new genes can be selected for testing.

4. Click "Run BLAST". The analysis usually requires a few minutes to complete.
5. The Blast results page displays different (meta)genomes in which similar sequences were hit. The E-value and percent identity scores are shown in the rightmost columns.
6. To access detailed information for a hit of interest, click on the corresponding "Gene ID". A new page will open with metadata such as the predicted gene product name, source genome, genomic coordinates, EC number, etc.
7. The section under "Evidence for protein prediction" displays the gene-neighborhood arrangement. Open reading frames (ORFs) are represented as light yellow arrows; rolling the cursor over an arrow reveals the annotated protein name. To compare neighborhood organization across related genomes, click "Show neighborhood regions with the same top COG hit (via top homolog)". This will open a new table
8. The Gene Neighborhoods page initially shows five genomes. Increase the number of genomes to the maximum and click "Redisplay". Each neighborhood highlights the marker (query) gene in red, and conserved neighboring genes are represented as colored arrows.
9. Hover over each arrow to view the annotated protein name. Click on an arrow to view detailed information.
10. To retrieve the sequence of a protein of interest, click on "Amino Acid Sequence Length" field (#aa). Functional domains should be verified using tools such as the 'NCBI Conserved Domain Search' https://www.ncbi.nlm.nih.gov/Structure/cdd/wrpsb.cgi. Save the protein sequences of interest for subsequent analysis.
11. Retrieved sequences can be used for various *in silico* analyses, such as phylogenetic reconstruction (Dereeper et al., 2008), solubility (Slabinski et al., 2007), and structure prediction (Jumper et al., 2021). Selected candidates may be codon-optimized for heterologous expression in *Escherichia coli* or *Saccharomyces cerevisiae* and engineered via rational design or directed evolution (Fig. 1).

2.2 Plant cys- and noncys-THI4s

A search for plant *THI4* genes across plant genomes revealed the presence of the conserved cysteine residue in plant *THI4s,* indicating that they are cys-THI4s and suicidal enzymes (Sun et al., 2019). However, in several cereal crops (oat, barley, wheat, maize), there are two *THI4* genes, one matching the canonical cys-*THI4s* present in yeast and other plant genomes, and a paralog that lacks the conserved active site cysteine and instead has histidine or aspartate at this position (Joshi et al., 2020). The cereal *THI4* paralogs lacking the conserved active site cysteine are also expressed under microoxic conditions found in developing grains and they successfully complemented an *E. coli* Δ*thiG* mutant (THIG is the *E. coli* functional analog of THI4) in the presence of an exogenous sulfur source (Rolletschek et al., 2004; Joshi et al., 2020).

Sequence analysis across a diverse set of cys- and noncys-*THI4s* revealed that, apart from the substitution of the active site cysteine, noncys-*THI4s* exhibit distinguishing features relative to plant cys-*THI4s*, including six substitutions at positions otherwise fully conserved in plants, one- or two-residue insertions after specific residues, and the absence of an approximately 20-residue C-terminal portion conserved in monocot and eudicot cys-*THI4s* (Joshi et al., 2020). Given the overall similarity between plant cys- and noncys-*THI4s* at the nucleotide level (~70 % sequence identity) and amino acid level (~72 % sequence identity), it appears feasible that one to several amino acid changes in cys- or noncys-*THI4s* may (1) make a suicidal enzyme truly catalytic, and (2) allow anaerobic enzymes to function under aerobic conditions. To this end, plant cys-*THI4s* could be engineered for true catalysis using rational design to swap the key active site residue (see Section 3.2), and plant noncys-*THI4s* could be evolved via continuous directed evolution for aerotolerance, as understanding what residues may be involved cannot be rationally determined based on sequence or protein structure alone (see Section 4.3).

2.3 Bacterial cys- and noncys-THI4s

Microbial THI4s lacking the active site cysteine, generally have a histidine residue at this position and are called noncys-THI4s or sulfide-dependent THI4s as they acquire the sulfur atom from an external sulfide (HS^-) source (Joshi et al., 2021). Noncys-THI4s are generally found in microorganisms inhabiting anoxic, sulfide-rich environments (e.g., hydrothermal vents, leech guts, and marine mud) and in microaerophilic and aerotolerant microorganisms, in environments exposed to lower sulfide levels (Eser et al., 2016; Joshi et al., 2021). Several noncys-THI4s with diverse ecophysiology complemented the *E. coli* Δ*thiG* mutant under aerobic conditions with an external sulfide source, indicating no apparent ecophysiological linkage between THI4s that could and could not tolerate oxygen (Sun et al., 2019; Joshi et al., 2021). Three of these diverse bacterial *THI4* genes were selected for directed evolution: *Thermovibrio ammonificans*, an anerobic thermophile, *Mucinivorans hirdunis*, an anerobic mesophile, and *Saccharicrinis fermentans*, a facultative aerobic mesophile. *M. hirduns* THI4 (*Mh*THI4) and *S. fermentans* (*Sf*THI4) both have histidine replacing the active site cysteine, and *T. ammonificans* (*Ta*THI4) has methionine (see Section 4, Joshi et al., 2021). However, it was later learned that a better metric for finding successful candidates for directed evolution was a combination of the average oxidation state of carbon (Z_C), and the presence of cytochrome oxidase-encoding sequences, which identified noncys-THI4 candidates from aerobic, mild-temperature prokaryotes suitable for directed evolution (Campos-Silva et al., 2026).

3. Rational design of THI4s

3.1 Constructing designer THI4s

3.1.1 Before you begin

Ensure that a crystal structure of your enzyme is available or that of a similar protein that can be used to model the structure of your enzyme (Fig. 2). Software is available online to create protein models (AlphaFold2, Phyre2, SWISS-MODEL, etc.). In the case of plant cys-THI4s, the *A. thaliana* THI4 (*At*THI4, PBD: 1RP0, Uniprot: Q38814) crystal structure has been published as have several noncys-THI4s (e.g., *Ta*THI4, PBD: 7RK0, Uniprot: E8T346).

3.1.2 Designing and cloning of THI4 sequences

1. Determine the amino acid residue to be altered and what amino acid will be the substitute. *At*THI4 has the conserved active site cysteine at position 216 and will be substituted for either histidine, alanine, or aspartate.
2. The three new *AtTHI4* sequences (Table 1) were codon optimized for expression in yeast, synthesized by a third party, amplified with primers *At*THI4_*Eco*RI_F (5′-GAATACGGGTTCGAATTCATGACCGCTGGTTACGACTTGAACG−3′) and *At*THI4_*Sph*I_R (5′-GCTGGTGATCACGCATGCTTAAGCATCTACGGTTTCAGCTGAATC−3′) and cloned into the ArEC plasmid under control of the TDH3 promoter and ADH1 terminator via restriction enzyme cloning with *Eco*RI (NEB Cat. No. R3101S) and *Sph*I (NEB Cat. No. R3182S). Constructs were transformed into *E. coli* TOP10 cells for plasmid generation and verified by Sanger sequencing (Fig. 2). Media for selecting positive transformants is listed in Table 2.

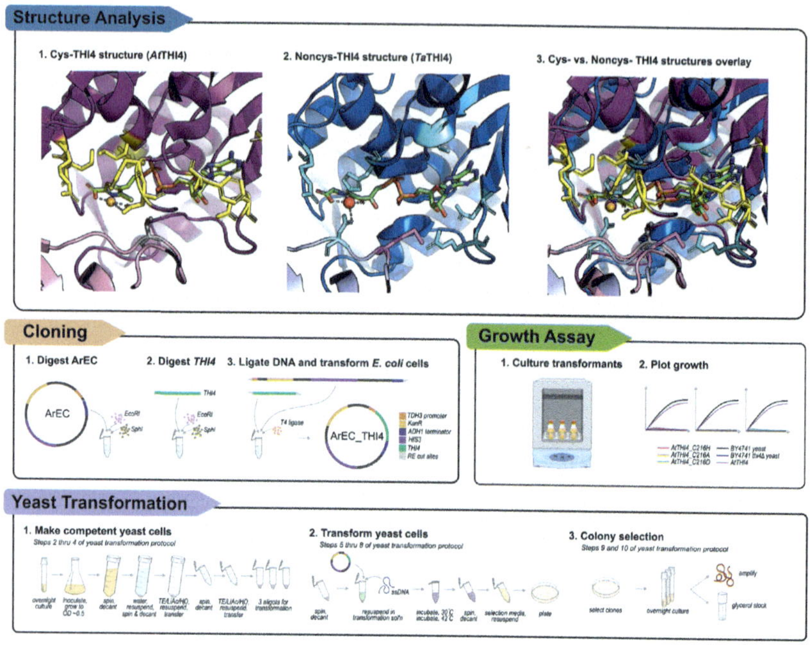

Fig. 2 Schematic of protocols to engineer proteins using information about the protein structure. Key residues involved in catalysis or binding to cofactors can be determined based on the structure, alongside residues that may be key in protein-protein interactions. Overlaying *At*THI4 with *Ta*THI4 indicates some shared and varied aspects of two nearly functionally identical proteins. For example, *Ta*THI4, a noncys-THI4 has a much more open binding pocket than *At*THI4, a cys-THI4. Residues in yellow and cyan indicate those involved in catalysis or metal cofactor binding. The active site cysteine of *At*THI4 is gray and the equivalent residue in noncys-THI4s is pink. The zinc metal cofactor of *At*THI4 is orange and the iron metal cofactor of *Ta*THI4 is red. The enzyme product thiazole is green. Residues identified for replacement can be cloned into the ArEC plasmid and transformed into *E. coli* or *S. cerevisiae*. Growth assay can be completed to determine how well the engineered enzymes function in the host when the native host gene is knocked out. Enzymatic assays may also be completed when feasible.

3. The constructs were transformed into yeast BY4741 *thi4*Δ following the standard yeast transformation protocol which can also be found in Section 4.4 (Gietz & Schiestl, 2007).
4. Wild type yeast (BY4741 *thi4*Δ) was grown alongside the three constructs, BY4741 *thi4*Δ-*At*THI4_C216H, BY4741 *thi4*Δ-*At*THI4_C216A, and BY4741 *thi4*Δ-*At*THI4_C216D in selection media without thiamin (SC – thiamin, Table 2) over ten days (Fig. 2). Growth was monitored by OD_{600}.

Table 1 List of designed truncated *At*THI4 sequences. Residue numbering is based on non-truncated sequences.

*At*THI4 name	Sequence (5–3′)
*At*THI4_C216H	MTAGYDLNAFTFDPIKESIVSREMTRRYMTDMIT YAETDVVVVGAGSAGLSAAYEISKNPNVQVAIIEQ SVSPGGGAWLGGQLFSAMIVRKPAHLFLDEIGVAY DEQDTYVVVKHAALFTSTIMSKLLARPNVKLFNA VAAEDLIVKGNRVGGVVTNWALVAQNHHTQS**H** MDPNVMEAKIVVSSCGHDGPFGATGVKRLKSIGM IDHVPGMKALDMNTAEDAIVRLTREVVPGMIVT GMEVAEIDGAPRMGPTFGAMMISGQKAGQLALK ALGLPNAIDGTLVGNLSPELVLAAADSAETVDA
*At*THI4_C216A	MTAGYDLNAFTFDPIKESIVSREMTRRYMTDMIT YAETDVVVVGAGSAGLSAAYEISKNPNVQVAIIEQ SVSPGGGAWLGGQLFSAMIVRKPAHLFLDEIGVAY DEQDTYVVVKHAALFTSTIMSKLLARPNVKLFNA VAAEDLIVKGNRVGGVVTNWALVAQNHHTQS**A** MDPNVMEAKIVVSSCGHDGPFGATGVKRLKSIGM IDHVPGMKALDMNTAEDAIVRLTREVVPGMIVT GMEVAEIDGAPRMGPTFGAMMISGQKAGQLALK ALGLPNAIDGTLVGNLSPELVLAAADSAETVDA
*At*THI4_C216D	MTAGYDLNAFTFDPIKESIVSREMTRRYMTDMIT YAETDVVVVGAGSAGLSAAYEISKNPNVQVAIIEQ SVSPGGGAWLGGQLFSAMIVRKPAHLFLDEIGVAY DEQDTYVVVKHAALFTSTIMSKLLARPNVKLFNA VAAEDLIVKGNRVGGVVTNWALVAQNHHTQS**D** MDPNVMEAKIVVSSCGHDGPFGATGVKRLKSIGM IDHVPGMKALDMNTAEDAIVRLTREVVPGMIVT GMEVAEIDGAPRMGPTFGAMMISGQKAGQLALK ALGLPNAIDGTLVGNLSPELVLAAADSAETVDA

3.2 Structure analysis

THI4s are homo-octameric, metal-dependent enzymes. Their structure has a characteristic central six-stranded β-sheet surrounded by a three-stranded β-sheet and several α-helices that together form the active site (Godoi et al., 2006; Zhang et al., 2016; Joshi et al., 2021). In cys-THI4s, the conserved active site cysteine is located at the bottom of the active site pocket and is converted to dehydroalanine (DHAla) upon donation of the sulfur atom to thiazole (Fig. 2) (Godoi et al., 2006; Chatterjee et al., 2011). In noncys-THI4s, the cysteine is generally replaced by histidine, or to a lesser degree by other amino acids (Met, Leu, Pro, Ala, Ser, Glu,

Table 2 List of media for cloning and growth assays.

Name	Recipe	Quantity	Volume	Sterilization	Organism
LB	10 g/L tryptone 5 g/L yeast extract 171 mM NaCl pH 7.0–7.5	10 g tryptone 5 g yeast extract 10 g NaCl 1 mL kanamycin	1 L	Autoclave	*E. coli*
Kan	100 mg/mL kanamycin	0.1 g kanamycin	1 mL	Filter sterilize	
SC – thiamin	6.7 g/L YNB w/o thiamin 1.4 g/L DO-Leu 38 mM $(NH_4)SO_4$ pH 6.1 2 % glucose	6.7 g YNB w/o thiamin 1.4 g DO-Leu 5 g $(NH_4)SO_4$ 10 mL 100x Leu 50 mL 40 % glucose	1 L	Autoclave	*S. cerevisiae*
100x Leu	100x Leucine 5 M HCl	0.38 g Leucine few drops HCl (till Leu dissolves)	10 mL	Filter sterilize	
40 % glucose	40 % glucose	40 g glucose	100 mL	Filter sterilize	
Agar plates[a]	15 g/L agar	15 g agar	1 L	Autoclave	

[a]To make any of the media into agar plates, add 15 g/L agar and autoclave.

Asp, Tyr, Trp) (Fig. 2). Noncys-THI4s also generally require sulfide-rich and anaerobic conditions for catalysis to occur (Joshi et al., 2021). Consistent with this observation, when cys-THI4s had their active site cysteine replaced by alanine or serine, the enzyme was still active, as long as the reaction proceeded in the presence of an external sulfide source under anaerobic conditions (Zhang et al., 2016; Sun et al., 2019). It has been hypothesized that residues replacing the catalytic cysteine help position sulfide within the active site (Zhang et al., 2016). Noncys-THI4s may have additional structural differences compared with the suicidal THI4s, including a more open active site architecture to allow access to sulfide from the solvent (Fig. 2) (Zhang et al., 2016; Joshi et al., 2021). In addition, metal ions are thought to contribute to catalysis by activating sulfide and stabilizing reaction intermediates (Table 3).

3.3 Yeast *THI4* knockout complementation

With the goal of engineering cys-THI4s to function as true catalysts (i.e., as noncys-THI4s), we designed three *A. thaliana* THI4 variants with either alanine, aspartate, or histidine to replace the conserved active site cysteine. Aspartate and histidine were chosen because the noncys-THI4s from cereal crops have either Asp or His in place of cysteine. The N-terminus was also truncated to remove the predicted transit peptide sequence. While the wildtype *A. thaliana* THI4 can complement a metabolic thiazole synthase knockout in yeast (Godoi et al., 2006), the *A. thaliana* THI4 with histidine, alanine, or aspartate do not restore growth (unpublished data). The background strain, BY4741 has *met15*Δ (Table 4), is considered a high sulfur strain, so external sulfide was not supplied. Previous attempts to convert suicide THI4s to true catalysts that employed similar changes to the conserved active site cysteine required anaerobic conditions with sulfide supplementation (Zhang et al., 2016; Sun et al., 2019). Additionally, the structure of plant cys-THI4s differs in some aspects from noncys-THI4s see (Section 2.2). These observations suggest that converting suicidal THI4 enzymes into fully catalytic variants in plants may require not only substitution of the active-site cysteine but also additional amino-acid changes and structural insertions that support the entry and use of sulfide for the reaction mechanism and support catalysis under aerobic conditions.

Table 3 List of key reagents for cloning, yeast transformation, protoplast fusion, culturing and genomic DNA extraction.

Reagent	**Vendor**	**Identifier**	**Protocol**
*Nsi*I restriction enzyme	NEB	R3127S	Cloning Mp1iv
*Sph*I restriction enzyme	NEB	R3182S	Cloning Mp1iv
*Sca*I restriction enzyme	NEB	R3122S	Yeast Transformation
Lithium acetate (LiAc)	ThermoFisher	0.1317. A1	Yeast Transformation
ssDNA	MilliporeSigma	D7646	Yeast Transformation
Poly(ethylene glycol), 3350	MilliporeSigma	202444	Yeast Transformation/ Protoplast Fusion
Zymolyase (R) 20 T	AMSBio	120491−1	gDNA extraction/ Protoplast Fusion
Proteinase K	ThermoFisher	AM2548	gDNA extraction
RNase A	ThermoFisher	R1253	gDNA extraction
Yeast Nitrogen Base (YNB) w/o ammonium sulfate, w/o amino acids	ThermoFisher	H26271.22	Yeast Transformation/ Protoplast Fusion
Drop-out mix synthetic minus HIS, LEU, TRP, URA, w/o YNB	US Biological	D9540	Protoplast Fusion/ Culturing
Drop-out mix synthetic minus LEU, w/o YNB	US Biological	D9525	Yeast Transformation/ Protoplast Fusion/ Culturing
Yeast Nitrogen Base (YNB) w/o ammonium sulfate, w/o amino acids, w/o thiamin	US Biological	Y2036	Culturing

Table 4 List of strains.

Strain name	Organism	Genotype
GA-Y319[a]	*S. cerevisiae*	MATa; *can1*; *his3*; *leu2Δ0*; *ura3Δ0*; *trp1Δ0*; *flo1*; p1; p2
BY4741 *thi4Δ*[b]	*S. cerevisiae*	MATa; *his3Δ1*; *leu2Δ0*; *met15Δ0*; *ura3Δ0*; *thi4Δ*::KanMX
TOP10	*E. coli*	F^{-} *mcr*A Δ(*mrr-hsd*RMS-*mcr*BC) φ80*lac*ZΔM15 Δ*lac*X74 *rec*A1 *ara*D139 Δ(*ara-leu*)7697 *gal*U *gal*K λ–*rps*L (Str^{R}) *end*A1 *nup*G

[a]Also known as F102–2 and can be purchased from ATCC (catalog no. 200585), mating type a
[b]Can be purchased from ATCC (catalog no. 201388), mating type a

4. Continuous directed evolution of THI4s

4.1 Experimental design

If the goal is to evolve bacterial and plant noncys-THI4s that can be expressed in plants under aerobic conditions, it is relevant to choose a host for evolution that is closely related to plants. Therefore, OrthoRep was chosen as the CDE platform for evolving bacterial sulfide-dependent *THI4s* as it is a well established eukaryotic based CDE platform, and is ideal for evolving enzymes for plant applications (Oliveira et al., 2023). OrthoRep is a yeast-based CDE platform that uses cytoplasmic terminal-protein linear plasmids (p1 and p2) that are replicated by a protein-primed DNA polymerase specific to each linear plasmid (TP-DNAP1 and TP-DNAP2). Several error-prone TP-DNAP1s were engineered to hypermutate a GOI at a rate of 10^{-4} to 10^{-6} substitutions per base (s.p.b) without increasing the mutation rate of the host genome ($\sim10^{-10}$ s.p.b), (Lang & Murray, 2008; Ravikumar et al., 2014; Ravikumar et al., 2018; Rix et al., 2024). Selection in OrthoRep is completed in vivo by linking expression of the GOI to the growth of the *S. cerevisiae* host. The knockout of the yeast cysteine-dependent *THI4* is lethal without thiamin supplementation or expression of a *THI4* homolog. Evolving the bacterial *THI4*s with OrthoRep requires complementation of the bacterial *THI4*s in the *S. cerevisiae* BY4741 *thi4Δ* strain with or without minimal thiamin supplementation to support growth. As beneficial mutations accumulate and are fixed in the bacterial *THI4*s, thiamin supplementation can be decreased or altogether removed as the evolved THI4 will be able to support thiamin biosynthesis for the yeast at a level similar to the native yeast THI4 or better when placed under tighter selection pressure (i.e. introduction of an inhibitor).

4.2 Before you begin: Reagents and microbial strains

In the tables below (Tables 3 and 4) are lists of key reagents and strains required for evolving bacterial sulfide-dependent *THI4*s in OrthoRep cells. For general cloning, several *E. coli* strains could work (DH101, DH5α, etc.) but our preference is for *E. coli* TOP10.

4.3 Cloning

The method presented here is one of several ways in which to clone genes that will be recombined into p1. This approach uses a multi-purpose p1 integration vector (Mp1iv) and restriction enzyme cloning (Fig. 3). Other approaches include Golden Gate cloning and Gibson Assembly (Javanpour & Liu, 2021; Pisera et al., 2025).

4.3.1 Buffers, media, and solutions for cloning

The tables below (Tables 5, 6) list the plasmids and primers required for cloning. Plasmid ArEC_611 and ArEC_633 can be purchased from Addgene, ID 130872 and ID 130873, respectively. Both plasmids contain an error-prone TP-DNAP1 with mutation rates of $\sim 7 \times 10^{-6}$ s.p.b.(611) and $\sim 1 \times 10^{-5}$ s.p.b.(633) and are biased for transitions over transversions (Ravikumar et al., 2018). New error-prone versions of TP-DNAP1 have been generated recently with mutation rates of $\sim 10^{-4}$ (Rix et al., 2024). However, the work evolving bacterial *THI4*s was completed using 611 and 633 before the newer error-prone TP-DNAPs were engineered. ArEC_611 and ArEC_633 are CEN6/ARS yeast plasmid, with the error-prone TP-DNAP1s under the control of the REV1 promoter and ADH1 terminator, with HIS3 for prototrophic selection in yeast and KanR resistance gene for antibiotic selection in *E. coli*.

The Mp1iv_A72 plasmid was built from 172-YTK-P4 (for the poly (A)/RZ self-cleaving ribozyme cassette) and GR306 (backbone with mkate2 fluorescent protein; García-García et al., 2022) where mkate2 is the place holder for the GOI. Mp1iv_A24 and Mp1iv_A0 were built from Mp1iv_A72 by site-directed mutagenesis (Van Gelder et al., 2025) and will not be discussed further except in terms of cloning *MhTHI4* evolved variants into these plasmids. All Mp1iv plasmids contain upstream and downstream homology arms to the linear p1, a 10B2 promoter taken from

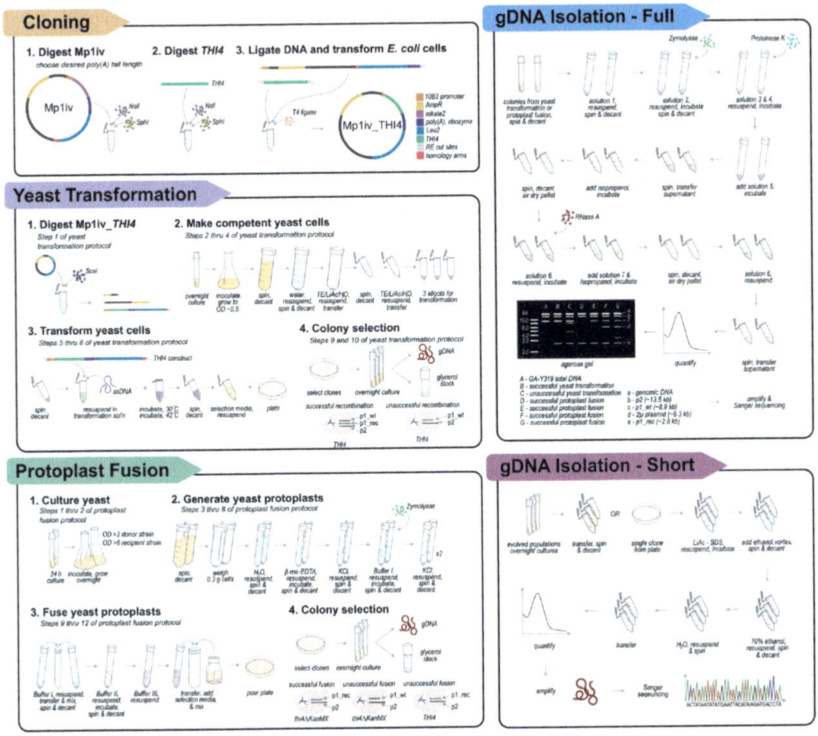

Fig. 3 Schematic of protocols to establish OrthoRep cells for continuous directed evolution. Colors used in the schematic are meant to denote different solutions, buffers, media, enzymes, etc., being used and are not indicative of their actual colors. Successful yeast transformation will result in the *THI4* construct recombining into the p1 (p1_rec) of GA-Y319 yeast cells. Unsuccessful yeast transformation will fail to recombine the *THI4* construct into p1 (p1_wt) and may result in aberrant p1_rec bands of incorrect sizes. Successful protoplast fusion will result in the fusion of the nucleus of BY4741 *thi4*ΔKanMX cells with the cytoplasmic fraction of GA-Y319 cells containing p1_rec. Successful protoplast fusion may also still retain some p1_wt and the 2 μ plasmid which is present in BY4741 cells. Unsuccessful protoplast fusions may fuse the nucleus of BY4741 *thi4*ΔKanMX cells with the cytoplasmic fraction of GA-Y319 cells containing only p1_wt. Unsuccessful protoplast fusions may fail to fuse BY4741 *thi4*ΔKanMX cells with GA-Y319 cells. Unsuccessful fusions may have additional aberrant p1_rec bands of various sizes. Thus, it is important to chose the correct selection media for the protoplast fusion that will select for both the BY4741 genotype and the linear plasmids from GA-Y319 cells.

the linear p2, a synthetic poly(A) tail of various lengths (A72, A24, A0), a ribozyme except when the poly(A) tail is absent, and LEU2 for prototrophic selection of positively transformed yeast cells (Zhong et al., 2018).

Table 5 List of plasmids.

Name	Backbone	Origin	Gene	Resistance	Organism
ArEC_611	CEN6/ARS	ColE1	TP-DNAP1−4−1	KanR	*E. coli*/yeast
ArEC_633	CEN6/ARS	ColE1	TP-DNAP1−4−2	KanR	*E. coli*/yeast
Mp1iv_A72[a]	pMB1	ColE1	mkate2	AmpR	*E. coli*
Mp1iv_A24[b]	pMB1	ColE1	mkate2	AmpR	*E. coli*
Mp1iv_A0[c]	pMB1	ColE1	mkate2	AmpR	*E. coli*

[a]A72 references the synthetic poly(A) tail comprised of 72 adenines
[b]A24 references the synthetic poly(A) tail comprised of 24 adenines
[c]A0 references the absence of a poly(A) tail and ribozyme

The AmpR resistance gene in Mp1iv is used for selection in *E. coli* TOP10 cells during the cloning and propagation steps (Fig. 3).

The cereal noncys-THI4s chosen to evolve were *Hordeum vulgare (HvTHI4)*, *Triticum aestivum (TraTHI4)* and *Avena sativa (AsTHI4)*. The bacterial THI4s that complemented the *E. coli* Δ*thiG*, *TaTHI4*, *MhTHI4*, and *SfTHI4* were also selected for directed evolution. All *THI4s* were codon optimized for yeast and synthesized by a third party. All genes were synthesized with the addition of a 'CAT' codon after the 'ATG' start codon to form the *Nsi*I cut site for cloning into the Mp1iv plasmids.

4.3.2 Cloning protocol

1. Amplify *HvTHI4*, *TraTHI4*, *AsTHI4*, *TaTHI4*, *MhTHI4*, *SfTHI4*, and *ScTHI4* with the primers listed in Table 6 and clean up the amplicons with a PCR purification kit according to the manufacturer's guidelines. Quantify DNA with a NanoDrop or Qubit.
2. Miniprep Mp1iv_A72 from 3−5 mL LB plus ampicillin overnight cultures (list of media in Table 7). Quantify DNA with NanoDrop or Qubit.
3. Digest 1−3 μg of each amplicon and Mp1iv plasmid with fast digest *Nsi*I and *Sph*I in a double digest for 30 min at 37 °C. Clean up digested amplicons with a PCR purification kit according to the manufacturer's guidelines. Clean up digested plasmids with a gel extraction kit according to the manufacturer's guidelines. Quantify all digested DNA with a NanoDrop or Qubit.
4. Ligate the digested amplicons with the digested plasmids with T4 ligase for 45 min at room temperature.

Table 6 List of primers. Table is adapted from Garcia-Garcia et al. (2022), Table S2.

Name	Sequence (5′ → 3′)	Purpose
*HvTHI4*_*Nsi*I_F	GAATACGGGTTCATGCATACCCCTCCATAC	Cloning *HvTHI4* into Mp1iv
*HvTHI4*_*Sph*I_R	GCTGGTGATCACGCATGCTCAACTCAATCTTC	Cloning *HvTHI4* into Mp1iv
*TraTHI4*_*Nsi*I_F	GAATACGGGTTCATGCATACTCCTCCATACGATCTGTCATC	Cloning *TraTHI4* into Mp1iv
*TraTHI4*_*Sph*I_R	GCTGGTGATCACGCATGCTTAAGCTGGAACAGTTTCAGAGTC	Cloning *TraTHI4* into Mp1iv
*AsTHI4*_*Nsi*I_F	GAATACGGGTTCATGCATACTCCACCATACGATTTCAACGC	Cloning *AsTHI4* into Mp1iv
*AsTHI4*_*Sph*I_R	GCTGGTGATCACGCATGCTCACAATCTGGAGGACCTAGTAC	Cloning *AsTHI4* into Mp1iv
*TaTHI4*_*Nsi*I_F	ACGATATGCATCAAAATTTGTCTGAAGTTG	Cloning *TaTHI4* into Mp1iv
*TaTHI4*_*Sph*I_R	TACTGCATGCTTATGCTTCCTTATTTTGGTTC	Cloning *TaTHI4* into Mp1iv
*MhTHI4*_*Nsi*I_F	CGATATGCATGAAAAGATTGTTTCTGCTGGTATTG	Cloning *MhTHI4* into Mp1iv
*MhTHI4*_*Sph*I_R	ATCGGCATGCTTACAATGAATCACAAATCATTTGTGC	Cloning *MhTHI4* into Mp1iv
*SfTHI4*_*Nsi*I_F	CGATATGCATGAACAAATCGTTTCAGTTGGTATCG	Cloning *SfTHI4* into Mp1iv
*SfTHI4*_*Sph*I_R	ATCGGCATGCTTACTTATTCAAATTATCTGCGATCAAATTAG	Cloning *SfTHI4* into Mp1iv
*ScTHI4*_*Nsi*I_F	CGATATGCATTCTGCTACCTCTACTGCTA	Cloning *ScTHI4* into Mp1iv
*ScTHI4*_*Sph*I_R	ATCGGCATGCCTAAGCAGCAAAGTGTTTC	Cloning *ScTHI4* into Mp1iv

(continued)

Table 6 List of primers. Table is adapted from Garcia-Garcia et al. (2022), Table S2. (*cont'd*)

Name	Sequence (5′ → 3′)	Purpose
10B2_upstream_F	ACCTCTGACACATGCAGCTC	Sequencing in Mp1iv
Leu_short_R	GCTGTGATTTCTTGACCAACGTGG	Sequencing in Mp1iv and p1
p1_F	TTATTGGAAGATTAGTACGTCTCC	Sequencing in p1
Ribozyme_R	CCTCACGGACTCATCAGACC	Sequencing in Mp1iv and p1
*MhTHI4*_Int_F	GCACCAGTTATTAGAGAGGGTATG	Sequencing *MhTHI4*
*MhTHI4*_Int_R	GCTTTAGCCAAGTAATATGCAGC	Sequencing *MhTHI4*
*SfTHI4*_Int_F	GAACTGGGCACCAGTTCATAG	Sequencing *SfTHI4*
*SfTHI4*_Int_R	CCTTGTCTTGCTAAGTAGTATGCAG	Sequencing *SfTHI4*
*ScTHI4*_Int_F	GCACCAGGTGGTGGTAGTTGG	Sequencing *ScTHI4*
*ScTHI4*_Int_R	GGTTCATGTCCAGACCCTTCATACCG	Sequencing *ScTHI4*

Underlined sequence indicates restriction enzyme cut sites (*Nsi*I – ATGCAT, *Sph*I – GCATGC)

Table 7 List of media required for growing *E. coli* and *S. cerevisiae*. Table is adapted from Garcia-Garcia et al. (2022), Table S3.

Name	Recipe	Quantity	Volume	Sterilization	Organism
LB	10 g/L tryptone 5 g/L yeast extract 171 mM NaCl pH 7.0–7.5	10 g tryptone 5 g yeast extract 10 g NaCl ± appropriate antibiotics	1 L	Autoclave	*E. coli*
YPD	20 g/L tryptone 10 g/L yeast extract 2 % glucose	20 g tryptone 10 g yeast extract 50 mL 40 % glu[a]	1 L	Autoclave	*S. cerevisiae*
DOL	6.7 g/L YNB[b] 1.4 g/L DO-Leu[c] 38 mM $(NH_4)SO_4$ pH 6.1 2 % glucose	6.7 g YNB[b] 1.4 g DO-Leu[c] 5 g $(NH_4)SO_4$ 50 mL 40 % glu[a]	1 L	Autoclave	*S. cerevisiae*
DOHT	6.7 g/L YNB[b] 1.4 g/L DO-His, Leu, Trp, Ura[d] 38 mM $(NH_4)SO_4$ pH 6.1 2 % glucose	6.7 g YNB[b] 1.4 g DO-His, Leu, Trp, Ura[d] 5 g $(NH_4)SO_4$ 10 mL 100x Leu[e] 10 mL 100x Ura[e] 50 mL 40 % glu[a]	1 L	Autoclave	*S. cerevisiae*
DOHLT	6.7 g/L YNB[b] 1.4 g/L DO-His, Leu, Trp, Ura[d] 38 mM $(NH_4)SO_4$ pH 6.1 2 % glucose	6.7 g YNB[b] 1.4 g DO-His, Leu, Trp, Ura[d] 5 g $(NH_4)SO_4$ 10 mL 100x Ura[e] 50 mL 40 % glu[a]	1 L	Autoclave	*S. cerevisiae*

(continued)

Table 7 List of media required for growing *E. coli* and *S. cerevisiae*. Table is adapted from Garcia-Garcia et al. (2022), Table S3. (*cont'd*)

Name	Recipe	Quantity	Volume	Sterilization	Organism
DOHLT – thiamin	6.7 g/L YNB w/o thiamin[f] 1.4 g/L DO-His, Leu, Trp, Ura[d] 38 mM $(NH_4)SO_4$ pH 6.1 2 % glucose	6.7 g YNB w/o thiamin[f] 1.4 g DO-His, Leu, Trp, Ura[d] 5 g $(NH_4)SO_4$ 10 mL 100x Ura[e] 50 mL 40 % glu[a]	1 L	Autoclave	*S. cerevisiae*
100x Leu	100x Leucine 5 M HCl	0.38 g Leucine few drops HCl (till Leu dissolves)	10 mL	Filter sterilize	
100x Ura	100x Uracil 5 M NaOH	0.076 g Uracil few drops NaOH (till Ura dissolves)	10 mL	Filter sterilize	
40 % glucose	40 % glucose	40 g glucose	100 mL	Filter sterilize	
Thiamin stock	10 mM thiamin[g]	0.0337 g thiamin	10 mL	n/a	
Thiamin sub-stock	100 μM thiamin[h]	100 μL 10 mM thiamin stock	10 mL	Filter sterilize	
Agar Plates[i]	15 g/L agar	15 g agar	1 L	Autoclave	

[a]50 mL of 40 % glucose is added after autoclaving the other media components
[b]YNB refers to yeast nitrogen base w/o ammonium sulfate, w/o amino acids
[c]DO-Leu refers to drop-out synthetic mix minus LEU, w/o YNB
[d]DO-His, Leu, Trp, Ura refers to drop-out synthetic mix minus HIS, LEU, TRP, URA, w/o YNB
[e]100x amino acid solutions are added after autoclaving the other media components
[f]YNB w/o thiamin refers to yeast nitrogen base w/o ammonium sulfate, w/o amino acids, w/o thiamin
[g]Determine exact concentration by spectrophotometer at 247 nm, pH < 5.5, molar extinction coefficient 14,200 M^{-1} cm^{-1}
[h]Working stock, store aliquots at −20 °C
[i]To make any of the media into agar plates, add 15 g/L agar and autoclave.

5. Transform 2 μL of each ligation reaction into chemical or electro-competent *E. coli* TOP10 cells using established heat shock or electroporation protocols. Recover cells in 950 μL LB at 37 °C for 1 h. Plate 50–100 μL of recovered cells on LB plus ampicillin and incubate overnight at 37 °C.
6. Select 5–10 colonies from the plates and resuspend in 10 μL sterile water. Take 5 μL for amplification with sequencing primers and the other 5 μL can be resuspended in 300 μL LB plus ampicillin, incubated at 37 °C, 250 rpm and used to inoculate overnight cultures and make glycerol stocks. Run amplicons on a 1 % agarose gel and look for clones with the appropriate band size (varies based on *THI4* and length of poly(A) tail) and verify positive clones with the correct band size by Sanger sequencing.

4.3.3 Cloning troubleshooting and optimization

The addition of a 'CAT' codon after the start codon to form the *Nsi*I cut site for restriction enzyme cloning introduces an artifact into the GOI. This codon, if mutated and fixed may be a result of mitigating potential negative impacts of this additional amino acid rather than any improvement to enzyme function, stability, turnover, etc. To this end, re-designing the vector to use a different restriction enzyme upstream of the start codon or to proceed with cloning via Gibson Assembly may be the better option.

4.4 Yeast transformation

4.4.1 Buffers, media, and solutions for yeast transformation

In Table 7 (Section 4.3.1) and the tables below (Tables 8, 9) are recipes for the buffers, solutions, and media required for the yeast transformation protocol and genomic DNA isolation (full) protocol used to verify positive transformants.

4.4.2 Yeast transformation protocol

Mp1iv constructs are transformed into *S. cerevisiae* GA-Y319 cells (which house the p1 and p2 linear plasmids), (Table 2) as a linear fragment according to the highly-efficient yeast transformation protocol (Gietz & Schiestl, 2007). The integration of the GOI into the cytoplasmic linear plasmids is completed by homologous recombination with homology arms (Fig. 3), as described here, or through the use of landing pads as described elsewhere (Wellner et al., 2021; Paulk et al., 2024). Plasmid ArEC_611 is also transformed into *S. cerevisiae* BY4741 *thi4*Δ cells as circular plasmids according to (Gietz & Schiestl, 2007).

Table 8 List of key buffers and solutions for yeast transformation.

Name	Recipe	Quantity	Volume	Sterilization
10x TE	100 mM Tris-HCl, pH 7.5–8.0 10 mM EDTA[a]	1.21 g Tris 0.37 g EDTA[a]	100 mL	Autoclave
0.9 % NaCl	0.9 % NaCl	0.9 g NaCl	100 mL	Autoclave
10x LiAc[b]	1 M LiAc[b]	10.2 g LiAc[a]	100 mL	Autoclave
50 % PEG, 3350[c]	50 % PEG, 3350[c]	50 g PEG, 3350[b]	100 mL	Filter sterilize
TE/LiAc/H_2O mix	1x TE 1x LiAc[b]	150 μL 10x TE 150 μL 10x LiAcb 1.2 mL ddH_2O	1.5 mL	Make fresh day of use
Transformation Solution	80 % of 50 % PEG[c] 10 % TE 10 % LiAc[b]	1.6 mL 50 % PEG[c] 200 μL 10x TE 200 μL 10x LiAc[b]	2 mL	Make fresh day of use

[a]EDTA – Ethylenediaminetetraacetic acid
[b]LiAc – Lithium acetate
[c]PEG, 3350 – Poly (ethylene glycol), with molecular weight (mw) averaging 3350 g/mol

Table 9 List of key buffers and solutions for genomic DNA isolation (gDNA isolation – full).

Name	Recipe	Quantity	Volume	Sterilization
Solution 1	0.9 % NaCl	0.9 g NaCl	100 mL	Autoclave
Solution 2	1 M sorbitol 0.1 M EDTA[a]	18.2 g sorbitol 2.9 g EDTA[a]	100 mL	Autoclave
Solution 3	50 mM Tris pH 7.4 20 mM EDTA[a]	0.61 g Tris 0.58 g EDTA[a]	100 mL	Autoclave
Solution 4	10 % SDS[b]	10 g SDS[b]	100 mL	Autoclave
Solution 5	5 M Potassium acetate (KAc)	49.1 g KAc	100 mL	Autoclave
Solution 6	10 mM Tris pH 7.4 1 mM EDTA	0.12 g Tris 0.03 g EDTA	100 mL	Autoclave
Solution 7	3 M Sodium acetate (NaAc)	24.6 g NaAc	100 mL	Autoclave
Zymolyase storage buffer	0.1 M Na_2HPO_4 pH 7.5 50 % glycerol 3 mM β-me[c]	1.42 g Na_2HPO_4 50 g glycerol 21 µL β-me[c]	100 mL	Filter sterilize
Isopropanol	n/a	n/a	50 mL	n/a

[a]EDTA – Ethylenediaminetetraacetic acid
[b]SDS – Sodium dodecyl sulfate
[c]β-me – β-mercaptoethanol

1. For transforming Mp1iv plasmids: Digest 10 μg of each Mp1iv construct with *Sca*I (total reaction volume of 50 μL) at 37 °C for 1 h followed by 65 °C for 20 min to denature the enzyme. Verify complete digest by running 2 μL of the digest on a 1 % agarose gel. Complete digest will reveal three bands at approximately 2.5 kb, 1.6 kb and 0.6 kb as the Mp1iv plasmids have three *Sca*I cut sites. The ~2.5 kb band contains the GOI and LEU2. Band sizes will vary depending on what *THI4* and poly(A) tail is present in the construct. For transforming the ArEC plasmid: Skip this step as this plasmid is transformed as the whole circular DNA.
2. The day before yeast transformation, start 3 mL YPD overnight cultures of GA-Y319 or BY4741 *thi4Δ* and incubate at 30 °C.
3. In the morning, start 30 mL YPD cultures with a 1:35 dilution of the overnight culture. One 30 mL culture is enough for three transformations. Incubate for approximately 3–6 h till OD_{600} reaches ~0.5–0.6.
4. Prep GA-Y319 cells for transformation with a series of washes.
   a. Centrifuge cells at 1500 x g for 5 min, decant and resuspend gently in 30 mL ddH_2O.
   b. Centrifuge cells at 1500 x g for 5 min, decant, and resuspend gently in 900 μL TE/LiAc/H_2O solution. Transfer to a 1.5 mL tube.
   c. Centrifuge at high speed for 15 s, decant, and resuspend gently in 300 μL TE/LiAc/H_2O solution. Aliquot 100 μL into three separate 1.5 mL tubes.
5. Centrifuge at high speed for 15 s, decant, and add 10 μL of pre-boiled (5–10 min) ssDNA (10–11 mg/mL) and the remaining 48 μL of the digest (for ArEC plasmid, add ~500–1000 ng of plasmid DNA). Mix gently by pipet. Add 600 μL of the transformation solution to the tube and mix by gently rotating the tube. Incubate at 30 °C for 30–45 min at 200–250 rpm.
6. Incubate at 42 °C for 20 min in a heat block without shaking.
7. Centrifuge at 700 x *g* for 5 min, decant, and resuspend in 1 mL DOL for Mp1iv transformants or DOHT for ArEC transformants.
8. Plate 20 μL and 100 μL of transformed cells onto DOL/DOHT plates. Incubate at 30 °C for up to five days.
9. Select 10 colonies and start 7 mL DOL cultures for Mp1iv transformants and 3 mL DOHT for ArEC transformants at 30 °C, 250 rpm for 1–2 days (Mp1iv transformants) or overnight (ArEC transformants).
10. ArEC transformants are verified by their ability to grow on media lacking histidine (HIS3 prototrophic marker in the ArEC plasmid will

allow positive transformants to synthesize histidine) and overnight cultures can be used to make glycerol stocks. Mp1iv transformants are verified first by their ability to grow on media lacking leucine (LEU2 marker recombined into p1 allows positive transformants to synthesize leucine) and by analyzing the band pattern of the linear plasmids (see Section 4.4.3).

4.4.3 gDNA isolation – full protocol

1. Take 1–2 mL of the 7 mL DOL culture and reserve for making glycerol stocks when needed. Centrifuge the remaining culture at 1500 x g for 5 min, decant, and resuspend in 1 mL solution 1. Samples can be stored at this step at 4 °C for a short period (<24 h) if need be.
2. Centrifuge cells at 1500 x g for 5 min, decant, and resuspend in 0.5 mL solution 2. Add 63 μL (25 units) of Zymolyase (R) 20 T and incubate at 37 °C for 1 h.
   a. Zymolyase (R) 20 T – 100 mg resuspended in 5 mL Zymolyase storage buffer for final concentration of 20 mg/mL or 0.4 units/μL. Aliquots can be made and stored at −20 °C.
3. Centrifuge cells at 1500 x g for 5 min, decant, and suspend in 0.5 mL solution 3. Add 50 μL of solution 4 and 5 μL of 20 mg/mL Proteinase K. Incubate at 57 °C for 30 min
4. Add 200 μL solution 5 and incubate on ice for minimum 1 h.
5. Centrifuge cells at top speed for 5 min and recover the supernatant into a new tube (~500–700 μL of supernatant).
6. Add 700 μL isopropanol and mix by inversion five times. Incubate at room temperature for 10 min
7. Centrifuge at top speed for 5 min and pipette off supernatant leaving a small white pellet undisturbed. Air dry the pellet for a minimum of 15 min
8. Resuspend the pellet in 300 μL solution 6 and add 1.5 μL of 10 mg/mL RNase A. Incubate at 37 °C for 30 min (no shaking).
9. Add 30 μL solution 7 and mix by inversion. Add 200 μL isopropanol and mix by inversion five times. Incubate at room temperature for 10 min
10. Centrifuge at top speed for 5 min. Pipette off supernatant leaving a small white pellet undisturbed. Air dry the pellet for a minimum of 15 min
11. Resuspend the pellet in 50 μL solution 6. The mixture can be frozen overnight at this step if necessary.

12. Centrifuge at top speed for 10 min. Pipette the supernatant (which contains the gDNA) into a new tube and quantify with a NanoDrop or Qubit.
13. Load approximately 1000–1200 ng of gDNA onto a 0.8 % agarose gel and separate the gDNA at 85 V for 1 h 20 min. Band pattern as is follows: p2 at ~13.5 kb, p1 without GOI recombined (p1_wt) at ~8.9 kb, and p1 with GOI recombined (p1_rec) at ~2.8 kb (but will vary depending on *THI4* and poly(A) tail). Select positive transformants that contain p2, p1_wt, p1_rec, and no other aberrant p1_rec bands. Amplify p1_rec from these clones with the sequencing primers (Table 6) to further verify the sequence of p1_rec with Sanger sequencing.

Three sequence verified clones from each transformant (i.e., three clones from Mp1iv_*HvTHI4*_A72 in GA-Y319, three clones from Mp1iv_*TraTHI4*_A72 in GA-Y319, etc.) were carried forward for the protoplast fusion.

4.4.4 Yeast transformation troubleshooting and optimization

The integration of the GOI into p1 relies on yeast homologous recombination (HR), wherein a DNA fragment can be swapped into a locus based on shared sections of DNA (homology arms) at the 5′ and 3′ ends of the DNA fragment. The ~2.5 kb linear fragment from the *Sca*I digest of Mp1iv constructs has homology arms that target p1 just inside of the inverted repeat regions at either end of the linear plasmid. In general, homology arms between 30–50 bp have been effective (David & Siewers, 2015). However, HR between the Mp1iv linear fragment and p1 often results in aberrant p1_rec of various lengths. Potential reasons for less efficient HR when targeting p1 could be the shorter left homology arm (34 bp compared to 155 bp of the right homology arm) or the specific target locus in p1. Testing various homology arm lengths and different regions in p1 for HR could improve the rate of positive clones from the yeast transformation. The use of 'landing pads' has also greatly increased the efficiency of HR into p1 (Wellner et al., 2021; Paulk et al., 2024).

In addition to the presence of aberrant p1_rec, p1_wt should be present in positive transformants. There can be between 3 and 10 copies of p1 in the cell, and not all copies necessarily integrate the GOI and thus p1_wt may persist (Ravikumar et al., 2018; Zhong et al., 2018). Only one copy of p1 that successfully integrated the GOI and LEU2 is required for growth on DOL media. However, still selecting for p1_wt at this stage is necessary to ensure replication of p1_rec as there is no other TP-DNAP1 in GA-Y319 besides the wild type TP-DNAP1 on p1_wt.

4.5 Protoplast fusion

4.5.1 Buffers, media, and solutions for protoplast fusion

In Table 7 (Section 4.3.1) and in the table below (Tables 10, 11) are buffers, media, and solutions for the protoplast fusion and short gDNA isolation protocol. The protoplast fusion is followed by gDNA isolation (full) protocol which can be found in Section 4.4.3, and thereafter, any sequencing of individual clones or populations from OrthoRep campaigns can use gDNA isolated from the short gDNA isolation protocol.

4.5.2 Protoplast fusion protocol

The protoplast fusion combines the cytoplasmic fraction of Mp1iv GA-Y319 transformants including p1_rec and p2 (donor strain) with the nucleus of ArEC BY4741 *thi4Δ* transformants (recipient strain) (Fig. 3). The presence of LEU2 on p1_rec and HIS3 on ArEC will be used as a prototrophic selection for positive fusions by selecting on media lacking leucine and histidine. Additionally, GA-Y319 cells lack TRP1 (Table 4) which can be used as secondary selection for the nucleus of ArEC BY4741 *thi4Δ* transformants (which have TRP1 and can synthesize tryptophan) by selecting positive fusions on media also lacking tryptophan.

1. 48 h before protoplast fusion, start 3 mL pre-cultures of GA-Y319 transformants (donor strain) and BY4741 *thi4Δ* transformants (recipient strain) in DOL and DOHT media, respectively, at 30 °C, 250 rpm.
2. The evening before protoplast fusion, dilute 1:50 pre-cultures into 50 mL YPD and incubate overnight at 30 °C, 250 rpm. It will take approximately 15−16 h for BY4741 *thi4Δ* cells to reach an OD_{600} of 6−7 and GA-Y319 cells to reach an OD_{600} of 2−4. For protoplast fusion, cultures should not have an OD_{600} of below 2 or over 8.
3. Measure the OD_{600} of cultures in the morning and calculate the volume of culture needed to get approximately 0.3 g of cells. (20 mL of OD_{600} = 4.5 is ~0.45 g of cells). Weigh 50 mL tubes, add the required volume and then centrifuge at 3000 x g for 5 min. Decant and re-weigh tubes to determine weight of cells (over 0.3 g is better than under).
4. Resuspend the cells in 10 mL ddH_2O and transfer to a 15 mL tube.

Table 10 List of key buffers and solutions for protoplast fusion.

Name	Recipe	Quantity	Volume	Sterilization
Citrate phosphate buffer (CPB)	solution A – 0.1 M citric acid solution B – 0.2 M Na_2HPO_4	1.06 g citric acid 5.36 g Na_2HPO_4	34.8 mL sol'n A 65.3 mL sol'n B 100 mL ddH_2O	Filter sterilize
3 M KCl	3 M KCl	22.36 g KCl	100 mL	Autoclave
EDTA[a]	0.5 M EDTA[a] pH 8.0	7.31 g EDTA[a]	50 mL	Autoclave
$CaCl_2$	0.5 M $CaCl_2$	3.68 g $CaCl_2$	50 mL	Autoclave
50 % PEG, 3350b[b]	50 % PEG, 3350[b]	20 g PEG, 3350[b]	40 mL	n/a
β-me[c] – EDTA[a]	0.06 M EDTA[a] 0.2 % β-me[c]	6 mL 0.5 M EDTA[a] 100 μL β-me[c] 43.9 mL sterile ddH_2O	50 mL	n/a
0.6 M KCl	0.6 M KCl	20 mL 3 M KCl 80 mL sterile ddH_2O	100 mL	n/a
Buffer I	CPB 0.6 M KCl 0.01 M EDTA[a]	39 mL CPB 10 mL 3 M KCl 1 mL 0.5 M EDTA[a]	50 mL	Make fresh day of use, filter sterilize

Buffer II	33 % PEG[b] 0.06 M KCl 50 mM $CaCl_2$	33.3 mL 50 % PEG[b] 10 mL 3 M KCl 5 mL 0.5 M $CaCl_2$ 1.7 mL sterile ddH_2O	50 mL	Make fresh day of use, filter sterilize
Buffer III	0.6 M KCl 50 mM $CaCl_2$	10 mL 3 M KCl 5 mL 0.5 M $CaCl_2$ 35 mL sterile ddH_2O	50 mL	Make fresh day of use, filter sterilize
Zymolyase storage buffer	0.1 M Na_2HPO_4 pH 7.5 50 % glycerol 3 mM β-me[c]	1.42 g Na_2HPO_4 50 g glycerol 21 μL β-me[c]	100 mL	Filter sterilize

[a]EDTA – Ethylenediaminetetraacetic acid
[b]PEG, 3350 – Poly (ethylene glycol), with molecular weight (mw) averaging 3350 g/mol
[c]β-me – β-mercaptoethanol

Table 11 List of key buffers and solutions for genomic DNA isolation (gDNA isolation – short).

Name	Recipe	Quantity	Volume	Sterilization
10 % SDS[a]	10 % SDS[a]	10 g SDS[a]	100 mL	Autoclave
LiAc[b] – SDS[a]	200 mM LiAc[b] 1 % SDS[a]	0.66 g LiAc[b] 5 mL 10 % SDS[a]	50 mL	n/a
100 % EtOH	n/a	n/a	50 mL	n/a
70 % EtOH	70 % EtOH	35 mL 100 % EtOH 15 mL ddH_2O	50 mL	n/a

[a]SDS – Sodium dodecyl sulfate
[b]LiAc – Lithium acetate

5. Centrifuge at 3000 x g for 5 min, decant, and resuspend in 1.8 mL β-me-EDTA by gently tapping the side of the tube. Do not vortex or resuspend with a pipet! Incubate at 30 °C for 30 min with gentle inversion of the tube five times at the half way point.
6. Centrifuge at 3000 x g for 5 min, decant, and gently resuspend in 3 mL 0.6 M KCl with a serological pipet. Limit the creation of bubbles while resuspending.
7. Centrifuge at 3000 x g for 5 min, decant, and gently resuspend in 4.8 mL Buffer I with 75 μL (30 units) of Zymolyase (R) 20 T. Incubate at 30 °C for 1 h with occasional inverting of the tube (every 10–15 min).
   a. Zymolyase (R) 20 T – 100 mg resuspended in 5 mL Zymolyase storage buffer for final concentration of 20 mg/mL or 0.4 units/μL. Aliquots can be made and stored at −20 °C.
   b. During the incubation period, make the DOHLT media with agarose in 100 mL aliquots and autoclave. Keep in a 60 °C water bath until step 11.
8. Centrifuge at 700 x *g* for 10 min at 0 °C, decant, and wash twice with 3 mL 0.6 M KCl. Use the serological pipette to resuspend. The pellet will be very sticky so avoid lifting the pellet off the side of the tube all at once.
9. Centrifuge at 700 x *g* for 10 min at 0 °C, decant, and resuspend in 3 mL Buffer I. Mix 1.5 mL of donor strain with 1.5 mL of recipient strain in a new 15 mL tube. Mix gently by inverting 2–3 times.

10. Centrifuge at 700 x g for 10 min at 0 °C, decant, and resuspend in 5 mL Buffer II. Incubate at 30 °C for 30 min with occasional gentle inversions.
11. Centrifuge at 700 x g for 10 min at 0 °C, decant, and resuspend in 5 mL Buffer III.
    a. Add uracil and glucose to the DOHLT media and let rest in the 60 °C water bath.
12. Place 250 μL and 500 μL of cells in two separate 50 mL tubes. Place DOHLT media at 42 °C for 2–3 min. Add 25 mL of DOHLT media to each of the 50 mL tubes with the cells. Mix gently my inverting the tube, and then pour into a petri dish. Allow to dry completely before covering with the lid, and incubating at 30 °C for up to 10 days.
    a. Use a fresh 50 mL tube for each fusion to avoid cross-contamination.
13. Colonies should be embedded in the agar. If all colonies are only on top of the agar, the protoplast fusion has failed. Dig 10 colonies from the plates and start 7 mL DOHLT cultures at 30 °C, 250 rpm, for 1–2 days. Double check for the presence of p1_rec and p2 with the gDNA isolation (full protocol) in Section 4.4.3. Select clones to carry forward that have no to minimal p1_wt and no aberrant p1_rec. Use the sequencing primers (Table 6) to further verify the sequence of p1_rec in BY4741 *thi4Δ* cells with Sanger sequencing. Three clones from each fusion should be at a minimum carried forward for the evolution campaign.

4.5.3 gDNA isolation – Short protocol

This protocol can be used when the band pattern of p1_wt, p1_rec, and p2 is not relevant. This protocol is generally used to isolate gDNA from evolved OrthoRep populations (Fig. 4).

1. Start 3 mL overnight DOHLT cultures of individual clones or have available, up to 1 mL of culture from a population in the OrthoRep campaign. Incubation should occur at 30 °C, 250 rpm.
2. Spin down 100–200 μL of yeast culture at 5000 x g for 1–2 min, decant, and resuspend the pellet in 100 μL of the LiAc – SDS solution. Incubate for 5 min at room temperature.
3. Add 300 μL of 100 % ethanol and vortex.
4. Centrifuge at 15,000 x g for 3 min, decant the supernatant, and resuspend the pellet in 70 % ethanol.

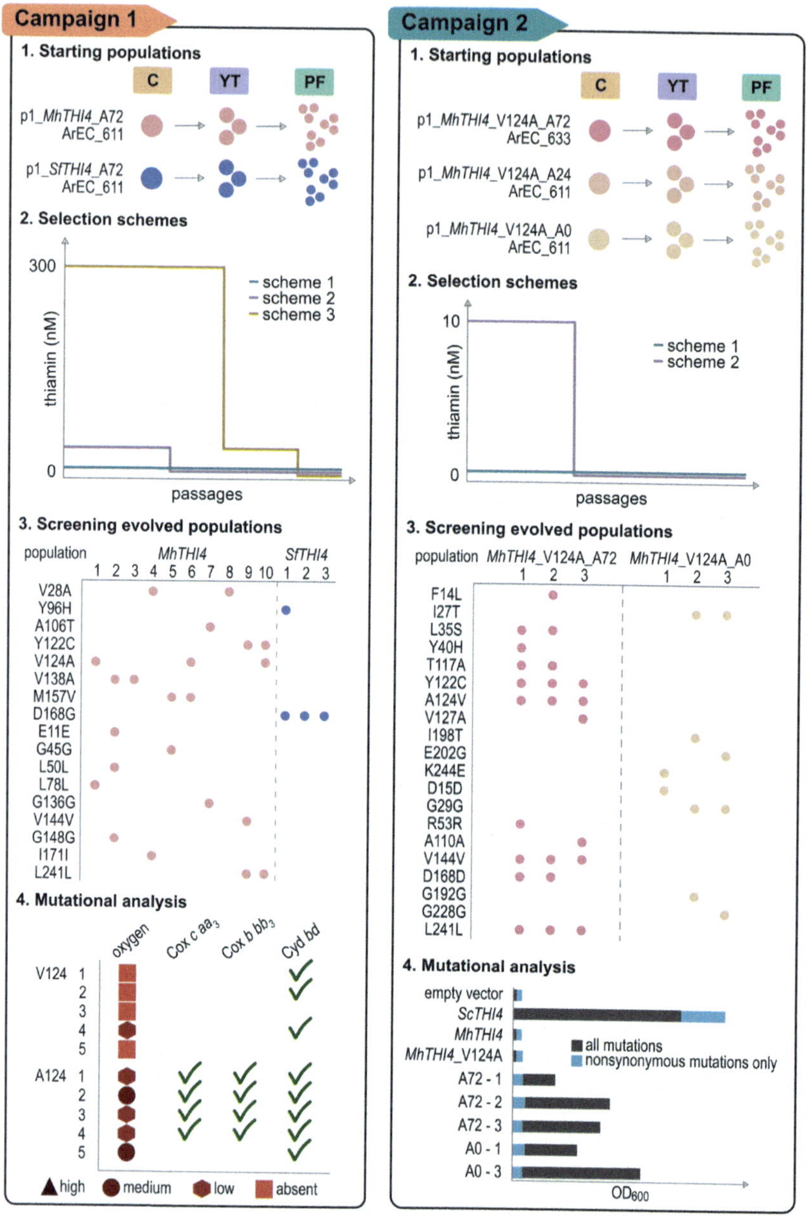

Fig. 4 Evolution campaigns. *Figures were used and/or modified with permission from García-García et al. (2022) Fig. S3 and 4, Van Gelder et al. (2023) Figs. 3 and 4 (https://pubs.acs.org/doi/10.1021/acssynbio.2c00512), and Van Gelder et al. (2025) Fig. 1. Further permissions related to Van Gelder et al. (2023) Figs. 3 and 4 should be directed to ACS.*

5. Centrifuge at 15,000 x *g* for 3 min, decant, and resuspend the pellet in 100 μL of sterile water.
6. Brief centrifuge at 15,000 x g for ~15 sonds.
7. Transfer 80 μL of supernatant into a new sterile tube and use 1 μL for PCR with appropriate sequencing primers (Table 6). PCR amplicons can be sent for Sanger sequencing.

4.5.4 Troubleshooting and optimization

In addition to p2 and p1_rec in clones from the protoplast fusion, p1_wt may or may not be present. As the recipient strain houses the error-prone TP-DNAP1, there is no need for maintaining the wild type TP-DNAP1 on p1_wt. With no selection pressure (i.e. no LEU2) for p1_wt, over time, after multiple passages, the p1_wt should naturally disappear. Therefore, it is not of major concern but we would recommend avoiding the selection of positive transformants with abundant p1_wt as it still houses a wild type copy of TP-DNAP1 which could contribute to GOI replication without mutagenesis.

Beyond looking for the correct band patterns in an agarose gel and sequence verifying p1_rec, proteomics can be a useful tool for determining metabolic changes in the host strain with expression of the GOI on p1_rec (Belt et al., 2025). Changes in protein expression of key enzymes in the biosynthetic pathway to which your GOI belongs can indicate how the host might compensate for poor GOI expression and/or function and circumvent any selection pressure that may be applied. Proteomics can also identify how much protein from p1 is being produced, an amount that may change based on p1 copy number in a single clone (Ravikumar et al., 2018; Zhong et al., 2018; Belt et al., 2025). It also highlights why having multiple populations derived from different clones in one evolution campaign is so important, as biology can vary from one clone to the next (Belt et al., 2025).

Lastly, of utmost importance is the selection of an appropriate yeast host strain (recipient strain in the protoplast fusion). Strains carrying a *met15Δ* allele, such as BY4741 *thi4Δ*, can be advantageous because they have elevated intracellular sulfide levels which may, in principle, support the activity and evolution of bacterial THI4 functional analogs that have naturally adapted to high-sulfide environments (Joshi et al., 2021; Van Oss et al., 2022; Yu et al., 2022). However, for plant or other eukaryotic THI4 enzymes, excessive intracellular sulfide can become cytotoxic, leading to poor growth, stalled evolution trajectories, or accumulation of deleterious mutations (Ono et al., 1991). An alternative is to use BY4742 *thi4Δ* as a host, which maintains lower basal sulfide levels. However, the MATα genotype of BY4742 raises an additional complication: fusion with MATa donor strains (like GA-Y319) can result in the formation of unwanted diploid (MATa/MATα) cells (Gibeaux & Knop, 2013). Diploidization disrupts the genetic stability required for continuous evolution and interferes with the selection dynamics of the p1/p2-plasmid system. To prevent nuclear fusion and maintain a haploid host following protoplast fusion, a p1-donor strain harboring the KAR1-Δ15 allele can be used. The Δ15 mutation introduces an in-frame deletion in KAR1, a gene required for karyogamy, and effectively suppresses nuclear fusion while still permitting cytoplasmic fusion and plasmid transfer (Vallen et al., 1992; Spencer et al., 1994; Georgieva & Rothstein, 2002; Pisera et al., 2025). Nonetheless, the absence of the native GOI in the final strain should still be verified (Oliveira-Filho et al., 2026).

4.6 Selection pressure and mutational analysis

4.6.1 Selection schemes for evolution campaigns

At a minimum, nine 3 mL cultures of each construct should be started for the evolution campaign. For example, *MhTHI4*_A72, has three sequence-verified clones from the yeast transformation used in the protoplast fusion. From the protoplast fusion, three sequence-verified clones of each fusion should be carried forward for the evolution campaign for a total of nine independent populations (Fig. 4). Campaigns can be conducted in test tubes or in multi-well plates depending on the equipment available. It is recommended that the incubator used for the evolution campaign have humidity control to minimize evaporation from the small culture volumes.

Cultures are grown in DOHLT – thiamin media (Table 7) with various supplementation of thiamin according to the selection scheme. There are three selection schemes: (1) library building in which OrthoRep cells are supplied with luxury amounts of thiamin (300 nM) allowing the bacterial *THI4*s to accumulate a diverse set of mutations, before selection is applied by decreasing the amount of thiamin supplied to the cultures (10 nM, 0 nM) to fix the beneficial mutations; (2) gradual in which OrthoRep cells are initially cultured with minimal amounts of thiamin (10 nM) to support minimal growth (OD_{600} of ~1) before moving towards no supplementation; and (3) cold turkey in which OrthoRep cells are started from the onset without any thiamin supplementation and thus there is full reliance on the bacterial THI4s to produce thiazole (Fig. 4). To the end of any one of these selection schemes, when a plateau is reached, where improved bacterial THI4s operate in yeast as good as the native yeast THI4, the inhibitor 2-amino−5-(2-hydroxymethyl)−4-methylthiazole (2-amino-HET) can be given in increasing concentrations to push the selection pressure beyond the threshold imposed by the yeast host's native THI4.

4.6.2 First evolution campaign

The cereal noncys-THI4s were unsuccessful in sustaining growth under selection scheme one. Under selection scheme two and three, populations were viable as long as thiamin was supplemented, even at minimal amounts to achieve minimal growth ($OD_{600} < 1$). However, once thiamin supplementation was ended, all populations failed to grow. Similar results were achieved with *Ta*THI4. These THI4s will not be discussed further.

MhTHI4 and *SfTHI4* with a full-length poly(A) tail, A72, were evolved under each of the three selection schemes (Fig. 4) (García-García et al., 2022; Van Gelder et al., 2023). In total, 16 unique variants were identified carrying one or two nonsynonymous mutations and up to three synonymous mutations. Mutations were identified on a population or single clone level by first isolating gDNA with the full (Table 9) or short (Table 11) protocol followed by Sanger sequencing of p1 amplicons. Several mutations were identified in multiple independent populations, including V28A, Y122C, V124A, and V138A in two or three independent *MhTHI4* populations each, and D168G in three independent *SfTHI4* populations (Fig. 4). These re-occurring mutations are an example of convergent evolution. They are also all located in the active-site cleft, near the adenine ring of the bound intermediate, and can be considered first shell mutations (Van Gelder et al., 2023; Goldsmith & Tawfik, 2017). These mutants were re-cloned into fresh OrthoRep cells (following protocols in Sections 4.3, 4.4, and 4.5) to determine whether they improved growth of the yeast host independent of the synonymous mutations found alongside them, and all were identified as beneficial i.e. improved growth compared to yeast expressing *ScTHI4*_A72 in p1; (García-García et al., 2022; Van Gelder et al., 2023).

The most independently re-occurring mutation, V124A, was also one of the most prevalent mutant variants across ~200 comparative bacterial genomes. The presence of alanine at this position was found at a frequency of ~8 %. Bacteria with alanine at position 124 were also found to inhabit environments with greater oxygen exposure and contained most, if not all, of the core oxygen metabolism genes (aa_3-type and ccb_3-type cytochrome c oxidases, cytochrome bd) compared to those species with valine at position 124 (Fig. 4). Therefore, the conclusion was that the mutation to alanine at position 124 may impart aerotolerance to *Mh*THI4 as *M. hirudinis* is found in environments with little to no oxygen and as such wild type *Mh*THI4 is considered an aerobic/microoxic enzyme (Van Gelder et al., 2023).

4.6.3 Second evolution campaign: MhTHI4_V124A

We were interested in pushing the selection pressure further in an attempt to accumulate more beneficial mutations in the V124A variant, particularly those outside the active site and that may be in the second or third shell, which generally are mutations outside the active site that can support the stabilizing of mutations in the first shell (Goldsmith & Tawfik, 2017). There were four means by which to increase selection pressure: (1) by decreasing the expression of *MhTHI4* at the gene level through weaker promoters, (2) by decreasing the expression of *MhTHI4* at the mRNA level through shortening the poly(A) tail, (3) by supplying an inhibitor to decrease the availability of active enzyme or (4) increasing the rate of mutagenesis (ArEC_633 instead of ArEC_611, Table 5). We attempted to increase the selection pressure through destabilizing the mRNA by having a poly(A) tail of 24 As (A24) or removing the poly(A) tail (A0) altogether or by increasing the rate of mutagenesis.

The protocols from Sections 4.3, 4.4, and 4.5 were repeated with MhTHI4_V124A with A24 or A0 (Table 5) (construction of Mp1iv with A24 or A0 are discussed in Van Gelder et al., 2025) for expression in ArEC_611 BY4741 thi4Δ cells (mutation rate of $\sim 10^{-6}$ s.p.b) and MhTHI4_V124A_A72 was expressed in ArEC_633 BY4741 thi4Δ cells (mutation rate of $\sim 10^{-5}$ s.p.b). Evolution campaigns were conducted under selection schemes one (cold turkey) and two (gradual) (Fig. 4). The only populations that persisted and accumulated both synonymous and nonsynonymous mutations included MhTHI4_V124A_A0 (with ArEC_611) under selection scheme one and MhTHI4_V124A_A72 (with ArEC_633) under selection scheme two. Mutants were identified at the population level using either the full (Table 9) or short (Table 11) gDNA isolation protocol and Sanger sequencing to verify mutants. Five unique mutants with two to four nonsynonymous mutations and two to four synonymous mutations were identified (Fig. 4). When only the nonsynonymous mutations were cloned back into MhTHI4_V124A in fresh OrthoRep cells, growth was comparable to wild type MhTHI4 and MhTHI4_V124A (both of which grew similarly).

However, with the synonymous mutations maintained, the five unique mutants grew better than wild type MhTHI4 and MhTHI4_V124A but not better than wild type ScTHI4, indicating that these synonymous mutations contribute to growth improvements as effectively as nonsynonymous mutations (Fig. 4, Van Gelder et al., 2025).

In the original experimental set up, evolution campaigns were conducted in test tubes (30 °C, 220–250 rpm, no humidity control), and *MhTHI4*_V124A grew significantly better than wild type *MhTHI4* and similar to *ScTHI4* (García-García et al., 2022; Van Gelder et al., 2023). However, when our set-up was improved to process more cultures at one time by using multi-well plates in a humidity-controlled incubator with very high shaking speeds (30 °C, 800 rpm, 80 % relative humidity), growth of *MhTHI4*_V124A significantly decreased and *ScTHI4* growth improved with the conclusion that growth rates had changed because aeration had increased under the new set up (Van Gelder et al., 2023, Van Gelder et al., 2025).

Acknowledgment

This manuscript is dedicated to the late Dr. Andrew Hanson who was a wonderful mentor and an even greater scientist.

References

Belt, K., Obe, D., Wilson, M. A., Millar, A. H., & Bathe, U. (2025). Harnessing mass spectrometry-based proteomics for continuous directed evolution. *Synthetic Biology, 10*(1), ysaf017. https://doi.org/10.1093/synbio/ysaf017.

Campos-Silva, R., Rahimi, F., Joshi, J., Voiniciuc, C., Zhou, J., & Hanson, A. D. (2026). Mining bacterial (meta)genomes for enzymes active in aerobic, mesophilic conditions. *Analytical Biochemistry, 771*, 116045. https://doi.org/10.1016/j.ab.2026.116045.

Chatterjee, A., Abeydeera, N. D., Bale, S., Pai, P. J., Dorrestein, P. C., Russel, D. H., ... Begley, T. P. (2011). *Saccharomyces cerevisiae* THI4p is a suicide thiamine thiazole synthase. *Nature, 478*, 542–546. https://doi.org/10.1038/nature10503.

David, F., & Siewers, V. (2015). Advances in yeast genome engineering. *FEMS Yeast Research, 15*(1), 1–14. https://doi.org/10.1111/1567-1364.12200.

Dereeper, A., Guignon, V., Blanc, G., Audic, S., Buffet, S., Chevenet, F., ... Gascuel, O. (2008). Phylogeny.fr: Robust phylogenetic analysis for the non-specialist. *Nucleic Acids Research, 36*, W465–W469. https://doi.org/10.1093/nar/gkn180.

Dorrestein, P. C., Zhai, H., McLafferty, F. W., & Begley, T. P. (2004). The biosynthesis of the thiazole phosphate moiety of thiamin: The sulfur transfer mediated by the sulfur carrier protein ThiS. *Chemistry & Biology, 11*(10), 1373–1381. https://doi.org/10.1016/j.chembiol.2004.08.009.

Eser, B. E., Zhang, X., Chanani, P. K., Begley, T. P., & Ealick, S. E. (2016). From suicide enzyme to catalyst: The iron-dependent sulfide transfer in *Methanococcus jannaschii* thiamin thiazole biosynthesis. *Journal of the American Chemical Society, 138*(11), 3639–3642. https://doi.org/10.1021/jacs.6b00445.

Esvelt, K. M., Carlson, J. C., & Liu, D. R. (2011). A system for the continuous directed evolution of biomolecules. *Nature, 472*, 499–503. https://doi.org/10.1038/nature09929.
García-García, J. D., Van Gelder, K., Joshi, J., Bathe, U., Leong, B. J., Burner, S. D., ... Hanson, A. D. (2022). Using continuous directed evolution to improve enzymes for plant applications. *Plant Physiology, 188*(2), 971–983. https://doi.org/10.1093/plphys/kiab500.
Georgieva, B., & Rothstein, R. (2002). *kar*-mediated plasmid transfer between yeast strains: Alternative to traditional transformation methods. *Methods in Enzymology, 350*, 278–289. https://doi.org/10.1016/S0076-6879(02)50969-1.
Gibeaux, R., & Knop, M. (2013). When yeast cells meet, karyogamy!. *Nucleus, 4*(3), 182–188. https://doi.org/10.4161/nucl.25021.
Gietz, R. D., & Schiestl, R. H. (2007). High-efficiency yeast transformation using the LiAc/SS carrier DNA/PEG method. *Nature Protocols, 2*(1), 31–34. https://doi.org/10.1038/nprot.2007.13.
Godoi, P. H. C., Galhardo, R. S., Luche, D. D., Van Sluys, M. A., Menck, C. F. M., & Oliva, G. (2006). Structure of the thiazole biosynthetic enzyme THI1 from *Arabidopsis thaliana. Journal of Biological Chemistry, 218*(41), 30957–30966. https://doi.org/10.1074/jbc.M604469200.
Goldsmith, M., & Tawfik, D. S. (2017). Enzyme engineering: Reaching the maximal catalytic efficiency peak. *Current Opinion in Structural Biology, 47*, 140–150. https://doi.org/10.1016/j.sbi.2017.09.002.
Halperin, S. O., Tou, C. J., Wong, E. B., Modavi, C., Schaffer, D. V., & Dueber, J. E. (2018). CRISPR-guided DNA polymerases enable diversification of all nucleotides in a tunable window. *Nature, 560*, 248–252. https://doi.org/10.1038/s41586-018-0384-8.
Hanson, A. D., Amthor, J. S., Sun, J., Niehaus, T. D., Gregory III, J. F., Bruner, S. D., & Ding, Y. (2018). Redesigning thiamin synthesis: Prospects and potential payoffs. *Plant Scince, 273*, 92–99. https://doi.org/10.1016/j.plantsci.2018.01.019.
Hwang, S., Cordova, B., Chavarria, N., Elbanna, D., McHugh, S., Roja, J., ... Maupin-Furlow, J. (2014). Conserved active site cysteine residue of archaeal THI4 homolog is essential for thiamine biosynthesis in *Haloferax volcanii. BMC Microbiology, 14*(260), 1–13. http://www.biomedcentral.com/1471-2180/14/260.
Javanpour, A. A., & Liu, C. C. (2021). Evolving small-molecule biosensors with improved performance and reprogrammed ligand preference using OrthoRep. *ACS Synthetic Biology, 10*(10), 2705–2714. https://doi.org/10.1021/acssynbio.1c00316.
Joshi, J., Beaudoin, G. A. W., Patterson, J. A., García-García, J. D., Belisle, C. E., Chang, L. Y., ... Hanson, A. D. (2020). Bioinformatic and experimental evidence for suicidal and catalytic plant THI4s. *Biochemical Journal, 477*(11), 2055–2069. https://doi.org/10.1042/BCJ20200297.
Joshi, J., Li, Q., García-García, J. D., Leong, B. J., Hu, Y., Bruner, S. D., & Hanson, A. D. (2021). Structure and function of aerotolerant, multiple-turnover THI4 thiazole synthases. *Biochemical Journal, 478*(17), 3265–3279. https://doi.org/10.1042/BCJ20210565.
Jumper, J., Evans, R., Pritzel, A., Green, T., Figurnov, M., Ronneberger, O., et al. (2021). Highly accurate protein structure prediction with AlphaFold. *Nature, 596*, 583–589. https://doi.org/10.1038/s41586-021-03819-2.
Kuchner, O., & Arnold, F. H. (1997). Directed evolution of enzyme catalysts. *Trends in Biotechnology, 15*(12), 523–530. https://doi.org/10.1016/S0167-7799(97)01138-4.
Lang, G. I., & Murray, A. W. (2008). Estimating the per-base-pair mutation rate in the yeast *Saccharomyces cerevisiae. Genetics, 178*(1), 67–82. https://doi.org/10.1534/genetics.107.071506.
Muller, Y. A., Lindqvist, Y., Furey, W., Schulz, G. E., Jordon, F., & Schneider, G. (1993). A thiamin diphosphate binding fold revealed by comparison of the crystal structures of transketolase, pyruvate oxidase and pyruvate decarboxylase. *Structure (London, England: 1993), 1*(2), 95–103. https://doi.org/10.1016/0969-2126(93)90025-c.

Oliveira, E. R., Voiniciuc, C., & Hanson, A. D. (2023). Adapting enzymes to improve their functionality in plants: Why and how. *Biochemical Society Transactions, 51*(5), 1957–1966. https://doi.org/10.1042/BST20230532.

Oliveira-Filho, E. R., Gayen, A. K., Leong, B. J., Belt, K., Voiniciuc, C., Millar, H. A., & Hanson, A. D. (2026). Continuous directed evolution of a short-lived histidinol dehydrogenase. *ACS Synthetic Biology, 15*(3), 985–992. https://doi.org/10.1021/acssynbio.5c00468.

Ono, B. I., Ishii, N., Fugino, S., & Aoyama, I. (1991). Role of hydrosulfide ions (HS^-) in methylmercury resistance in *Saccharomyces cerevisiae. Applied and Environmental Microbiology, 57*(11), 3183–3186. https://doi.org/10.1128/aem.57.11.3183-3186.1991.

Packer, M. S., & Liu, D. R. (2015). Methods for the directed evolution of proteins. *Nature Reviews. Genetics, 16*, 379–394. https://doi.org/10.1038/nrg3927.

Park, H., & Kim, S. (2021). Gene-specific mutagenesis enables rapid continuous evolution of enzymes in vivo. *Nucleic Acids Research, 49*(6), e32–e42. https://doi.org/10.1093/nar/gkaa1231.

Paulk, A. M., Williams, R. L., & Liu, C. C. (2024). Rapidly inducible yeast surface display for antibody evolution with OrthoRep. *ACS Synthetic Biology, 13*(8), 2629–2634. https://doi.org/10.1021/acssynbio.4c00370.

Pisera, A. O., Yu, Y., Williams, R. L., & Liu, C. C. (2025). Ultra-efficient integration of gene libraries onto yeast cytosolic plasmids. *ACS Synthetic Biology, 14*(4), 1002–1008. https://doi.org/10.1021/acssynbio.4c00786.

Ravikumar, A., Arrieta, A., & Liu, C. C. (2014). An orthogonal DNA replication system in yeast. *Nature Chemical Biology, 10*, 175–177. https://doi.org/10.1038/nchembio.1439.

Ravikumar, A., Arzumanya, G. A., Obadi, M. K. A., Javanpour, A. A., & Liu, C. C. (2018). Scalable, continuous evolution of genes at mutation rates above genomic error thresholds. *Cell, 175*(7), 1946–1957. https://doi.org/10.1016/j.cell.2018.10.021.

Rix, G., Williams, R. L., Hu, V. J., Spinner, A., Pisera, A. O., Marks, D. S., & Liu, C. C. (2024). Continuous evolution of user-defined genes at 1 million times the genomic mutation rate. *Science (New York, N. Y.), 386*(6722), eadm9073. https://doi.org/10.1126/science.adm9073.

Rolletschek, H., Weschke, W., Weber, H., Wobus, U., & Borisjuk, L. (2004). Energy state and its control on seed development: Starch accumulation is associated with high ATP and steep oxygen gradients within barley grains. *Journal of Experimental Botany, 55*(401), 1351–1359. https://doi.org/10.1093/jxb/erh130.

Slabinski, L., Jaroszewski, L., Rychlewski, L., Wilson, I. A., Lesley, S. A., & Godzik, A. (2007). XtalPred: A web server for prediction of protein crystallizability. *Bioinformatics (Oxford, England), 23*(24), 3403–3405. https://doi.org/10.1093/bioinformatics/btm477.

Spencer, F., Hugerat, Y., Simchen, G., Hurko, O., Connelly, C., & Hieter, P. (1994). Yeast *kar1* mutants provide an effective method for YAC transfer to new hosts. *Genomics, 22*(1), 118–126. https://doi.org/10.1006/geno.1994.1352.

Sun, J., Sigler, C. L., Beaudoin, G. A. W., Joshi, J., Patterson, J. A., Cho, K. H., et al. (2019). Parts-prospecting for a high-efficiency thiamin thiazole biosynthesis pathway. *Plant Physiology, 179*(3), 958–968. https://doi.org/10.1104/pp.18.01085.

Tian, R., Zhae, R., Guo, H., Yan, K., Wang, C., Lu, C., ... Liu, Y. (2023). Engineered bacterial orthogonal DNA replication system for continuous evolution. *Nature Chemical Biology, 19*, 1504–1512. https://doi.org/10.1038/s41589-023-01387-2.

Vallen, E. A., Hiller, M. A., Scherson, T. Y., & Rose, M. D. (1992). Separate domains of KAR1 mediate distinct functions in mitosis and nuclear fusion. *Journal of Cell Biology, 117*(6), 1277–1287. https://doi.org/10.1083/jcb.117.6.1277.

Van Gelder, K., Gayen, A. K., & Hanson, A. D. (2025). Mirages in continuous directed enzyme evolution: A cautionary case study with plantized bacterial THI4 enzymes. *Plant Biotechnology Journal, 23*(4), 1070–1072. https://doi.org/10.1111/pbi.14563.

Van Gelder, K., Oliveira-Filho, E. R., García-García, J. D., Hu, Y., Bruner, S. D., & Hanson, A. D. (2023). Directed evolution of aerotolerance in sulfide-dependent thiazole synthases. *ACS Synthetic Biology, 12*(4), 963–970. https://doi.org/10.1021/acssynbio.2c00512.

Van Oss, S. B., Parikh, S. Bipin, Castilho Coelho, N., Wacholder, A., Belashov, I., Zdancewicz, X., et al. (2022). On the illusion of auxotrophy: *met15Δ* yeast cells can grow on inorganic sulfur, thanks to the previously uncharacterized homocysteine synthase YII058w. *Journal of Biological Chemistry, 298*(12), 102697. https://doi.org/10.1016/j.jbc.2022.102697.

Wang, Y., Xue, P., Cao, M., Yu, T., Lane, S. T., & Zhao, H. (2021). Directed evolution: Methodologies and applications. *Chemical Reviews, 121*(20), 12384–12444. https://doi.org/10.1021/acs.chemrev.1c00260.

Wellner, A., McMahon, C., Gilma, M. S. A., Clements, J. R., Clark, S., Nguygen, K. M., et al. (2021). Rapid generation of potent antibodies by autonomous hypermutation in yeast. *Nature Chemical Biology, 17*, 1057–1064. https://doi.org/10.1038/s41589-021-00832-4.

Yu, J. S. L., Heineike, B. M., Hartl, J., Aukakh, S. K., Correia-Melo, C., Lehmann, A., et al. (2022). Inorganic sulfur fixation via a new homocysteine synthase allows yeast cells to cooperatively compensate for methionine auxotrophy. *PLoS Biology, 20*(12), e3001912. https://doi.org/10.1371/journal.pbio.3001912.

Zhang, X., Eser, B. E., Chanani, P. K., Begley, T. P., & Ealick, S. E. (2016). Structural basis for iron-mediated sulfur transfer in archaeal and yeast thiazole synthases. *Biochemistry, 55*(12), 1826–1838. https://doi.org/10.1021/acs.biochem.6b00030.

Zhong, Z., & Liu, C. C. (2019). Probing pathways of adaptation and continuous evolution. *Current Opinion in Systems Biology, 14*, 18–24. https://doi.org/10.1016/j.coisb.2019.02.002.

Zhong, Z., Ravikumar, A., & Liu, C. C. (2018). Tunable expression systems for orthogonal DNA replication. *ACS Synthetic Biology, 7*(12), 2930–2934. https://doi.org/10.1021/acssynbio.8b00400.

CHAPTER TWO

Evolutionary reconstruction of thiamine biosynthesis pathway: An integrative bioinformatics workflow

Naiara Almeida de Toledo[a], Marie-Anne Van Sluys[a,*], and Henrique Moura Dias[b,*]

[a]Departamento de Botânica, Instituto de Biociências, Universidade de São Paulo, Sao Paulo, SP, Brazil
[b]Department of Agronomy, Horticulture and Plant Science, South Dakota State University, Brookings, SD, United States
*Corresponding authors. e-mail address: mavsluys@usp.br; henrique.dias@sdstate.edu

Contents

Methods in Enzymology, Volume 735
ISSN 0076-6879, https://doi.org/10.1016/bs.mie.2026.06.014

Abstract

Thiamine (vitamin B1) biosynthesis involves enzymatic activities that exhibit remarkable diversity across the tree of life, including deep sequence divergence, domain rearrangements, lineage-specific duplications and losses, and horizontal gene transfer events. These features complicate computational inference based on single evidence types. This chapter describes a reproducible, modular bioinformatics workflow for pathway-scale comparative analysis of thiamine biosynthesis genes across Bacteria, Archaea, and Eukarya. The workflow integrates four complementary evidence streams: (i) profile-based homolog discovery using Hidden Markov Models (HMMs) with domain architecture validation; (ii) per-family phylogenetic reconstruction to evaluate evolutionary relationships, distinguish orthologs from paralogs, and identify lineage-specific patterns; (iii) detection of TPP riboswitches using covariance models, with an overview of upstream promoter motif discovery as a complementary observational approach; and (iv) transcriptome integration to assess context-dependent gene expression. We further demonstrate how variation in protein domain architecture, particularly single versus multi domain configurations, shape sequence recovery, phylogenetic inference, and evolutionary interpretation. Step-by-step protocols cover computational environment setup, parameter optimization, quality control checkpoints, and interpretation guidelines. The workflow addresses common bioinformatic challenges and annotation inconsistencies across databases. All scripts, HMM profiles, and example datasets are available through GitHub, enabling researchers to apply these methods to thiamine pathway analysis or adapt them for other metabolic pathways with similar evolutionary diversity.

1. Introduction

1.1 Thiamine: An essential cofactor with complex biosynthetic origins

Vitamin B1 (thiamine) is an important cofactor for all living cells, as it plays an essential role in carbohydrate and amino acid metabolism, stress responses (Fitzpatrick et al., 2012) and even can reduce mtDNA damage (Machado et al., 1997). *De novo* thiamine biosynthesis results from the condensation of hydroxymethylpyrimidine (HMP) and thiazole moiety, which is universally conserved. The genetic and enzymatic strategies underlying their biosynthesis, however, vary widely across lineages (Gonçalves & Gonçalves, 2019; Rodionov et al., 2002), as illustrated in Fig. 1 (plants: green circles; fungi: purple outlined circles; archaea: blue squares; bacteria: pink triangles). *De novo* thiamine biosynthesis requires ATP, carbon and nitrogen-based intermediates, and a source of sulfur incorporated into the thiazole ring (Jurgenson et al., 2009), highlighting the metabolic cost and biochemical integration underlying precursor formation.

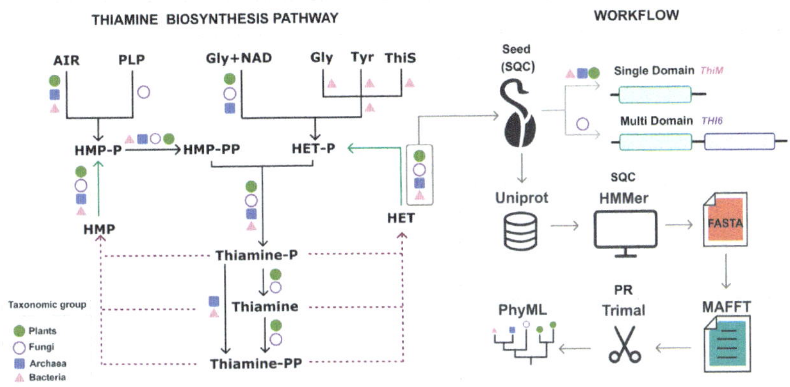

Fig. 1 General vitaminB1 biosynthetic pathway and workflow. (I) Known thiamine biosynthetic pathway in archaea, bacteria, plants and fungi. Solid arrows indicate *de novo* biosynthesis, dashed arrows indicate thiamine breakdown, and green arrows represent *salvage* pathways. Metabolites are labeled according to biochemical convention and converge into thiamine monophosphate (Thiamine-P) and thiamine pyrophosphate (Thiamine-PP). **(II)** Overview of the modular protocol presented in this chapter. Each step corresponds to a section of the protocol based on a computational framework for the identification and comparative analysis of genes involved in the biosynthesis and activation of thiamine (Thiamine/ThMP/ThDP), based on curated gene families and associated Pfam/HMM profiles. It does not include experimental validation, *de novo* genome assembly, or extensive manual re-annotation of primary genome resources. Optional modules (promoter analysis and transcriptome integration) depend on the availability and comparability of genome coordinates, gene models, and associated metadata. Green arrows indicate of the salvage pathway to recycle HET and HMP moieties. Acronyms and EC numbers when available.

The hydroxymethyl pyrimidine (HMP) synthesis varies across organisms. In bacteria, archaea, and plants, it is mainly produced by phosphomethylpyrimidine synthase (THIC EC:4.1.99.17), an [Fe–S] radical SAM enzyme that converts 5-aminoimidazole ribonucleotide (AIR) into hydroxymethyl pyrimidine phosphate (HMP -P). In fungi, phosphorylated HMP is generated by pyrimidine precursor biosynthesis enzyme (THI5 EC not assigned) using pyridoxal-5′-phosphate (PLP) as substrate via a single-turnover mechanism that leads to irreversible enzyme inactivation (Jurgenson et al., 2009). This alternative pathway also occurs in some bacteria, such as *Legionella pneumophila* (Paxhia et al., 2021).

In bacteria, the thiazole moiety, hydroxyethyl thiazole (HET) is synthesized via a multistep pathway, which names and functional descriptions are summarized in Table 1, including a sulfur-carrier protein ThiS, ThiF (EC 2.7.7.73), ThiI (EC 2.8.1.4), ThiH (EC 4.1.99.19), ThiO (EC 1.4.3.19),

Table 1 Seed query and profile summary for thiamine-form synthesis of genes and proteins.

Gene/protein	Function	Taxonomic distribution (KEGG)	Domain	Pfam domain (s)	Expected length (aa)	References
Thiazole branch						
Thiamine thiazole synthase (THI4, THI1)	Thiazole precursor of thiamine biosynthesis; iron-dependent suicide enzyme.	Plants, Fungi and Archaea	Thi4 family	PF01946	240–300	Chatterjee et al., Hwang et al., Godoi et al.
Sulfur carrier protein (ThiS)	Sulfur donor in the synthesis of the thiazole phosphate moiety.	Bacteria	ThiS family	PF02597	60–70	Duda et al., Lehmann et al.
Sulfur carrier protein ThiS adenylyltransferase (ThiF)	Catalyzes the adenylation by ATP of the carboxyl group of the C-terminal glycine of sulfur carrier protein ThiS.	Bacteria	Thif	PF00899	Around 250	Duda et al., Lehmann et al.
Glycine oxidase (ThiO)	FAD-dependent oxidase producing dehydroglycine for thiazole ring formation in thiamine biosynthesis	Bacteria	(DAO) FAD dependent oxidoreductase	PF01266	300–700	Job et al.
2-iminoacetate synthase (ThiH)	Radical-mediated cleavage of tyrosine to 2-iminoacetate and 4-cresol, required for thiazole formation in thiamine biosynthesis	Bacteria and Archaea	Radical SAM	PF04055	200–300	Kriek et al.

Pyrimidine branch						
Phosphomethylpyrimidine synthase (THIC)	HMP-P moiety synthesis of thiamine; radical SAM-dependent enzyme.	Plants Archaea, diatoms and bacteria	Radical SAM ThiC family	PF01964	Around 600	Palmer and Downs, Martinez-Gomez et al., Rodionov et al. (2002)
Pyrimidine precursor biosynthesis enzyme (THI5)	Required for biosynthesis of the pyrimidine moiety of thiamine (HMP-P); thiamine-regulated protein.	Fungi	NMT1/THI5 like	PF09084	Around 200	
Thiamine-phosphate synthase/TenI (ThiE)	Catalyzes condensation of thiazole and pyrimidine intermediates to form thiamine monophosphate (TMP).	Bacteria and Archaea	TMP-TENI	PF17792	250–300	Horn et al., Muller et al.
Thiamine phosphate synthase (ThiN or ThiDN)	Catalyzes condensation of thiazole and pyrimidine intermediates to form thiamine monophosphate (TMP).	Archaea	ThiN	PF10120	100–450	Morett et al. (2003), Hayashi et al. (2014)

(continued)

Table 1 Seed query and profile summary for thiamine-form synthesis of genes and proteins. (*cont'd*)

Gene/protein	Function	Taxonomic distribution (KEGG)	Domain	Pfam domain (s)	Expected length (aa)	References
Thiamine phosphorilation						
Thiamin phosphate kinase (ThiI or Ncs6)	Catalyzes sulfur transfer to tRNA (4-thiouridine) and to ThiS during thiazole synthesis in thiamine biosynthesis.	Archaea and bacteria	ThiI	PF02568	250–400	Liu et al., Kambampati and Lauhon
Salvage branch						
Hydroxyethylthiazo-le kinase (ThiM)	Unique enzyme responsible for thiazole salvage in thiamine biosynthesis.	Plants, Archaea and Bacteria	HK	PF02110	Around 270	Mizote et al. (1999), Yazdani et al. (2013)
HMP/HMP-P kinase (ThiD)	Unique enzyme responsible for HMP salvage in thiamine biosynthesis.	Bacteria and Archaea	Phos_pyr_kin	PF08543	Around 250	Hayashy et al., Morett et al. (2003)
Commonly Associated Domains or Fused or Co-occurring Domains						
TH1 or ThiDE	Bifunctional enzyme catalyzing HMP-P phosphorylation and TMP formation.	Plants	Phos_pyr_kin	PF08543	Around 500	Komeda et al. (1988)
			TMP-TENI	PF02581		

THI6	A bifunctional enzyme that catalyzes the condensation of thiazole and pyrimidine intermediates to form TMP and mediates thiazole salvage in thiamine biosynthesis	Fungi	HK	PF02110	Around 500	Nosaka et al. (1994)
			TMP-TENI	PF02581		
Hydroxymethylpyrimidine/phosphomethylpyrimidine kinase (THI20)	Catalyzes HMP phosphorylation to HMP-PP and thiamine degradation via thiaminase II activity	Fungi	Phos_pyr_kin	PF08543	500	Onozuka et al. (2008)
			TENA_THI-4	PF03070		

ThiG (EC 2.8.1.10), and 1-deoxy-D-xylulose-5-phosphate synthase (Dxs EC:2.2.1.7); collectively, this system is often referred to as the ThiSFGO complex (Dias et al., 2024; Gonçalves & Gonçalves 2019). In plants, fungi, and archaea, thiazole (HET) biosynthesis is catalyzed by Thiamine thiazole synthase (THI1/THI4 EC 2.4.2.60), a single-turnover enzyme that donates sulfur from an internal conserved cysteine residue, becoming irreversibly inactivated. Thiazole formation in this pathway derives from glycine (Gly) and NAD^+, contrasting with the recyclable sulfur transfer system observed in bacteria (Dias et al., 2024).

Although hydroxymethylpyrimidine phosphate (HMP-P) and hydroxyethyl thiazole phosphate (HET-P) are distinct precursors, they are condensed to form thiamine monophosphate and are co-regulated to balance metabolite availability and energy demand. In some organisms, genes from both branches are controlled by thiamine pyrophosphate (TPP) riboswitches, RNA elements located in untranslated regions (UTRs) that bind metabolites and regulate gene expression through conformational changes (Serganov & Nudler, 2013). When TPP levels are high, binding to the riboswitch represses biosynthetic genes such as thiC and thi1 (Dias et al., 2024), acting as a negative feedback mechanism. In bacteria, this typically occurs through sequestration of the Shine-Dalgarno sequence, inhibiting translation initiation (Rinaldi et al., 2016), whereas in eukaryotes regulation is often mediated by alternative splicing (Bocobza et al., 2007). The TPP riboswitch (THI-box), which responds to thiamine pyrophosphate and related intermediates, is widely distributed, but not universal, across bacteria, archaea, fungi, and plants (Jurgenson et al., 2009).

In some organisms, gene and domain fusions generate multifunctional enzymes that catalyze consecutive steps, whereas others retain modular pathways with dedicated enzymes, potentially increasing regulatory flexibility (Gerdes et al., 2012; Henry et al., 2016). This diversity extends to key steps linking the pyrimidine and thiazole branches, where enzymes such as ThiD catalyze sequential phosphorylation reactions in both biosynthesis and salvage pathways (Mizote et al., 1999; Onozuka et al., 2008). In several prokaryotes, fusion events (e.g., ThiD–ThiE or ThiD–ThiN) further integrate pathway steps and may enhance metabolic efficiency by limiting intermediate diffusion (Komeda et al., 1988; Maupin-Furlow, 2018).

A similar organizational variation is observed in the thiazole branch, where ThiM functions as a thiazole kinase responsible for phosphorylating hydroxyethyl thiazole (HET), including forms acquired from the environment, during salvage as indicated by the green arrows in Fig. 1 (Mizote et al., 1999).

While ThiM occurs as a standalone enzyme in many bacteria and plants, in fungi it is fused to ThiE, forming the bifunctional enzyme THI6, which catalyzes both HET phosphorylation and thiamine phosphate synthesis (Nosaka et al., 1994). More broadly, thiamine salvage pathways enable the uptake and recycling of HMP and HET derivatives and often share enzymes with the *de novo* pathway (Maupin-Furlow, 2018; Yazdani et al., 2013), while additional enzymatic activities provided by TenA contribute to the regeneration of pyrimidine intermediates (Zallot et al., 2014). These variations illustrate the modularity of thiamine synthetic pathways across evolutionary lineages and highlight the complexity of integrating diverse enzymatic activities across organisms, motivating pathway-scale comparative analyses.

1.2 The challenge of pathways-scale comparative analysis

Thiamine biosynthesis relies on protein-coding gene families that can exhibit sequence divergence, domain rearrangements, lineage-specific duplication and loss across the tree of life (Dias et al., 2024; Gonçalves & Gonçalves, 2019; Morett et al., 2003). These features complicate inference based on single evidence types. Annotation-based approaches often suffer from inconsistencies across databases, and similarity thresholds may perform unevenly across protein families. Distinguishing orthologs from paralogs frequently requires explicit phylogenetic context (Altenhoff & Dessimoz, 2009; Promponas et al., 2015). Likewise, inference of transcriptional regulation from promoter regions depends on genome annotation quality and varies with regulatory architecture across taxa, while transcriptome-based analyses require careful control of dataset comparability and normalization (Rangannan & Bansal, 2010).

Fig. 1 summarizes the workflow that integrates multiple evidence streams can therefore improve biological interpretability and reduce false inference. Exploring the protein sequence establishes a tractable candidate set; domain and motif checks provide orthogonal validation; per-family phylogenies contextualize homolog relationships and highlight lineage-specific patterns; promoter analyses identify putative regulatory conservation under explicitly stated assumptions; and transcriptome summaries offer complementary information on when and where genes are active, provided that datasets are curated and normalized appropriately.

1.3 Scope and objectives of this chapter

This chapter describes computational workflow to identify, curate, and compare homologs of genes and proteins involved in the biosynthesis of thiamine across Eukarya, Bacteria and Archaea. The reproducible workflow supports pathway-scale analyses by integrating complementary evidence streams, including: (i) sequence-based discovery and curation of candidate protein homologs; (ii) phylogenetic reconstruction for each protein family to evaluate evolutionary relationships, paralog and lineage-specific diversification; (iii) TPP riboswitch detection using covariance models and an overview of promoter cis-regulatory motif discovery; and (iv) access to curated public transcriptome repositories to assess context-dependent gene expression across taxa.

The workflow emphasizes transparency, reproducibility, and modularity. Each step generates outputs that can be inspected, filtered, and reanalyzed as hypotheses evolve. Analytical choices, including thresholds, coverage cutoffs, alignment trimming parameters, and tree-building methods, are explicitly stated, allowing users to adapt the protocol to their specific questions and taxonomic scope.

2. Materials, equipment and reagents

This workflow is entirely computational and does not require wet-laboratory reagents or specialized equipment beyond standard research computing infrastructure. The following subsections describe the hardware configurations, software dependencies, and publicly available databases required to execute each module of the pipeline. All software tools can be installed via the Conda package manager using the environment specification file provided. No commercial licenses are required.

Note: *All scripts, HMM profiles, seed sequence sets, and Conda environment YAML files is available at GitHub repository [*https://github.com/Chapter-TPP-Evolution/TPP-evo-pipeline*]. Users are encouraged to clone the repository before proceeding, as several steps in the Methods section reference scripts and data files organized within its directory structure.*

2.1 Hardware requirements

The workflow can be executed on a standard research workstation for small to moderate datasets (up to ~1000 sequences per gene family). For analyses spanning hundreds of genomes or involving computationally intensive steps such as maximum-likelihood phylogenetic inference, access to a high-performance computing (HPC) cluster with job scheduling (e.g., SLURM) is recommended. Detailed hardware specifications for each workflow module are available in the companion repository (https://github.com/Chapter-TPP-Evolution/TPP-evo-pipeline).

2.2 Software dependencies

All software dependencies, including recommended versions, purposes, and installation channels, are documented in the companion repository (https://github.com/Chapter-TPP-Evolution/TPP-evo-pipeline/blob/main/software_tools.md). All tools are open-source and freely available for academic use. Installation via Conda is recommended to ensure version consistency and avoid dependency conflicts. A fully tested environment file is provided (envs/evo_thiamine_env.yml) to install all dependencies simultaneously.

2.3 Databases and data sources

Before data retrieval, appropriate genome and annotation databases should be selected according to the taxonomic scope of the analysis (Table 2). In this workflow, UniProtKB/Swiss-Prot serves as the primary source for seed sequences due to its high-quality manual curation, consistent functional annotation, and reliable domain assignments, making it well suited for initial homolog identification and parameter optimization. To expand taxonomic coverage beyond curated entries, sequence retrieval may be complemented with additional organism-specific databases as needed. Regardless of the source, users should ensure the availability of protein FASTA files compatible with HMMER-based searches.

Domain architecture validation is performed using Pfam and InterPro, either via their web interfaces or through locally available HMM profiles. Other resources, such as KEGG, are intended for downstream biological interpretation and are not directly used in sequence retrieval or filtering steps.

2.4 Local database organization and metadata

Before initiating sequence searches, it is essential to organize downloaded data into a structured local directory and build a unified metadata table that links each protein file to its taxonomic and assembly-level information. This step prevents errors in sequence tracking and annotation consistency during downstream filtering and reporting and is required by several companion scripts. A directory structure is provided in the companion repository as a recommended starting point but can be freely adapted to suit the organizational needs of each project.

Note: *For organisms with multiple genome assemblies or annotation versions available (e.g., model organisms in Ensembl), prioritize the most recent reference assembly with a complete chromosome-level assembly. For non-model organisms, genome assembly quality (N50, completeness) should be evaluated using BUSCO prior to inclusion in comparative analyses.*

Table 2 Publicly available genome and annotation databases recommended for sequence retrieval and taxonomic sampling. summarizes publicly available genome and annotation databases recommended for sequence retrieval and taxonomic sampling. All listed resources are freely accessible for academic and non-commercial use, and the database statistics correspond to the date of last access indicated. Because database content, assembly counts, and gene model annotations are frequently updated, it is essential to record the exact database version and date of access for all data downloads to ensure reproducibility. Whenever possible, users should also download and retain a local copy of the proteome FASTA files used in the analysis. For organisms with multiple genome assemblies or annotation versions available (e.g., model organisms in Ensembl), the most recent reference assembly with chromosome-level completeness should be prioritized. In the case of non-model organisms, genome assembly quality metrics (e.g., N50 and completeness) should be evaluated—-preferably using BUSCO—before inclusion in comparative analyses.

Name	Integrated tools and resources	Taxonomic coverage	Genomes	Last access (M/D/Y)	References
NCBI- Refseq	Genome annotation; BLAST; comparative genomics; functional and variation resources	All domains of life (including viruses)	170,401	01/15/26	Pruitt et al., 2007
Swiss-Prot and Uniprot	Functional annotation; GO terms; protein domains; manual curation	All domains of life (including viruses)	~570,000	01/15/26	Bairoch and Apweiler (1997)
KEGG	Metabolic pathways; functional modules; genome mapping; pathway visualization	All domains of life (including viruses)	11,476	01/15/26	Kanehisa et al. (2016)

Phytozome v14	Comparative genomics; gene family analysis; functional annotation for plant genomes	Land plants	439	01/15/26	Goodstein et al. (2012)
Phycocosm (JGI)	Genome annotation; comparative genomics; metabolic reconstruction; gene family analysis; functional annotation.	Algae (emphasis on chlorophytes)	348	03/16/26	Grigoriev et al. (2021)
Ensembl	Genome annotation; comparative genomics; orthology and gene family inference; synteny analysis.	Vertebrates and selected eukaryotes	348	01/15/26	Yates et al. (2019)
Ensembl Bacteria	Genome annotation; comparative genomics; orthology and gene family inference; synteny analysis.	Bacteria	31,332	01/15/26	Kersey et al. (2010)
Ensembl plants	Genome annotation; comparative genomics; orthology and gene family inference; synteny analysis.	Land plants	240	01/15/26	Kersey et al. (2010)

(continued)

Table 2 Publicly available genome and annotation databases recommended for sequence retrieval and taxonomic sampling. summarizes publicly available genome and annotation databases recommended for sequence retrieval and taxonomic sampling. All listed resources are freely accessible for academic and non-commercial use, and the database statistics correspond to the date of last access indicated. Because database content, assembly counts, and gene model annotations are frequently updated, it is essential to record the exact database version and date of access for all data downloads to ensure reproducibility. Whenever possible, users should also download and retain a local copy of the proteome FASTA files used in the analysis. For organisms with multiple genome assemblies or annotation versions available (e.g., model organisms in Ensembl), the most recent reference assembly with chromosome-level completeness should be prioritized. In the case of non-model organisms, genome assembly quality metrics (e.g., N50 and completeness) should be evaluated—-preferably using BUSCO—before inclusion in comparative analyses. (*cont'd*)

Name	Integrated tools and resources	Taxonomic coverage	Genomes	Last access (M/D/Y)	References
Ensembl Metazoa	Genome annotation; comparative genomics; orthology and gene family inference; synteny analysis.	Metazoa	386	01/15/26	Kersey et al. (2010)
Ensembl Fungi	Genome annotation; comparative genomics; orthology and gene family inference; synteny analysis.	Fungi and yeast	1502	01/17/26	Kersey et al. (2010)

3. Methods

This section describes the complete step-by-step protocol for the computational workflow. Each subsection corresponds to a discrete analytical module and can be executed independently, provided that its required inputs, as described at the beginning of each subsection, have been generated by the preceding steps. All commands of reference scripts, parameter files, and directory structures as organized in the companion repository (see Section 2.4).

3.1 Environment setup

a. Clone or download the repository containing all scripts, HMM profiles, seed sequences, and configuration files:

```
git clone https://github.com/Chapter-TPP-Evolution/TPP_
evo_pipeline.git
cd TPP_evo_pipeline
```

b. Verify the directory structure before proceeding

```
ls -d scripts/databases/uniprot/uniprot_fasta/envs/
|| echo "ERROR: Missing directories!"
echo "Checking scripts..."
```

for script in fasta_to_phylip.py rename_uniprot.py hmmer_query_to_fasta_multi_v3.py run_riboswitch_detection.py; do[-*f* "scripts/$script"] && echo " ✓ $script" || echo " ✗ $script MISSING"

```
done
```

Expected files include:

```
scripts/                    # Python workflow scripts
databases/uniprot/          # UniProt FASTA databases
envs/                       # Conda environment files
README.md
```

Note: *The repository URL is available at* https://github.com/Chapter-TPP-Evolution/TPP-evo-pipeline.Check *the repository README for a full description of the directory hierarchy.*

c. Create the environment using the pinned YAML specification file distributed with the repository and activate the Conda environment:

```
# For Linux:
conda env create -f envs/evo_thiamine_env_linux.yml
```

```
# For macOS:
conda env create -f envs/evo_thiamine_env_macOs.yml

# Activate the environment
conda activate evo_thiamine_env
```

d. Verify that all required tools are available and report versions:

```
conda activate evo_thiamine_env
hmmsearch -h | head
python -version
```

e. Confirm that key tools are available and report the expected versions:

```
echo "===Tool Versions==="
hmmsearch -version     # Expected: HMMER 3.4
mafft -version         # Expected: MAFFT v7.526
trimal -version        # Expected: trimAl 1.5.1
infernal -version      # Expected: Infernal 1.1.5 (for
                         riboswitch)
iqtree -version        # Expected: IQ-TREE version 3.0.1
python -version        # Expected: Python 3.9+
```

Note: *If any tools are missing or show unexpected versions, the conda environment was not properly activated. Troubleshoot with:*

```
conda env list                     # Verify environment
                                     exists
conda activate evo_thiamine_env    # Re-activate
which hmmsearch                    # Check tool path
```

3.2 Defining a starting point: Seed query curation (SQC)

A critical prerequisite for pathway-scale comparative analyses is an explicit, curated definition of the seed query set used to retrieve homologs of genes involved in the synthesis of thiamine, including its phosphorylated forms. Across taxa, these genes can vary in sequence, domain architecture, copy number, and regulatory context (Dias et al., 2024). Without a documented seed set and family-specific expectations, homology discovery and filtering become difficult to reproduce and can introduce untracked bias into downstream phylogenetic inference.

In this protocol, seed queries follow four roles (Fig. 1 and Table 1): they define the target families to be analyzed (the pathway gene set), provide protein sequences for similarity, or profile-based discovery, anchor expected domain architecture and motif constraints for quality control, and establish a consistent reference framework for reporting identity/coverage and constructing comparative tables.

1. For each gene family listed in Table 1, begin by identifying a high-quality reference entry in UniProtKB/Swiss-Prot. Search for the gene or protein name of interest using the UniProt search interface, filtering by reviewed status to restrict results to manually curated Swiss-Prot entries: https://www.uniprot.org/
2. From the search results, select the entry that best represents the canonical, experimentally characterized form of the protein. Prioritize entries based on two quality indicators visible directly on the entry page (Fig. 2; I and II): Entries meeting these criteria have been manually curated and supported by primary literature, making them reliable anchors for HMM profile selection and downstream interpretation.

   Note: *If no entry with annotation score 5/5 is available for a given gene family, select the highest-scoring reviewed entry available. For families with few experimentally characterized members* (e.g., *ThiS, ThiN), entries with scores of 3/5 or 4/5 may be the best available option and document this choice explicitly in the methods.*

3. Within the selected Swiss-Prot entry, scroll to the "Family & Domains" section. This section lists all protein family and domain database cross-references for the entry, including InterPro, Pfam, HAMAP, Gene3D, and others. Locate the Pfam subsection, which displays the Pfam accession detected in the canonical sequence (e.g., PF01946 for the Thi4 family in THI1/THI4 entries). Click on the Pfam accession or the "View protein in Pfam" link, which redirects to the corresponding Pfam family page within the InterPro portal (Fig. 2; III).
4. On the InterPro/Pfam family page, review the following information before proceeding to download:

   First, confirm that the family description and short name match the expected function of the target gene; Second, Inspect the taxonomic distribution panel to verify that the family encompasses the domains of life relevant to the analysis; and; Finally, locate the "Profile HMM" section in the left navigation sidebar (Fig. 2; IV), which provides access to the downloadable HMM profile and displays the gathering threshold values (sequence-level and domain-level scores) used during profile-based searches.

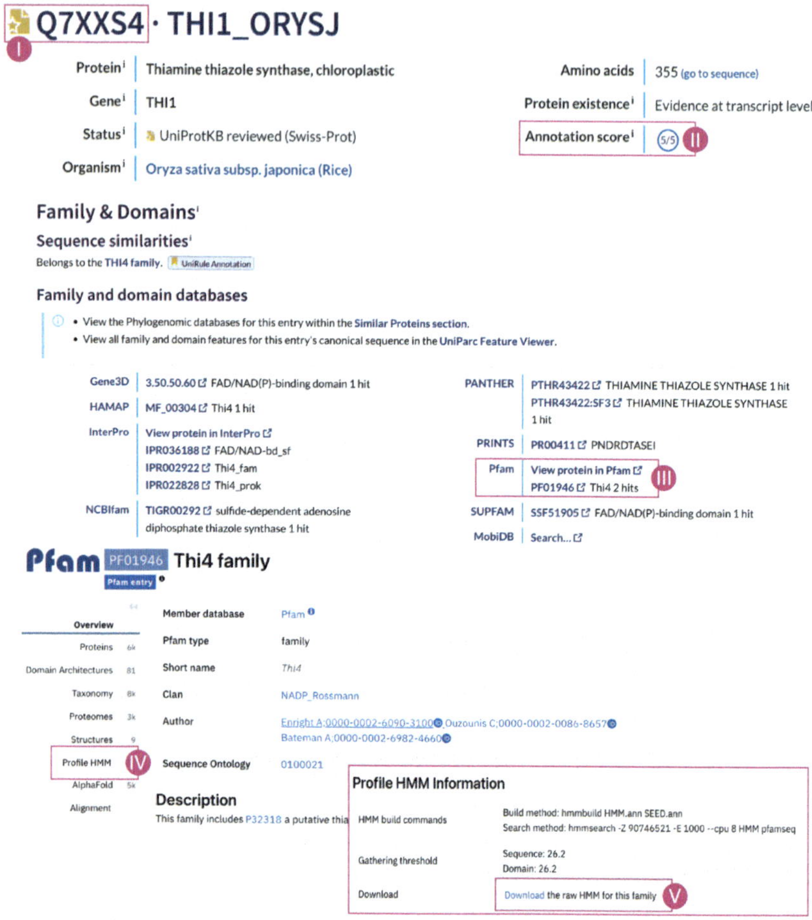

Fig. 2 Schematic overview of the UniProtKB portal illustrating the key steps to access curated entries and retrieve relevant information. (I–V) indicate the recommended navigation order to locate and download the required data. (I) the gold star icon in the top-left corner of the entry, which confirms UniProtKB reviewed (Swiss-Prot) status and manual expert curation; and (II) the annotation score of 5/5, displayed in the entry header, which reflects the highest level of functional and structural annotation confidence available in the database. (III) Inspect the protein domain section, which links to the corresponding Pfam interface; (IV) Locate the "Profile HMM" section; and (V) download the HMM profile.

Note: *Some Swiss-Prot entries contain multiple Pfam domain hits, particularly for multi-domain or fusion proteins (e.g., TH1/ThiDE in plants carries both PF08543 and PF02581; THI6 in fungi carries PF02110 and PF02581). For these entries, repeat steps 2 and 3 for each Pfam accession listed, downloading one HMM profile per domain, which will be used together in the compound hmmsearch step.*

5. Within the "Profile HMM" section of the Pfam entry page on InterPro, click "Download the raw HMM for this family" (Fig. 2; V) to download the profile file in HMMER3 format (.hmm). Save each downloaded profile to the hmm_profiles/ directory of the companion repository, using the Pfam accession as the filename.

The examples below illustrate how profiles can be organized; however, the specific families used will depend on the genes of interest in each analysis:

```
#Example:
hmm_profiles/
  PF02110.hmm              # THIM – HK family
  PF09084.hmm              # THI5 - NMT1/THI5-like family
  PF17792.hmm              # ThiE - TMP-TenI family
```

Note: *Representative HMM profiles were selected for parameter optimization, including single-domain (THI1/THI4) and multi-domain (THIM + THI6) proteins, whose domain architectures are well characterized. These cases were used to validate script behavior and refine filtering thresholds based on expected outputs. Additional tests were performed using TH1; however, as this family is largely restricted to plants, its representation in Swiss-Prot is limited, resulting in a reduced number of sequences for robust benchmarking.*

3.3 Profile-based homolog discovery (PHD)

Profile-based homolog discovery using Hidden Markov Models (HMMs) is the primary sequence search strategy in this workflow. HMM searches are substantially more sensitive than pairwise BLAST for detecting divergent homologs across deep evolutionary distances, making them particularly suited for cross-kingdom analyses of thiamine biosynthesis genes. For each gene family, the validated Pfam HMM profile downloaded in Section 3.2 (SQC) is searched against the target protein database(s) using the companion script hmmer_query_to_fasta_multi_v3.py. This script automates the full process: it runs hmmsearch for one or more HMM profiles against one or multiple FASTA databases, applies user-defined domain-centric filters, handles multi-domain architecture requirements, and produces filtered tabular reports and a ready-to-use FASTA file of passing sequences.

The script produces the following outputs per run, all written to the specified –outdir:

```
#Per-database raw outputs (one set per HMM x database combination):
<PREFIX>__<HMM_ID>__<DB_ID>.tblout.tsv       # per-sequence hit table
<PREFIX>__<HMM_ID>__<DB_ID>.domtblout.tsv    # per-domain hit table
<PREFIX>__<HMM_ID>__<DB_ID>.report.txt       # full hmmsearch report

#Combined outputs across all databases:
<PREFIX>__allDB__domains.filtered.tsv        # all passing domain hits
<PREFIX>__allDB__targets.filtered.tsv        # one row per passing target
<PREFIX>__allDB__filtered.faa                # filtered FASTA (final output)
<PREFIX>__allDB__run_parameters.txt          # full provenance record
```

a. For gene families encoded by a single Pfam domain (e.g., THIC, THI4/THI1, ThiS, ThiE), provide a single HMM profile with --hmm and point the script to the directory containing the target proteome FASTA files with --db_dir. The script will automatically detect all FASTA files in the directory (extensions: .fa, .faa, .fasta, .fna, and their .gz variants) and run hmmsearch against each one:

```
#Example for single-domain gene family:
python scripts/hmmer_query_to_fasta_multi_v3.py \
  --hmm hmm_profiles/PF02110.hmm
  --db_dir data/proteomes/\
  --e_seq 1e-5 \
  --e_dom 1e-3 \
  --min_query_cov 0.80 \
  --cpu 8 \
  --outdir results/hmmer/THIM/
```

b. The key filtering parameter is --min_query_cov, which sets the minimum fraction of the HMM profile length that must be covered by each reported domain alignment. This domain-centric coverage threshold (computed as (hmm_to-hmm_from+1)/ profile_length) is preferred over target sequence coverage (--min_target_cov) because it does not penalize multi-domain proteins where the target is substantially longer than the queried domain. By default, the script operates in domain-only mode (--domain_only), meaning target coverage is not applied and filtering relies exclusively on --e_seq, --e_dom, and --min_query_cov. Refer to Table 3 for recommended threshold values and rationale for each parameter.

Note: *An important distinction: the "domains" reported in the domtblout output correspond to HMMER alignment segments matching the query HMM profile, not to Pfam/InterPro domain annotations. Refer to* Table 3 *for recommended parameter values and rationale for each HMMER search parameter.*

c. For gene families encoded by two or more Pfam domains, or for confirmed fusion proteins (e.g., TH1/ThiDE in plants, which combines PF08543 and PF02581; THI6 in fungi, which combines PF02110 and PF02581), provide all required HMM profiles simultaneously using --hmms and enforce co-occurrence of all domains in the same target sequence using --require and --min_domains:

```
#Example for THIM/THI6 bifunctional enzyme:
python scripts/hmmer_query_to_fasta_multi_v3.py
  --hmms hmm_profiles/PF02110.hmm hmm_profiles/
    PF02581.hmm \
  --require PF02110,PF02581 \
  --min_domains PF02110 = 1 PF02581 = 1 \
  --db_dir data/proteomes/ \
  --e_seq 1e-5 \
  --e_dom 1e-3 \
  --min_query_cov 0.80 \
  --cpu 8 \
  --prefix THI6\
  --outdir results/hmmer/THI6 /
```

d. The --require flag specifies which HMM IDs (derived from the filename stem, e.g., PF08543 from PF08543.hmm) must both be present in a target sequence for it to pass. The --min_domains flag sets the minimum number of passing domain hits per required HMM profile.

Table 3 HMMER search parameter guidelines. Recommended values are validated for thiamine pathway gene families spanning Bacteria, Archaea, and Eukarya. Adjust parameters for highly divergent families or specialized taxonomic scopes.

Parameter	Recommended value	Acceptable range	Rationale
--E (sequence E-value)	≤1e−5	1e−3–1e−10	Controls the per-sequence E-value threshold. Values >1e−3 increase false positives substantially for divergent families.
--domE (domain E-value)	≤1e−3	1e−2–1e−5	Controls per-domain reporting. Use together with –domtblout for domain-level filtering in the validation step.
--min_query_cov	0.8	0.70–0.80	Minimum fraction of the HMM profile that must be covered by the alignment. Lower values recover partial sequences (e.g., gene fragments); higher values are more stringent.
--min_domains	≥1 per expected domain	Protein family specific	Ensures that retrieved sequences contain the minimum number of expected Pfam domains. Critical for multi-domain and fusion proteins.
--require	Protein-family specific		For multi-domain families (e.g., TH1/ThiDE, THI6), require all expected domain HMMs to match simultaneously to avoid recovering single-domain partial hits.

Only sequences that satisfy all --require constraints at the specified --min_domains thresholds will appear in the final filtered FASTA output. This architecture enforcement step is critical for distinguishing genuine bifunctional enzymes from single-domain partial hits that would otherwise pass individual profile searches.

Note: *For multi-domain searches, use --prefix to assign a meaningful name to the output files (e.g., --prefix THI6). Without --prefix, the script defaults to the stem of the first HMM file provided, which may be ambiguous when multiple profiles are used. The prefix appears in all output filenames and in the run_parameters.txt provenance file.*

e. After each run, inspect the tabular outputs generated by the workflow before proceeding to domain validation. The domains.filtered.tsv file reports per-domain hits, including query coverage (query_cov), target coverage (target_cov), domain E-value (dom_ievalue), and full-sequence E-value (full_seq_evalue), whereas targets.filtered.tsv summarizes domain architecture per sequence, reporting the number of passing domain hits per HMM and the minimum domain E-value.

Caution: *FASTA record IDs in the target proteome files must match the target names reported in the HMMER domtblout output (the first whitespace-delimited token in each sequence header). If IDs do not match, the script will report the correct number of passing targets in the TSV tables but will extract zero sequences into the filtered FASTA, with a message indicating the discrepancy. Ensure that proteome FASTA files use clean, unambiguous identifiers before running the search.*

f. (Optional) A list of organisms can be generated from UniProt sequences, and unique codenames assigned to each protein to facilitate downstream analyses and improve phylogenetic visualization. These identifiers link FASTA headers to taxonomic classification via the UniProt REST API. In this study, codes were assigned based on domain, kingdom, or clade. Sequences were renamed using the following format:

```
<GroupLetter>_<SpeciesCode>-<GeneName>
```

GroupLetter being:

```
A = Archaea; Cy = Cyanobacteria B = Bacteria; P = Plantae;
F = Fungi; M = Metazoa
```

SpeciesCode:

`First two letters of the genus + First letter of the species`

Example:

`Oryza glumipatula → P_Org-GeneName`

Although optional, this step enhances sequence traceability and supports more accurate interpretation of evolutionary relationships in phylogenetic analyses.

3.4 Phylogenetic reconstruction (PR)

3.4.1 Reconstruction of evolutionary tree

Per-family phylogenetic reconstruction contextualizes homolog relationship identified by HMMER searches, distinguishes orthologs from paralogs, and reveals lineage-specific duplication, loss, and horizontal gene transfer events that are otherwise invisible solely from sequence similarity. The workflow follows a standard maximum-likelihood approach: multiple sequence alignment > alignments trimming > model selection and tree inference > bootstrap support evaluation > visualization.

a. Align the non-redundant, domain-validated sequence set for each gene family using MAFFT in auto mode:

```
mafft --auto --thread 8
/thiamine_synthesis_pipeline/results/hmmer/THIM/
THIM__allDB__filtered.fasta
/thiamine_synthesis_pipeline/results/msa/
msa_THIM__allDB__filtered.fasta
```

b. For gene families with highly variable loop regions or insertions (e.g., THIC, which contains a large radical SAM insertion domain), inspect the alignment visually in AliView (Larsson, 2014) or UGENE (Okonechnikov et al., 2012) before trimming. Excessively long insertions present in only a few sequences may warrant either removal of those sequences or manual masking of the insertion region.

Note: *MAFFT --auto selects FFT-NS−1 for datasets larger than ~2000 sequences and L-INS-i for smaller datasets. L-INS-i is significantly more accurate but computationally intensive. For datasets between 200 and 1000 sequences, consider forcing L-INS-i explicitly (-localpair --maxiterate 1000) if computation time permits.*

c. Remove poorly aligned, gap-rich columns from the alignment using trimAl in automated mode:

```
trimal -in
"/thiamine_synthesis_pipeline/results/msa/
msa_THIM__allDB__filtered.fasta"
-out
"/thiamine_synthesis_pipeline/results/msa/msa_
THIM_allDB__filtered_trimmed.fasta" \
-automated1
```

Note: *An alternative approach to running trimAl is through the Galaxy Europe service (*https://usegalaxy.eu/root?tool_id=trimal*), using a multiple sequence alignment (MSA) as input.*

d. After trimming, compare the number of columns retained against the original using the summary printed by trimAl to stderr. With -automated1, substantial column removal (50–70 %) is expected and methodologically appropriate for cross-kingdom datasets, where divergent loop regions and lineage-specific insertions are common (Capella-Gutiérrez et al., 2009). The relevant diagnostic is not the fraction of columns removed, but whether the remaining columns are informative: open the trimmed alignment in AliView or UGENE and confirm that conserved functional motifs for the target family (e.g., the radical SAM CxxCxxxxC motif for THIC, Chatterjee et al., 2008) are still present and well-aligned across representatives of all major clades. If conserved motifs are absent or disrupted in the trimmed output, this indicates that one or more highly divergent or fragmentary sequences are collapsing the conserved core during trimming. In this case, identify the outlier sequences, typically visible as rows with excessive gaps in the pre-trimmed alignment, remove them, and re-run the alignment and trimming steps before proceeding to phylogenetic inference.

e. Maximum-likelihood phylogenetic trees are inferred using PhyML, which implements efficient SPR (subtree pruning and regrafting) moves combined with automatic substitution model selection via SMS (smart model selection; Lefort et al., 2017). PhyML can be used either its online web server (ATGC – Montpelier Bioinformatics Platform), which is suitable for moderate sized datasets and requires no local installation, or through the command-line binary for larger datasets or automated batch processing. Only the web server mode is described in this chapter.

f. Navigate to the PhyML online execution interface at http://www.atgc-montpellier.fr/phyml/ and upload the trimmed alignment file. The input must be in PHYLIP format; convert the FASTA output from trimAl using the companion script (fasta_to_phylip.py) before uploading:

```
python fasta_to_phylip.py \
  --input results/msa/msa_THIM__allDB__filtered_trimmed
  .fasta
  --output results/msa/msa_THIM__allDB__filtered_trimmed.
  phy
```

g. Configure the run using the following settings on the web interface (Fig. 3).

1. Upload the data file;
2. Select "amino-acids" as the data type;
3. Substitution Model: select "Automatic model selection by SMS" and set the selection criterion to BIC (Bayesian Information Criterion), which penalizes model complexity more strongly than AIC and is generally preferred for phylogenetic model selection;
4. Optimization options: set Starting tree to BioNJ, enable topology optimization and edge length optimization, and enable random starting trees (recommended: 5 replicates) to reduce the risk of converging on a local likelihood maximum;
5. Branch Supports: enable "Fast likelihood-based methods" and select aLRT-SH (Shimodaira-Hasegawa-like approximate likelihood ratio test). aLRT-SH computes branch support by comparing the likelihood of the best ML tree against the best alternative topology for each internal branch, then applies an SH correction for multiple testing. This produces support values on a scale of 0–1 that are computationally very efficient (no resampling required) and conservative relative to standard bootstrap, making them well suited for large cross-kingdom datasets. Disable standard and transfer bootstrap for this run;
6. Provide your email address to receive a notification with the results link when the analysis is complete.

Note: *aLRT-SH support values are interpreted on a scale of 0–1, where values ≥ 0.9 indicate Strong support and values ≥ 0.95 are considered highly supported. These values are not directly comparable to bootstrap percentages and tend to be more conservative.*

Fig. 3 Interface of online software PhyML. Step by step procedure used in this study. (I) Input for Phylip file, this step is necessary because these formats are flexible among software's for phylogenetics. (II) Type of data, for larger and evolutionary approaches we recommend using the amino-acid sequence. (III) Smart model selection, compared to AIC, BIC more reliably recovers the correct model under large sample sizes. (IV) Tree reconstruction using BioNJ, a distance-based method that improves Neighbor-Joining (NJ). (V) Branch support estimation, among available methods (aLRT-SH, aLRT-Chi^2, and aBayes), SH-aLRT is more conservative and robust to model violations.

h. Import the PhyML output tree directly into iTOL (https://itol.embl.de/) for simultaneous topological inspection and figure preparation. Navigate to the upload page, select the Newick tree files and click upload as shown in Fig. 4. I.

i. Once loaded, apply initial display settings in the "Basic" tab (Fig. 4. III):

1. Select "Rectangular" display for datasets up to ~150 sequences or "Circular" for larger datasets;
2. Enable branch length display to distinguish short from long branches;
3. Set label font size to a legible value;
4. Root the tree on the most appropriate outgroup for the local family using "Tree > Reroot": for most thiamine pathway families, archaeal sequences (clade_code_A) or deep-branching bacterial lineages are appropriate outgroups.
5. In the "Advanced" tab, set the aLRT-SH display threshold to 0.70 to suppress weakly supported nodes and reduce visual clusters. Nodes with aLRT-SH ≥ 0.90 should be visually distinct.
6. Apply annotation for a specific gene clade, to better resolution.

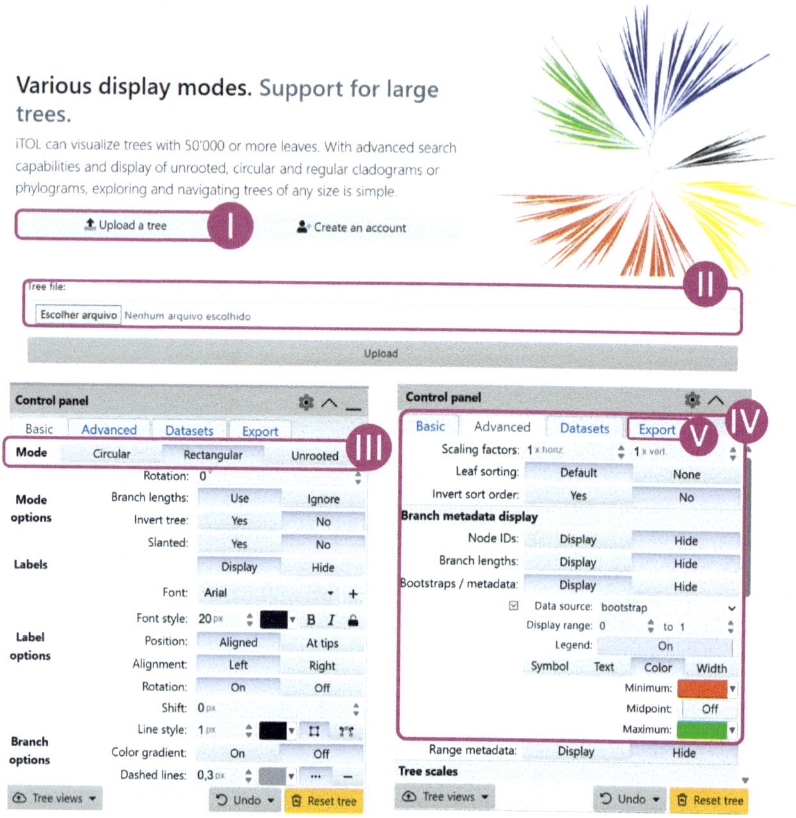

Fig. 4 Interface of online software ITOL. (I) File upload section; no login is required, but uploaded files are not stored, so modifications must be downloaded manually. (II) Input file settings, supporting formats such as Newick, Nexus, and PhyloXML, with optional naming for easier identification, as we suggested in PHD step j. (III) Basic options, including tree display mode and customization of size, fonts, and colors for labels and branches. (IV) Advanced options for metadata editing, such as visualizing bootstrap values using customizable color gradients. (V) Options of export image files.

j. With clade_code colors applied, inspect the tree for the following patterns:
   1. Monophyletic clades composed exclusively of sequences sharing the same clade_code and supported by aLRT-SH ≥ 0.90 represent high-confidence ortholog groups;
   2. Record their membership and supported value. If two or more sequences from the same organism appear in well-supported, separate parts of the tree (aLRT-SH ≥0.90 at the separating node), these are likely paralogs from lineage-specific duplication;

3. Weak support (aLRT-SH < 0.70) at a separating node makes the duplication signal ambiguous and insufficient to report without corroborating evidence (e.g., synteny).

Note: *For families where, bifunctional fusions are expected (TH1/ThiDE in plants, THI6 and THI20 in fungi), fused and non-fused forms performing the same catalytic step may appear in different parts of the tree. These are not paralogs, they represent structurally distinct solutions to the same enzymatic function.*

k. HGT candidates appear as sequences whose clade_code label is discordant with their position in the tree: for example, an eukaryotic sequence (P or F) embedded within a well-supported bacterial clade, or a bacterial sequence branching inside an archaeal clade. Inspect those carefully and evaluate each candidate using the following criteria directly in iTOL:

1. Confirm the clade_code is genuinely discordant, not simply sister to the expected clade. A plant sequence (P) branching within cyanobacteria (Cy) is a potential endosymbiotic gene transfer (EGT) candidate, not HGT, if the pattern is consistent across all plant sequences in the dataset;
2. Verify that the discordant placement is supported by aLRT-SH ≥ 0.90. Weakly supported placements may reflect long-branch attraction or sparse taxon sampling rather than genuine HGT
3. Check the branch length of the candidate sequence in the iTOL display: an unusually long branch relative to its neighbours is a warning sign for long-branch attraction. Use "Tree > Prune" in iTOL to remove the candidate and re-examine topology; an artifactual placement due to long-branch attraction (LBA) typically shifts or collapses when the outlier is removed;
4. Cross-check sequence length and domain completeness in the domains.filtered.tsv output from Section 3.3: unusually short sequences or those with partial domain coverage are more likely misassembled contigs than genuine HGT events.

Caution: *Long-branch attraction (LBA) is a systematic artifact that causes rapidly evolving sequences to cluster artifactually with other long-branch sequences, mimicking HGT. Before reporting an HGT event, verify: if the candidate does not carry a disproportionately long branch; the placement is supported by aLRT-SH ≥ 0.90; and the placement is robust to removal of other long-branch sequences.*

1. Export the final annotated tree using "Export" in the iTOL toolbar (Fig. 4. V). SVG format is recommended for publication figures as it preserves vector quality and allows post-processing in Inkscape or Adobe Illustrator. PDF is suitable for direct submission; PNG at 300 dpi is acceptable for journal pre-submission checks.

3.5 Regulatory sequence analyses (RSA)

3.5.1 TPP-riboswitch detection using covariance models

Thiamine pyrophosphate (TPP) riboswitches are conserved RNA secondary structures in mRNA molecules that regulate gene expression post-transcriptionally in response to intracellular TPP concentration (Winkler et al., 2002). Depending on the organism, TPP binding leads to transcription termination, translation inhibition, or alternative splicing. For thiamine pathway genes, particularly *Thi1/Thi4* and *ThiC*, the presence or absence of a riboswitch reflects whether the gene is regulated in a TPP-responsive manner (Dias et al., 2024; Jurgenson et al., 2009). Covariance model-based detection with Infernal (Nawrocki & Eddy, 2013) is the gold standard for riboswitch identification because it simultaneously models sequence conservation and RNA secondary structure through base-pair covariation, which is not captured by sequence homology methods alone. This section describes the protocol for detecting TPP riboswitches (RF00059) across the full set of homologs identified in Section 3.3.

a. Activate the workflow Conda environment before running any commands in this section:

```
conda activate evo_thiamine_env
```

b. Navigate to the directory where the covariance model file will be stored. Create the directory if it does not already exist:

```
mkdir -p [cm_profiles/]
cd cm_profiles
```

c. Download the covariance model (CM) file for RF00059 (TPP riboswitch) from the Rfam database (Ontiveros-Palacio et al., 2024). The CM file encodes both the sequence and secondary structure consensus of the TPP riboswitch family and is required as input for cmsearch:

```
wget -q "https://rfam.org/family/RF00059/cm" -O RF00059.cm
head -5 RF00059.cm
```

Note: *The head command prints the first five lines of the CM file. Confirm that the output begins with INFERNAL1/a and that the name field reads TPP. This verifies the download is complete and the file corresponds to the correct family.*

d. Index the CM files using cmpress to create the binary index files required by cmmsearch. This step must be completed before any search can be run:

```
cmpress RF00059.cm
ls -lh RF00059.cm*
```

Note: *After successful indexing, it should show four additional files: RF00059.cm.i1f, RF00059.cm.i1i, RF00059.cm.i1m, and RF00059.cm.i1p. If any of these files are missing, re-run cmpress before proceeding. cmsearch will fail with an error if the index is absent or incomplete.*

e. Inspect the gathering threshold (GA) embedded in the CM file. The GA is the minimum bit score used by Rfam curators to define true family members; it is the threshold enforced by the --cut_ga flag used in Step 6:

```
Grep "GA " RF00049.cm
```

Note: *The output will show two numbers (e.g., GA 37.60 37.60). The first value is the sequence-level gathering threshold; the second is the domain-level threshold. Only the sequence-level value (first number) is enforced by --cut_ga in cmsearch. Record this value, it will appear in the summary report produced in Step 6 and should be reported in the methods section of the manuscript.*

f. Run cmsearch and parse the results using the run_riboswitch_detection.py script. The script searches all FASTA files in –seqdir against RF00059.cm in a single step and produces a consolidated hit table, an annotated TSV with UTR context assessment, and a human-readable summary report:

```
BASE=thiamine_synthesis_pipeline/
  python scripts/run_riboswitch_detection.py \\
    --seqdir $BASE/results/hmmer/THI1/
    --outdir $BASE/results/cmsearch/THI1/
    --cm cm_profiles/RF00059.cm
    --cpu 8
```

> ***Note:*** *The --seqdir argument should point to the directory containing the filtered FASTA files produced by Section 3.3 (one FASTA per gene family per database). For each FASTA file, the script writes a per-sample.tblout and.report file, then consolidates all hits into summary_report.tsv, hits_annotated.tsv, and summary_report.txt in --outdir.*

g. Inspect the output files to assess detection results. Open summary_report.txt for an overview of hits per sample. For detailed per-hit inspection, open hits_annotated.tsv in a spreadsheet application. The key columns for interpretation are:

1. (score_bits): bit score of the hit; must exceed the GA threshold (Step 5) to be significant under --cut_ga;
2. (bits_above_GA): score margin above the threshold; values ≥ 10 bits above GA indicate high-confidence hits. Values within 1–5 bits of the threshold warrant manual inspection;
3. (mdl_coverage_pct): percentage of the RF00059 consensus (79 nt) covered by the hit alignment. Values below 70% may indicate a partial or degraded riboswitch element;
4. (utr_context): estimated position of the hit within the sequence window (5UTR_likely, 3UTR_likely, or gene_body). Functional TPP riboswitches in eukaryotes are invariably located in UTRs; hits in gene body positions are unlikely to be functional and should be flagged for manual review;
5. (evalue): expectation value. For genome-scale searches, use bit score and the GA threshold as the primary filter; E-values are less reliable when sequence databases vary in size across samples;
6. (strand): strand orientation of the hit. Riboswitches are always on the same strand as the mRNA; hits on the antisense strand are non-functional and should be excluded.

h. If known positive and negative controls are available, for example, sequences from taxa where TPP riboswitch presence and absence has been experimentally confirmed (Bocobza et al., 2007; Dias et al., 2024), rerun the script with the –positives and –negatives arguments to compute sensitivity.

```
python $BASE/workflow/scripts/run_riboswitch_detection
.py \
    --seqdir $BASE/results/hmmer/THI1/ \
    --outdir $BASE/results/cmsearch/THI1/ \
```

```
--cm [cm_profiles/]RF00059.cm \
--cpu 8 \
--positives Creinhardtii,Vcarteri,Ppatens,
Smoellendorffii \
--negatives Athaliana,Osativa,Zmays,Slycopersicum
```

Note: *Sample names passed to --positives and --negatives must match the FASTA filename stems exactly* (e.g., *Ppatens for Ppatens.fa). The script prints sensitivity (fraction of known positives recovered), false positive rate (fraction of known negatives with unexpected hits), a pass/fail verdict. For Thi1/Thi4, the expected result under --cut_ga is recovery of all chlorophyte (5′-UTR) and non-seed embryophyte (3′-UTR) sequences, with zero hits in seed plant sequences where the riboswitch has been lost* (Dias et al., 2024).

Caution: *TPP riboswitches in eukaryotes can be located in either the 5′-UTR or the 3′-UTR. Detection therefore requires that input sequences include adequate flanking sequence on the correct side of the CDS. If input sequences were prepared from genomic coordinates, confirm that --upstream and --downstream windows (default: 1500 bp each) are sufficient to capture UTR regions for the taxa under study. CDS-only FASTA files will produce false negatives for all riboswitch-positive taxa, regardless of the bit score threshold used.*

3.5.2 Promoter motif analysis in eukaryotes

In addition to post-transcriptional regulation via riboswitches, the promoter regions of thiamine pathway genes may harbor conserved cis-regulatory motifs that coordinate transcriptional responses to TPP availability or metabolic signals. Motif discovery tools such as MEME suite (*De novo* Motif Discovery; Bailey et al., 2009) can identify overrepresented sequence elements in the upstream regions of co-regulated genes. For plant genomes, two conserved promoter motifs shared across all plant *Thi1/Thi4* lineages are identified by extracting a fixed upstream window relative to the annotated start codon and applying MEME to the resulting sequence set (Dias et al., 2024). The general approach is applicable to any eukaryotic thiamine pathway family for which a curated genome annotation is available.

Note: *This approach is not included as a formal protocol in this chapter because its applicability is contingent on three conditions that are not universally met across the taxonomic breadth of this workflow. First, it requires a well-curated genome annotation with reliable start codon coordinates; misannotated or partial gene models* *will produce artifactual upstream windows that undermine motif discovery. Second, the choice of window size upstream of the start codon is not standardized and varies substantially across studies and organisms, typical values range from 500 to 2000 bp, but the biologically meaningful regulatory window may be shorter (in compact genomes) or longer (in organisms with complex promoter architectures). Because MEME is sensitive to input window size, results obtained with different window sizes are not directly comparable across gene families or taxa. Third, MEME produces purely statistical motifs with no built-in significance threshold equivalent to the Rfam gathering threshold (GA) used in RF00059-based riboswitch detection; all candidate motifs require independent experimental or comparative validation before biological interpretation. Researchers wishing to apply this approach should consult the MEME Suite documentation (*https://meme-suite.org*) and define window size and filtering criteria explicitly in their methods.*

3.6 Transcriptome integration

In this work, sequence homology searches and phylogenetic inference establish the evolutionary framework of thiamine biosynthesis genes but do not directly address whether these genes are expressed across conditions, tissues, or developmental stages. To extend the analysis, public transcriptome repositories can be used as complementary evidence to increase the depth of functional interpretation, supporting the validation of candidate genes and the identification of lineage or condition specific expression patterns. As summarized in Table 4, resources such as Expression Atlas provide broad cross-species comparisons and baseline versus differential expression analyses, while the Bio-Analytic Resource for Plant Biology (BAR) eFP Browser (Electronic Fluorescent Pictograph) is particularly suited for model plants, enabling spatially resolved expression analyses across tissues and conditions. Phytozome is especially valuable for non-model plant lineages, offering genome-integrated expression data (e.g., RNA-seq TPM/FPKM) linked to curated gene models and functional annotations, and facilitating direct exploration of homologs identified in the workflow, particularly in algae and basal land plants (Goodstein et al., 2012). For taxa or conditions not covered by these platforms, additional datasets can be retrieved from repositories such as NCBI GEO, although variability in metadata and annotation requires careful manual curation prior to interpretation.

***Note:** When querying expression databases with homologs identified by HMMER, confirm that the gene identifier used for the query matches the identifier system of the target database. HMMER output typically uses sequence accessions*

Table 4 Public transcriptome repositories recommended for querying thiamine pathway gene expression. Summarizes publicly available transcriptome repositories recommended for querying the expression of thiamine pathway genes. In addition to general-purpose resources, specialized databases such as FungiDB and the Eukaryotic Pathogen Database Resource provide valuable expression data for fungi and other eukaryotic pathogens. However, despite these additions, comprehensive and standardized coverage of thiamine pathway gene expression across prokaryotes and fungi remains limited. For bacterial datasets and complementary fungal expression data, the NCBI Gene Expression Omnibus (GEO) remains a primary resource. GEO DataSets can be queried using combinations of gene names and organism identifiers as keywords (e.g., "thiC Bacillus subtilis" or "THI4 Saccharomyces cerevisiae"). Neither GEO nor specialized resources such as FungiDB and EuPathDB enforce fully standardized annotation or experimental metadata across studies. As a result, dataset quality, normalization approaches, and experimental context may vary substantially. For these groups, careful manual curation including inspection of associated publications, experimental design, and data processing methods is essential before interpreting gene expression patterns.

Database	URL	Taxonomic coverage	Data types	Key features
Expression Atlas	https://www.ebi.ac.uk/gxa/home	Multiple eukaryotes including plants and fungi	RNA-seq, microarray, baseline and differential expression	Manually curated gene expression datasets, visualization across tissues and experimental conditions, cross-species comparisons
BAR (Bio-Analytic Resource for Plant Biology)	https://bar.utoronto.ca	Mainly *Arabidopsis thaliana* and some crop species	Transcriptomics, co-expression networks	Interactive tools for exploring gene expression patterns and functional associations in plants
NCBI Gene Expression Omnibus	https://www.ncbi.nlm.nih.gov/geo/	Broad taxonomic coverage including plants and fungi	RNA-seq, microarray, functional genomics datasets	Public repository for curated gene expression studies with extensive metadata and experimental descriptions

(continued)

Table 4 Public transcriptome repositories recommended for querying thiamine pathway gene expression. Summarizes publicly available transcriptome repositories recommended for querying the expression of thiamine pathway genes. In addition to general-purpose resources, specialized databases such as FungiDB and the Eukaryotic Pathogen Database Resource provide valuable expression data for fungi and other eukaryotic pathogens. However, despite these additions, comprehensive and standardized coverage of thiamine pathway gene expression across prokaryotes and fungi remains limited. For bacterial datasets and complementary fungal expression data, the NCBI Gene Expression Omnibus (GEO) remains a primary resource. GEO DataSets can be queried using combinations of gene names and organism identifiers as keywords (e.g., "thiC Bacillus subtilis" or "THI4 Saccharomyces cerevisiae"). Neither GEO nor specialized resources such as FungiDB and EuPathDB enforce fully standardized annotation or experimental metadata across studies. As a result, dataset quality, normalization approaches, and experimental context may vary substantially. For these groups, careful manual curation including inspection of associated publications, experimental design, and data processing methods is essential before interpreting gene expression patterns. (*cont'd*)

Database	URL	Taxonomic coverage	Data types	Key features
FungiDB	https://fungidb.org	Diverse fungal species including model and pathogenic fungi	Transcriptomics, genomics, proteomics	Integrated fungal omics resource with curated RNA-seq datasets and tools for comparative and functional analysis
Eukaryotic Pathogen Database Resource	https://eupathdb.org	Eukaryotic microbes including fungi and protists	Transcriptomics, genomics, functional datasets	Platform integrating multiple organism-specific databases (including FungiDB) with curated expression data and analysis tools
Phytozome	https://phytozome-next.jgi.doe.gov	Green plants (Viridiplantae)	Genome assemblies, gene annotations, transcript evidence	Comparative plant genomics platform integrating transcriptomic evidence for gene annotation and functional analysis

from the source database (e.g., UniProt, RefSeq, or genome-specific locus tags), which may differ from the Ensembl gene IDs expected by Expression Atlas or the AGI identifiers used by BAR. Use the identifier mapping tools provided by each database (e.g., Ensembl BioMart, UniProt ID mapping) to convert between accession systems before querying expression data.

Caution: *Expression values from different databases, platforms, or normalization pipelines are not directly comparable. Microarray signal intensities (BAR/ATH1, primarily for Arabidopsis thaliana), RNA-seq FPKM/TPM values (Phytozome), and normalized counts from Expression Atlas each reflect different technical and biological assumptions; BAR datasets for non-Arabidopsis species are often RNA-seq-based. Cross-database comparisons should be qualitative (e.g., expressed vs. not expressed, tissue-enriched vs. ubiquitous) rather than quantitative. For quantitative cross-species comparisons, re-processing raw reads from a common pipeline with unified normalization is require.*

4. Case study: Navigating domain architecture variation in ThiM and THI6

Thiamine biosynthesis genes vary not only in sequence but in the structural organization of the proteins that carry out each pathway step. As introduced in Section 1.1, ThiM and THI6 occupy the same biochemical position that is the phosphorylation of hydroxyethyl thiazole (HET), but differ fundamentally in domain architecture and taxonomic distribution. ThiM is a standalone HK-domain enzyme in bacteria and plants, while THI6 is a bifunctional HK + TMP-TENI fusion restricted to fungi. By tracing how this fundamental difference propagates through each stage of the workflow, from search strategy and parameter choice to phylogenetic interpretation, this section illustrates how domain-aware search strategies are essential for accurate homolog recovery, and how the same methodological decisions yield different information depending on whether the target is a single-domain or multi-domain protein.

4.1 Search strategy and parameter choices: What you search for determines what you find

The shared HK domain between ThiM and THI6 creates a fundamental ambiguity in profile-based sequence searches: a query anchored exclusively on PF02110 will recover both proteins without providing any basis for distinguishing them at the retrieval stage. To make this effect explicit, four complementary search strategies are applied sequentially, each recovering a different subset of the evolutionary space occupied by these two proteins (Fig. 5, panel I).

In the first search, an HMM constructed from ThiM sequences (HK.hmm) is used as the query against all target proteomes. ThiM sequences are recovered at near-complete target coverage (target_cov ~1.0), consistent with the query profile spanning the full length of a single-domain protein. THI6 sequences from fungal proteomes are also retrieved, but at substantially lower target coverage (target_cov < 0.5). This asymmetry is a direct signal of multidomain architecture: the HK profile aligns across its full length against the N-terminal kinase domain of THI6, while the C-terminal TMP-TENI domain (PF02581) falls outside the scope of the query and remains undetected (Fig. 5, panel II). The recovery of *Saccharomyces cerevisiae* THI6 (UniProt P41835) at target_cov < 0.5 illustrates how target coverage values in the domtblout output can flag multidomain candidates without prior knowledge of their architecture.

In the second search, an HMM is built *de novo* from confirmed full-length THI6 sequences (THI6.hmm), using the fungal hits recovered in search one as a seed alignment. This profile captures sequence variation within the HK domain in the specific context of the fusion protein, improving sensitivity for divergent THI6 homologs in poorly characterized fungal lineages. The recovery pattern inverts relative to search one: THI6 sequences are now retrieved at full target coverage (target_cov ~1.0), while ThiM sequences appear at partial coverage (target_cov < 0.5), reflecting the absence of the TMP-TENI region in the single-domain hits.

In the third search, a co-occurrence filter requires simultaneous detection of both HK.hmm and TMP-TENI.hmm within the same target sequence (Section 3.3, Step c). This restricts the output to architecturally confirmed THI6 sequences and excludes ThiM entirely. The trade-off is reduced coverage: THI6 homologs where the TMP-TENI domain has diverged below the detection threshold, or where gene models are incomplete, will not satisfy the co-occurrence criterion. Sequences recovered in searches one and two but absent from search three are treated as candidates for manual inspection, potentially representing annotation gaps or partial gene models.

In the fourth search, the outputs of searches one and two are merged, dereplicated, and annotated by architectural class based on target_cov values and co-occurrence results. Full-length sequences are aligned using

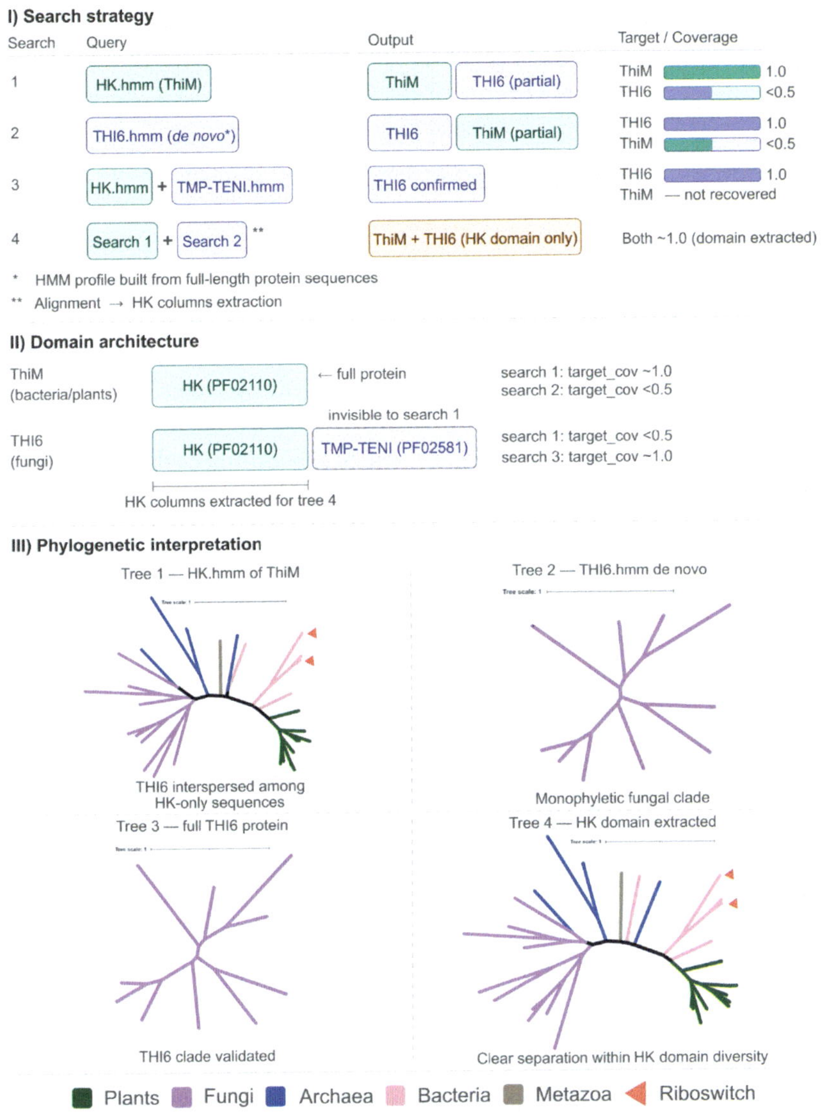

Fig. 5 HK-containing protein domain case study. (I) Distinct query strategies sample different portions of sequence space depending on domain composition, resulting in systematic differences in target coverage between ThiM and THI6. (II) ThiM proteins contain a single HK domain, whereas THI6 proteins exhibit a dual-domain architecture (HK + TMP-TENI), explaining partial versus full-length matches across searches. (III) Phylogenetic trees reflect these differences: HK-based searches capture broader diversity, full-length approaches restrict retrieval to THI6-like proteins, and domain-focused alignment improves comparability between architectures. Color coding indicates taxonomic groups as described below; red triangles mark branches corresponding to organisms in which TPP riboswitch elements were detected.

MAFFT, and the alignment columns corresponding to the HK domain are extracted using the hmm_from and hmm_to coordinates from the domtblout output. This domain-delimited alignment places ThiM and THI6 on equal terms, same region, same alignment, without the phylogenetic signal contributed by the TMP-TENI domain in THI6, and constitutes the input for the fourth phylogenetic analysis described in Section 4.2.

4.2 Phylogenetic reconstruction: Four trees, four perspectives

The resulting phylogenetic analyses based distinct SQC search strategies differ how they sample sequence space and domain architecture (Fig. 5, panels I–II). Because ThiM and THI6 share the HK domain but diverge in overall protein organization, the SQC directly determines which sequences are recovered and how they are represented in the resulting trees. Single-domain searches recover a broader and more diverse set of HK-containing sequences, including both standalone proteins and domains embedded within multi-domain architectures, whereas full-length or co-occurrence-based profiles restrict retrieval to proteins sharing the same dual-domain organization as THI6. These differences are reflected in systematic variation in target coverage values, where partial matches signal domain-level overlap rather than full-length similarity. By combining these approaches and restricting one analysis to the shared HK region, the workflow enables direct comparison between proteins of distinct architectures while maintaining explicit control over sensitivity and specificity. This framework underlies the phylogenetic patterns observed in Fig. 5, panel III.

Tree 1, built from the ThiM-biased search (HK.hmm), captures the broadest sequence diversity, retrieving HK-containing proteins across bacteria, plants, archaea, fungi, and metazoa. It includes ThiM-like standalone kinases and HK domains embedded within multi-domain proteins such as THI6. THI6 sequences appear interspersed among HK-only sequences rather than forming a discrete clade, a placement that reflects their shared domain ancestry and illustrates directly how a single-domain query maximizes sensitivity at the expense of architectural resolution. The riboswitch-associated sequences distribution indicated by red triangles in Fig. 5, suggests that TPP-responsive regulation is not restricted to a single lineage within the broader HK domain radiation, consistent with the cross-kingdom distribution of RF00059 described in Section 3.5.

Despite starting from distinct inputs, trees 2 and 3 converge on the same sequence set and produce similar topologies. Tree 2, built from the

THI6.hmm *de novo* profile, recovers a well-supported monophyletic fungal clade based on full-length sequence similarity, while tree 3, built from the co-occurrence filter requiring simultaneous detection of HK and TMP-TENI domains (Section 3.3, Step c), enforcing the same dual-domain architecture explicitly. The agreement between these two independent retrieval strategies confirms both the sensitivity and the robustness of the workflow for THI6 protein family and supports the inference of a single domain fusion event early in the fungal lineage. The internal branching pattern shared by both trees reflects the diversification of THI6 across fungal lineages following the fusion event.

Tree 4, built from the HK domain alignment extracted from the combined ThiM and THI6 sequence sets, enables direct comparison of the HK domain across both architectural contexts. The overall topology is broadly consistent with tree 1, but the domain-restricted alignment improves the resolution of divergent representatives, including sequences from *Amphimedon queenslandica* (Metazoa), *Bacillus subtilis* (Bacteria), and *Pyrococcus furiosus* (Archaea). The presence of metazoan sequences in this tree reflects the broad phylogenetic distribution of HK-like domains rather than the conservation of a complete thiamine biosynthesis pathway, which is absent in metazoans. The clear separation in the domain-based tree suggests that restricting the alignment to homologous columns reduces noise introduced by structurally heterogeneous flanking regions and improves comparability across distantly related sequences.

The partial topological convergence observed across the four trees reflects both the internal consistency of the workflow and the limited taxonomic breadth of the sequence set used here. Rather than indicating redundancy, this pattern confirms that independent retrieval strategies consistently identify the same core sequence space when applied to a well-defined protein family. Broader sampling across fungi, bacteria, plants and protists would likely reveal additional variation in domain architecture and provide greater resolution for questions concerning the timing and genomic context of the HK domain fusion event.

5. Summary

This study presents a modular and reproducible workflow for pathway-scale identification and comparative analysis of thiamine biosynthesis proteins, integrating sequence retrieval, domain architecture validation, and phylogenetic reconstruction. By combining profile-based searches, multi-domain filtering, and domain-focused alignments, the pipeline allows systematic control over sensitivity, specificity, and dataset composition. The HK domain was used as a case study to illustrate how different retrieval strategies shape downstream analyses. Single-domain searches recovered a broader and more diverse set of homologs, whereas full-length and multi-domain approaches restricted the dataset to structurally consistent THI6-like proteins. Independent strategies based on full-length similarity and domain architecture converged on the same candidate set, supporting the robustness of the workflow. Domain extraction further improved alignment comparability while preserving evolutionary signal.

Beyond these results, effective application of the workflow requires careful evaluation at each step. Although the pipeline is designed to be modular and partially automated, parameter selection, domain composition, and gene-specific features should be critically assessed in light of the biological context. In addition, cellular organization must be considered if organellar genomes are present. Awareness of the structural and evolutionary characteristics of the target gene family remains essential to guide filtering decisions, interpret outputs, and ensure reliable downstream inference. In this sense, the workflow provides a flexible framework, but its performance ultimately depends on informed and iterative use.

Acknowledgments

The authors acknowledge financial support from grants FAPESP 2016/17545–8 and CNPq 310779/2017–0 to Marie-Anne Van Sluys, FAPESP 2019/08239–9 to Henrique Moura Dias and CAPES 88887.900872/2023–00 to Naiara Almeida de Toledo. The funders had no role in study design, data collection, analysis, publication decision, or manuscript preparation.

Appendix A. Supporting information

Supplementary data associated with this article can be found in the online version at https://doi.org/10.1016/bs.mie.2026.06.014.

References

Altenhoff, A. M., & Dessimoz, C. (2009). Phylogenetic and functional assessment of orthologs inference projects and methods. *PLoS Computational Biology, 5*(1), e1000262. https://doi.org/10.1371/journal.pcbi.1000262.

Bailey, T. L., Boden, M., Buske, F. A., Frith, M., Grant, C. E., Clementi, L., ... Noble, W. S. (2009). MEME SUITE: Tools for motif discovery and searching. *Nucleic Acids Research, 37*, W202–W208. https://doi.org/10.1093/nar/gkp335.

Bairoch, A., & Apweiler, R. (1997). The SWISS-PROT protein sequence data bank and its supplement TrEMBL. *Nucleic Acids Research, 25*, 31–36. https://doi.org/10.1093/nar/25.1.31.

Bocobza, S., Adato, A., Mandel, T., Shapira, M., Nudler, E., & Aharoni, A. (2007). Riboswitch-dependent gene regulation and its evolution in the plant kingdom. *Genes & Development, 21*, 2874–2879. https://doi.org/10.1101/gad.443907.

Capella-Gutiérrez, S., Silla-Martínez, J. M., & Gabaldón, T. (2009). trimAl: A tool for automated alignment trimming in large-scale phylogenetic analyses. *Bioinformatics (Oxford, England), 25*, 1972–1973. https://doi.org/10.1093/bioinformatics/btp348.

Chatterjee, A., Li, Y., Zhang, Y., Grove, T. L., Lee, M., Krebs, C., et al. (2008). Reconstitution of ThiC in thiamine pyrimidine biosynthesis expands the radical SAM superfamily. *Nature Chemical Biology, 4*, 758–765. https://doi.org/10.1038/nchembio.121.

Dias, H. M., De Toledo, N. A., Mural, R. V., Schnable, J. C., & Van Sluys, M.-A. (2024). *THI1* gene evolutionary trends: A comprehensive plant-focused assessment via data mining and large-scale analysis. *Genome Biology and Evolution, 16*, evae212. https://doi.org/10.1093/gbe/evae212.

Fitzpatrick, T. B., Basset, G. J. C., Borel, P., Carrari, F., DellaPenna, D., Fraser, P. D., et al. (2012). Vitamin deficiencies in humans: Can plant science help? *The Plant Cell, 24*(2), 395–414. https://doi.org/10.1105/tpc.111.093120.

Gerdes, S., Lerma-Ortiz, C., Frelin, O., Seaver, S. M., Henry, C. S., de Crécy-Lagard, V., & Hanson, A. D. (2012). Plant B vitamin pathways and their compartmentation: A guide for the perplexed. *Journal of Experimental Botany, 63*(15), 5379–5395. https://doi.org/10.1093/jxb/ers208.

Gonçalves, C., & Gonçalves, P. (2019). Multilayered horizontal operon transfers from bacteria reconstruct a thiamine salvage pathway in yeasts. *Proceedings of the National Academy of Sciences of the United States of America, 116*, 22219–22228. https://doi.org/10.1073/pnas.1909844116.

Goodstein, D. M., Shu, S., Howson, R., Neupane, R., Hayes, R. D., Fazo, J., et al. (2012). Phytozome: A comparative platform for green plant genomics. *Nucleic Acids Research, 40*, D1178–D1186. https://doi.org/10.1093/nar/gkr944.

Grigoriev, I. V., Hayes, R. D., Calhoun, S., Kamel, B., Wang, A., Ahrendt, S., et al. (2021). PhycoCosm, a comparative algal genomics resource. *Nucleic Acids Research, 49*, D1004–D1011. https://doi.org/10.1093/nar/gkaa898.

Henry, C. S., Lerma-Ortiz, C., Gerdes, S. Y., Mullen, J. D., Colasanti, R., Zhukov, A., et al. (2016). Systematic identification and analysis of frequent gene fusion events in metabolic pathways. *BMC Genomics, 17*, 473. https://doi.org/10.1186/s12864-016-2782-3.

Jurgenson, C. T., Begley, T. P., & Ealick, S. E. (2009). The structural and biochemical foundations of thiamin biosynthesis. *Annual Review of Biochemistry, 78*, 569–603. https://doi.org/10.1146/annurev.biochem.78.072407.102340.

Kanehisa, M., Sato, Y., Kawashima, M., Furumichi, M., & Tanabe, M. (2016). KEGG as a reference resource for gene and protein annotation. *Nucleic Acids Research, 44*, D457–D462. https://doi.org/10.1093/nar/gkv1070.

Kersey, P. J., Lawson, D., Birney, E., Derwent, P. S., Haimel, M., Herrero, J., et al. (2010). Ensembl genomes: Extending ensembl across the taxonomic space. *Nucleic Acids Research, 38*, D563–D569. https://doi.org/10.1093/nar/gkp871.

Komeda, Y., Tanaka, M., & Nishimune, T. (1988). A *th-1* Mutant of *Arabidopsis thaliana* is defective for a thiamin-phosphate-synthesizing enzyme: Thiamin phosphate pyrophosphorylase. *Plant Physiology, 88*, 248–250. https://doi.org/10.1104/pp.88.2.248.

Larsson, A. (2014). AliView: A fast and lightweight alignment viewer and editor for large datasets. *Bioinformatics (Oxford, England), 30*, 3276–3278. https://doi.org/10.1093/bioinformatics/btu531.

Lefort, V., Longueville, J.-E., & Gascuel, O. (2017). SMS: Smart model selection in PhyML. *Molecular Biology and Evolution, 34*, 2422–2424. https://doi.org/10.1093/molbev/msx149.

Machado, C. R., Praekelt, U. M., De Oliveira, R. C., Barbosa, A. C. C., Byrne, K. L., Meacock, P. A., & Menck, C. F. M. (1997). Dual role for the yeast THI4 gene in thiamine biosynthesis and DNA damage tolerance. *Journal of Molecular Biology, 273*(1), 114–121. https://doi.org/10.1006/jmbi.1997.1302.

Maupin-Furlow, J. A. (2018). Vitamin B1 (Thiamine) metabolism and regulation in archaea. In J. G. LeBlanc, & G. S. D. Giori (Eds.). *B group vitamins - current uses and perspectives*. InTech. https://doi.org/10.5772/intechopen.77170.

Mizote, T., Tsuda, M., Smith, D. D. S., Nakayama, H., & Nakazawa, T. (1999). Cloning and characterization of the thiD/J gene of Escherichia coli encoding a thiamin-synthesizing bifunctional enzyme, hydroxymethylpyrimidine kinase/phosphomethylpyrimidine kinase. *Microbiology (Reading, England), 145*(Pt 2), 495–501. https://doi.org/10.1099/13500872-145-2-495.

Morett, E., Korbel, J. O., Rajan, E., Saab-Rincon, G., Olvera, L., Olvera, M., Schmidt, S., et al. (2003). Systematic discovery of analogous enzymes in thiamin biosynthesis. *Nature Biotechnology, 21*, 790–795. https://doi.org/10.1038/nbt834.

Nawrocki, E. P., & Eddy, S. R. (2013). Infernal 1.1: 100-fold faster RNA homology searches. *Bioinformatics (Oxford, England), 29*, 2933–2935. https://doi.org/10.1093/bioinformatics/btt509.

Nosaka, K., Nishimura, H., Kawasaki, Y., Tsujihara, T., & Iwashima, A. (1994). Isolation and characterization of the THI6 gene encoding a bifunctional thiamin-phosphate pyrophosphorylase/hydroxyethylthiazole kinase from Saccharomyces cerevisiae. *Journal of Biological Chemistry, 269*, 30510–30516. https://doi.org/10.1016/S0021-9258(18)43843-4.

Okonechnikov, K., Golosova, O., & Fursov, M. the UGENE team. (2012). Unipro UGENE: A unified bioinformatics toolkit. *Bioinformatics (Oxford, England), 28*, 1166–1167. https://doi.org/10.1093/bioinformatics/bts091.

Onozuka, M., Konno, H., Kawasaki, Y., Akaji, K., & Nosaka, K. (2008). Involvement of thiaminase II encoded by the *THI20* gene in thiamin salvage of *Saccharomyces cerevisiae*. *FEMS Yeast Research, 8*, 266–275. https://doi.org/10.1111/j.1567-1364.2007.00333.x.

Paxhia, M. D., Swanson, M. S., & Downs, D. M. (2021). Functional characterization of the HMP-P synthase of Legionella pneumophila (Lpg1565). *Molecular Microbiology, 115*(4), 539–553. https://doi.org/10.1111/mmi.14622.

Promponas, V. J., Iliopoulos, I., & Ouzounis, C. A. (2015). Annotation inconsistencies beyond sequence similarity-based function prediction – phylogeny and genome structure. *Standards in Genomic Sciences, 10*(1), 108. https://doi.org/10.1186/s40793-015-0101-2.

Pruitt, K. D., Tatusova, T., & Maglott, D. R. (2007). NCBI reference sequences (RefSeq): A curated non-redundant sequence database of genomes, transcripts and proteins. *Nucleic Acids Research, 35*, D61–D65. https://doi.org/10.1093/nar/gkl842.

Rangannan, V., & Bansal, M. (2010). High-quality annotation of promoter regions for 913 bacterial genomes. *Bioinformatics (Oxford, England), 26*(24), 3043–3050. https://doi.org/10.1093/bioinformatics/btq577.

Rodionov, D. A., Vitreschak, A. G., Mironov, A. A., & Gelfand, M. S. (2002). Comparative genomics of thiamin biosynthesis in procaryotes. *Journal of Biological Chemistry, 277*(50), 48949–48959. https://doi.org/10.1074/jbc.M208965200.

Serganov, A., & Nudler, E. (2013). A decade of riboswitches. *Cell, 152*(1–2), 17–24. https://doi.org/10.1016/j.cell.2012.12.024.

Winkler, W., Nahvi, A., & Breaker, R. R. (2002). Thiamine derivatives bind messenger RNAs directly to regulate bacterial gene expression. *Nature, 419*, 952–956. https://doi.org/10.1038/nature01145.

Yates, A. D., Achuthan, P., Akanni, W., Allen, J., Allen, J., Alvarez-Jarreta, J., et al. (2019). Ensembl 2020. *Nucleic Acids Research*, gkz966. https://doi.org/10.1093/nar/gkz966.

Yazdani, M., Zallot, R., Tunc-Ozdemir, M., De Crécy-Lagard, V., Shintani, D. K., & Hanson, A. D. (2013). Identification of the thiamin salvage enzyme thiazole kinase in Arabidopsis and maize. *Phytochemistry, 94*, 68–73. https://doi.org/10.1016/j.phytochem.2013.05.017.

Zallot, R., Yazdani, M., Goyer, A., Ziemak, M. J., Guan, J.-C., McCarty, D. R., et al. (2014). Salvage of the thiamin pyrimidine moiety by plant TenA proteins lacking an active-site cysteine. *Biochemical Journal, 463*(1), 145–155. https://doi.org/10.1042/BJ20140522.

Further reading

Cock, P. J. A., Antao, T., Chang, J. T., Chapman, B. A., Cox, C. J., Dalke, A., et al. (2009). Biopython: Freely available Python tools for computational molecular biology and bioinformatics. *Bioinformatics (Oxford, England), 25*, 1422–1423. https://doi.org/10.1093/bioinformatics/btp163.

Finn, R. D., Clements, J., & Eddy, S. R. (2011). HMMER web server: Interactive sequence similarity searching. *Nucleic Acids Research, 39*, W29–W37. https://doi.org/10.1093/nar/gkr367.

Guindon, S., Dufayard, J.-F., Lefort, V., Anisimova, M., Hordijk, W., & Gascuel, O. (2010). New algorithms and methods to estimate maximum-likelihood phylogenies: Assessing the performance of PhyML 3.0. *Systematic Biology, 59*, 307–321. https://doi.org/10.1093/sysbio/syq010.

Hunter, S., Apweiler, R., Attwood, T. K., Bairoch, A., Bateman, A., Binns, D., et al. (2009). InterPro: The integrative protein signature database. *Nucleic Acids Research, 37*, D211–D215. https://doi.org/10.1093/nar/gkn785.

Katoh, K., & Standley, D. M. (2013). MAFFT multiple sequence alignment software version 7: Improvements in performance and usability. *Molecular Biology and Evolution, 30*, 772–780. https://doi.org/10.1093/molbev/mst010.

Mistry, J., Chuguransky, S., Williams, L., Qureshi, M., Salazar, G. A., Sonnhammer, E. L. L., et al. (2021). Pfam: The protein families database in 2021. *Nucleic Acids Research, 49*, D412–D419. https://doi.org/10.1093/nar/gkaa913.

Nancy, O.-P., Emma, C., Eric, P.N., Sandra, T., Manja, M., Elena, R., … Blake, S. (2024). *Nucleic Acids Research*, https://doi.org/10.1093/nar/gkae1023.

Wightman, R., & Meacock, P. A. (2003). The THI5 gene family of Saccharomyces cerevisiae: Distribution of homologues among the hemiascomycetes and functional redundancy in the aerobic biosynthesis of thiamin from pyridoxine. *Microbiology (Reading, England), 149*, 1447–1460. https://doi.org/10.1099/mic.0.26194-0.

CHAPTER THREE

Generation and characterization of thiamin-biofortified plants

Aymeric Goyer*, Keonny Cosek, and David Handy
Department of Botany and Plant Pathology, Oregon State University, Corvallis, OR, United States
*Corresponding author. e-mail address: aymeric.goyer@oregonstate.edu

Contents

Methods in Enzymology, Volume 735
ISSN 0076-6879, https://doi.org/10.1016/bs.mie.2026.05.041

Abstract

Thiamin biofortification of food crops by genetic engineering has the potential to solve human health issues related to thiamin deficiencies. Rice is an obvious target for thiamin biofortification because white rice, which is obtained after removing the thiamin-rich outer layer of the grain, is a poor source of thiamin. To increase the levels and stability of thiamin in white rice, engineering strategies that boost the expression of biosynthesis genes and/or thiamin binding proteins have been reported. In this chapter, we describe protocols for generating and characterizing thiamin-engineered plants: (1) design and building of plant expression vectors, (2) introduction of DNA constructs into plant genome by *Agrobacterium*-mediated transformation, (3) analysis of transgene expression by quantitative reverse-transcription polymerase chain reaction, (4) quantification of thiamin biosynthesis proteins by Western blot, and (5) quantification of thiamin by HPLC.

1. Introduction

Thiamin (vitamin B1) is an essential micronutrient for proper growth, development, and health of humans. Humans obtain thiamin from dietary sources. Whole-grain cereals like rice are rich sources of thiamin. However, most of the thiamin found in rice grain is located in the bran, the outer layer of the grain that is removed during polishing to produce white rice (Goyer, 2017). Low-income populations whose diets are mainly based on white rice have inadequate intake of thiamin and may suffer from thiamin deficiency. Severe thiamin deficiency is associated with neurological and cardiovascular symptoms, referred to as beriberi, a disease that can be lethal.

Thiamin biofortification of food crops like rice has been attempted to increase the thiamin content of rice grain endosperm, also known as white rice. One strategy aimed to increase thiamin production in rice grain by overexpressing thiamin biosynthesis genes (Dong et al., 2015; Dong et al., 2016), while another focused on the stabilization of thiamin by overexpressing thiamin binding protein (Fitzpatrick et al., 2024). In the first approach, the thiamin biosynthesis genes *THI1* and *THIC* were placed under the control of the constitutive *ubiquitin* promoter. Thiamin concentrations increased up to fivefold in the unpolished rice grain. However, after polishing, the rice endosperm of engineered plants had the same thiamin content than that of the control plants. In the second approach, endosperm-specific expression of a thiamin binding protein from *Sesamum indicum* increased thiamin concentration 3- to 4-fold in the polished grains compared to wild-type polished grains. Both approaches used similar

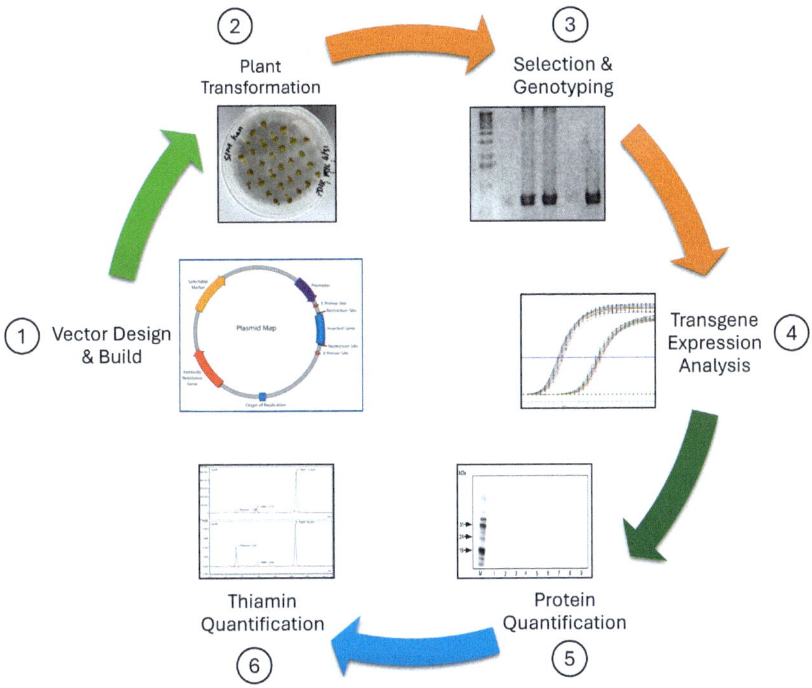

Fig. 1 Schematic of the experimental pipeline used to generate and characterize transgenic plants.

experimental pipelines to generate and characterize transgenic plants (Fig. 1). In this chapter, we outline steps required to generate transgenic plants engineered for increased thiamin levels by overexpression of thiamin biosynthesis genes and describe protocols for each of the required steps.

2. Plant transformation

2.1 Rationale

Plant scientists have used the ability of *Agrobacterium tumefaciens* to transfer a piece of DNA (transfer DNA or T-DNA) from the tumor-inducing (Ti) plasmid to plant genomes, thus introducing new DNA elements to the genomes of plants. Disarmed Ti plasmids that retain virulence genes necessary for nicking the T-DNA region at left and right borders but lacking phytohormone synthesis genes responsible for tumorigenesis have been developed. While different plant species have different requirements for successful transformation, the general principles are the same and involve the following steps: preparation of plant explants, induction of callus formation, co-cultivation of explants with *A. tumefaciens*, and regeneration of plants from calli. Here, we describe a protocol for *Agrobacterium*-mediated transformation of potato stem explants, which is based on a previously published protocol (Chetty, Narváez-Vásquez, & Orozco-Cárdenas, et al., 2015) (Fig. 2). We invite readers to refer to a previously published paper for details on *Agrobacterium*-mediated rice embryo transformation (Cheng, Sardana, & Altosaar, et al., 1998).

2.2 Plant binary vectors

Plant binary vectors are used for transformation of plants mediated by *A. tumefaciens*. A diversity of vectors is available for cloning genes of interest. Binary vectors are composed of the following main elements: left and right borders of T-DNA, replication operons for *Escherichia coli* and *A. tumefaciens*, and selectable marker genes for bacteria and plants. Various cloning strategies can be used to assemble binary vectors, including restriction digestion, Gateway recombination (Walhout et al., 2000), and Golden Gate shuffling (Engler et al., 2009). For thiamin engineering, the T-DNA region will contain a promoter, e.g., Cauliflower Mosaic Virus 35S (CaMV35S) for dicotyledonous species and Ubiquitin for monocotyledonous species, the coding sequence of the gene of interest, and a terminator. Correct assembly must be verified by PCR, restriction digestion, or whole plasmid sequencing. A map of the plasmid used for Arabidopsis transformation (Dong et al., 2015) is shown in Fig. 3.

2.3 Equipment

- Electroporator 2510 (Eppendorf, Hamburg, Germany)
- Shaking incubator MaxQ 4000 (Thermo Fisher Scientific, Waltham, MA, USA)
- Spectrophotometer Genesys 10UV (Thermo Fisher Scientific, Waltham, MA, USA)

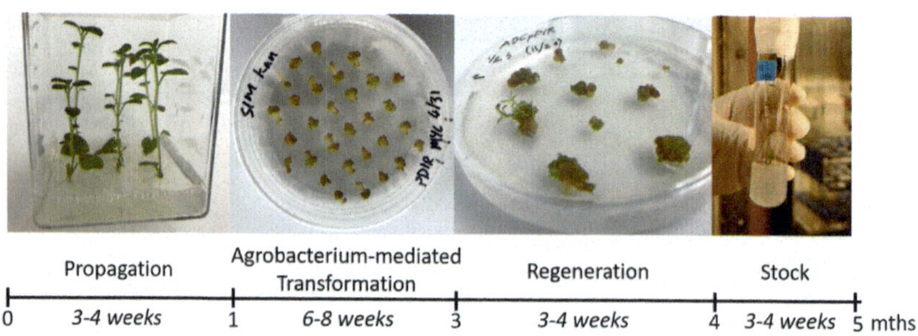

Fig. 2 Timeline of *Agrobacterium*-mediated transformation of potato stem explants.

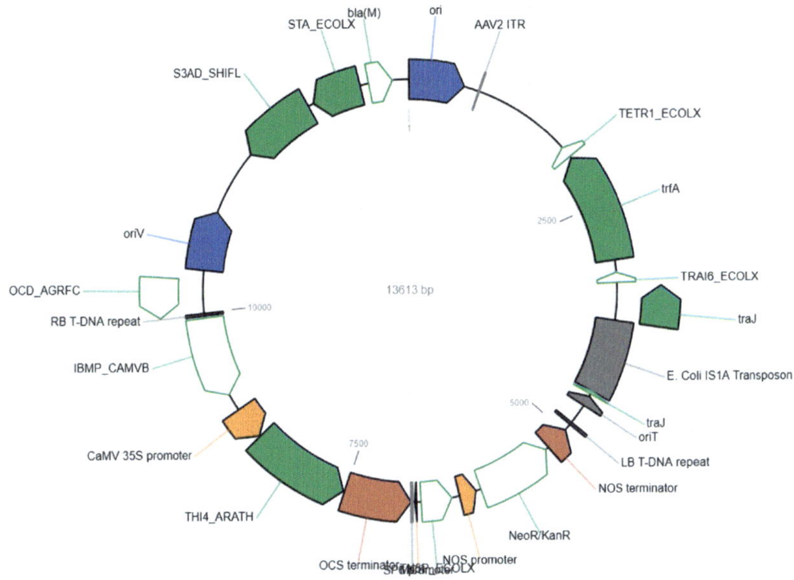

Fig. 3 Map of the plasmid containing THI1 (THI4) under the control of the CaMV35S promoter (Dong et al., 2015).

- Thermal cycler C1000 (BioRad, Hercules, CA, USA)
- Agarose gel electrophoresis apparatus (VWR, Radnor, PA, USA)
- Laminar flow hood
- Sorvall Legend 23R centrifuge (Fisher Scientific, Hampton, NH, USA)
- Growth chamber
- Autoclave

2.4 Reagents and supplies

2.4.1 Agrobacterium

- Electrocompetent *A. tumefaciens* strain EHA105 (Hood et al., 1993). This strain contains a rifampicin resistance gene (rif) and the Ti plasmid pEHA105 (pTiBo542DT0DNA). We obtained it from Stanton Gelvin (Purdue University).
- Luria-Bertani (LB) agar plates (per liter: 10 g tryptone, 5 g yeast extract, 10 g NaCl, 15 g agar)
- Yeast Extract Peptone (YEP) medium (per liter: 10 g peptone, 10 g yeast extract, 5 g NaCl)
- Agrobacterium Resuspension Medium (ARM) (Table 1)
- Rifampicin (PhytoTech Labs, Lenexa, KS, USA)

Table 1 Plant tissue culture recipe per liter.

Component	SPM	ARM	CIM	SIM
MS modified BC potato medium	4.4 g	4.4 g	4.4 g	4.4 g
Sucrose	30 g	20 g	20 g	20 g
Myo-inositol	100 mg	100 mg	100 mg	100 mg
MSV1 (1000X)[a]	1 mL	1 mL	1 mL	1 mL
MES	0.5 g	0.5 g	0.5 g	0.5 g
Phytoagar	8 g	—	8 g	8 g
NAA (1 g L^{-1})[b]	20 μL	—	200 μL	20 μL
GA_3 (1 g L^{-1})[b]	—	—	20 μL	20 μL
Trans-ZR (1 g L^{-1})[b]	—	—	2.5 mL	2 mL
Acetosyringone (100 mM)[b]		2 mL	—	—
Cefotaxime (50 g L^{-1})[b]	—	—	—	5 mL
Carbenicillin (100 g L^{-1})[b]	—	—	—	0.4 mL
Antibiotic (based on vector)[b]	—	—	—	x mL

pH adjusted to 5.6 with NaOH before adding phytoagar.
All media should be autoclaved at 121 °C for 20 min.
Preparation of stock solutions:
NAA (1 g L^{-1}): Dissolve 10 mg in 0.5 mL 1 N NaOH, then adjust to 10 mL with deionized water.
trans-Zeatine riboside (1 g L^{-1}): Dissolve 10 mg in 0.5 mL 1 N NaOH, then adjust to 10 mL with deionized water.
GA3 (1 g L^{-1}): Dissolve 10 mg in 0.5 mL ethanol, then adjust to 10 mL with deionized water.
Cefotaxime (50 g L^{-1}): Dissolve 500 mg in 10 mL deionized water.
Acetosyringone (100 mM): Dissolve 196 mg in 4 mL ethanol, then adjust to 10 mL with deionized water.
Filter all the solutions with 0.2 μm filters.
[a]MSV1 (1000X) = 2 g L^{-1} glycin, 0.5 g L^{-1} nicotinic acid, 0.5 g L^{-1} pyridoxine, 0.1 g L^{-1} thiamin
[b]Add hormones and antibiotics after autoclaving.

- Kanamycin (Sigma-Aldrich, St. Louis, MO, USA)
- Disposable electroporation cuvettes, 2 mm, 400 μL (VWR, Radnor, PA, USA)
- Spectrophotometer plastic cuvettes, 4.5 mL (VWR, Radnor, PA, USA)

2.4.2 Potato tissue culture

- Petri dishes (15 × 100 mm) (VWR, Radnor, PA, USA)

- Test tubes round-bottom (25 mm × 150 mm) (PhytoTech Labs, Lenexa, KS, USA)
- Closures for culture tubes, 25 mm (PhytoTech Labs, Lenexa, KS, USA)
- Culture boxes, 77 mm × 77 mm × 95 mm (PhytoTech Labs, Lenexa, KS, USA)
- Closures for culture boxes (PhytoTech Labs, Lenexa, KS, USA)
- 50-mL conical tubes (VWR, Radnor, PA, USA)
- MS modified BC potato medium (PhytoTech Labs, Lenexa, KS, USA)
- Sucrose (PhytoTech Labs, Lenexa, KS, USA)
- Myo-inositol (PhytoTech Labs, Lenexa, KS, USA)
- MES monohydrate (PhytoTech Labs, Lenexa, KS, USA)
- Phytoagar (PhytoTech Labs, Lenexa, KS, USA)
- Napthalene acetic acid (NAA) (PhytoTech Labs, Lenexa, KS, USA)
- Gibberrelic acid 3 (GA_3) (Sigma-Aldrich, St. Louis, MO, USA)
- Trans-Zeatin riboside (transZR) (PhytoTech Labs, Lenexa, KS, USA)
- Glycine (Sigma-Aldrich, St. Louis, MO, USA)
- Nicotinic acid (PhytoTech Labs, Lenexa, KS, USA)
- Pyridoxine (PhytoTech Labs, Lenexa, KS, USA)
- Thiamin hydrochloride (Sigma-Aldrich, St. Louis, MO, USA)
- Shoot propagation medium (SPM) (Table 1)
- Callus inducing medium (CIM) (Table 1)
- Shoot inducing medium (SIM) (Table 1)
- MSV1 solution (1000x) (Table 1)
- Sterile Whatman filter paper (VWR, Radnor, PA, USA)
- Acetosyringone (PhytoTech Labs, Lenexa, KS, USA)
- Cefotaxime (PhytoTech Labs, Lenexa, KS, USA)
- Carbenicillin (PhytoTech Labs, Lenexa, KS, USA)

2.4.3 Genomic DNA extraction and PCR

- Chloroform (VWR, Radnor, PA, USA)
- Extraction buffer (200 mM Tris-HCl pH 7.5, 250 mM NaCl, 25 mM EDTA pH 8.0, 0.5% SDS)
- Tris EDTA (10 mM Tris, 0.1 mM EDTA, pH 7.5)
- PrimeStar HS DNA Polymerase (premix) (Takara Bio, Shiga, Japan)
- Agarose (IBI Scientific, Dubuque, IA, USA)

2.5 Protocols

All procedures of plant transformation should be done under sterile conditions.

2.5.1 *Agrobacterium preparation*

1. Electroporate 50 µL of electrocompetent EHA105 cells with 1 µL plant binary vector (1 ng/µL) in an electroporation cuvette at 2500 V. Add 1 mL pre-warmed LB and incubate for 3 h at 30 °C with shaking at 220 rpm.
2. Plate bacteria on selective LB agar (15 mg/L rifampicin + 50 mg/L kanamycin) and incubate at 30 °C for 2 days.
3. Start 5-mL YEP + 15 mg/L rifampicin + 50 mg/L kanamycin preculture from a single isolated colony. Grow overnight with shaking at 220 rpm.
4. Inoculate 50-mL YEP + 15 mg/L rifampicin + 50 mg/L kanamycin with the preculture. Grow overnight at 30 °C with shaking at 220 rpm.
5. Measure optical density (OD) at 600 nm, pellet 10 mL in 50-mL conical tube, and resuspend in ARM to obtain an OD_{600} about 0.4–0.6.

2.5.2 *Potato plantlets and internodes preparation*

1. Cultivate potato plantlets in test tubes in 15-mL SPM for 3–4 weeks at 20–22 °C (16 h photoperiod, 65–84 µmoL m^{-2} s^{-1}) (see Note 1).
2. Cut 4–6-mm long internodes and place onto sterile Whatman filter paper over 30-mL CIM medium in 100 × 15 mm petri dishes (about 30 explants per dish).
3. Cultivate explants for 2 days at 20–22 °C (16 h photoperiod, 65–84 µmol m^{-2} s^{-1}).

2.5.3 *Transformation*

1. In a 50-mL conical tube, inoculate internodes with the *Agrobacterium* solution and shake slowly for 15 min at room temperature.
2. Gently blot-dry explants between two sterile filter papers and transfer onto sterile Whatman filter paper over 30-mL CIM + 200 µM acetosyringone medium in 100 × 15 mm petri dishes (about 30 explants per dish). Seal plates with surgical tape and co-cultivate at 20 °C for 2 days in the dark.
3. In a 50-mL conical tube, wash explants for 5 min with 40 mL water supplemented with 250 mg/L cefotaxime, blot dry on sterile filter papers, transfer explants to 30-mL CIM + 250 mg/L cefotaxime + 200 mg/L carbenicillin + 50 mg/L kanamycin (no Whatman filter paper). Place in the light (16 h photoperiod, 65–84 µmol m^{-2} s^{-1}).
4. After 14 days, transfer to culture boxes containing 50-mL SIM.
5. Transfer explants to SIM every 14 days. Callus and shoot formation occurs after 4 weeks (Fig. 2).

6. Carefully excise developing shoots (see Note 2) and transfer to SPM + 250 mg/L cefotaxime + 200 mg/L carbenicillin + 50 mg/L kanamycin.

2.5.4 Genomic DNA extraction and PCR analysis of putative transgenic potato plants

1. When plantlets have produced sufficient leaf tissue, collect a piece of leaf (5 × 5 mm) from putative transgenic plants in a 1.7-mL tube.
2. Add 450 μL of extraction buffer (200 mM Tris-HCl pH 7.5, 250 mM NaCl, 25 mM EDTA pH 8.0, 0.5 % SDS) and grind with a micropestle.
3. Add 450 μL of chloroform, vortex vigorously, and centrifuge at 12,000 *x g* for 5 min
4. Transfer upper aqueous phase (300 μL) to a new tube and add 1 vol of cold isopropanol. Invert several times to mix and place on ice for 5 min. Centrifuge at 12,000 *x g* for 10 min
5. Wash the DNA pellet with 70 % ethanol and resuspend in 200 μL of Tris EDTA buffer.
6. Set up a PCR reaction as follows: 5 μL PrimeStar HS DNA Polymerase Premix, 0.8 μL forward primer (5 μM), 0.8 μL reverse primer (5 μM), 1.5 μL DNA template, 1.9 μL deionized water. PCR Cycle: 1 cycle of 30 s at 98 °C, 35 cycles of 10 s at 98 °C, 5 s at 55 °C, 15 s at 72 °C. Forward primer: 5′-CGCTATTGTGAGGTTGACCAGA-3′; Reverse primer: 5′-GTTTCTGCGGACTGGCT-3′.

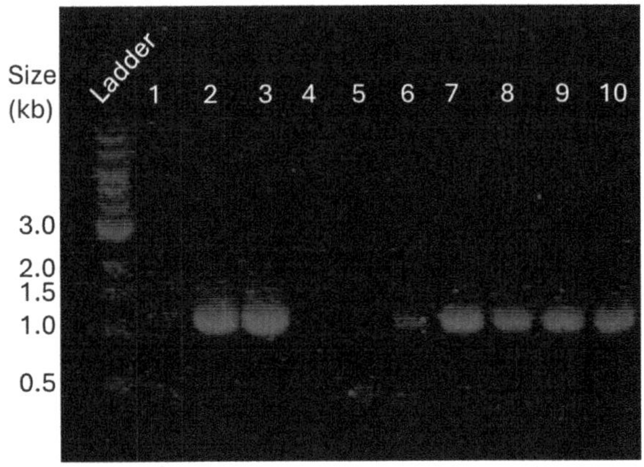

Fig. 4 Agarose gel electrophoresis showing a 1.2-kb amplicon in positive transgenic plants (2, 3, 6–10) and the absence of an amplicon in non-transgenic plants (4, 5). Well 1 contains a non-transformed negative control.

7. Run PCR amplicons on a 0.8 % agarose gel for 30 min at 100 Volts.
8. Visualize amplicons under UV light. A band at 1.2 kb should be visible in positive transformants (Fig. 4).

2.6 Notes

1. For each transformation, we usually grow 40 plantlets to obtain 200–300 internodal explants.
2. Several shoots usually form from each end of the explants. To ensure that independent transformation events are moved forward for analysis, we take one developing shoot per explant end.

3. Transgene expression analysis

3.1 Rationale

Prior to moving forward with experiments using the newly generated transgenic plants, it is standard practice to validate the expected change in expression of the targeted gene(s). After isolation from leaves of transgenic potato plants, RNAs are reverse-transcribed to cDNA, then real-time quantitative PCR (qPCR) is used to measure the level of overexpression of target genes (transgene + endogenous gene). Proper design and analysis of qPCR experiments are essential for accurate interpretations of biological differences across samples. It is essential to include two endogenous reference genes (housekeeping genes) for normalization of the data. Universal housekeeping genes such as *18S rRNA* and *Actin-2* are often used; however, literature research should guide the selection of alternative and validated reference genes. For guidelines on qPCR statistical analysis, we refer to Taylor et al. (2019) on normalization of qPCR data using the geometric mean of multiple reference genes. The expression of genes of interest is normalized to the expression of reference genes to reduce variation due to technical errors.

3.2 Equipment

- Microcentrifuge Prism (LabNet, Corning, NY, USA)
- Hot water bath (VWR, Radnor, PA, USA)
- Stratagene Mx3005P qPCR instrument (Thermo Fisher Scientific, Waltham, MA, USA)
- Pipettors
- Barrier tips

3.3 Reagents

- Hot phenol RNA extraction (see Note 1)
  - 8 M lithium chloride
  - 1 M Tris-HCl pH 8.5
  - 0.5 M EDTA
  - 10 % sodium dodecyl sulfate (SDS)
  - 1 M dithiothreitol (DTT)
  - Diethyl pyrocarbonate (DEPC)-treated water, or other RNase-free water
  - Phenol solution saturated with 0.01 M citrate buffer, pH 4.3, for molecular biology (Sigma-Aldrich, Saint Louis, MO, USA)
  - Chloroform:isoamylalcohol (IAA) (24:1) (Sigma-Aldrich, Saint Louis, MO, USA)
- DNase Treatment
  - DNA-*free*™ DNA Removal Kit (Thermo Fisher Scientific, Waltham, MA, USA)
- cDNA Synthesis
  - M-MuLV reverse transcriptase (New England Biolabs, Ipswich, MA)
  - 100 μM dT18 oligonucleotide (MilliporeSigma, Burlington, MA, USA)
  - 10 mM dNTP mix (New England Biolabs, Ipswich, MA)
  - 10X reverse transcriptase buffer (New England Biolabs, Ipswich, MA)
  - Ultra-pure water
- qPCR
  - 2x Brilliant III Ultra-Fast SYBR® Green QPCR Master Mix (Cat# 600882) (Agilent, Santa Clara, CA)
  - Reference dye, 1 mM (included in the Brilliant III Ultra-Fast SYBR® Green QPCR Master Mix) (Agilent, Santa Clara, CA)
  - 1.5 μM THI1 Forward primer 5′-CGCTATTGTGAGGTTGACC AGA-3′
  - 1.5 μM THI1 Reverse primer 5′-CAAAAGTTGGTCCCATTC TCG-3′
  - 1.5 μM 18S rRNA Forward primer 5′-GGGCATTCGTATTTCAT AGTCAGAG-3′
  - 1.5 μM 18S rRNA Reverse primer 5′-CGGTTCTTGATTAATGA AAACATCCT-3′
  - 1.5 μM L2 Forward primer 5′-GGCGAAATGGGTCGTGTTAT-3′
  - 1.5 μM L2 Reverse primer 5′-CATTTCTCTCGCCGAAATCG-3′

3.4 Protocols

3.4.1 Hot phenol RNA extraction

RNAs are extracted using a hot phenol method previously described (Goyer et al., 2015; Xu et al., 2011).

1. Prepare RNA extraction buffer (Table 2).
2. Add 0.8 mL of extraction buffer and 0.8 mL of phenol pH 4.3 to a 15-mL conical tube for each sample. Place tubes in a 60 °C water bath for at least 5 min before adding plant material.
3. Remove tube from water bath immediately before use; do not take other tubes out before that specific sample is ready.
4. Add 0.1 g of frozen, finely ground leaf tissue to the tube. Use a spatula chilled in liquid nitrogen to transfer material.
5. Immediately vortex for 60 s
6. Put tube back in water bath.
7. Add 0.8 mL of chloroform:IAA (24:1) and vortex for 30 s
8. Centrifuge for 10 min at 1800 *xg* at 4 °C.
9. Add 0.4 mL of phenol and 0.4 mL chloroform:IAA (24:1) to a fresh 2 mL screw cap tube.
10. After centrifuging, transfer aqueous phase to the new tube containing phenol:chloroform:IAA and vortex for 30 s immediately.
11. Centrifuge for 10 min at 12,000 × *g* at 4 °C.

Table 2 RNA extraction buffer.

Compound	Molecular weight (g/mol)	Stock concentration	Amount/ volume per 20 mL (1X)	Final concentration (1X)
LiCl	42.394	8 M	0.5 mL	100 mM
Tris pH 8.5	121.14	1 M	2.0 mL	100 mM
EDTA	292.24	0.5 M	0.4 mL	10 mM
SDS	288.28	10 %	2.0 mL	1 %
Dithiothreitol (DTT)[a]	154.25	1 M	0.3 mL	15 mM
Nuclease free water			14.8 mL	

[a]Add just before use.

12. Repeat steps 9–11 once.
13. Add 0.8 mL chloroform:IAA (24:1) to a fresh tube.
14. Transfer aqueous phase to the new tube containing chloroform:IAA and vortex immediately for 30 s
15. Centrifuge for 10 min at 12,000 × *g* at 4 °C.
16. Repeat steps 13–15 two more times.
17. Transfer aqueous phase to a fresh tube and add an equivalent volume of 4 M LiCl. Mix by vortexing.
18. Store overnight at 4 °C.
19. Centrifuge for 20 min at 12,000 × *g* at 4 °C.
20. Decant the tube then rinse the pellet in 1 mL of 70 % ethanol, vortex to help wash.
21. Centrifuge for 15 min at 12,000 × *g* at 4 °C.
22. Remove the ethanol by decanting and/or using a pipette to remove as much ethanol as possible.
23. Let tube air dry for ~30 min.
24. Resuspend in 50 μL of DEPC water (can use more for large pellets) by vortexing.
25. To ensure all RNA is off the sides of the tubes, do a pulse spin in a benchtop centrifuge for 1–2 min.
26. Measure RNA concentrations (concentrations usually range between 100 and 600 ng/μL) and ratios A260/A280 and A260/A230 using a Nanodrop (acceptable ratio > 1.8).
27. Remove traces of genomic DNA by DNase treatment following the manufacturer's recommendations (see Note 2).
    - Dilute an aliquot of extracted RNA to 200 ng/μL.
    - Add 3 μL 10X DNase buffer to 25 μL of 200 ng/μL RNA.
    - Add 1 μL of DNase I, incubate for 30 min at 37 °C.
    - Add 3 μL of DNase inactivation reagent, vortex, and incubate at room temperature for 3 min.
    - Centrifuge for 2 min.
    - Transfer the supernatant (~25 μL) into a new tube being careful not to disturb the beads.
    - Run a 5-μL aliquot on a 1 % agarose electrophoresis gel to check RNA integrity (Fig. 5).

3.4.2 cDNA synthesis

1. Add 1 μg of DNase-treated RNA to a PCR tube, then add nuclease free water to bring the total volume to 10.75 μL.

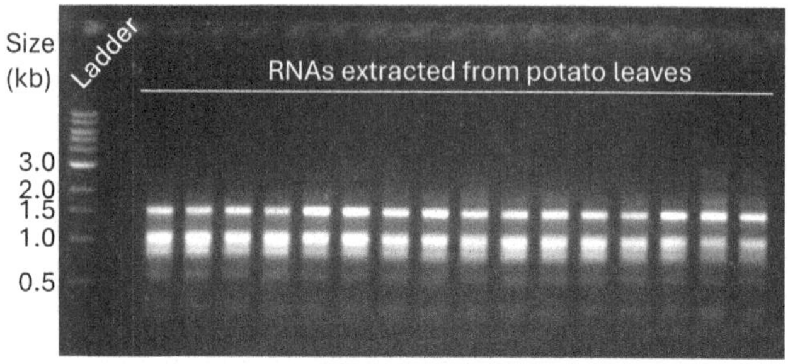

Fig. 5 Agarose gel (1 %) electrophoresis of RNAs extracted from potato leaves using the hot phenol method. Electrophoresis was run for 45 min at 100 volts. The first well on the left contains 1 kb ladder.

2. Add 5 μL of 100 μM dT18 oligonucleotide.
3. Denature the RNA by incubating at 70 °C in a thermocycler for 10 min, then cool on ice.
4. Add 2 μL of 10X reverse transcriptase buffer, 2 μL of 10 mM dNTPs, and 0.25 μL of M-MuLV reverse transcriptase (200,000 U/mL) (see Note 3).
5. Incubate RT reactions for 60 min at 42 °C in a thermocycler and inactivate M-MuLV reverse transcriptase for 20 min at 65 °C.
6. Store at −20 °C until ready for qPCR.

3.4.3 Quantitative PCR

1. Design primers to amplify the gene of interest. The sequences of THI1 primers and housekeeping genes *18S rRNA* and *L2* primers are provided above.
   - Generate primers pairs using NCBI primer tool (Primer designing tool, https://www.ncbi.nlm.nih.gov/tools/primer-blast/index.cgi?ORGANISM=1235996&INPUT_SEQUENCE=JX869059.1). An ideal primer pair will:
     - Amplify a target range between 70–200 base pairs.
     - Have no off targets, high specificity to the target gene.
     - Have a T_m range between 60 °C to 64 °C.
   - Test primer efficiencies by generating standard curves of Ct (cycle threshold) values plotted against log cDNA dilution.
     - Example: dilute cDNA 1:10, 1:50, 1:100, 1:200 and 1:500.
     - Run qPCR with a minimum of two technical replicates per primer set as outlined in steps 3 and 4 below.

- In an Excel spreadsheet calculate the average Ct value of each cDNA dilution per primer set (see Note 4).
- Create a second column entry with the log cDNA dilution (1:10 is log10 (0.1) = −1, repeat for the other four dilutions) and plot the average Ct values vs the log cDNA dilution.
- For each primer set, generate a regression line with the equation (Ct = slope (log cDNA) + intercept) and corresponding R2 value (an R^2 value approaching 1 shows a strong fit between the data and the regression line).
- Use the slope of the standard curve to estimate the PCR amplification efficiency.
- PCR efficiency (%) is calculated using the following equation

$$(10^{1/slope} - 1) \times 100$$

which indicates how effectively each primer set amplifies its target gene across cDNA dilutions.

2. Prepare cDNAs at the appropriate dilution (see Note 5).
3. For each primer set, prepare a master mix based on the number of reactions needed (see Note 6). For each reaction (1X), use the following reagents:
   - 3.7 μL ultra-pure water
   - 10 μL 2x Brilliant III SYBR® Green QPCR Master Mix
   - 0.3 μL of 1 mM Reference dye diluted to 1:500
   - 2 μL Forward primer (1.5 μM)
   - 2 μL Reverse primer (1.5 μM)
4. To each reaction, add 2 μL diluted cDNAs (final reaction volume = 20 μL) (see Note 7).
5. Place PCR reactions in a qPCR thermal cycler and use the following PCR cycle: 1 cycle at 95 °C for 3 min, 40 cycles (95 °C for 10 s, 60 °C for 20 s), 1 cycle (95 °C for 1 min, 55 °C for 30 s, 95 °C for 30 s) (see Note 8).
6. Report expression levels as normalized fold expression and run statistical analysis using the log transformed normalized expression as described in (Taylor et al., 2019) (see note 9).

3.5 Notes

1. High quality total RNA can also be obtained using other methods like the PureLink™ RNA Mini Kit (Thermo Fisher Scientific, Waltham, MA, USA).

2. Procedures may vary depending on manufacturers. Refer to manufacturer's recommendations.
3. For each RNA sample, run a reaction without M-MuLV RT (NoRT control) to verify that qPCR amplifications are from cDNAs and not from genomic DNA.
4. Ct values should be determined at a threshold set within the exponential phase of amplification (Fig. 6).

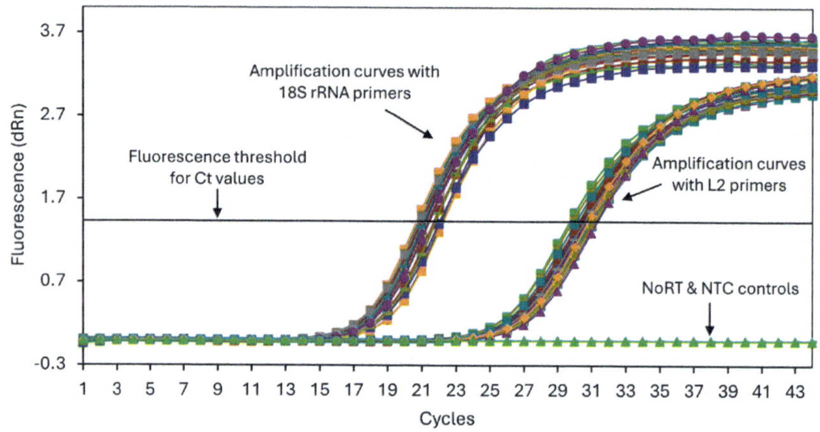

Fig. 6 Representative qPCR amplification curves.

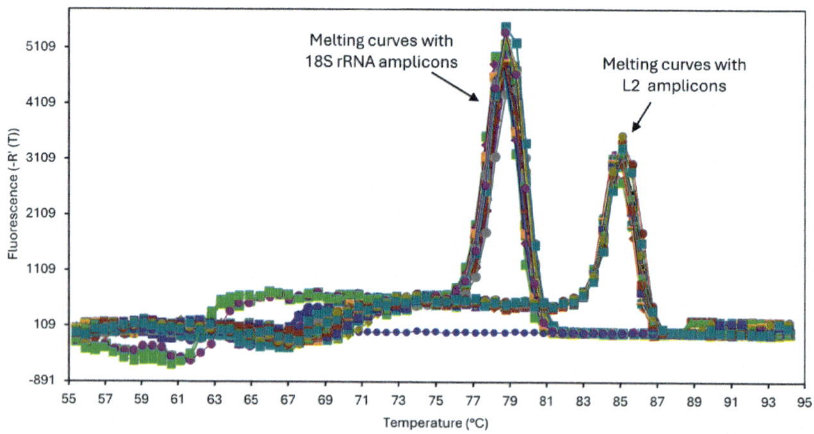

Fig. 7 Representative dissociation curves for amplicons generated using 18S rRNA (first peak) and L2 (second peak) primers. All PCR reactions containing 18S rRNA primers produced amplicons with a melt temperature of 78.5 °C while PCR reactions containing L2 primers produced amplicons with a melt temperature of 85 °C.

5. cDNA dilutions should be determined based on primers' efficiency curves and should fall within the upper and lower limits of the linear range.
6. For each experimental treatment, a minimum of three biological replicates per treatment must be used. For each biological replicate, three technical replicates should be used, ideally. However, if the sample size is large, then one or two technical replicates may be sufficient. Costs should be taken into consideration.
7. For each primer set, include a No Template Control (NTC) to verify that there is no contamination or primer dimers.
8. The last PCR cycle is to generate melt curves which are used to determine the specificity of primers. For each pair of primers, there should be one single peak which corresponds to a single amplicon (Fig. 7).

=AVERAGE(F2:F3) =AVERAGE(G2:G8) =H2-G2 =2^I2 =GEOMEAN(J2.J18) =J34/K34 =LOG(L34,2) =STEDV(M34:M40) =GEOMEAN(L34:L40)

	A	B	C	D	E	F	G	H	I	J	K	L	M	N	O
1	Treatment	Well	Well Name	Well Type	Threshold (dRn)	Ct (dRn)	Average tech reps	Ctrl group Ave Ct per target	ΔCt	RQ 2^(ΔCt)	Norm. factor GEOMEAN	Normalized Expression	Log2 Normalized Expression	GEOMEAN	STDEVI
2	Control	A1	L2 Rep 1	Unknown	0.9893	28.3	28.45	29.54	1.09	2.128740365					
3		B1	L2 Rep 1	Unknown	0.9893	28.6									
4		C1	L2 Rep 2	Unknown	0.9893	29.96	29.79		-0.25	0.840896415					
5		D1	L2 Rep 2	Unknown	0.9893	29.62									
6		E1	L2 Rep 3	Unknown	0.9893	28.95	29.345		0.195	1.144724161					
7		F1	L2 Rep 3	Unknown	0.9893	29.74									
8		G1	L2 Rep 4	Unknown	0.9893	30.02	30.575		-1.035	0.48801588					
9		H1	L2Rep 4	Unknown	0.9893	31.13									
10	Treatment	A4	L2 Rep 1	Unknown	0.9893	30.1	30.01		-0.47	0.721964598					
11		B4	L2 Rep 1	Unknown	0.9893	29.92									
12		C4	L2 Rep 2	Unknown	0.9893	29.06	29.3		0.24	1.180992661					
13		D4	L2 Rep 2	Unknown	0.9893	29.54									
14		E4	L2 Rep 3	Unknown	0.9893	28.08	28.035		1.505	2.838246712					
15		F4	L2 Rep 3	Unknown	0.9893	27.99									
16		G4	L2 Rep 4	Unknown	0.9893	28.07	28.51		1.03	2.042024251					
17		H4	L2 Rep 4	Unknown	0.9893	28.95									
18	Control	A2	18s Rep 1	Unknown	0.9893	18.45	18.385	18.4275	0.0425	1.029896957					
19		B2	18s Rep 1	Unknown	0.9893	18.32									
20		C2	18s Rep 2	Unknown	0.9893	18.14	18.295		0.1325	1.096191612					
21		D2	18s Rep 2	Unknown	0.9893	18.45									
22		E2	18s Rep 3	Unknown	0.9893	18.4	18.4		0.0275	1.019244379					
23		F2	18s Rep 3	Unknown	0.9893	18.4									
24		G2	18s Rep 4	Unknown	0.9893	18.61	18.63		-0.2025	0.86904332					
25		H2	18s Rep 4	Unknown	0.9893	18.65									
26	Treatment	A5	18s Rep 1	Unknown	0.9893	19.07	18.745		-0.3175	0.802459229					
27		B5	18s Rep 1	Unknown	0.9893	18.42									
28		C5	18s Rep 2	Unknown	0.9893	No Ct	19.07		-0.6425	0.640601908					
29		D5	18s Rep 2	Unknown	0.9893	19.07									
30		E5	18s Rep 3	Unknown	0.9893	19.4	19.31		-0.8825	0.542426663					
31		F5	18s Rep 3	Unknown	0.9893	19.22									
32		G5	18s Rep 4	Unknown	0.9893	18.98	19.075		-0.6475	0.638385594					
33		H5	18s Rep 4	Unknown	0.9893	19.17									
34	Control	A3	GOI Rep 1	Unknown	0.9893	32.52	32.53	32.69375	0.1638	1.120195082	1.480669856	0.756546152	-0.4025	1	0.50283
35		B3	GOI Rep 1	Unknown	0.9893	32.54									
36		C3	GOI Rep 2	Unknown	0.9893	31.99	32.02		0.6738	1.595214026	0.960095619	1.661515786	0.7325		
37		D3	GOI Rep 2	Unknown	0.9893	32.05									
38		E3	GOI Rep 3	Unknown	0.9893	32.57	32.695		-0.0012	0.999133941	1.080163722	0.924983798	-0.1125		
39		F3	GOI Rep 3	Unknown	0.9893	32.82									
40		G3	GOI Rep 4	Unknown	0.9893	33.32	33.53		-0.8363	0.560097541	0.651234935	0.860054507	-0.2175		
41		H3	GOI Rep 4	Unknown	0.9893	33.74									
42	Treatment	A6	GOI Rep 1	Unknown	0.9893	32.05	31.42		1.2738	2.417892327	0.761148576	3.176636471	1.6675	10.689519	1.26681
43		B6	GOI Rep 1	Unknown	0.9893	30.79									
44		C6	GOI Rep 2	Unknown	0.9893	29.19	29.415		3.2788	9.705146563	0.869796615	11.15794933	3.48		
45		D6	GOI Rep 2	Unknown	0.9893	29.64									
46		E6	GOI Rep 3	Unknown	0.9893	28.6	28.52		4.1738	18.04778649	1.24078229	14.54549007	3.8625		
47		F6	GOI Rep 3	Unknown	0.9893	28.44									
48		G6	GOI Rep 4	Unknown	0.9893	27.88	27.84		4.8538	28.91507622	1.141752541	25.32516915	4.6625		
49		H6	GOI Rep 4	Unknown	0.9893	27.8									

Fig. 8 Example of calculations of relative expression of a gene of interest (GOI) compared to two housekeeping genes, *18S rRNA* and *L2*.

9. Normalization of qPCR data using the geometric mean of multiple reference genes ensures accurate comparisons of gene expression between samples (Fig. 8). It is essential to verify that the expression of reference genes is stable across treatments.

4. Protein quantification by Western blot

4.1 Rationale

Increased accumulation of thiamin biosynthesis proteins must be verified in transgenic plants that overexpress the thiamin biosynthesis or thiamin binding protein genes. The most common method for protein quantification is Western blot. Total proteins are extracted from plant tissues and then separated on a polyacrylamide gel under denaturing conditions. Proteins are then transferred to a nitrocellulose membrane. Transfer quality is verified by staining with Ponceau solution. Proteins of interest are detected with primary polyclonal antibodies. Secondary antibodies conjugated with horseradish peroxidase (HRP) are then used to bind to primary antibodies. Different substrates of HRP can then be used for detection to produce a colored or chemiluminescent product. Difference in color or chemiluminescence intensity between non-transgenic and transgenic materials is then assessed to estimate the degree of protein accumulation.

4.2 Equipment

- Microcentrifuge Prism (LabNet, Corning, NY, USA)
- Spectrophotometer Genesys 10UV (Thermo Fisher Scientific, Waltham, MA, USA)
- Protein gel electrophoresis system Invitrogen XCell SureLock Mini-Cell (Thermo Fisher Scientific, Waltham, MA, USA)
- Mini Trans-Blot Cell (Bio-Rad, Hercules, CA, USA)
- Rocking shaker

4.3 Reagents

- Bradford Protein Assay Kit (Sigma-Aldrich, St. Louis, MO, USA)
- NuPAGE 4–12% Bis-Tris Gel 1.5 mm × 10 well (Thermo Fisher Scientific, Waltham, MA, USA)
- Nitrocellulose membrane (0.2 μm, 7.5 × 8 cm) (GenScript, Nanjing, China)
- Liquid nitrogen

- Bovin serum albumin (Sigma-Aldrich, St. Louis, MO, USA)
- THI1 antibodies raised by immunization of rabbits with the peptide CEVAEIDGAPRMGPT (GenScript, Nanjing, China)
- Conjugated rabbit antibodies (GenScript, Nanjing, China)
- Protein molecular markers (Thermo Fisher Scientific, Waltham, MA, USA)
- Tris-HCl (Thermo Fisher Scientific, Waltham, MA, USA)
- Tris base (Thermo Fisher Scientific, Waltham, MA, USA)
- MOPS (Sigma-Aldrich, St. Louis, MO, USA)
- Sodium phosphate monobasic (Sigma-Aldrich, St. Louis, MO, USA)
- Sodium phosphate dibasic (Sigma-Aldrich, St. Louis, MO, USA)
- Tween 20 (Sigma-Aldrich, St. Louis, MO, USA)
- Sodium chloride (Sigma-Aldrich, St. Louis, MO, USA)
- Glycerol (Thermo Fisher Scientific, Waltham, MA, USA)
- Triton X-100 (Thermo Fisher Scientific, Waltham, MA, USA)
- Phenylmethylsulfonyl fluoride (Sigma-Aldrich, St. Louis, MO, USA)
- Dithiothreitol (Sigma-Aldrich, St. Louis, MO, USA)
- Lithium dodecyl sulfate (LDS) (Sigma-Aldrich, St. Louis, MO, USA)
- Sodium dodecyl sulfate (SDS) (Sigma-Aldrich, St. Louis, MO, USA)
- EDTA (Sigma-Aldrich, St. Louis, MO, USA)
- Coomassie Brilliant Blue G250 (VWR, Radnor, PA, USA)
- Phenol Red 1 % (w/v) (Thermo Fisher Scientific, Waltham, MA, USA)
- Ponceau S staining solution (Thermo Fisher Scientific, Waltham, MA, USA)
- Glycine (Sigma-Aldrich, St. Louis, MO, USA)
- Ethanol (Sigma-Aldrich, St. Louis, MO, USA)
- Non-fat dry milk (Walmart)
- 4-chloro-1-naphthol (Sigma-Aldrich, St. Louis, MO, USA)
- Hydrogen peroxide (Sigma-Aldrich, St. Louis, MO, USA)

4.4 Protocols

4.4.1 Protein extraction and quantification

1. Collect leaves and flash freeze in liquid nitrogen.
2. Grind frozen leaves in mortar with pestle in liquid nitrogen.
3. Weigh out 100 mg of frozen, powdered leaves into a pre-frozen 1.7-mL centrifuge tube.
4. Add 500 μL of Protein extraction buffer (Table 3).
5. Vortex thoroughly and place tubes on ice for 10 min
6. Centrifuge at 15,000 × *g* for 15 min at 4 °C.

Table 3 Protein extraction buffer.

Compound	Molecular weight (g/mol)	Amount/ volume for 10 mL (1X)	Final concentration (1X)
1 M Tris-HCl pH 8.0	157.6	0.5 mL	50 mM
NaCl	58.44	0.0701 g	120 mM
Glycerol		1 g	10%
Triton X-100		0.02 g	0.2%
Phenylmethylsulfonyl fluoride (1% solution)	174.19	1 mL	5 mM
Dithiothreitol	154.25	0.0154 g	10 mM

7. Transfer the supernatant to a new centrifuge tube, flash freeze in liquid nitrogen and store at −80 °C until analysis.
8. Determine the protein concentration of each sample using the Bradford method (Bradford, 1976), using serial dilutions of bovine serum albumin standards to generate standard curves.

4.4.2 Protein separation by gel electrophoresis

1. Mix sample (30 μg) (see Note 1) with LDS sample buffer (Table 4) (2.5 μL 4X) and reducing agent (1 μL β-mercaptoethanol) in a final volume of 10 μL.
2. Heat at 70 °C for 10 min
3. Load on 4−12% polyacrylamide gel Nu-PAGE along with protein molecular markers.
4. Separate proteins in 1X Running buffer (Table 5) at 200 V for 50 min or until the blue dye front is at the end of the gel.
5. Separate the plastic casting plates with a knife. Cut the top (wells) and bottom of the gel with a razor blade. Cut one corner of the gel to keep track of orientation. Place gel in 1X Transfer buffer (Table 6).

4.4.3 Transfer of proteins from gel to membrane

1. Transfer proteins to a nitrocellulose membrane.
   a. Soak nitrocellulose membrane in deionized water.
   b. Soak fiber pads and Whatman papers in 1X Transfer buffer (Table 6).

Table 4 LDS sample buffer.

Compound	Molecular weight (g/mol)	Amount/volume for 10 mL (4X)	Final concentration (1X)
Tris-HCl	157.6	0.666 g	106 mM
Tris base	121.14	0.682 g	141 mM
Lithium dodecyl sulfate	272.33	0.800 g	2 %
Glycerol		4 g	10 %
EDTA	292.2	0.006 g	0.51 mM
Blue G250 (1 % solution)	854.02	0.75 mL	0.22 mM
Phenol Red (1 % solution)	354.38	0.25 mL	0.175 mM

The pH of the 1X solution is 8.5. Do not use acid or base to adjust the pH.

Table 5 Running buffer.

Compound	Molecular weight (g/mol)	Amount/volume per liter (20X)	Final concentration (1X)
MOPS	209.26	210 g	50 mM
Tris base	121.14	120 g	50 mM
SDS	288.38	20 g	0.1 %
EDTA	292.2	5.8 g	1 mM

The pH of the 1X solution is 7.7. Do not use acid or base to adjust the pH.
Store at 4 °C. For electrophoresis, dilute the 20X buffer to 1X with water.

c. Prepare the transfer sandwich by assembling one fiber pad, two Whatman papers, SDS gel, nitrocellulose membrane, two Whatman papers, one fiber pad. Make sure no air bubbles are trapped in the sandwich.
d. Secure the sandwich in a cassette and place it in a Mini Trans-Blot Cell tank. The cathode is positioned behind the gel, and the anode is behind the membrane.
e. Fill up the tank with 1X Transfer buffer.
f. Transfer overnight at 4 °C at 15 V.

2. Briefly rinse the blot in water.

Table 6 Transfer buffer.

Compound	Molecular weight (g/mol)	Amount/ volume per liter (1X)	Final concentration (1X)
Tris base	121.14	3.03 g	25 mM
Glycine	75.07	14.4 g	190 mM
Methanol (see Footnote)		200 mL	20 %

Store at 4 °C.
We often replace methanol with ethanol as a less toxic alternative.

Table 7 Phosphate buffer saline (PBS).

Compound	Molecular weight (g/mol)	Amount per liter (1X)	Final concentration (1X)
Na_2HPO4	141.96	1.4 g	9.8 mM
NaH_2PO4	119.98	0.2 g	1.7 mM
NaCl	58.44	8.5 g	145 mM

Adjust to pH 7.4 with HCl or NaOH.
Store at 4 °C

4.4.4 Immunodetection

1. Stain in Ponceau solution to check transfer quality. Signal of similar intensity between wells indicates even loading and transfer.
2. Rinse off the Ponceau stain with 3 washes with Washing buffer (PBS + 0.5 % Tween 20).
3. Block by incubating the membrane in 5 % non-fat dry milk in PBS (Table 7) on a shaker for 1 h at room temperature.
4. Dilute primary antibody in Dilution buffer (PBS + 0.5 % Tween 20 + 1 % BSA) (1:1000) (see Note 2). Incubate the membrane with the diluted primary antibody on a shaker for 1 h 30 at room temperature.
5. Wash the membrane 3–5 times for 5 min in Washing buffer (PBS + 0.5 % Tween 20).
6. Dilute secondary antibody in 5 % non-fat dry milk in PBS (1:3000) (see Note 2) and incubate the membrane with the diluted secondary antibody on a shaker for 1 h 30 at room temperature.

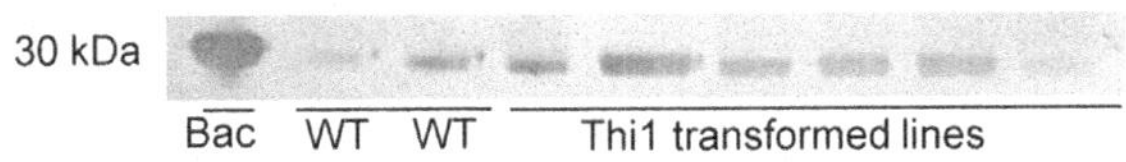

Fig. 9 Western blot of THI1 protein expression in non-transformed and transformed lines.

7. Wash the membrane 3–5 times for 5 min in Washing buffer (PBS + 0.5 % Tween 20).
8. Incubate the membrane with horseradish peroxidase substrate (30 mg 4-chloro-1-naphthol dissolved in 10 mL ethanol + 30 μL 30 % hydrogen peroxide + 50 mL PBS) (see Note 3). (Fig. 9).

4.5 Notes

1. Thiamin biosynthesis proteins are expressed at low levels. Depending on the sensitivity of the protein detection system, total amount of proteins loaded on SDS-PAGE can be adjusted.
2. Optimal dilutions should be tested depending on antibody quality and performance.
3. For higher sensitivity, use chemiluminescent substrate, e.g., ECL (Amersham ECL Plus).

5. Thiamin quantification

5.1 Rationale

Thiamin and its phosphate esters thiamin monophosphate (ThMP) and thiamin diphosphate (ThDP) are the main forms of thiamin found in plant tissues. In photosynthetic tissues, ThDP represents over 80 % of total thiamin, while thiamin is the main form in storage tissues like seeds and tubers (Dong et al., 2015). Cellular ThDP is almost entirely bound to proteins that use it as a cofactor (Hanson et al., 2018). Thiamin and its derivatives are stable under acidic conditions, but labile in alkaline conditions (Voelker et al., 2021). Therefore, extraction from plant tissues under acidic conditions enables to free ThDP from proteins, with little hydrolysis to thiamin (Ollilainen et al., 1993), while maintaining stability. After extraction, thiamin and its phosphate esters are oxidized to fluorescent thiochrome derivatives in the presence of the oxidizing agent potassium hexacyanoferrate (III) (see Note 1). This step is done under alkaline conditions because thiochrome forms at pH values greater than 8.0 and its fluorescence is maximal at pH above 9–13

(Edwards et al., 2017; Ishii et al., 1979; Pérez-Ruiz et al., 2004). The addition of methanol during the oxidation step has been reported to increase yield and fluorescence intensity (Bubeshko et al., 2011). Thiochrome derivatives of thiamin, ThMP, and ThDP are then separated by HPLC on a Capcell Pak NH_2 column with a mobile phase composed of acetonitrile and potassium phosphate buffer (Ishii et al., 1979). Quantification is done by reference to standard curves prepared by injecting commercially available thiamin, ThMP, and ThDP after oxidation to thiochromes.

5.2 Equipment

- UltiMate 3000 HPLC system (Thermo Fisher Scientific, Waltham, MA, USA)
- FLD-3000 fluorescence detector (Thermo Fisher Scientific, Waltham, MA, USA)
- Microcentrifuge Prism (LabNet, Corning, NY, USA)
- Branson 3200 ultrasonic cleaner (Branson Ultrasonics, Brookfield, CT, USA)
- Mortar and pestle

5.3 Reagents

- Capcell Pak NH_2 (5 μm, 4.6 × 150 mm) (Osaka Soda, Kita-ku, Osaka, Japan)
- Cartridge column NH_2 UG80 S5 (4 × 10 mm) (Parma CnS Inc., Tustin, CA, USA)
- 2-mL amber HPLC vials (Microsolv, Lelend, NC, USA)
- Screw caps with pre-slit (Microsolv, Lelend, NC, USA)
- Vial inserts clear glass conical base (6 × 31 mm) (Pekybio, Shaoxing City, China)
- 1.7-mL centrifuge tubes (Avantor, Phillipsburg, NJ, USA)
- 10-mL glass assay tubes (PhytoTech Labs, Lenexa, KS, USA)
- Plastic lids (PhytoTech Labs, Lenexa, KS, USA)
- Liquid nitrogen
- HPLC-grade acetonitrile (Fisher Scientific, Waltham, MA, USA)
- HPLC-grade methanol (Sigma-Aldrich, St. Louis, MO, USA)
- Potassium phosphate monobasic (Thermo Fisher Scientific, Waltham, MA, USA)
- Potassium phosphate dibasic (Thermo Fisher Scientific, Waltham, MA, USA)
- Hydrochloric acid (Thermo Fisher Scientific, Waltham, MA, USA)

- Sodium hydroxide (Sigma-Aldrich, St. Louis, MO, USA)
- Thiamin (Sigma-Aldrich, St. Louis, MO, USA)
- Thiamin monophosphate (Sigma-Aldrich, St. Louis, MO, USA)
- Thiamin diphosphate (Sigma-Aldrich, St. Louis, MO, USA)
- Potassium hexacyanoferrate (III) (Sigma-Aldrich, St. Louis, MO, USA)

5.4 Protocols

5.4.1 Sample preparation

1. Collect leaves and flash freeze in liquid nitrogen.
2. Grind frozen leaves in mortar with pestle in liquid nitrogen (see Note 2).
3. Add 100 mg of frozen (see Note 2), powdered leaves to 4 mL of 0.1 N HCl in 10-mL glass assay tubes (see Notes 3, 4).
4. Vortex to homogenize and sonicate in ultrasonic cleaner for 30 min.
5. Vortex gently to homogenize and transfer 1-mL to 1.7-mL centrifuge tube.
6. Centrifuge at 15,000 $\times g$ for 10 min.
7. Transfer 100 μL of supernatant to a new tube.

5.4.2 Thiamin oxidation to thiochrome

1. Add 34 μL of methanol and 17 μL of oxidizing solution (10 mM potassium hexacyanoferrate (III) in 3.7 N sodium hydroxide) (16.5 mg in 5 mL NaOH).
2. Vortex thoroughly.
3. Centrifuge at 15,000 $\times g$ for 10 min.
4. Transfer supernatant to HPLC vial containing insert and close with a screw-cap.

5.4.3 Separation by HPLC, detection and quantification

1. Inject 5 μl of each sample into the HPLC for separation on a Capcell Pak NH_2 column.
2. Run HPLC at a flow of 1 mL min^{-1} in isocratic mode with the following mobile phase: 60% acetonitrile – 40% 90 mM potassium phosphate pH 8.2 (Table 8) (see Note 5).
3. Record fluorescence with excitation at 365 nm and emission at 435 nm.
4. Integrate areas of peaks corresponding to oxidized forms of thiamin, thiamin monophosphate, and thiamin diphosphate (see Note 6) (Fig. 10).
5. Calculate thiamin and its esters concentrations by relating peak areas to a calibration curve created with oxidized thiamin, thiamin monophosphate, and thiamin diphosphate standards (40–0.04 μM) (see Notes 7, 8).

Table 8 Preparation of the HPLC mobile phase.

Compound	Molecular weight (g/mol)	Amount per liter	Final concentration
K_2HPO_4	174.2	87.1 g	0.5 M
KH_2PO_4	136.08	68.04 g	0.5 M

1. Prepare 90 mM solutions of each of K_2HPO_4 and KH_2PO_4 buffers by diluting 180 mL of 0.5 M solutions with 820 mL of deionized water.
2. Mix the two 90 mM solutions until a pH of 8.2 is reached (you will use a lot of K_2HPO_4 and very little of KH_2PO_4).
3. Filter buffer with 0.2 μm cellulose filter.
4. Prepare mobile phase by adding acetonitrile to 90 mM potassium phosphate pH 8.2 slowly with constant stirring to prevent salt formation.

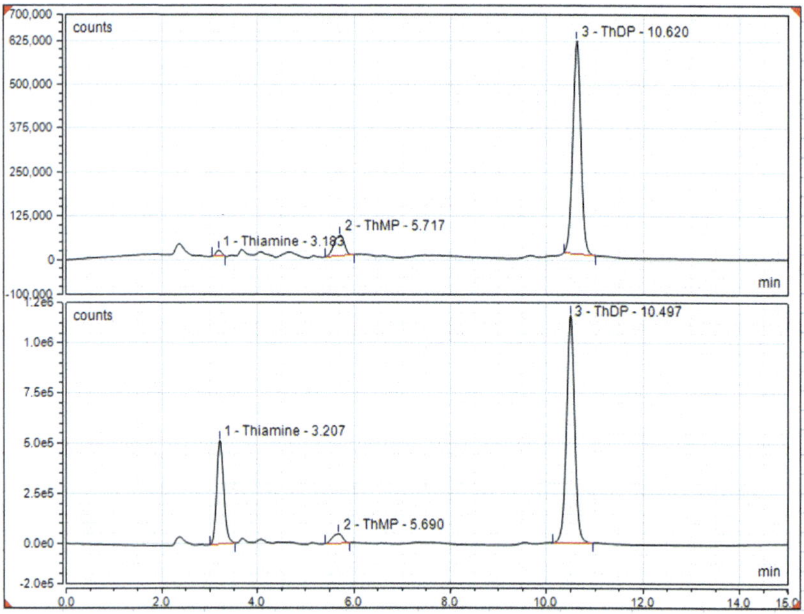

Fig. 10 Representative HPLC chromatograms of wild-type (top panel) and engineered (bottom panel) Arabidopsis leaves.

5.5 Notes

1. Other oxidizing agents have been used and reported in the literature, e.g., mercury(II) chloride, cyanogen bromide (Edwards et al., 2017).
2. Leaf samples can be lyophilized. In this case, smaller amounts of dried powdered sample should be used, e.g., 15 mg.

Table 9 Limit of detection (LOD) and limit of quantification (LOQ) of thiamin and its esters by HPLC coupled with fluorescence detection.

Compound	LOD (μM)	LOQ (μM)
Thiamin	0.0244	0.0812
Thiamin monophosphate	0.0782	0.2605
Thiamin diphosphate	0.0529	0.1762

3. We have found that smaller extraction volumes, e.g., 1 mL, decrease recoveries.
4. This procedure is well adapted for Arabidopsis leaves. However, it should be verified for other plant species. For instance, extraction of thiamin from fresh potato leaves in hydrochloric acid leads to poor recoveries (<20 %), probably because of the presence of compounds, e.g., ascorbic acid, polyphenols, which compete for ferricyanide. We were able to increase recovery (>90 %) by extracting thiamin from potato leaves in 1 % polyvinylpyrrolidone, as was done with wines (Liddicoat et al., 2015), but in this case, all thiamin monophosphate and thiamin diphosphate were hydrolyzed to thiamin. Extraction from lyophilized potato leaf samples in hydrochloric acid enabled us to quantify all thiamin species with ~50 % recoveries.
5. We prefer to pre-mix acetonitrile and phosphate buffer because we have experienced salt precipitation at the pump head when pumping solutions separately and mixing them with the HPLC system.
6. We prepare standard solutions fresh because thiamin monophosphate and thiamin diphosphate get hydrolyzed to thiamin relatively quickly in solution. Degradation of these standards can be checked by injecting them independently into the HPLC.
7. Concentrations of thiamin and its esters in leaf sample (nmol/g fresh weight) can be calculated by multiplying concentration in extract (μM) by 0.04.
8. Thiamin diphosphate is the dominant form in leaves, followed by thiamin. In our procedures, thiamin monophosphate is usually not quantifiable (Table 9).

6. Conclusions

In this chapter, we describe procedures to generate plants that overexpress THI1, the first committed step of the thiazole branch of thiamin biosynthesis. These procedures include designing and building a binary vector, using *Agrobacterium* to transform plants, and regenerating and characterizing transformed plants for gene (qPCR) and protein (Western blot) expression and thiamin levels (HPLC). Similar procedures can be used to overexpress THIC, the first committed step of the pyrimidine branch of thiamin biosynthesis, or other genes of interest.

Acknowledgements

Na.

References

Bradford, M. M. (1976). Rapid and sensitive method for quantitation of microgram quantities of protein utilizing principle of protein-dye binding. *Analytical Biochemistry, 72*(1-2), 248–254. https://doi.org/10.1016/0003-2697(76)90527-3.

Bubeshko, N. N., Stsiapura, V. I., & Stepuro, I. I. (2011). Fluorescent properties of thiochrome in solvents of different polarity. *Journal of Applied Spectroscopy, 78*(3), 337–343. https://doi.org/10.1007/s10812-011-9467-0.

Cheng, X., Sardana, R. K., & Altosaar, I. (1998). Rice transformation by agrobacterium infection. In C. Cunningham, & A. J. R. Porters (Eds.). *Recombinant proteins from plants: Production and isolation of clinically useful compounds* (pp. 1–9). Totowa, NJ: Humana Press.

Chetty, V., Narváez-Vásquez, J., & Orozco-Cárdenas, M. (2015). Potato. In K. Wang (Ed.). *Agrobacterium protocols. Methods in molecular biology* (pp. 1–1224). New York: Springer.

Dong, W., Stockwell, V. O., & Goyer, A. (2015). Enhancement of thiamin content in *Arabidopsis thaliana* by metabolic engineering. *Plant and Cell Physiology, 56*(12), 2285–2296. https://doi.org/10.1093/pcp/pcv148.

Dong, W., Thomas, N., Ronald, P. C., & Goyer, A. (2016). Overexpression of thiamin biosynthesis genes in rice increases leaf and unpolished grain thiamin content but not resistance to *Xanthomonas oryzae* pv. *oryzae*. *Frontiers in Plant Science, 7*. https://doi.org/10.3389/fpls.2016.00616.

Edwards, K. A., Tu-Maung, N., Cheng, K., Wang, B., Baeumner, A. J., & Kraft, C. E. (2017). Thiamine assays—Advances, challenges, and caveats. *ChemistryOpen, 6*(2), 178–191. https://doi.org/10.1002/open.201600160.

Engler, C., Gruetzner, R., Kandzia, R., & Marillonnet, S. (2009). Golden gate shuffling: A one-pot DNA shuffling method based on type IIs restriction enzymes. *PLoS One, 4*(5), e5553. https://doi.org/10.1371/journal.pone.0005553.

Fitzpatrick, T. B., Dalvit, I., Chang, F.-H., Wang, K., Fudge, J. B., Chang, S.-H., & Gruissem, W. (2024). Vitamin B1 enhancement in the endosperm of rice through thiamine sequestration. *Plant Biotechnology Journal, 22*(8), 2330–2332. https://doi.org/10.1111/pbi.14348.

Goyer, A. (2017). Thiamin biofortification of crops. *Current Opinion in Biotechnology, 44*, 1–7. https://doi.org/10.1016/j.copbio.2016.09.005.

Goyer, A., Hamlin, L., Crosslin, J. M., Buchanan, A., & Chang, J. H. (2015). RNA-Seq analysis of resistant and susceptible potato varieties during the early stages of Potato Virus Y infection. *BMC Genomics, 16*, 472.

Hanson, A. D., Amthor, J. S., Sun, J. Y., Niehaus, T. D., Gregory, J. F., Bruner, S. D., & Ding, Y. S. (2018). Redesigning thiamin synthesis: Prospects and potential payoffs. *Plant Science, 273*, 92–99. https://doi.org/10.1016/j.plantsci.2018.01.019.

Hood, E. E., Gelvin, S. B., Melchers, L. S., & Hoekema, A. (1993). NewAgrobacterium helper plasmids for gene transfer to plants. *Transgenic Research, 2*(4), 208–218. https://doi.org/10.1007/BF01977351.

Ishii, K., Sarai, K., Sanemori, H., & Kawasaki, T. (1979). Analysis of thiamine and its phosphate esters by high-performance liquid chromatography. *Analytical Biochemistry, 97*(1), 191–195. Retrieved from. http://www.ncbi.nlm.nih.gov/entrez/query.fcgi?cmd=Retrieve&db=PubMed&dopt=Citation&list_uids=484841.

Liddicoat, C., Hucker, B., Liang, H., & Vriesekoop, F. (2015). Thiamin analysis in red wine by fluorescence reverse phase-HPLC. *Food Chemistry, 177*, 325–329. https://doi.org/10.1016/j.foodchem.2015.01.009.

Ollilainen, V., Vahteristo, L., Uusi-Rauva, A., Varo, P., Koivistoinen, P., & Huttunen, J. (1993). The HPLC determination of total thiamin (Vitamin B1) in foods. *Journal of Food Composition and Analysis, 6*, 152–165.

Pérez-Ruiz, T., Martínez-Lozano, C., Sanz, A., & Guillén, A. (2004). Successive determination of thiamine and ascorbic acid in pharmaceuticals by flow injection analysis. *Journal of Pharmaceutical and Biomedical Analysis, 34*(3), 551–557. https://doi.org/10.1016/s0731-7085(03)00606-x.

Taylor, S. C., Nadeau, K., Abbasi, M., Lachance, C., Nguyen, M., & Fenrich, J. (2019). The ultimate qPCR experiment: Producing publication quality, reproducible data the first time. *Trends in Biotechnology, 37*(7), 761–774. https://doi.org/10.1016/j.tibtech.2018.12.002.

Voelker, A. L., Taylor, L. S., & Mauer, L. J. (2021). Effect of pH and concentration on the chemical stability and reaction kinetics of thiamine mononitrate and thiamine chloride hydrochloride in solution. *BMC Chemistry, 15*(1), 47. https://doi.org/10.1186/s13065-021-00773-y.

Walhout, A. J. M., Temple, G. F., Brasch, M. A., Hartley, J. L., Lorson, M. A., van den Heuvel, S., & Vidal, M. (2000). GATEWAY recombinational cloning: Application to the cloning of large numbers of open reading frames or ORFeomes. In J. Thorner, S. D. Emr, & J. N. Abelson (Vol. Eds.), *Methods in enzymology: 328*, (pp. 575–IN577). Academic Press.

Xu, X., Pan, S. K., Cheng, S. F., Zhang, B., Mu, D. S., Ni, P. X., & Potato Genome Sequencing, C. (2011). Genome sequence and analysis of the tuber crop potato. *Nature, 475*(7355), 189–195. https://doi.org/10.1038/nature10158.

CHAPTER FOUR

Incubation and physiological readouts *in vitro* for thiamine triphosphate and related compounds in plants

Kai Wang[a,b,*], **Jean Alric**[c], **and Teresa B. Fitzpatrick**[b,*]

[a]Faculty of Synthetic Biology & Institute of Emerging Agricultural Technology, Shenzhen University of Advanced Technology, Shenzhen, P.R. China

[b]Vitamins & Environmental Stress Responses in Plants, Department of Plant Sciences, University of Geneva, Geneva, Switzerland

[c]CEA, CNRS, BIAM, Photosynthesis & Environment, Aix Marseille University, Saint Paul-Lez-Durance, France

*Corresponding authors. e-mail address: wangkai@suat-sz.edu.cn; theresa.fitzpatrick@unige.ch

Contents

Abstract

Thiamine triphosphate (ThTP) is a low-abundance thiamine derivative the metabolism of which in plants remains poorly understood because of its dynamic behavior, apparent lability, and strong dependence on subcellular context. In this chapter, we describe an integrated experimental framework for investigating ThTP and related thiamine derivatives in plants. The workflow combines isolation and

Methods in Enzymology, Volume 735
ISSN 0076-6879, https://doi.org/10.1016/bs.mie.2026.07.013

incubation *in vitro* of chloroplasts, HPLC-based quantification of thiamine derivatives, enzymatic verification of ThTP peak identity, and electrochromic shift (ECS) measurements as a physiological readout of altered ThTP levels. Together, these methods provide a practical platform for examining the metabolic significance of this minor thiamine derivative in plant systems.

1. Introduction

Thiamine is essential for both plant performance and human nutrition (as vitamin B_1). Whereas humans depend on dietary acquisition of thiamine, plants biosynthesize this compound *de novo* and thereby serve as a major entry point of vitamin B_1 into food systems. In plants, thiamine status influences not only nutritional quality but also growth, stress tolerance, disease responses, and overall metabolic performance (Fitzpatrick & Chapman, 2020). However, thiamine metabolism cannot be adequately described by this compound alone (Fig. 1). Rather, the homeostasis of individual derivatives often referred to as vitamers (*i.e.* when they can fulfil the role of vitamin), including their abundance, balance, and interconversion, is increasingly recognized as a more informative framework for understanding biological function (Fitzpatrick, 2024).

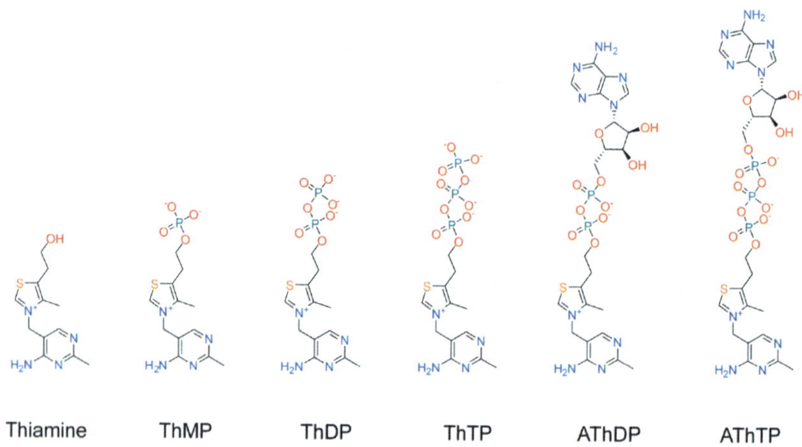

Fig. 1 Chemical structures of thiamine and its derivatives. The structures shown include thiamine, thiamine monophosphate (ThMP), thiamine diphosphate (ThDP), thiamine triphosphate (ThTP), adenosine thiamine diphosphate (AThDP), and adenosine thiamine triphosphate (AThTP). Phosphate groups are attached to the hydroxyethyl side chain of thiamine, while AThDP and AThTP contain an additional adenosine moiety linked through the phosphate chain.

Thiamine is non-phosphorylated but can exist as mono-, di- and triphosphorylated forms, as well as adenylated derivatives (Fig. 1). Among those, thiamine diphosphate (ThDP; also termed thiamine pyrophosphate, TPP) is the most renowned bioactive form as it is well established to serve as an essential coenzyme for central metabolic enzymes, including pyruvate dehydrogenase, 2-oxoglutarate dehydrogenase, and transketolase (Schellenberger, 1998; Bettendorff & Wins, 2013). Thiamine and thiamine monophosphate (ThMP) serve as intermediates to ThDP biosynthesis. By contrast, thiamine triphosphate (ThTP) remains far less well understood. Although its existence is conserved across biological kingdoms, ThTP generally occurs at much lower abundance than ThDP, and its biosynthesis, turnover, and physiological significance remain incompletely resolved (Bettendorff, 2021). This uncertainty is particularly pronounced in plants, where investigation of ThTP metabolism has only recently progressed from initial detection toward mechanistic analysis (Hofmann et al., 2020; Wang et al., 2025). Adenylated derivatives are also not well understood but have been reported to accumulate in bacteria during nutritional limitations and have been detected in plants, although it is unclear under which circumstances (Bettendorff et al., 2007; Gigliobianco et al., 2010). Here we will focus on methods that can be used to investigate ThTP in plants.

Current evidence indicates that ThTP in plants is a dynamically regulated vitamer closely linked to cellular energy metabolism rather than a by-product of thiamine metabolism (Fig. 2) (Wang et al., 2025). In Arabidopsis, its accumulation is strongly influenced by light and was originally associated with plastidial and mitochondrial function, consistent with a dependence on the proton motive force (*pmf*) and ThDP availability (Hofmann et al., 2020). More recent work supports the chloroplast as a major site of ThTP formation (Fig. 2A) and suggests that ThTP participates in the coordination of phosphate utilization, ThDP economy, and ATP synthase regulation in this organelle during the early light phase (Fig. 2B) (Wang et al., 2025). Together, these observations support the view that ThTP should be considered a responsive metabolic effector in plant cells, while also highlighting the need for experimental approaches that can resolve its formation, turnover, and functional consequences at subcellular resolution.

Despite this progress, major questions surrounding ThTP metabolism in plants remain unresolved. Its low abundance, apparent lability, and sensitivity to developmental stage, light regime, and subcellular compartmentation

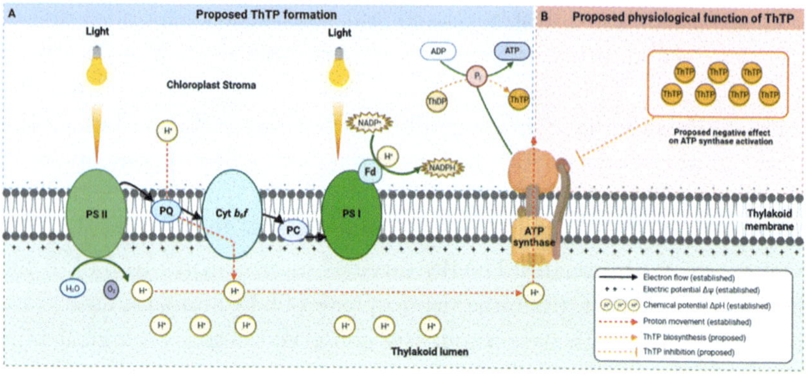

Fig. 2 Proposed model linking ThTP formation and physiological function. (A) Proposed model for ThTP formation in chloroplasts. Light-driven electron transport across the thylakoid membrane is shown as an established process involving photosystem II (PSII), plastoquinone (PQ), cytochrome b_6f (Cyt b_6f), plastocyanin (PC), photosystem I (PSI), ferredoxin (Fd), and $NADP^+$ reduction. Electron flow is indicated by solid black arrows. Water oxidation at PSII and proton translocation associated with the PQ/Cyt b_6f pathway promote proton accumulation in the thylakoid lumen, while photosynthetic electron transport establishes an electrochemical proton gradient across the thylakoid membrane. This proton motive force (*pmf*) is composed of an electric potential component Δψ, and a chemical proton gradient ΔpH. In the diagram, Δψ is indicated by the charge separation across the membrane, whereas ΔpH is indicated by lumenal H^+ accumulation and proton movement. In the proposed model, chloroplast ATP synthase uses the *pmf* to support ATP formation from ADP and inorganic phosphate (Pi). In parallel, ThTP formation from ThDP and Pi is proposed to occur in association with chloroplast ATP synthase, with the ΔpH component of the *pmf* being directly linked to ThTP formation. This model is supported by *in vitro* incubation assays using isolated chloroplasts. (B) Proposed physiological function of ThTP. A stromal ThTP pool is proposed to negatively influence the activation state of chloroplast ATP synthase, suggesting a potential feedback mechanism by which ThTP may modulate ATP synthase activity and *pmf* utilization. Solid black arrows indicate established electron flow; red dashed arrows indicate proton movement and ΔpH formation; orange dashed arrows indicate proposed ThTP biosynthesis; and orange inhibitory lines indicate the proposed negative effect of ThTP on ATP synthase activation.

make robust analysis inherently challenging. Moreover, the biosynthesis and turnover of ThTP are still not completely resolved. Biosynthesis is thought to operate through ATP synthase, although the mechanism is unclear (Hofmann et al., 2020; Wang et al., 2025), whereas no enzyme has been demonstrated to be involved in turnover in plants, despite the existence of such enzymes in mammalian systems (Pesquera et al., 2022). In particular, members of the CyaB–thiamine triphosphatase (CYTH) domain or triphosphate tunnel metalloenzyme (TTM) family in mammals are highly specific for degradation of ThTP, referred to as thiamine triphosphatase (ThTPase) (Iyer & Aravind, 2002; Lakaye et al., 2002). Although plant genomes encode CYTH/TTM family proteins related to the superfamily that includes mammalian ThTPases, convincing evidence that any of these proteins functions as *bona fide* ThTPases in plants is still lacking (Pesquera et al., 2022). Therefore, current approaches to characterize ThTP mainly rely on accurate quantification of the compound, as well as on experimental systems that enable testing of its function that may in the future be integrated with downstream metabolic outcomes.

In this chapter, we describe an experimental framework for investigating ThTP and related thiamine derivatives in plants, with a particular focus on chloroplast-associated ThTP metabolism and function. We outline strategies for isolating intact chloroplasts, performing substrate-based incubation assays *in vitro*, extracting thiamine derivatives, and quantifying major vitamers by fluorescence-based HPLC. We also describe the use of recombinant *Mus musculus* thiamine triphosphatase (*Mm*ThTPase) as an enzymatic control for ThTP peak verification. Finally, we include light-induced electrochromic shift (ECS) measurements of the trans-thylakoid membrane potential as a non-invasive physiological readout to assess how altered ThTP levels are associated with changes in chloroplast ATP synthase activity. Together, these approaches provide a practical platform for dissecting the formation, turnover, compartmentation, and physiological consequences of thiamine derivatives in plant systems.

2. Incubation strategies *in vitro* with isolated chloroplasts

2.1 Rationale

ThTP is a low-abundance thiamine derivative the metabolism of which is difficult to resolve in whole tissues because its pool size is small, dynamic, and strongly influenced by subcellular context. In *Arabidopsis*, ThTP has been linked to organellar bio-energetics and shown to accumulate transiently at the start of the light period, predominantly through a chloroplast-associated chemiosmotic mechanism (Wang et al., 2025). Assays with isolated chloroplast preparations therefore provide a tractable experimental system for testing substrate dependence and compartment-specific metabolism under controlled conditions, while minimizing the complexity of competing whole-cell processes (Fig. 3A).

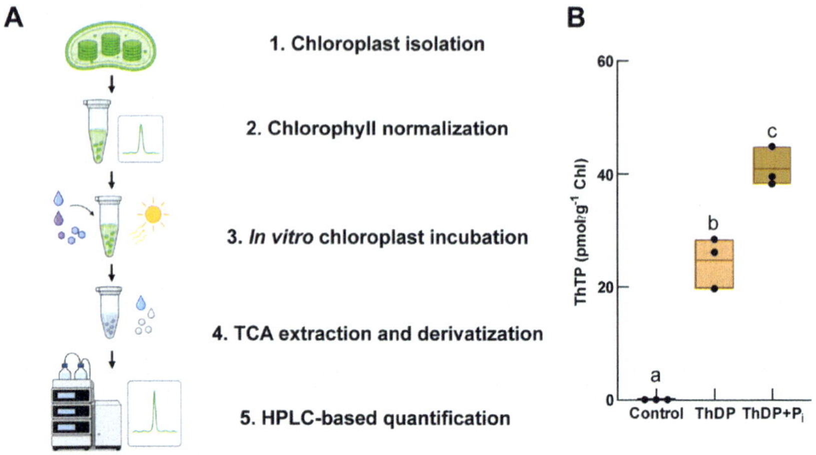

Fig. 3 The workflow for investigating ThTP formation with isolated chloroplasts. (A) Experimental workflow for assessing substrate-dependent ThTP formation in isolated chloroplasts. Intact chloroplasts are first isolated and normalized based on chlorophyll content, then incubated under light with defined substrate conditions. After incubation, chloroplasts are washed to remove external substrates, followed by trichloroacetic acid (TCA) extraction of thiamine derivatives and fluorescence-based HPLC quantification. This workflow provides a biochemical readout of ThTP accumulation after controlled chloroplast incubation *in vitro*. (B) A representative result of ThTP accumulation after incubation with selected substrate combinations. The control sample showed minimal ThTP accumulation, whereas addition of ThDP increased ThTP levels. The combined addition of ThDP and P_i further enhanced ThTP accumulation, supporting substrate- and phosphate-dependent ThTP formation in isolated chloroplasts. Individual data points are shown together with mean values and variation. Different letters indicate statistically significant differences between treatments. Data were extracted from Wang et al., 2025, reanalyzed, and replotted.

Because chloroplast yield and integrity vary between preparations, normalization to chlorophyll content is required to standardize organelle input across samples. Following incubation, acidic extraction coupled to fluorescence-based HPLC enables sensitive separation and quantification of the major thiamine derivatives, including thiamine, ThMP, ThDP, and ThTP. Since ThTP is typically present at low abundance and authentic standards may contain partially hydrolyzed ThDP, enzymatic confirmation of peak identity is highly advantageous. For this purpose, recombinant *Mm*ThTPase provides a specific analytical control by converting ThTP to ThDP (Hofmann et al., 2020). Together, chloroplast isolation, chlorophyll normalization, incubation *in vitro*, vitamer extraction, HPLC quantification, and

*Mm*ThTPase-based verification constitute an integrated workflow for dissecting the formation and turnover of minor thiamine derivatives in plant organelles (Fig. 3B).

2.2 Isolation of intact chloroplasts

2.2.1 Materials, equipment and reagents

- *Arabidopsis thaliana* leaf tissue, typically 15–30 g fresh weight per preparation; in our hands, ~20 g is sufficient for a standard isolation
- Cold room or ice buckets; the entire procedure should be performed at 4 °C or with samples and reagents continuously maintained on ice
- Pre-chilled tubes, bottles, centrifuge rotors, and pipettes
- Blender or tissue homogenizer suitable for rapid homogenization of leaf tissue
- Calbiochem/Merck Millipore/Sigma or equivalent rayon-polyester filtration material with an approximate pore size around 25 μm. Use as two layers for filtration of the leaf homogenate
- Soft paintbrush for gentle resuspension of the crude chloroplast pellet
- Truncated or wide-bore pipette tips; narrow tips should be avoided to minimize chloroplast breakage
- Centrifuge capable of 1500 g, 2070 g, 2600 g, and 16,100 g at 4 °C
- Vacuum aspirator for removal of the upper broken-chloroplast layer
- **Extraction buffer:** 15 mM MOPS-KOH (pH 7.4) containing 300 mM sorbitol and 2 mM EDTA; followed by adding 0.6 % (w/v) polyvinylpyrrolidone (PVP), 0.1 % (w/v) BSA, 1 mM DTT, 0.2 mM PMSF, and 0.2 % (w/v) protease inhibitor cocktail
- **Washing buffer:** 10 mM Tricine-KOH (pH 7.6), containing 400 mM sorbitol, 1.25 mM EDTA, and 2.5 mM $MgCl_2$
- Percoll gradients (Sigma): 45 % and 85 % Percoll layers prepared with 2 × washing buffer

2.2.2 Protocols

- **Pre-chill all equipment and reagents:** perform the entire procedure at 4 °C or continuously on ice. Pre-chill centrifuges, tubes, bottles, and buffers before starting. To minimize organelle damage, use truncated or wide-bore tips whenever possible.
- **Harvest and homogenize the tissue:** use approximately 20 g fresh Arabidopsis leaf tissue and homogenize in 250 mL ice-cold extraction buffer. Perform homogenization rapidly using short pulses in a prechilled blender or tissue homogenizer until the tissue is disrupted and the

suspension becomes uniformly green. Avoid prolonged blending, excessive foaming, or warming of the homogenate, as over-homogenization increases chloroplast breakage and reduces the proportion of intact chloroplasts.

- **Remove tissue debris:** filter the homogenate through two layers of Miracloth (pre-wetted with ice-cold extraction buffer) to remove residual tissue debris. If necessary, retained material may be re-extracted and filtered again with a reduced volume of extraction buffer. Gently squeeze only if necessary to recover the liquid fraction, avoiding excessive pressure that may increase chloroplast breakage.
- **Collect the crude chloroplast fraction:** centrifuge the filtrate at 1500 g for 10 min at 4 °C. The resulting pellet corresponds to the crude chloroplast fraction. Keep the pellet on ice until further processing.
- **Resuspend the crude chloroplast pellet:** gently resuspend the pellet in 6 mL washing buffer using a soft paintbrush. Mix by gentle inversion until a homogeneous suspension is obtained. Do not vortex.
- **Prepare and load the Percoll gradients:** prepare a discontinuous Percoll step gradient in a pre-chilled centrifuge tube, with the denser 85 % Percoll layer at the bottom and the lighter 45 % Percoll layer above it. Both Percoll solutions should be prepared using 2 × washing buffer to maintain the appropriate osmotic and ionic conditions. Carefully add the 85 % Percoll first, then slowly overlay the 45 % Percoll along the tube wall to form a sharp interface between the two layers. Finally, gently layer the crude chloroplast suspension on top of the 45 % Percoll layer. The gradient should therefore be arranged from bottom to top as 85 % Percoll/45 % Percoll/crude chloroplast suspension. Avoid disturbing the interfaces during loading, as mixing of the layers reduces separation efficiency and compromises recovery of intact chloroplasts.
- **Separate intact chloroplasts:** centrifuge the chloroplast loaded gradient at 2600 g for 10 min at 4 °C using slow acceleration and no brake. Intact chloroplasts should accumulate at the 45 %/85 % Percoll interface.
- **Remove the broken-chloroplast fraction:** carefully aspirate the upper portion of the gradient, which contains broken chloroplasts, using a vacuum aspirator (Fig. 4A). Avoid disturbing the intact chloroplast band at the interface.
- **Wash the intact chloroplast fraction:** carefully collect the intact chloroplast band from the 45 %/85 % Percoll interface using a wide-bore pipette tip or Pasteur pipette (Fig. 4A), and transfer it to a pre-chilled

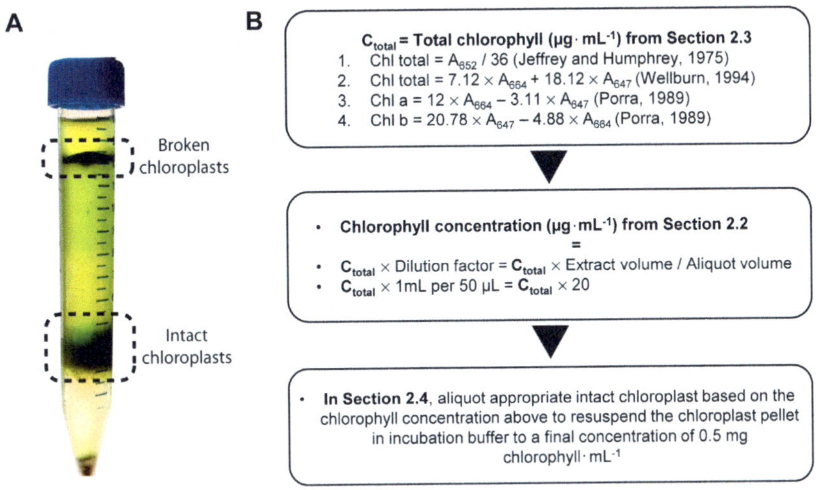

Fig. 4 Isolation of intact chloroplasts and chlorophyll-based normalization for *in vitro* incubation. (A) Isolation of intact chloroplasts by Percoll density-gradient centrifugation. After centrifugation, broken chloroplasts and partially disrupted chloroplast material are enriched in the upper green band, whereas intact chloroplasts form a distinct lower green band at the 45 %/85 % Percoll interface. The intact chloroplast band is carefully collected for washing and downstream assays. (B) Workflow for calculating and adjusting chlorophyll concentration before *in vitro* chloroplast incubation. Chlorophyll content is first determined from the isolated chloroplast preparation using absorbance-based equations described in **Section 2.3**. The measured chlorophyll concentration is then corrected for extraction volume, aliquot volume, and dilution factor to estimate the chlorophyll concentration of the intact chloroplast preparation obtained in **Section 2.2**. Based on this value, an appropriate amount of intact chloroplast suspension or pellet is resuspended in incubation buffer to reach the desired final chlorophyll concentration for the *in vitro* incubation assay described in **Section 2.4**.

centrifuge tube. Dilute the recovered suspension with 5–10 mL ice-cold washing buffer to reduce the Percoll concentration. Mix gently by slow inversion; do not vortex or pipette repeatedly. Centrifuge at 2070 g for 2 min at 4 °C once full speed is reached. Carefully remove the supernatant without disturbing the green chloroplast pellet. Repeat the wash once more with 5–10 mL washing buffer. After the final wash, gently resuspend the chloroplast pellet in 2–3 mL washing buffer using a soft paintbrush or wide-bore tip. Keep the suspension on ice and proceed immediately to chlorophyll normalization.

2.2.3 Notes

1. **Maintain low temperature throughout the isolation.** Chloroplast integrity declines rapidly if extracts, buffers, or pellets are allowed to warm. All steps should therefore be performed at 4 °C or on ice.
2. **Minimize mechanical damage.** Use wide-bore tips and gentle resuspension rather than repeated pipetting or vortexing, as excessive shear markedly increases chloroplast breakage.
3. **Avoid over-homogenization.** Prolonged grinding increases contamination with broken chloroplasts and other cellular debris, thereby reducing the quality of the final preparation.
4. **Leaf starch content can influence chloroplast recovery.** Preparations from starch-rich tissue may yield a higher proportion of damaged chloroplasts during centrifugation. When compatible with the biological question, dark or cold pre-treatment of plants may reduce starch accumulation and improve recovery.
5. **Percoll handling is critical.** Slow acceleration and omission of the brake during gradient centrifugation are important for maintaining a clean interface and maximizing recovery of intact chloroplasts.
6. **Fresh preparations are preferred.** Chloroplasts should ideally be used immediately for downstream assays. If storage is unavoidable, short-term freezing (less than 2 weeks) may be possible prior to incubation assays *in vitro*.
7. Although this section focuses on chloroplasts, a similar strategy could also be adapted for isolated mitochondria, to investigate specific aspects of ThTP metabolism therein (Gangolf et al., 2010).

2.3 Chlorophyll normalization

2.3.1 Materials, equipment and reagents

- Isolated intact chloroplasts from **Section 2.2**
- Centrifuge capable of 2600 g and 16,100 g at 4 °C
- Liquid nitrogen
- Dimethylformamide (DMF), analytical-grade with purity ≥ 99.8 % or equivalent (Sigma)
- A probe sonicator or bath sonicator
- Spectrophotometer or NanoDrop capable of absorbance measurements at 664 nm and 647 nm

2.3.2 Protocols

- **Prepare an aliquot of the chloroplast suspension:** transfer a 50 μL aliquot (**Aliquot volume**) of the isolated chloroplast preparation to a pre-chilled microcentrifuge tube while keeping the remaining sample on ice.
- **Pellet the chloroplasts:** centrifuge at 2600 g for 2 min at 4 °C and carefully remove the supernatant.
- **Disrupt the chloroplasts:** subject the pellet to two rapid freeze-thaw cycles using liquid nitrogen to disrupt chloroplast membranes and release chlorophyll.
- **Extract chlorophyll:** resuspend the disrupted pellet in 1 mL DMF (**Extract volume**) and mix thoroughly to extract chlorophyll. This protocol is adapted from published DMF-based chlorophyll extraction and quantification methods (Evans et al., 2012). Because DMF has relatively low volatility compared with acetone or methanol, absorbance measurement can be performed using a NanoDrop spectrophotometer.
- **Improve extraction by sonication:** sonicate the suspension 3 times for 10 s using a probe sonicator fitted with a microtip at low-to-moderate power, for example 20–30 % amplitude, with the tube kept on ice between pulses. Alternatively, a bath sonicator can be used for 3 × 30 s if a probe sonicator is not available, although extraction may be less efficient. Avoid foaming, prolonged sonication, or noticeable warming of the sample.
- **Clarify the extract:** centrifuge at 16,100 g for 2 min at 4 °C to remove insoluble material.
- **Measure absorbance:** record the absorbance of the clarified supernatant at 664 nm and 647 nm.
- **Calculate total chlorophyll concentration:** several published absorbance-based equations can be used to estimate total chlorophyll concentration (Fig. 4B). Although the absolute values may differ slightly between equations, they generally provide comparable estimates when applied consistently. Therefore, the same equation should be used throughout the entire protocol.
- **Normalize chloroplast input:** use the calculated chlorophyll concentration to adjust the chloroplast preparation to a defined working concentration for downstream assays. In our workflow, intact chloroplasts were typically adjusted to 0.5 mg chlorophyll·mL^{-1} prior to use.

2.3.3 Notes

1. **Proceed promptly after disruption.** Dimethylformamide extraction should be initiated immediately after freeze-thaw treatment to minimize variation between samples.
2. **Define one working concentration and use it consistently**. Once a chlorophyll-normalized concentration is selected for the assay series, it should be applied uniformly across all samples to minimize variation in downstream measurements.

2.3.4 Results part I

Following Percoll density-gradient centrifugation, two major chloroplast-containing fractions are typically observed (Fig. 4A). The upper green band contains broken chloroplasts and partially disrupted chloroplast material, whereas the lower, darker green band at the 45 %/85 % Percoll interface is enriched in intact chloroplasts. This intact chloroplast fraction should be carefully collected and washed before chlorophyll normalization and downstream incubation assays.

To ensure comparable chloroplast input between treatments, the isolated chloroplasts are normalized based on chlorophyll content before the *in vitro* incubation assay. Total chlorophyll concentration is first determined from an aliquot of the isolated chloroplast suspension after DMF extraction, as described in **Section 2.3**. Several published absorbance-based equations can be used to calculate total chlorophyll concentration (Fig. 4B, **upper panel; unit: μg·mL^{-1}**). Although different equations may give slightly different absolute values, they generally provide comparable estimates when used consistently (Jeffrey & Humphrey, 1975; Porra et al., 1989; Wellburn, 1994). Therefore, the same equation should be used throughout the protocol for all samples and experiments. The chlorophyll concentration measured in the DMF extract is then converted back to the chlorophyll concentration of the chloroplast suspension obtained in **Section 2.2** by correcting for the dilution factor, defined here as the extraction volume divided by the aliquot volume (Fig. 4B, **middle panel**):

$$\text{Chlorophyll suspension concentration in } \textbf{Section 2.2} = C_{\text{total}} \times \frac{\text{Extraction volume}}{\text{Aliquot volume}}$$

C_{total} is the total chlorophyll concentration directly measured from the DMF extract in **Section 2.3**, expressed as μg·mL^{-1}.

For example, if chlorophyll is extracted from a **50 μL** aliquot of chloroplast suspension (aliquot volume) using **1 mL DMF** (extraction volume), and no additional dilution is applied, the chlorophyll concentration of the original chloroplast suspension from **Section 2.2** is calculated as:

$$\frac{C_{total} \times 1\ mL}{50\ \mu L} = C_{total} \times 20(\mu g \cdot mL^{-1})$$

This calculated chlorophyll concentration is then used to prepare the chloroplast input for the *in vitro* incubation assay described in **Section 2.4** (Fig. 4B, **lower panel**). For example, if the desired target incubation concentration is **0.5 mg chlorophyll·mL^{-1}**, and the final incubation volume is **1 mL**, the amount of chlorophyll required is:

$$0.5\ mg \cdot mL^{-1} \times 1\ mL = 0.5\ mg = 500\ \mu g\ chlorophyll$$

The required volume of chloroplast suspension from **Section 2.2** is therefore:

$$\frac{\text{(The amount of chlorophyll required)}}{\text{(Chlorophyll suspension concentration in Section 2.2)}}$$

$$= \frac{500\ \ \mu g\ chlorophyll}{C_{total} \times 20}$$

$$= \frac{25}{C_{total}}\ (mL)$$

Thus, **25 per C_{total} (mL)** of the chloroplast suspension obtained in **Section 2.2** should be aliquoted, gently pelleted, and resuspended in incubation buffer to a final volume of **1 mL** to obtain a final chlorophyll concentration of **0.5 mg·mL^{-1}** for the **Section 2.4**.

2.4 Chloroplast incubation with reagents *in vitro* to assess effects on ThTP levels

2.4.1 Materials, equipment and reagents

- Chlorophyll-normalized chloroplast preparation from **Section 2.3**
- Centrifuge or microcentrifuge capable of 2600 g at 20 °C
- Plant growth chamber or equivalent illuminated incubator set to 20 °C and 250 μmol photons m^{-2} s^{-1} light
- Liquid nitrogen
- Tissue lyser suitable for 2 mL microcentrifuge tubes
- 2 mm glass beads

- Pre-chilled microcentrifuge tubes
- Pipettes with wide-bore or cut tips for gentle handling of chloroplast suspensions
- **Incubation buffer:** 50 mM HEPES-KOH (pH 7.8) containing 330 mM sorbitol, 10 mM NaCl and 2 mM $MgCl_2$
- **Substrates:** in this protocol 1 mM ThDP and 10 mM potassium phosphate (referred to as P_i) were used as representative concentrations for ThTP formation.

2.4.2 Protocols

- **Pellet the chloroplasts:** centrifuge the chlorophyll-normalized chloroplast preparation at 2600 g for 2 min and carefully remove the supernatant.
- **Resuspend in incubation buffer:** resuspend the chloroplast pellet in 1 mL incubation buffer to a final chloroplast concentration of 0.5 mg chlorophyll·mL^{-1}.
- **Add substrates:** add the desired substrates according to the experimental design. A minimal set of conditions includes a control containing incubation buffer alone and a treatment group (here containing ThDP plus P_i).
- **Incubate under defined conditions:** incubate the chloroplast suspension for 30 min at 20 °C under 250 μmol photons m^{-2} s^{-1} light. During incubation, gently mix the suspension every 10 min to maintain a homogeneous preparation.
- **Wash the chloroplasts:** after incubation, wash the chloroplasts twice with incubation buffer by centrifugation at 2600 g for 2 min to remove residual external substrates.

2.4.3 Notes

1. **Fresh chloroplast preparations are preferred.** This assay is intended to probe chloroplast-associated biosynthetic capacity, and freshly isolated chloroplasts generally provide the most reliable results. Samples can be stored at −80 °C but for less than 2 weeks.
2. **Mix gently during incubation.** Periodic resuspension every 10 min helps prevent settling without introducing excessive mechanical damage.
3. **Wash thoroughly after incubation.** This step is important to minimize carryover of external ThDP or P_i into the downstream extraction.
4. **Keep assay conditions constant across samples.** Because ThTP levels are low and dynamic, incubation time, light intensity, chloroplast input, and handling should be kept identical between treatments.

2.5 Thiamine derivative extraction and HPLC-based quantification

2.5.1 *Materials, equipment and reagents*

- Post-incubation chloroplast pellets from **Section 2.4**
- Liquid nitrogen
- Tissue lyser suitable for 2 mL microcentrifuge tubes
- 2 mm glass beads
- Vortex mixer
- Thermomixer set to 37 °C and 800 rpm
- Microcentrifuge capable of 16,100 g
- HPLC vials with 200 μL glass inserts
- Long pipette tips for careful loading into inserts
- Agilent 1260 Infinity HPLC system or equivalent
- Cosmosil π-NAP column (2.0 mm ID × 150 mm, 5 μm particle size; Nacalai Tesque)
- **Thiamine derivative extraction and derivatization reagents:** 1 % TCA (v/v), buffer E (1 M Tris-Cl (pH 9.0) containing 50 mM $MgCl_2$), freshly prepared 46 mM potassium ferricyanide ($K_3Fe(CN)_6$) dissolved in 23 % KOH (v/v), 1 M KOH, methanol
- **Standard solutions:** prepare standard curves in the range of 0–4 μM using 6–8 dilutions for the different thiamine derivatives. Based on the retention time, thiamine, ThMP and ThTP (this standard is not commercially available and can be chemically synthesized using published protocols (Bettendorff et al., 2003; Hofer et al., 2016)) can be mixed together for co-injection, whereas ThDP needs to be injected separately
- ThTP identity can be confirmed by treatment with purified *Mm*ThTPase, as described in next **Section 2.6**
- **HPLC mobile phases:** Solvent A (50 mM potassium phosphate (pH 7.2)), Solvent B (100% methanol, HPLC grade or LC-MS grade)

2.5.2 *Protocols*

- **Disrupt the chloroplast pellets:** after incubation and washing, subject the chloroplast pellet to two freeze-thaw cycles using liquid nitrogen.
- **Release the chloroplast contents:** disrupt the pellet in a pre-chilled 2 mL screw-cap tube containing 3 sterile 2 mm glass beads using a tissue lyser at 25–30 Hz for 2 × 30 s, with 1 min cooling on ice between cycles. Keep samples cold throughout.
- **Extract thiamine derivatives:** resuspend the disrupted material in 100 μL of 1 % TCA.

- **Extract the soluble fraction:** vortex vigorously for 30 min, then centrifuge at 16,100 g for 10 min at room temperature.
- **Prepare the sample for derivatization:** transfer 50 µL of the supernatant or standard solution to a fresh tube.
- **Adjust pH and set up verification reactions:** add 5 µL Buffer E and vortex briefly. For ThTPase verification, add 1 µL of purified *Mm*ThTPase (10 ng·µL^{-1}). For routine vitamer measurement, add 1 µL of diluted Buffer E (1:20) instead.
- **Incubate the reaction:** incubate for 30 min at 37 °C and 800 rpm in a thermomixer. In the verification reaction, this step converts ThTP to ThDP.
- **Derivatize the sample:** add 7 µL of freshly prepared 46 mM $K_3Fe(CN)_6$ in 23 % KOH (v/v), vortex briefly, and incubate for 10 min in the dark at room temperature.
- **Complete sample preparation:** add 12 µL of 1 M KOH and 25 µL methanol.
- **Clarify the derivatized sample:** centrifuge again at 16,100 g for 10 min at room temperature. If necessary, transfer the supernatant to a new tube and repeat the centrifugation to remove residual insoluble material.
- **Load the HPLC vials:** transfer 50−70 µL of the final supernatant into an HPLC vial containing a 200 µL glass insert, using a long pipette tip to dispense the sample to the bottom of the insert while avoiding bubble formation.
- **Analyze the samples by HPLC:** inject 10 µL of each derivatized sample onto a Cosmosil π-NAP column mounted on an HPLC system equipped with fluorescence detection.
- **Run the solvent gradient:** separate the derivatized vitamers using Solvent A and Solvent B with the following gradient: 0 min: 89 % A/11 % B; 9.5 min: 68 % A/32 % B; 13 min: 40 % A/60 % B; 18 min: 10 % A/90 % B; 20–22 min: 0 % A; 23–25 min: return to 89 % A/11 % B.
- **Set the chromatographic conditions:** run the column at a flow rate of 0.5 mL·min^{-1} and a column temperature of 35 °C. Detect the derivatized vitamers by fluorescence at 375 nm excitation and 450 nm emission. In the current workflow, the photomultiplier gain was set to 18, peak width to > 0.2 min, and equilibration conditions to 89 % A/11 % B.
- **Assign peaks and quantify vitamers.** Identify thiamine, ThMP, and ThDP by comparison with authentic standards (Fig. 5). Assign ThTP by retention time relative to the authentic standard and confirm its identity

A

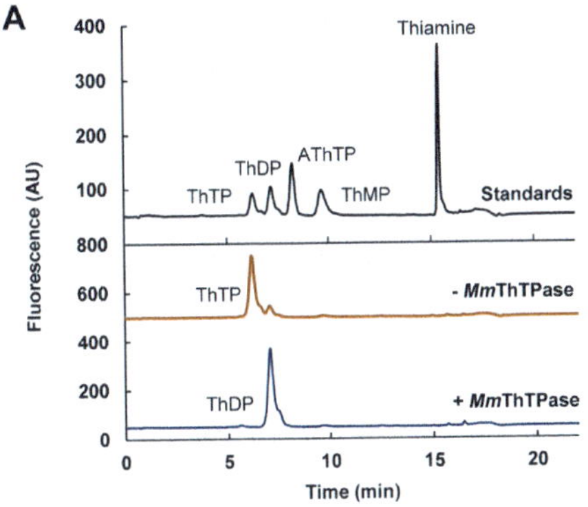

B

Vitamer	Approximate retention time (min)
ThTP	~6.8
ThDP	~7.8
AThTP	~8.4
ThMP	~9.6
Thiamine	~15.4

Fig. 5 **HPLC-based verification of the ThTP peak and retention time assignment of thiamine derivatives.** (A) Representative HPLC-fluorescence chromatograms of thiamine derivative standards, before and after *Mm*ThTPase treatment. The upper chromatogram shows the separation of authentic standards, including ThTP, ThDP, AThTP, ThMP, and thiamine. In the untreated sample, a peak corresponding to ThTP is detected at the expected retention time. After treatment with recombinant *Mm*ThTPase, the ThTP peak is reduced or abolished, with a corresponding increase in the ThDP peak. This enzymatic conversion provides an important specificity control for confirming the identity of the ThTP peak. (B) Approximate retention times of the major thiamine derivatives under the HPLC conditions used in this protocol. Retention times are method-specific and should be verified using authentic standards under the same column, mobile phase, gradient, flow rate, and fluorescence detection settings.

by comparison of chromatograms obtained before and after *Mm*ThTPase treatment. Quantification of ThTP should be based on both standard calibration and the signal decrease specifically attributable to enzymatic hydrolysis.

2.5.3 Notes

1. The protocol described in this section, including *Mm*ThTPase treatment for ThTP peak verification and pre-column thiochrome derivatization prior to HPLC-fluorescence detection, was adapted and modified from previously published methods for plant vitamin B_1/thiamine derivative analysis (Hofmann et al., 2020; Wang et al., 2025).
2. **Use freshly prepared $K_3Fe(CN)_6$ reagent.** The derivatization reagent should be prepared immediately before use and protected from light during the 10 min incubation.
3. **Handle ThTP-containing samples consistently.** Because ThTP is low-abundance and is labile, extraction and derivatization should be performed identically across all treatments.

4. ***Mm*ThTPase treatment is recommended for peak verification.** Comparison of samples before and after *Mm*ThTPase treatment provides an important control for confirming the identity of the ThTP peak.
5. **Sample cleanliness is critical for HPLC performance.** Residual insoluble material can compromise peak quality and column performance; an additional clarification spin is recommended when necessary.
6. **ThTP quantification requires caution.** Co-eluting contaminants may contribute to the apparent ThTP signal. For this reason, ThTP should be quantified by combining authentic standard-based retention time assignment with enzymatic confirmation through *Mm*ThTPase treatment.
7. **Run standards under identical chromatographic conditions.** Authentic standards for thiamine, ThMP, ThDP, and ThTP should be analyzed in parallel with the samples to support peak assignment and quantification.
8. **Maintain consistent instrument settings across experiments.** Gain, peak width, equilibration conditions, and injection volume should be kept constant within an assay series to enable robust comparison between treatments.

2.5.4 Results part II

After *in vitro* chloroplast incubation and extraction of thiamine derivatives, fluorescence-based HPLC provides both peak identification information and quantitative readouts of ThTP formation. Under the HPLC conditions used in this protocol, authentic standards are separated in the order of ThTP, ThDP, AThTP, ThMP, and thiamine (Fig. 5A). The corresponding approximate retention times are also summarized (Fig. 5B). These retention times are method-specific and should be used only under the same column, mobile phase, gradient, flow rate, derivatization, and fluorescence detection conditions. For each batch of measurements, authentic standards should be run in parallel to verify retention-time stability.

Because ThTP is usually present at low abundance and may be difficult to distinguish from nearby minor peaks, enzymatic verification provides an important specificity control. Recombinant *Mm*ThTPase hydrolyzes ThTP to ThDP. Therefore, treatment of the sample or standard mixture with *Mm*ThTPase should reduce or abolish the ThTP peak and increase the ThDP peak (Fig. 5A). This peak conversion confirms that the signal assigned as ThTP corresponds to an authentic ThTP-derived peak rather than to an unrelated compound with a similar retention time.

Once peak identity is confirmed, the same HPLC method can be used to quantify ThTP formation after *in vitro* incubation of isolated chloroplasts with defined substrates. As shown in Fig. 3B, control incubations without added substrate show minimal ThTP accumulation. Addition of ThDP increases ThTP levels, indicating that ThDP can serve as a substrate for ThTP formation in the isolated chloroplast system. The combined addition of ThDP and Pi further enhances ThTP accumulation, supporting substrate- and phosphate-dependent ThTP formation in isolated chloroplasts.

Together, the retention time assignment, *Mm*ThTPase-based peak verification, and substrate incubation results provide a coherent workflow for measuring ThTP formation *in vitro*. The HPLC chromatogram establishes the identity of the ThTP peak, while the chloroplast incubation assay provides the biochemical readout showing that ThTP accumulation depends on substrate availability. This workflow therefore links controlled chloroplast incubation with reliable HPLC-based quantification of ThTP production.

2.6 Purification of recombinant *Mus musculus* thiamine triphosphatase (*Mm*ThTPase) from *E. coli*

2.6.1 *Materials, equipment and reagents*

- *E. coli* BL21(DE3)-RIL expression strain (Invitrogen) with the plasmid pMH-HT-*Mm*ThTPase, which carries the codon-optimized full-length *Mm*ThTPase (UniProt ID Q8JZL3) (Hofmann et al., 2020).
- High-pressure homogenizer such as an Emulsiflex C-3 (Avestin)
- Ni-NTA agarose resin or Ni-NTA affinity column (Macherey-Nagel)
- SDS-PAGE equipment for monitoring protein expression and purification
- **LB medium:** 20 $g \cdot L^{-1}$ LB, autoclaved. Add the appropriate antibiotic for plasmid selection. In the current workflow, kanamycin was added to a final concentration of 50 $\mu g \cdot mL^{-1}$.
- Glycerol for preparing freezer stocks
- **Lysis buffer:** 50 mM sodium phosphate (pH 8.0) containing 300 mM NaCl, 10 mM imidazole, 1 mM DTT, 1 mM PMSF, 1x protease inhibitor cocktail (Sigma), lysozyme (0.5 $mg \cdot mL^{-1}$, w/v)
- Isopropyl β-D-1-thiogalactopyranoside (IPTG)
- 1 x Phosphate Buffered Saline (PBS) buffer (Sigma)
- 1 $mg \cdot mL^{-1}$ deoxyribonuclease I (DNase I) from bovine pancreas stock (Sigma)
- **Buffer A** (binding/wash buffer): 50 mM sodium phosphate (pH 8.0) containing 300 mM NaCl, 10 mM imidazole.

- **Buffer B** (elution buffer): 50 mM sodium phosphate (pH 8.0) containing 300 mM NaCl, 250 mM imidazole.
- A commercial bicinchoninic acid (BCA) protein assay kit (Sigma)

2.6.2 Protocols

- **Prepare the starter culture:** streak *E. coli* BL21(DE3)-RIL cells harboring pMH-HT-*Mm*ThTPase onto an LB agar plate containing 50 μg·mL^{-1} kanamycin and incubate overnight at 37 °C. The next day, inoculate 5–10 mL LB medium containing 50 μg·mL^{-1} kanamycin with a single freshly grown colony and incubate overnight at 37 °C with shaking at 180 rpm.
- **Express recombinant *Mm*ThTPase:** inoculate 1 L pre-warmed LB medium containing kanamycin with the overnight culture and grow at 37 °C, 180 rpm until the culture reaches OD600 ≈ 0.6. Collect a small aliquot (0.5−1 mL) before induction for later SDS-PAGE analysis.
- **Induce protein expression:** add IPTG to a final concentration of 0.25 mM and continue incubation at 16 °C for 16 h. A second aliquot may be collected after induction for verification of protein expression.
- **Harvest the cells:** transfer the induced culture to pre-chilled centrifuge bottles and collect the cells by centrifugation at 4500 g for 30 min at 4 °C. Discard the supernatant and gently resuspend the pellet in 80 mL ice-cold PBS buffer. Centrifuge again at 4500 g for 15 min at 4 °C. Discard the supernatant and repeat the wash once more with 40 mL ice-cold PBS buffer if residual medium remains. After the final centrifugation, remove the supernatant completely. The washed pellet can be processed immediately or snap-frozen in liquid nitrogen and stored at −80 °C until use.
- **Prepare the cell lysate:** resuspend the washed cell pellet in 20 mL ice-cold lysis buffer with bovine DNase I at a final concentration of 5–10 μg·mL^{-1}, and keep the suspension on ice. Lyse the cells using a high-pressure homogenizer (*e.g.* Emulsiflex C−3). Collect the homogenate and clarify it by centrifugation at 6000 g at 4 °C to remove insoluble material. Retain the supernatant as the soluble crude extract for Ni-NTA purification. Save a small aliquot of the crude extract and the insoluble pellet for SDS-PAGE analysis.
- **Bind the His-tagged protein to Ni-NTA resin or column:** prepare a gravity-flow Ni-NTA column using 4 mL of 50 % Ni-NTA slurry, corresponding to approximately 2 mL settled resin, in a disposable polypropylene column with a 10 mL reservoir and an internal diameter of

approximately 0.8–1.0 cm. Allow the resin to settle and equilibrate the column with approximately 20 mL Buffer A. Load approximately 18–20 mL clarified crude extract onto the equilibrated resin or column. Collect the flow-through for SDS-PAGE analysis if desired.

- **Wash the column:** wash the resin bed with 5–10 column volumes of Buffer A to remove weakly bound proteins. For a 2 mL settled Ni-NTA resin bed, wash the column with 10–20 mL Buffer A at least 2–3 times.
- **Elute *Mm*ThTPase:** elute the bound protein using Buffer B. For a 2 mL settled resin bed, perform elution with 4–6 column volumes of Buffer B in two fractions, for example 4 mL followed by 6 mL. Collect the elution fractions separately and analyze a small aliquot by SDS-PAGE before pooling.
- **Measure and adjust protein concentration:** a BCA protein assay kit is recommended because it is widely available and compatible with most routine protein purification workflows. Determine the protein concentration of the eluted *Mm*ThTPase fractions according to the manufacturer's instructions. If necessary, concentrate the pooled elution fractions using a centrifugal protein concentrator with an appropriate molecular-weight cutoff.

2.6.3 Notes

1. **Monitor purification by SDS-PAGE.** The recombinant *Mm*ThTPase should migrate at the expected size of approximately 24 kDa on SDS-PAGE. SDS-PAGE should be used to assess enrichment and purity rather than to determine the exact molecular mass. Representative fractions to assess may include before induction, after induction, crude extract, insoluble fraction, flow-through, wash fractions and elution fractions.
2. **Small aliquots are recommended.** Because the purified enzyme is subsequently used for sensitive ThTP-conversion assays, repeated freeze-thaw cycles should be avoided.
3. **Maintain cold conditions after induction.** Cell harvest, lysis, and purification should be carried out at 4 °C or on ice to preserve enzyme integrity.
4. **Use the same preparation consistently within an assay series.** When possible, the same purified *Mm*ThTPase batch should be used throughout a given experimental set to minimize variation in ThTP verification.

3. Electrochromic shift measurements as a physiological readout of altered ThTP levels

3.1 Rationale

Chloroplast ATP synthase activity is closely associated with ThTP accumulation in plants, raising the question of whether ThTP itself can reciprocally influence ATP synthase function. *Arabidopsis* lines in which organellar ThTP is depleted through targeted expression of recombinant *Mm*ThTPase provide a useful system to examine this possibility. By comparing these lines with wild-type plants and catalytically inactive *Mm*ThTPase controls, the physiological consequences of reduced organellar ThTP can be assessed under photosynthetically active conditions.

The ECS measurements provide a quantitative and non-invasive approach for monitoring thylakoid bioenergetics in intact photosynthetic tissues. The ECS signal, detected as carotenoid absorbance changes around 520 nm, reflects changes in the electric field across the thylakoid membrane and can be used to estimate *pmf*, ATP synthase proton conductivity ($g_{H}{}^{+}$), and the total proton efflux rate through ATP synthase ($\nu_{H}{}^{+}$) after appropriate calibration. In the workflow described here, a single-turnover saturating flash is used to normalize the ECS signal to the total amount of active photosystems, followed by short actinic illumination to induce ECS kinetics under steady-state conditions. This assay therefore provides a physiological readout linking altered ThTP levels to changes in chloroplast ATP synthase activity (Fig. 6).

3.2 ECS-based analysis of chloroplast ATP synthase activity and proton motive force

3.2.1 Materials, equipment and reagents

- *Arabidopsis* line with altered organellar ThTP level *Mm*ThTPase$_{plastid}$, the inactive form *Mm*ThTPase E157A E159A$_{plastid}$ were used as an example together with *Col-0* wild type (Hofmann et al., 2020).
- JTS-10 spectrophotometer (BioLogic) or equivalent instrument (DUAL-EP515/DUAL-DP515, Waltz, or MultispeQ) capable of measuring absorbance changes around 515−520 nm
- Single-turnover saturating flash system for ECS normalization
- Fine forceps for handling seedlings
- Computer and data acquisition software for recording and analyzing ECS kinetics

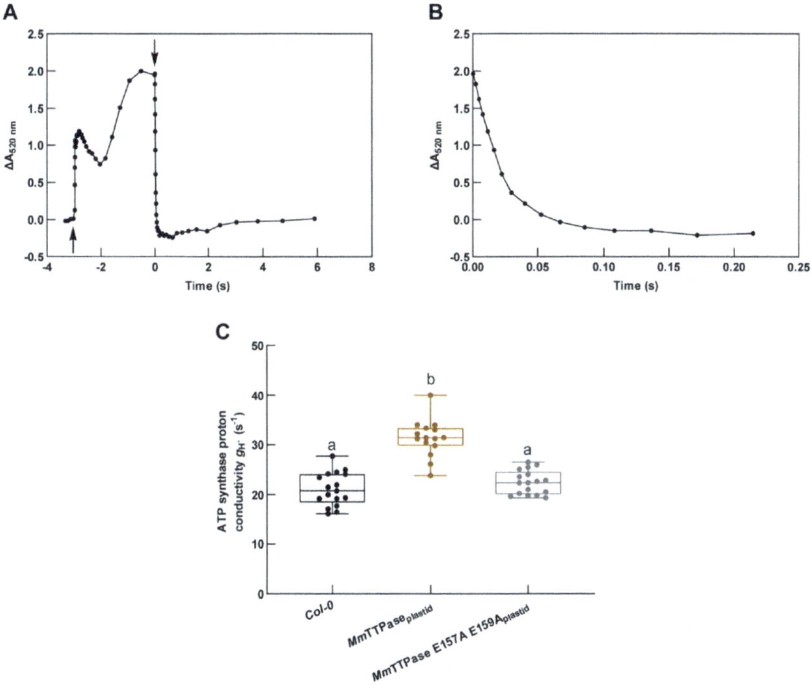

Fig. 6 ECS-based analysis of chloroplast ATP synthase proton conductivity in altered organellar ThTP plants. (A) Representative kinetics of light induced absorbance changes at 520 nm showing electrochromic shift (ECS). The ECS signal was induced by a 3 s continuous illumination (see arrows) and reflects changes in the electric field across the thylakoid membrane and provides a non-invasive readout of thylakoid bioenergetics in intact photosynthetic tissue. The signal was normalized to a single turnover saturating flash, so a vertical unit corresponds to one charge separated per PSI and PSII. Normalization allows for comparison between samples. (B) Enlarged view of the rapid post-illumination ECS decay (0–200 ms). This decay phase is used to estimate *pmf* and other proton flux parameters. (C) Quantification of ATP synthase proton conductivity ($g_H{}^+$) of altered organellar ThTP levels as an example. ECS measurements provide estimates of thylakoid bioenergetic parameters: *pmf*, $g_H{}^+$, and total proton efflux through ATP synthase ($v_H{}^+$). In Arabidopsis lines that are ThTP-depleted in the plastid by targeted ectopic expression of *Mus musculus* ThTPase (*Mm*ThTPase$_{plastid}$), ECS analysis indicates increased ATP synthase proton conductivity and proton efflux, together with reduced *pmf*, consistent with an increased activation state of chloroplast ATP synthase. A catalytically inactive version of *Mm*ThTPase E157A E159A$_{plastid}$ is also shown as a control. Statistical relevance was calculated by one-way ANOVA ($\alpha = 0.05$) with Tukey's *post hoc* multiple comparisons. Data were extracted from Wang et al., 2025, reanalyzed, and replotted.

3.2.2 Protocols

- **Prepare plant material:** grow *Arabidopsis* seedlings under defined photoperiodic conditions. In our workflow, 14-day-old sterile culture-grown seedlings were used. For comparison of altered ThTP levels, *Col-0* wild type, *Mm*ThTPase$_{\text{plastid}}$ line, catalytically inactive *Mm*ThTPase E157A E159A$_{\text{plastid}}$ line were used simultaneously.
- **Maintain steady-state photosynthetic conditions before measurement:** different experiments may apply different dark-adaptation periods before measurement. In the current experiment, do not dark-adapt seedlings before ECS measurement to preserve genotype-dependent differences in ThTP levels.
- **Set up the ECS instrument:** use the flash-induced spectrophotometer to monitor absorbance changes around 515−520 nm, corresponding to the electrochromic shift of thylakoid carotenoids.
- **Position the sample:** place an intact seedling or leaf in the sample holder. Ensure that the photosynthetic tissue is flat, stable, and properly aligned with the measuring beam. Avoid mechanical damage or dehydration during handling.
- **Record the baseline signal:** allow the signal to stabilize before applying the saturating flash or actinic illumination. Exclude samples with unstable baseline traces.
- **Apply a single-turnover saturating flash for normalization:** use a single-turnover saturating flash to normalize the ECS signal to the total amount of active photosystems (PSI+PSII). This normalization allows comparison of ECS amplitudes between genotypes and biological replicates.
- **Induce ECS kinetics by short illumination:** after normalization, illuminate the sample for 3 s using actinic light. In our workflow, an actinic light intensity of 210 μmol photons m^{-2} s^{-1} was used to approximate the growth-light condition.
- **Record post-illumination ECS decay:** after the actinic light is switched off, read the ECS decay kinetics at 520 nm. The full kinetic trace is shown in (Fig. 6A), the post-illumination rapid decay reflects relaxation of the thylakoid membrane electric field and is used to derive chloroplast bioenergetic parameters (Fig. 6B).
- **Define the fitting window:** set the time of actinic light switch-off as 0 ms. Use the initial 0–200 ms post-illumination decay phase for fitting the first-order exponential decay model (Fig. 6B).

- **Estimate *pmf* and proton flux parameters:** determine the *pmf* (H^+ per PSI+PSII) from the ECS amplitude and derive the total proton efflux rate through $\nu_H{}^+$ (H^+ s^{-1} per PSI+PSII) from the fitted decay. Calculate ATP synthase proton conductivity $g_H{}^+$ (s^{-1}) according to the relationship: $\nu_H{}^+ = pmf \times g_H{}^+$.
- **Compare genotypes:** compare *pmf*, $g_H{}^+$, and $\nu_H{}^+$ across wild-type, ThTP-depleted, and inactive-enzyme control.

3.2.3 Notes

1. **Avoid dark adaptation before steady-state ECS measurement.** In this workflow, plants are measured without prior dark adaptation to preserve genotype-dependent differences in ThTP levels and to maintain steady-state photosynthetic conditions.
2. **Measure samples at a consistent time of day.** Because ThTP levels are light-responsive and temporally dynamic, all genotypes should be measured within the same photoperiod window and in a balanced order.
3. **Use appropriate controls.** Wild-type plants provide the baseline reference, while catalytically inactive *Mm*ThTPase lines serve as control for effects of transgene expression.
4. **Keep sample positioning consistent.** Leaf or seedling placement strongly influences the absorbance signal. Samples should be flat, stable, and similarly aligned with the measuring beam across all measurements.
5. **Avoid sample dehydration.** Prolonged handling can alter photosynthetic performance. Samples should be measured promptly after removal from growth conditions, and dehydration should be minimized during mounting.
6. **Include sufficient biological replication.** Because ECS measurements are sensitive to sample positioning and physiological state, multiple independent seedlings or leaves should be measured for each genotype.

3.2.4 Results

A representative ECS kinetic trace is shown in Fig. 6A. After normalization with a single-turnover saturating flash, a 3 s actinic illumination at 210 μmol photons m^{-2} s^{-1} induced a rapid increase in the ECS signal, recorded as the absorbance change at 520 nm. This light-induced ECS rise reflects the formation of the electric field across the thylakoid membrane, which is generated by photosynthetic electron transport and proton translocation. When the actinic light was switched off, the ECS signal rapidly decayed, reflecting relaxation of the thylakoid electric field and proton efflux from the lumen mainly through chloroplast ATP synthase. For quantitative analysis, the time point of light-off was defined as 0 ms. The early post-illumination decay phase from 0 to 200 ms was extracted and replotted as an enlarged kinetic trace (Fig. 6B). This initial decay phase was used because it represents the rapid relaxation of light-induced *pmf* before slower baseline drift or secondary relaxation components become dominant. Each decay curve was fitted with a first-order exponential decay model:

$$Y = A \times e^{-k \cdot t} + offset$$

Where Y is the normalized ECS signal at time t, A is the fitted decay amplitude, k is the fitted decay-derived rate parameter, and *offset* represents the residual baseline after relaxation. The fitted amplitude A was used to estimate the total light-induced *pmf*, while the fitted decay-derived rate k was used to estimate the total proton efflux rate $\nu_H{}^+$. And the ATP synthase proton conductivity $g_H{}^+$ is calculated based on the formula:

$$g_H{}^+ = \frac{\nu_H{}^+}{pmf}$$

For each biological replicate, the following calculation workflow was applied. First, the raw ECS kinetic trace was normalized to the single-turnover flash amplitude. Second, the 0–200 ms post-illumination decay phase was extracted. Third, the decay curve was fitted using the first-order exponential model to obtain the decay amplitude and decay-derived proton efflux parameter. Fourth, $g_H{}^+$ was calculated by dividing $\nu_H{}^+$ by *pmf*. Finally, the calculated $g_H{}^+$ values from all biological replicates were plotted and compared among genotypes (Fig. 6C).

Using this ECS-based analysis, *Col-0* wild-type plants showed a basal level of ATP synthase proton conductivity. The plastid-targeted active *Mm*ThTPase line displayed a significantly higher $g_H{}^+$ value than *Col-0*, indicating enhanced proton conductivity through chloroplast ATP synthase. In contrast, the catalytically inactive *Mm*ThTPase E157A E159A$_{plastid}$ line showed $g_H{}^+$ values similar to *Col-0*. This result indicates that the increased ATP synthase proton conductivity depends on the catalytic activity of *Mm*ThTPase and is therefore associated with the depletion of plastidial ThTP rather than with the presence of the transgene itself.

4. Summary and conclusion

This chapter brings together complementary biochemical and biophysical approaches for investigating ThTP and related thiamine derivatives in plants. The chloroplast incubation workflow enables substrate-dependent ThTP formation to be examined under controlled organellar conditions (Fig. 3), while chlorophyll normalization improves comparability between independent preparations (Fig. 4). Coupling this workflow to acidic extraction and fluorescence-based HPLC permits sensitive detection of thiamine, ThMP, ThDP, and ThTP, and recombinant *Mm*ThTPase provides an important specificity control for confirming ThTP peak identity through conversion to ThDP (Fig. 5).

Beyond vitamer quantification, ECS measurements provide a non-invasive physiological readout of thylakoid bioenergetics, including *pmf*, ATP synthase proton conductivity, and total proton efflux through ATP synthase (Fig. 6). This biophysical assay is particularly useful for evaluating whether changes in organellar ThTP levels are associated with altered chloroplast ATP synthase activity. Thus, the combination of chloroplast-based biochemical assays and ECS measurements allows ThTP metabolism to be linked more directly with photosynthetic energy regulation.

These approaches nevertheless have inherent limitations. Organelle isolation and *in vitro* incubation simplify the native cellular environment and may not fully reproduce biochemical conditions *in vivo*. For low-abundance metabolites such as ThTP, robust interpretation depends on careful sample handling, reliable chromatographic separation, authentic standards, and orthogonal validation of peak identity. ECS measurements also require strict control of growth conditions, measurement timing, sample positioning, and data fitting, because the derived parameters are sensitive to physiological state and technical variation.

Overall, the methods presented here provide a practical foundation for mechanistic analysis of ThTP and related thiamine derivatives in plants. Their application to additional tissues, organelles, genotypes, and physiological conditions should help define more precisely how these metabolites are formed, turned over, and linked to plant metabolic regulation.

Acknowledgments

This research was supported by the Swiss National Science Foundation (Grant 310030_192466/1 to T.B.F.) and the University of Geneva.

References

Bettendorff, L. (2021). Update on thiamine triphosphorylated derivatives and metabolizing enzymatic complexes. *Biomolecules, 11*, 1645. https://doi.org/10.3390/biom11111645.

Bettendorff, L., Nghiêm, H. O., Wins, P., & Lakaye, B. (2003). A general method for the chemical synthesis of γ-^{32}P-labeled or unlabeled nucleoside 5′-triphosphates and thiamine triphosphate. *Analytical Biochemistry, 322*, 190–197. https://doi.org/10.1016/j.ab.2003.08.013.

Bettendorff, L., & Wins, P. (2013). Biological functions of thiamine derivatives: Focus on non-coenzyme roles. *OA Biochemistry, 1*, 10. https://orbi.uliege.be/bitstream/2268/160643/1/1381918674.pdf.

Bettendorff, L., Wirtzfeld, B., Makarchikov, A. F., Mazzucchelli, G., Frédérich, M., Gigliobianco, T., ... Wins, P. (2007). Discovery of a natural thiamine adenine nucleotide. *Nature Chemical Biology, 3*, 211–212. https://doi.org/10.1038/nchembio867.

Evans, T., Song, J., & Jameson, P. E. (2012). Micro-scale chlorophyll analysis and developmental expression of a cytokinin oxidase/dehydrogenase gene during leaf development and senescence. *Plant Growth Regulation, 66*(1), 95–99. https://doi.org/10.1007/s10725-011-9627-5.

Fitzpatrick, T. B. (2024). B vitamins: An update on their importance for plant homeostasis. *Annual Review of Plant Biology, 75*, 67–93. https://doi.org/10.1146/annurev-arplant-060223-025336.

Fitzpatrick, T. B., & Chapman, L. M. (2020). The importance of thiamine (vitamin B_1) in plant health: From crop yield to biofortification. *Journal of Biological Chemistry, 295*, 12002–12013. https://doi.org/10.1074/jbc.REV120.010918.

Gangolf, M., Wins, P., Thiry, M., El Moualij, B., & Bettendorff, L. (2010). Thiamine triphosphate synthesis in rat brain occurs in mitochondria and is coupled to the respiratory chain. *Journal of Biological Chemistry, 285*, 583–594. https://doi.org/10.1074/jbc.M109.054379.

Gigliobianco, T., Lakaye, B., Wins, P., El Moualij, B., Zorzi, W., & Bettendorff, L. (2010). Adenosine thiamine triphosphate accumulates in *Escherichia coli* cells in response to specific conditions of metabolic stress. *BMC Microbiology, 10*, 148. https://doi.org/10.1186/1471-2180-10-148.

Hofer, A., Marques, E., Kieliger, N., Gatter, S. K., Jordi, S., Ferrari, E., ... Jessen, H. J. (2016). Chemoselective dimerization of phosphates. *Organic Letters, 18*, 3222–3225. https://doi.org/10.1021/acs.orglett.6b01466.

Hofmann, M., Loubéry, S., & Fitzpatrick, T. B. (2020). On the nature of thiamine triphosphate in *Arabidopsis*. *Plant Direct, 4*, e00258. https://doi.org/10.1002/pld3.258.

Iyer, L. M., & Aravind, L. (2002). The catalytic domains of thiamine triphosphatase and CyaB-like adenylyl cyclase define a novel superfamily of domains that bind organic phosphates. *BMC Genomics, 3*, 33. https://doi.org/10.1186/1471-2164-3-33.

Jeffrey, S. T., & Humphrey, G. F. (1975). New spectrophotometric equations for determining chlorophylls a, b, c1 and c2 in higher plants, algae and natural phytoplankton. *Biochemie und Physiologie der Pflanzen, 167*(2), 191–194. https://doi.org/10.1016/S0015-3796(17)30778-3.

Lakaye, B., Makarchikov, A. F., Antunes, A. F., Zorzi, W., Coumans, B., De Pauw, E., ... Bettendorff, L. (2002). Molecular characterization of a specific thiamine triphosphatase widely expressed in mammalian tissues. *Journal of Biological Chemistry, 277*, 13771–13777. https://doi.org/10.1074/jbc.M111241200.

Pesquera, M., Martinez, J., Maillot, B., Wang, K., Hofmann, M., Raia, P., ... Fitzpatrick, T. B. (2022). Structural and functional studies of *Arabidopsis thaliana* triphosphate tunnel metalloenzymes reveal roles for additional domains. *Journal of Biological Chemistry, 298*, 102438. https://doi.org/10.1016/j.jbc.2022.102438.

Porra, R. J., Thompson, W. A., & Kriedemann, P. E. (1989). Determination of accurate extinction coefficients and simultaneous equations for assaying chlorophylls a and b extracted with four different solvents: Verification of the concentration of chlorophyll standards by atomic absorption spectroscopy. *Biochimica et Biophysica Acta (BBA)-Bioenergetics, 975*(3), 384–394. https://doi.org/10.1016/S0005-2728(89)80347-0.

Schellenberger, A. (1998). Sixty years of thiamin diphosphate biochemistry. *Biochimica et Biophysica Acta - Protein Structure and Molecular Enzymology, 1385*, 177–186. https://doi.org/10.1016/S0167-4838(98)00067-3.

Wang, K., Alric, J., & Fitzpatrick, T. B. (2025). Thiamine triphosphate puts the brake on the activation state of chloroplast ATP synthase. *Plant Physiology, 199*, kiaf424. https://doi.org/10.1093/plphys/kiaf424.

Wellburn, A. R. (1994). The spectral determination of chlorophylls a and b, as well as total carotenoids, using various solvents with spectrophotometers of different resolution. *Journal of Plant Physiology, 144*(3), 307–313. https://doi.org/10.1016/S0176-1617(11)81192-2.

ThiN as a versatile domain of transcriptional repressors and catalytic enzymes of thiamine biosynthesis in archaea

Nicholas Cox[a], Katherine R. Weber[a], and Julie A. Maupin-Furlow[a,b,*]
[a]Department of Microbiology and Cell Science, Institute of Food and Agricultural Science, University of Florida, Gainesville, FL, United States
[b]Genetics Institute, University of Florida, Gainesville, FL, United States
*Corresponding author. e-mail address: jmaupin@ufl.edu

Contents

Abstract

Thiamine pyrophosphate (TPP), the active form of vitamin B1, is a cofactor required for catalyzing key reactions in carbohydrate and amino acid biosynthesis. TPP can be acquired through environmental salvage or *de novo* biosynthesis. Two separate pathways are used for the *de novo* biosynthesis of thiamine: one produces a sulfur-containing thiazole ring and the other a pyrimidine moiety, which are then joined and phosphorylated to form TPP. While the synthesis and regulation of thiamine are well characterized in certain bacteria, plants and fungi, these pathways are comparatively understudied in archaea. Recent work reveals archaeal thiamine metabolism to be chimeric, combining eukaryotic THI4-dependent thiazole formation with bacterial ThiC-mediated pyrimidine synthesis. Moreover, archaeal thiamine synthesis can be regulated by a ThiN-domain-dependent transcriptional repression mechanism distinct from riboswitch-based control. This chapter describes the experimental techniques that have been utilized for defining catalytic enzymes and transcriptional regulators of

Methods in Enzymology, Volume 735
ISSN 0076-6879, https://doi.org/10.1016/bs.mie.2026.05.039

thiamine biosynthesis in archaea, using *Haloferax volcanii* as a model organism. Approaches include targeted gene deletion, complementation, site-directed mutagenesis (SDM), 3D-structural modeling, DNA regulatory motif identification, transcriptional reporter assays, and bioinformatic comparisons providing a versatile toolbox for functional and regulatory analysis of thiamine metabolism in extremophilic microorganisms.

Protein (UniProt):

Escherichia coli ThiS (O32583), ThiG (P30139), and ThiC (B5Z090); *Haloferax volcanii* Thi4 (HvThi4; HVO_0665), ThiR (HVO_0662), ThiC (HVO_2154), ThiE (HVO_2668) and UbaA (HVO_0558).

1. Introduction to TPP reactions

Thiamine pyrophosphate (TPP) is an essential cofactor required for central metabolic processes, serving as the catalytic prosthetic group for enzymes involved in glycolysis, the tricarboxylic acid (TCA) cycle, and amino acid biosynthesis (Luo et al., 2025; Pohl et al., 2004). Although TPP dependence is conserved across all domains of life, the enzymatic pathways and regulatory mechanisms governing thiamine biosynthesis vary substantially between eukaryotes and prokaryotes (Chatterjee et al., 2011; Lehmann et al., 2006). In particular, formation of the thiazole ring, a five-membered sulfur-containing heterocycle that serves as a precursor to thiamine (vitamin B1), represents one of the most chemically complex steps in the biosynthetic pathway and exhibits pronounced mechanistic diversity (Jurgenson et al., 2009).

Distinct biochemical strategies for thiazole-ring formation have been characterized in eukaryotic and bacterial systems, providing a comprehensive framework for investigating archaeal thiamine biosynthesis. In the yeast *Saccharomyces cerevisiae*, the enzyme THI4p is required for thiazole ring formation and contains an active-site cysteine residue (Chatterjee et al., 2011). *In vitro*, the sulfur atom of the cysteine sulfhydryl group is incorporated into the thiazole biosynthetic intermediate, resulting in the irreversible modification and inactivation of THI4p after a single catalytic cycle (Chatterjee et al., 2011). The completed thiazole intermediate is further processed through condensation with a pyrimidine moiety to form thiamine, which is then activated via phosphorylation to thiamine monophosphate (TMP) and ultimately TPP (Chatterjee et al., 2011; Tazuya et al., 1989). While this mechanism provides a direct solution for sulfur incorporation *in vitro*, it is suggested to be metabolically costly (Sun et al., 2019). In contrast, many bacteria synthesize the thiazole moiety of thiamine though a multienzyme, catalytic pathway that relies on dedicated sulfur relay systems. This bacterial pathway is extensive and complex, varying among species. In *Escherichia coli*, for example, sulfur is mobilized from cysteine (or provided *in vitro* using H_2S) and transferred via the ThiF-activated sulfur-carrier protein ThiS before being incorporated into the thiazole ring by ThiG, which catalyzes the rearrangement of 1-deoxy-D-xylulose 5-phosphate to produce the thiazole phosphate (Lehmann et al., 2006). This modular strategy allows repeated catalytic turnover and efficient sulfur utilization, albeit at the cost of increased pathway complexity and the requirement for multiple accessory proteins. Together, the mechanistic and organizational differences learned from the study of thiazole biosynthesis in model yeast and bacteria highlight the flexibility of this pathway and provide a comparative framework for probing thiamine biosynthesis in archaea.

In archaea, thiamine biosynthesis proceeds through a hybrid strategy that combines a ThiC-dependent pathway for pyrimidine synthesis with a THI4-dependent pathway for thiazole-ring formation, with these

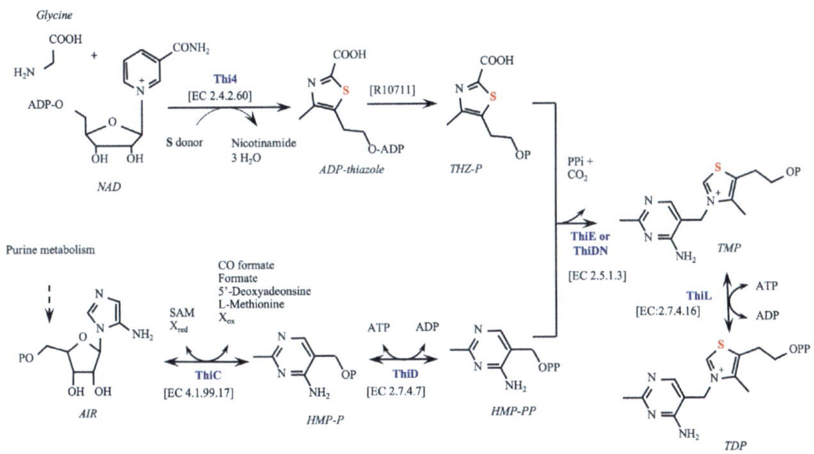

Fig. 1 Thiamine biosynthetic pathway in archaea. Enzymes in blue with EC or reaction numbers in brackets. Abbreviations: AIR, 5-aminoimidazole ribotide; SAM, S-adenosyl-methionine; HMP-P, 4-aminohydroxymethyl-2-methylpyrimidine phosphate; HMP-PP, 4-aminohydroxymethyl-2-methylpyrmidine diphosphate; TMP, thiamine monophosphate; TDP, thiamine diphosphate; THZ-P, 4-methyl-5-(β-hydroxyethyl)thiazolium phosphate; X, electron carrier.

intermediates used to ultimately form TPP (Maupin-Furlow, 2018) (Fig. 1). ThiC, the phosphomethylpyrimidine synthase catalyzing the formation of the pyrimidine precursor (hydroxymethylpyrimidine phosphate) that is joined with the thiazole ring, is conserved across domains of life (Hwang et al., 2014; Sharma et al., 2024). Archaea do not use the bacterial ThiF-ThiS pathway for sulfur mobilization to form the thiazole ring and instead use eukaryotic-like THI4 homologs (Hwang et al., 2014). Many archaea, such as *Haloferax volcanii*, have THI4 homologs with a cysteine residue that is required for thiamine biosynthesis and is conserved with the sulfur-donating cysteine residue of *S. cerevisiae* THI4 (Hwang et al., 2014). Other archaea – such as *Methanocaldococcus jannaschii* – encode THI4 homologs with a histidine residue in this position. These archaeal non-Cys Thi4 homologs purify with Fe as a metal cofactor and efficiently use exogenous sulfur sources (e.g., sodium sulfide or thioacetate) to support catalytic turnover *in vitro* (Eser et al., 2016; Zhang et al., 2016). Non-Cys Thi4 homologs are also found in bacteria and archaea that inhabit aerobic, low-sulfide environments. For example, the non-Cys Thi4 homolog from the aerotolerant bacterium *Thermovibrio ammonificans* supports multiple catalytic turnover using metals other than Fe under low sulfide conditions (Joshi et al., 2021).

Most archaea are predicted to regulate thiamine biosynthesis via transcriptional control using a ThiR protein that contains an N-terminal helix-turn-helix DNA binding domain and ThiN-related sensor domain (Hwang et al., 2017; Rodionov et al., 2017). By contrast, bacteria, plants and fungi commonly regulate thiamine biosynthesis via TPP-sensing riboswitches (Donovan et al., 2018; Hübenthal et al., 2025), with archaea of the order *Thermoplasmatales* also suggested to use this type of mechanism (Rodionov et al., 2017). The regulatory role of archaeal ThiR proteins is experimentally demonstrated in *H. volcanii*, where ThiR is shown to repress the transcription of *thi4* and *thiC* genes, encoding enzymes responsible for thiazole and pyrimidine biosynthesis, respectively (Hwang et al., 2017). Although the ThiN domain of the ThiR regulatory protein shares an overall structural fold with ThiN enzymes that catalyze ATP-dependent phosphorylation of thiamine to TPP, it lacks the conserved catalytic residues and corresponding enzymatic activity (Hwang et al., 2017). Instead, ThiR represses the transcription of *thi4* and *thiC* under conditions of thiamine or thiamine derivative abundance, thereby preventing unnecessary gene expression (Hwang et al., 2017).

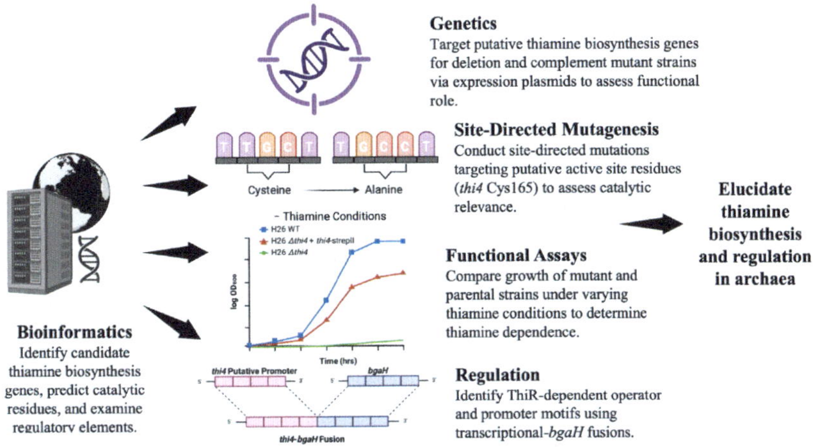

Fig. 2 Experimental workflow for investigating thiamine biosynthesis and regulation in archaea. A multi-disciplinary approach was used to investigate genes involved in archaeal thiamine biosynthesis and their regulatory mechanisms. Bioinformatic analysis was first performed to identify candidate thiamine biosynthetic genes (*thi4, thiC, thiE, thiN/R, ubaA*), predict conserved catalytic residues (*thi4* Cys165), and examine potential regulatory elements using sequence and structural tools. These predictions guided experimental studies that included genetic analysis through targeted gene deletion and complementation to determine function. Site-directed mutagenesis was then applied to conserved catalytic residues (e.g, substitution of the Thi4 Cys165 residue for alanine) to assess their role in thiazole formation and enzymatic activity. Functional assays, including phenotypic growth analysis under varying thiamine conditions, were used to evaluate the physiological consequences of these mutations. Finally, transcriptional regulation was examined using promoter-reporter (*thi4-bgaH*) fusions to assess ThiR-mediated repression and promoter activity. Together, these integrated approaches provide a comprehensive framework for understanding thiamine biosynthesis and its regulation in archaea.

Despite recent advances in understanding thiamine biosynthesis in archaea, several key questions remain. These include the physiological roles of single-turnover versus catalytic THI4 mechanisms in vivo, the extent to which exogenous sulfur sources contribute to thiazole formation, and how ThiN-sensor domains in ThiR regulators control transcriptional repression in response to pathway metabolites.

The overall methodological workflow, which employes a combination of bioinformatic, genetic and biochemical approaches, used to address these questions, is summarized in Fig. 2. Initial analyses used bioinformatic tools to identify thiamine biosynthesis-related proteins, predict conserved catalytic residues, and examine regulatory elements. Protein homologs were

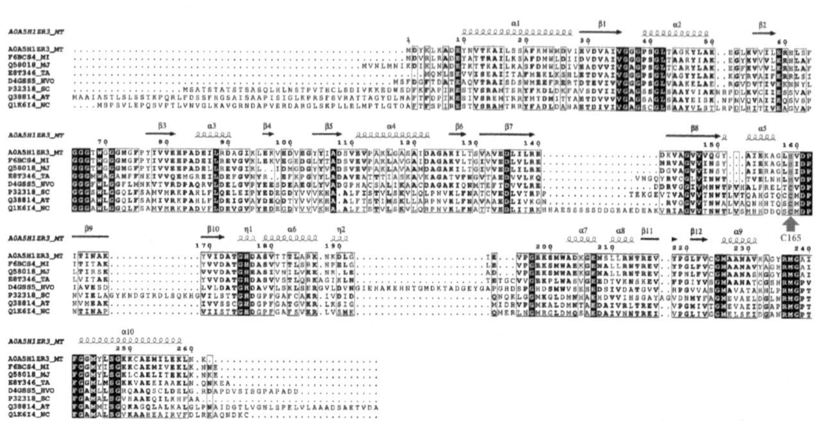

Fig. 3 Identification of conserved residues in Thi4 using bioinformatic analysis. Bioinformatic approaches were used to identify conserved residues of potential functional significance in the *Haloferax volcanii* Thi4 protein. Structural homology and putative crystal structures were predicted using Phyre2, allowing identification of related thiazole synthase proteins and overall fold conservation. Multiple sequence alignment of Thi4 homologs was then performed and visualized using ESPript 3, highlighting conserved amino acid residues across species. Secondary structure elements, including α-helices and β-sheets derived from the structural model of *Methanothermococcus thermolithotrophicus* (pdb_00006hk1), are displayed above the amino acid alignment. A highly conserved cysteine residue (Cys165), implicated in catalytic sulfur donation, is highlighted in purple. Sequences are displayed in the format UniProt ID_Organism. Abbreviations: MT, *Methanothermococcus thermolithotrophicus*; MI, *Methanococcus igneus*; MJ, *Methanocaldococcus jannaschii*; TA, *Thermovibrio ammonificans*; HVO, *Haloferax volcanii*; SC, *Saccharomyces cerevisiae*; AT, *Arabidopsis thaliana*; NC, *Neurospora crassa.*

identified using InterPro, Pfam, and BLASTP, while conserved active-site residues were analyzed through multiple sequence alignments using ClustalW and visualized with ESPript 3 (Fig. 3). Structural insights were obtained through 3D modeling with ChimeraX, and potential regulatory DNA motifs were identified using the MEME suite. These strategies allow for developing a comprehensive metabolic profile of relevant thiamine-biosynthetic genes and key active site residues, as well as important regulatory domains and DNA binding site motifs for transcriptional repressors.

Building upon this foundation, experimental work followed a gene deletion-complementation and sited-directed mutagenesis framework. Because *H. volcanii* can be readily grown in defined media, the contribution of individual genes to thiamine biosynthesis could be evaluated using targeted gene deletions. Genes of interest *(thi4, thiR, thiC, thiE*, and *ubaA*)

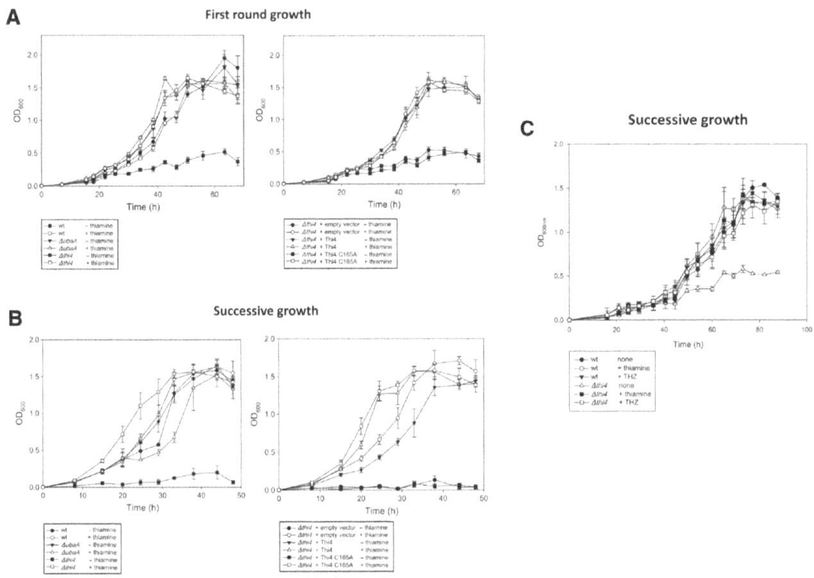

Fig. 4 Growth analysis identifies Thi4, rather than UbaA, as essential for thiamine biosynthesis. (A) First-round growth assays of wild-type (wt), ΔubaA, and Δthi4 strains cultured in defined medium with (+) or without (−) thiamine or thiazole (THZ) supplementation. Complementation of Δthi4 with wild-type thi4 restored growth in the absence of thiamine, whereas strains expressing the Thi4 C165A protein variant remained thiamine-dependent, demonstrating the importance of the C165 residue. (B) Successive growth assays following passage of cells into fresh medium confirmed the stability of the observed phenotypes and further demonstrated that growth defects associated with Δthi4 could be rescued by wild-type thi4 but not by catalytic variants. In contrast, ΔubaA strains exhibited growth comparable to wt, indicating that UbaA is not essential for thiamine biosynthesis. (C) Chemical complementation assays of wt and Δthi4 strains grown in the absence of supplementation, with thiamine, or with THZ. Restoration of growth by THZ demonstrated that the Δthi4 growth defect results from impaired thiazole biosynthesis. Collectively, these results demonstrate that Thi4, and its conserved C165 residue, is essential for thiamine biosynthesis in H. volcanii, whereas UbaA does not play a direct role in this pathway. This figure is generously provided by Hwang et al., 2014, under open-access licensing.

were deleted using a pop-in/pop-out strategy and subsequently reintroduced on an overexpression plasmid containing a C-terminal StrepII tag (Hwang et al., 2017). If these genes were required for thiamine biosynthesis, deletion mutants were expected to exhibit impaired growth in non-supplemented media, with growth restored upon gene complementation. Consistent with these expected outcomes, mutants lacking '*thi*' genes displayed thiamine-dependent growth phenotypes, confirming their roles in thiamine biosynthesis and regulation (Hwang et al., 2017) (Fig. 4).

To further investigate catalytic mechanisms, site-directed mutagenesis (SDM) was performed. Initial efforts focused on the conserved cysteine residue of Thi4 (Cys165), predicted to function as the sulfur donor for thiazole-ring formation based on analogy to yeast THI4 (Chatterjee et al., 2011). Exchange of this residue to alanine was expected to abolish sulfur incorporation and therefore prevent growth in the absence of thiamine. In parallel, transcriptional reporter assays were used to investigate the regulatory control of thiamine biosynthetic genes. The β-galactosidase gene (*bgaH*) derived from *Haloferax alicantei* (Patenge et al., 2000) was used as a transcriptional reporter to measure the promoter activity of regions upstream of thiamine metabolic genes under different thiamine metabolic conditions (Hwang et al., 2017). These promoter assays enabled the assessment of ThiR-mediated transcriptional repression, allowing the activity of ThiR regulators containing the inactive ThiN-domain helix-turn-helix (HTH) fusion to be evaluated in response to thiamine pathway intermediates (Hwang et al., 2017; Rodionov et al., 2017).

Growth assays and reporter measurements were therefore used collectively to assess the functional consequences of gene deletions, catalytic mutations, and regulatory truncations affecting archaeal thiamine biosynthesis. Together, these complementary approaches provide a holistic framework for dissecting both the enzymatic mechanisms underlying thiazole formation and the transcriptional regulatory networks that control thiamine biosynthesis in archaea.

2. Key resources tables

Resource	Source or description	Identifier or reference
Chemicals:		
4-amino-5-hydroxymethyl-2-methyl-pyrmidine (HMP)	TCI Chemicals	A2730
4-methyl-5-(2-hydroxyethyl) thiazole (THZ)	TCI Chemicals	H0527
5-Fluoroorotic acid	USBiological	F5050

Agar	Fisher	BP2641
Agarose	Fisher	BP160
Ampicillin	Sigma	A9518
Biotin	Acros Organics	230090010
β-mercaptoethanol	Sigma	M3148
$CaCl_2 \cdot 2H_2O$	Fisher	BP510
Casamino acids	Fisher	BP1424
$CuSO_4 \cdot 5\ H_2O$	Acros Organics	315281000
Ethidium bromide	Biorad	1610433
$FeSO_4 \cdot 7\ H_2O$	Alfa Aesar	12967
Glycerol	Fisher	J61059. AP
HCl	Supelco	HX0603
Isopropyl β-D-1-thiogalactopyranoside (IPTG)	Acros	302970100
K_2HPO_4	Fisher	P288
K_2SO_4	Fisher	424220010
KCl	Fisher	P217
KH_2PO_4	Acros	205925000
KOH	Acros Organic	206075000
$MgCl_2 \cdot 6H_2O$	Fisher	197530050
$MgSO_4 \cdot 7\ H_2O$	Fisher	BP213
$MnCl_2 \cdot 4\ H_2O$	Fisher	S80070
NaCl	Fisher	S271
NH_4Cl	Fisher	A661
Novobiocin	Fisher	74675
o-nitrophenyl-β-D-galactopyranoside (ONPG)	Sigma	48712-1GM
PEG600	Sigma	P3390
Sucrose	Fisher	BP220

Thiamine	Sigma	T-4625
Tris-base	Fisher	BP152
Tryptone	Fisher	BP1431
Uracil	Sigma	U0750
5-bromo-4-chloro-3-indolyl-beta-D-galactopyranoside (X-gal)	Fisher	BP1615
Yeast extract	Fisher	BP1422
$ZnSO_4 \cdot 7\, H_2O$	Fisher	Z68
Plasmid generation		
2x Quick Ligase	New England Biolabs (NEB) (Ipswitch, MA, USA)	M2200
Antarctic Phosphatase and its 10X Reaction Buffer	New England Biolabs	M0289
*Bam*HI-HF	New England Biolabs	R3136
Custom DNA oligonucleotides	Eurofins Genomics (Louisville, KY, USA)	
DNA Gel Extraction Kit	New England Biolabs	T1120
*Dpn*I	New England Biolabs	R0176
*Hind*III-HF	New England Biolabs	R3104
KLD (T4 polynucleotide kinase, T4 DNA ligase, and *Dpn*I) enzyme mix	New England Biolabs	M0554
Monarch PCR and DNA Clean-up Kit	New England Biolabs	T4130
Phusion High-Fidelity DNA polymerase	New England Biolabs	M0530
PureLink Quick Plasmid Miniprep Kit	Invitrogen, ThermoFisher Scientific (Waltham, MA, USA)	K210010
Sanger or Oxford Nanopore Technologies long-read DNA sequencing service	Eurofins Genomics or Plasmidsaurus (Louisville, KY, USA)	
Equipment		
0.22 μM filter sterilization units	Gen Clone	25–239
37 °C and 42 °C Incubators with rotary shaking	Varying	—

Chemglass beads	Fisher	1804412
DNA gel electrophoresis system	Biorad	1704401
Microcentrifuge (>12,000×*g* at 4 °C)	Fisher	Sorvall Legend Micro 21
Microcentrifuge tubes	1.5 mL; 2.0 mL	C3017; C3020
Microplate reader with double-orbital shaking and capable of maintaining temperature for 99 h	BioTek Epoch 2 Microplate Spectrophotometer	Epoch2
MilliQ water (dH_2O)	Varying	—
Nanodrop	Varying	—
Pipettors and tips	Varying	—
Thermocycler	BioRad (Hercules, CA, USA)	Model T100

E. coli strains		
TOP10	*F– recA1 endA1 hsdR17(r$_K$–m$_K$+) supE44 thi-1 gyrA relA1*	Invitrogen
GM2163	*F– ara-14 leuB6 fhuA31 lacY1 tsx78 glnV44 galK2 galT22 mcrA dcm-6 hisG4 rfbD1rpsL136 dam13::Tn9 xylA5 mtl-1 thi-1 mcrB1 hsdR2*	New England Biolabs
H. volcanii strains		
DS70	wild-type isolate DS2 cured of plasmid pHV2	Allers et al. (2004)
H26	*DS70 ΔpyrE2*	Allers et al. (2004)
NC1011	H26 Δ*thi4 (HVO_0665)*	Hwang et al. (2014)
HM1052	H26 Δ*ubaA (HVO_0558)*	Miranda et al. (2011)
SH101	H26 Δ*thiC (HVO_2154)*	Hwang et al. (2017)
SH105	H26 Δ*thiE (HVO_2668)*	Hwang et al. (2017)

SH100	H26 Δ*thiR (HVO_0662)*	Hwang et al. (2017)
Deletion plasmids		
pTA131	Ap^r; pBluescript II carries P_{fdx}-*pyrE2* with MCS	Allers et al. (2004)
pJAM960	Ap^r; pTA131-based plasmid for *ubaA* deletion	Miranda et al. (2011)
pJAM3119	Ap^r; pTA131-based plasmid for *thiR* deletion	Hwang et al. (2017)
pJAM3126	Ap^r; pTA131-based plasmid for *thiC* deletion	Hwang et al. (2017)
pJAM3127	Ap^r; pTA131-based plasmid for *thiE* deletion	Hwang et al. (2017)
pJAM2820	Ap^r; pTA131-based plasmid for *thi4* deletion	Hwang et al. (2014)
Expression plasmids		
pJAM202c	Ap^r; Nv^r; *Hfx. volcanii-E. coli* shuttle plasmid with $P2_{rrnA}$ promoter	Zhou et al. (2008)
pJAM2821	Ap^r; Nv^r; *expresses *thi4-strepII*	Hwang et al. (2014)
pJAM2822	Ap^r; Nv^r; *expresses *thi4* (C165A)-*strepII*	Hwang et al. (2014)
pJAM3143	Ap^r; Nv^r; *expresses *thi4* (C165H)-*strepII*	Hwang et al. (2017)
pJAM3141	Ap^r; Nv^r; *expresses *thiC*	Hwang et al. (2017)
pJAM3142	Ap^r; Nv^r; *expresses *thiE*	Hwang et al. (2017)
pJAM3130	Ap^r; Nv^r; *expresses *thiR-strepII*	Hwang et al. (2017)
pJAM3155	Ap^r; Nv^r; *expresses *thiRa-strepII*	Hwang et al. (2017)
pJAM3156	Ap^r; Nv^r; *expresses *thiDN-strepII*	Hwang et al. (2017)
pJAM3163	Ap^r; Nv^r; *expresses *thiDN* (E297A)-*strepII*	Hwang et al. (2017)

pJAM3166	Ap^r; Nv^r; *expresses *thiDN* (N301A)-*strepII*	Hwang et al. (2017)
pJAM3168	Ap^r; Nv^r; *expresses *thiDN* (N362A)-*strepII*	Hwang et al. (2017)
pJAM3172	Ap^r; Nv^r; *expresses *thiDN* (R321A)-*strepII*	Hwang et al. (2017)
pJAM3173	Ap^r; Nv^r; *expresses *thiDN* (H342A)-*strepII*	Hwang et al. (2017)
pJAM3174	Ap^r; Nv^r; *expresses *thiDN* (K434A)-*strepII*	Hwang et al. (2017)
pJAM3185	Ap^r; Nv^r; *expresses *thiDN* (E396A)-*strepII*	Hwang et al. (2017)
pJAM3186	Ap^r; Nv^r; *expresses *thiDN* (E398A)-*strepII*	Hwang et al. (2017)
pJAM3190	Ap^r; Nv^r; *expresses *thiDN* (M400A)-*strepII*	Hwang et al. (2017)
pJAM3192	Ap^r; Nv^r; *expresses thiDN *(E396A E398A)-strepII*	Hwang et al. (2017)
β-galactosidase (bgaH) reporter plasmids		
pJAM2678	Ap^r; Nv^r; *carries $P2_{rrnA}$-*bgaH*	Sudarsan et al. (2003)
pJAM2714	Ap^r; Nv^r; *carries *bgaH* lacking $P2_{rrnA}$ and SD site	Sudarsan et al. (2003)
pJAM3110	Ap^r; Nv^r; *carries $P_{thi4-91\ bp}$-*bgaH*	Hwang et al. (2017)
pJAM3111	Ap^r; Nv^r; *carries $P_{thi4-36\ bp}$-*bgaH*	Hwang et al. (2017)
pJAM3129	Ap^r; Nv^r; *carries $P_{thi4-45\ bp}$-*bgaH*	Hwang et al. (2017)
pJAM3131	Ap^r; Nv^r; *carries $P_{thi4-80\ bp}$-*bgaH*	Hwang et al. (2017)
pJAM3132	Ap^r; Nv^r; *carries $P_{thi4-70\ bp}$-*bgaH*	Hwang et al. (2017)
pJAM3133	Ap^r; Nv^r; *carries $P_{thi4-60\ bp}$-*bgaH*	Hwang et al. (2017)

pJAM3134	Ap^r; Nv^r; *carries $P_{thi4-50\ bp}$-*bgaH*	Hwang et al. (2017)
pJAM3139	Ap^r; Nv^r; *carries $P_{thiC-113\ bp}$-*bgaH*	Hwang et al. (2017)
pJAM3140	Ap^r; Nv^r; *carries $P_{thiE-92\ bp}$-*bgaH*	Hwang et al. (2017)
pJAM3180	Ap^r; Nv^r; *carries $P_{thi4\text{-}91\ bp}$ (TATCAC → cgTCga)-*bgaH*	Hwang et al. (2017)
pJAM3181	Ap^r; Nv^r; *carries $P_{thi4\text{-}91\ bp}$ (GTAGT → acAGc)-*bgaH*	Hwang et al. (2017)
pJAM3182	Ap^r; Nv^r; *carries $P_{thi4\text{-}91\ bp}$ (TATCAC → cgTCga, GTAGT → acAGc)-*bgaH*	Hwang et al. (2017)

Abbreviations: pJAM202c related Hv expression plasmids marked by asterisk (*); Ap^r, ampicillin resistance; Nv^r, novobiocin resistance; MCS, multiple cloning site; -*strepII*, C-terminal StrepII tag fusion coding sequence (preceded by a *Kpn*I site that facilitated generation of the C-terminal StrepII fusion with GlyThr linker); His_6-, N-terminal His_6 tag fusion coding sequence; *ubaA*, HVO_0558; *thi4*, HVO_0665; *thiC*, HVO_2154; *thiR*, HVO_0662; *thiE*, HVO_2668; SD, Shine Dalgarno ribosomal binding site; *thiRa*, Nmag_2413; *thiDN*, Nmag_1282 (encodes single protein with ThiD and ThiN domains); amino acid and nucleic acid substitutions in parentheses.

E. coli growth medium:

Luria-Bertani (LB) medium: Dissolve 10 g of NaCl, 10 g of tryptone, and 5 g of yeast extract in 700 mL of dH_2O. For LB agar plates, add 15 g of agar (1.5 % w/v). Adjust volume to 1 L and autoclave the medium. Once cooled to 55 °C, add ampicillin (100 µg/mL) as needed for plasmid-containing strains.

LB plates supplemented with 100 µg/mL ampicillin were evenly surface-treated with 50 µL of 5-bromo-4-chloro-3-indolyl-β-D-galactopyranoside (X-gal) (50 mg/mL prepared in 100 % N,N-dimethylformamide) and 100 µL of isopropyl β-D-1-thiogalactopyranoside (IPTG) (23.8 mg/mL prepared in deionized water).

H. volcanii growth media:

ATCC974 rich medium: Dissolve 125 g of NaCl, 50 g of $MgCl_2{\cdot}6H_2O$, 5 g of K_2SO_4, 0.134 g of $CaCl_2{\cdot}2H_2O$, 5 g of tryptone, and 5 g of yeast extract in 700 mL dH_2O. For ATCC agar plates, add 15 g of agar (1.5 % w/v). Adjust pH to 6.8 using 1 N KOH. Adjust volume to 1 L and autoclave the medium. Include a stir bar during autoclaving for mixing. Once cooled to 55 °C, add novobiocin (0.2 µg/mL) or 5-fluoroorotic acid (5-FOA, 50 µg/mL) as required. Dissolve 5-FOA in 100 % DMSO at 50 mg/mL before adding to the medium or directly add as a powder.

Prepare novobiocin as a 10 mg/mL stock in dH_2O, perform a 1:10 dilution, and then add 200 μL of the 1 mg/mL stock to the 1 L medium.

Hv-Ca$^{\pm}$ semi-defined medium: Mix 708 mL of 30 % salt-water (SW), 256 mL of dH_2O and 35 mL of 1 M Tris-HCl pH 7.5. For Hv-Ca+ plates, add 17.71 g of agar (1.5 % w/v). Autoclave base medium. Once cosoled to 55 °C, add 117 mL of 10X casamino acids, 30 mL of 2 M glycerol, 14 mL of Hv-min salts, 11.5 mL of 0.1 M KPO_4 buffer, pH 7.5, and 1.06 mL of thiamine and biotin stock for a final volume of 1.170 L.

Defined Hv-minimal (Hv-min) medium: Add 600 mL of 30 % SW, 280 mL of dH_2O, and 30 mL of 1 M Tris-HCl pH 7.5. For Hv-min plates, add 15 g of agar (1.5 % w/v). Autoclave base media and once cooled to 55 °C, add 10 mL of 2 M glycerol, 6 mL of 0.5 M $CaCl_2$, 1 mL of trace elements, 0.9 mL of thiamine-biotin stock, 10 mL of 1 M NH_4Cl, 51 mL of uracil (1 mg/mL) and 9.75 mL of 0.1 M KPO_4 buffer pH 7.5 (Dyall-Smith, 2009).

Medium supplements:

Concentrated 30 % SW (w/v): Into 700 mL of warm dH_2O, add 240 g of NaCl, 30 g of $MgCl_2{\cdot}6\,H_2O$, 35 g of $MgSO_4{\cdot}7\,H_2O$, 7 g of KCl, and 20 mL of 1 M Tris-HCl pH 7.5. Raise to 1 L and autoclave before storing at room temperature.

1 M Tris-HCl buffer: Into 300 mL of dH_2O, add 60.55 g of Tris-base. Adjust to desired pH with concentrated 1 N HCl and raise to 500 mL. Filter sterilize and store at room temperature.

10X Casamino acid (CA) stock: Into 300 mL of dH_2O, add 25 g of casamino acids and 11.762 mL of 1 M KOH. Raise to 500 mL and filter sterilize before storing at 4 °C.

Hv-Min salts: Into 40 mL of dH_2O, add 1.6 g of NH_4Cl, 2.65 g of $CaCl_2{\cdot}2\,H_2O$, and 6 mL of trace elements. Raise to 72 mL and filter sterilize before storing at 4 °C.

Trace elements: Into 100 mL of dH_2O, add 3 drops of 1 N concentrated HCl and dissolve the following metals: 36 mg of $MnCl_2{\cdot}4\,H_2O$, 44 mg of $ZnSO_4{\cdot}7\,H_2O$, 230 mg of $FeSO_4{\cdot}7\,H_2O$, and 5 mg of $CuSO_4{\cdot}5\,H_2O$. Filter sterilize and store in dark at 4 °C.

Thiamine and biotin stock: Add 9.6 mg of thiamine and 1.2 mg of biotin in 10.8 mL of dH_2O. Filter sterilize and store at 4 °C (Dyall-Smith, 2009). Add 0.9 mL of this stock to 1 L Hv-min medium for a final concentration of 2.38 μM thiamine and 0.41 μM biotin. Omit the thiamine when examining the capacity of *H. volcanii* mutant strains to synthesize thiamine. For more detailed analysis of the thiamine biosynthetic pathway, exchange thiamine for the thiazole-precursor 4-methyl-5-(2-hydroxyethyl) thiazole (THZ; 2 μM) or the pyrimidine precursor 4-amino-5-hydroxymethyl-2-methylpyrmidine (HMP; 2 μM) (Hwang et al., 2017).

Additional reagents:

0.8 % DNA agarose gel: Add 392 mL of dH_2O to 8 mL of 50X TAE buffer, 3.92 g of agarose, and 10 μL of ethidium bromide. Microwave until agarose is fully dissolved and mixture is clear. Alter concentration of agarose as needed to optimize separation of DNA products based on size (*e.g.*, smaller DNA fragments may require a higher % agarose).

β-galactosidase buffer B: Into 500 mL of dH_2O, add 25 mL of 1 M Tris-HCl (pH 7.2), 250 mL of 5 M NaCl, and 10 μL of 0.5 M $MnCl_2$. The final concentrations are as follows: 50 mM Tris-HCl (pH 7.2), 2.5 M NaCl, and 10 μM $MnCl_2$.

BgaH Buffer: Into 4.995 mL of β-galactosidase buffer B, add 5 μL of β-mercaptoethanol (0.1 % v/v).

Spheroplasting Solution: Into 70 mL of dH_2O, add 4.7 g of NaCl, 0.2 g of KCl, 15 g of sucrose, 5 mL of 1 M Tris-HCl, pH 8.2, and 15 mL of 100 % glycerol. Raise to 100 mL. Filter sterilize and store at 4 °C.

Polyethylene glycol solution (PEG): Mix 4 mL of spheroplasting solution with 6 mL of PEG600 (Sigma P3390) to a final volume of 10 mL. Filter sterilize and store at 4 °C.

Regeneration Buffer: Into 600 mL of dH_2O, add 206 g of NaCl, 37 g of $MgSO_4 \cdot 7\,H_2O$, 3.7 g of KCl, 50 mL of 1 M Tris-HCl pH 7.2, 5 mL of 10 % $CaCl_2 \cdot 2\,H_2O$, and 250 mL of 60 % (w/v) sucrose. Raise to 1 L and filter sterilize before storing at 4 °C (Dyall-Smith, 2009).

3. Polymerase chain reaction: Primer design and general methods

Polymerase chain reaction (PCR) is integral to microbial strain construction. PCR is employed to generate DNA fragments used in plasmid construction for ectopic gene expression and targeted gene knockout on the chromosome via homologous recombination. Halophilic archaea typically have GC-rich genomes as exemplified by *H. volcanii* (65 % GC) (Hartman et al., 2010) and *Natrialba magadii* (61 % GC) (Siddaramappa et al., 2012), species used to generate the constructs that formed the basis for providing insights into the regulatory role of ThiR and metabolic function of Thi4 and other thiamine biosynthetic enzymes

(Hwang et al., 2014, 2017; Siddaramappa et al., 2012). While GC-content can reduce PCR efficiency, we have found the following approaches facilitate yield of PCR product for amplifying the DNA of halophilic archaea.

Oligonucleotide Primer Specifications: Design primers to optimize PCR efficacy within the following parameters: GC content of 40–70 %; T_m of 65–85 °C and less than 5 °C difference in T_m between the primers in the pair; at least one G or C at the 3' end (and the 5' end if performing site-directed mutagenesis); and minimal hairpin structures. Calculate primer T_m values according to the New England Biolabs T_m calculator with settings for Phusion DNA polymerase in GC buffer and 500 nM primer concentration. Examine primer harpin structures using IDT Technologies OligoAnalyzer with settings at: Oligo Conc 0.5 μM, Na^+ Conc 50 mM, Mg^{2+} Conc 1.5 mM, and dNTPs Conc 0.2 mM (Jia et al., 2022).

Screening colonies by PCR (cPCR): Screen *H. volcanii* and *E. coli* strains for mutations, plasmids and other genetic features by PCR. Screen 6–10 original colonies per sample type. For each original colony, pick approximately three clonal isolates from fresh plates using a toothpick and transfer them into 30 μL of nuclease-free water in 500 μL microcentrifuge tubes. Boil samples at 95 °C for 5 min, incubate on ice for 5 min, and centrifuge at 12,000 g for 5 min at room temperature, leaving the cell debris undisturbed. Combine 2 μL of the supernatant with 18 μL of PCR master mix in thin-walled 0.2 mL polypropylene tubes. Prepare the PCR master mix from the components listed below, maintaining all reagents and mixtures on ice prior to PCR.

	PCR master mix	
Reagent	**1x Reaction**	**6.5 x reaction (screen 6)**
Nuclease free water	10.8 μL	64.8 μL
5X Phusion GC buffer	4 μL	26 μL
10 mM dNTPs	0.4 μL	2.6 μL
10 μM forward primer	1 μL	6.5 μL
10 μM reverse primer	1 μL	6.5 μL
DMSO	0.6 μL	3.9 μL
Phusion DNA polymerase	0.2 μL	1.3 μL

Genomic and plasmid DNA preparation and use in PCR-based amplification: Isolate plasmids from *E. coli* or *H. volcanii* using the PureLink Quick Plasmid Miniprep Kit. Extract genomic DNA (gDNA) from *H. volcanii* H26 (Allers et al., 2004) and *Natrialba magadii* ATCC 43099 T (Siddaramappa et al., 2012) by the spooling method as described in *The Halohandbook* (Dyall-Smith, 2009). Use the isolated gDNA or plasmid DNA as template in PCR mixtures as detailed below for a 100 μL reaction.

Reagent	1x Reaction
Nuclease free water	62 μL
5X Phusion GC buffer	20 μL
10 mM dNTPs	2 μL
10 μM forward primer	5 μL
10 μM reverse primer	5 μL
DMSO	3 μL
Phusion DNA polymerase	1 μL
Plasmid/gDNA (10 ng/μL or 20 ng total)	2 μL

PCR thermocycler conditions: Program the thermocycler with the PCR steps as outlined below. Initial a hot start by transferring the PCR reaction mixtures directly from ice to the preheated (98 °C) thermocycler wells. Vary the annealing step (*) according to the calculated T_m values of the primer pair (typically 5 °C above the lowest T_m). Optimize the PCR conditions as needed. Adjust the annealing step (*) by incorporating a thermal gradient that ranges from 5 °C below to 5 °C above the calculated average T_m of the primer pair. Use an extension time (**) of 15–30 s per kb, erroring on the side of 30 s per kb for complex reactions.

Steps	Temperature	Time	Description of step
1 Cycle	98 °C	4 min	Initial denaturation
30 Cycles	98 °C	10 s	Denaturation
	65–85 °C (variable)*	30 s	Annealing
	72 °C	15–30 s per kb**	Extension
1 Cycle	72 °C	5 min	Final extension

4. Bioinformatic analysis and structural modeling

Comparative bioinformatics and structural modeling approaches can be employed to guide experimental design, identify thiamine biosynthesis-related protein homologs, determine potential interacting partners, and map key amino acid residues within an evolutionary framework, as illustrated for Thi4 in Fig. 3. As the thiazole-ring formation represents one of the most chemically challenging steps in thiamine biosynthesis, prioritization of sequence and structure-based comparisons to optimize THI4 and related homologs would be beneficial to enhance thiamine production. Further expansion of this analysis to study archaeal ThiR transcriptional regulators and their cognate operator motifs would advance knowledge of how to regulate thiamine biosynthesis in extremophilic microorganisms. Overall, these strategies can provide critical insights into sulfur mobilization to form the thiazole ring, catalytic turnover potential, and regulatory divergence. In addition, conserved DNA elements upstream of thiamine biosynthetic operons that provide sites for transcriptional control may be identified. Collectively, this analysis can serve as the foundation for selecting candidate residues, domains, protein-homologs, and promoter regions for subsequent studies.

Analyze archaeal protein homologs for thiamine biosynthetic or regulatory capacity:

1. Retrieve archaeal protein sequences sharing homology with enzymes of thiamine biosynthesis. Orthologous protein groups available via InterPro, Pfam, AnnoTree, and arCOG are found useful (Hunter et al., 2009; Makarova et al., 2015; Mendler et al., 2019; Mistry et al., 2021). Basic Local Alignment Search Tool using BLASTP (protein-protein BLAST) is also an option (Jones et al., 2014). Select archaeal representatives for further analysis using the FASTA files with UniProtKB/Swiss-Prot accession numbers for reference (Altschul et al., 1990).
2. Examine conserved active site residues, domains and other relevant protein features by aligning the sequences of the archaeal homologs to the related thiamine biosynthetic enzymes using ClustalW (Thompson et al., 1994). Visualize this alignment using ESPript 3 (Robert & Gouet, 2014).
3. Perform 3D-structural modeling of the archaeal homolog (see Note 1). Using the matchmaker tool within the interface of ChimeraX, compare the 3D-model to experimentally derived 3D-structures of enzymes known to catalyze steps in thiamine biosynthesis or regulation. Highlight

the conserved active site residues and compare the overall structural fold (reporting the RMSD values of the structural comparison). For example, archaeal Thi4 homologs can be examined for the conserved active site histidine and cysteine residues, observed in thiazole-synthesizing enzymes (Kelley & Sternberg, 2009).

Analyze chromosomal elements for DNA sequences that may serve as operator regions for transcriptional regulation of thiamine metabolism:

4. Use MEME suite (https://meme-suite.org/meme/index.html) to identify DNA regulatory motifs of thiamine biosynthetic operons. Input DNA sequences upstream of thiamine biosynthetic genes (*thi4, thiC,* and *thiE*) in the MEME motif discovery mode to discover ungapped motifs of recurring, fixed-length patterns. To refine the motif and identify new motifs de novo, use this motif to scan archaeal genomes with the finding individual motif occurrences (FIMO) feature of the MEME suite (Bailey et al., 2015).

Publicly available databases and software can be used to guide downstream experimental design and prioritize specific genes of interest.

5. Archaeal strains and growth conditions

The following general guidelines are recommended for *H. volcanii* strain and plasmid construction. For routine plasmid construction, use an *E. coli* strain that displays high transformation efficiency, such as TOP10. To obtain plasmids devoid of methylation for subsequent transformation into *H. volcanii*, use a *dam*$^-$ *dcm*$^-$ strain such as *E. coli* GM2163 as a shuttle (Cline, Lam, et al., 1989). Use PEG-mediated transfer to transform *H. volcanii* (Cline, Schalkwyk, et al., 1989). Aerate liquid cultures by orbital shaking at 200 rpm. Culture *E. coli* strains at 37 °C in Luria-Bertani (LB) supplemented with ampicillin (100 μg/mL) when required. Grow *H. volcanii* strains at 42 °C in American Type Culture Collection 974 rich medium (ATCC974) or Hv-minimal media (Hv-MM) supplemented with a carbon source such as 20 mM glycerol, which is preferred over glucose (Sherwood et al., 2009). The advantage of this Hv-MM is that the concentration of thiamine can be controlled. Prepare media according to the protocols in *The Halohandbook* (Dyall-Smith, 2009) and described in the key resources section. Supplement media with novobiocin (0.2 μg/mL) to select for and maintain plasmids that carry a *gyrB** (DNA gyrase) mutation that renders *H. volcanii* cells resistant to novobiocin (Holmes et al., 1994). When generating *H. volcanii* mutant strains with gene deletions, use the HvCa+ uracil minus medium for the initial selection and then supplement ATCC974 medium with 5-fluoroortic acid (5-FOA, 50 μg/mL) for the counterselection (Allers et al., 2004). Strains and plasmids utilized for study of *H. volcanii* thiamine metabolism and regulation are listed in the key resources table.

6. Gene deletion and complementation

Gene deletion and complementation studies can be used to elucidate thiamine biosynthetic pathways in *H. volcanii* as the microorganism is amenable to growth on defined medium in which the concentration of thiamine and thiamine-related chemical supplements can be controlled (Dyall-Smith, 2009). Past studies have targeted gene homologs of thiamine metabolism and regulation for deletion in the *H. volcanii* genome, and the mutants have been genetically complemented by ectopic expression from plasmids to further validate and characterize the genes of interest (Hwang et al., 2014, 2017). Included among the *H. volcanii* mutants studied for their impact on thiamine biosynthesis are Δ*thi4* (HVO_0665), Δ*thiR* (HVO_0662), Δ*thiC* (HVO_2154), and Δ*thiE* (HVO_2668) (Hwang et al., 2014, 2017). In addition, *ubaA* (HVO_0558) encoding the E1-like enzyme of sampylation (Miranda et al., 2011) was targeted for deletion due to its relationship with bacterial ThiF used to mobilize sulfur to ThiS for incorporation into the thiazole ring (Butler et al., 2025). While the '*thi*' designated mutants were unable to grow in the absence of thiamine supplementation, the Δ*ubaA* mutation had no impact revealing the bacterial-like ThiF-ThiS mechanism is not needed for thiamine biosynthesis (Hwang et al., 2017). Use of the complementation system is important to distinguish the phenotypes observed from possible secondary regulatory or polar effects. Since thiamine biosynthesis genes are frequently co-regulated, deletion constructs should be designed with minimal disruption of adjacent genes in mind, as well as to allow for phenotypic rescue through complementation with wild-type or mutant alleles expressed in *trans*. Established genetic tools for gene deletion in *H. volcanii* via homologous recombination (Bitan-Banin et al., 2003) with general recommendations and modifications are outlined below.

Total time: Allocate 1.5 weeks for plasmid generation and passage through *E. coli* GM2163, and an additional 4 weeks for transformation and isolation of the *H. volcanii* mutant strains. While ensuring the mutant is fully isolated is time consuming, this latter stage of the project is important for downstream studies and avoids genetic ambiguity.

Designing pre-KO plasmid:

1. Design pre-knockout (pre-KO) primers with *Bam*HI and *Hin*dIII restriction endonuclease sites to amplify the deletion gene of interest with 500 bp upstream and downstream. See Section 3 for primer design constraints.
2. Perform PCR with 10 ng/μL (20 ng total) of *Hv* genomic DNA (gDNA) and pre-KO primers. Examine the quality of the PCR product via DNA gel electrophoresis. See Section 3 for Phusion PCR settings. This PCR typically involves a temperature gradient to identify ideal primer T_m.
3. If a single band is present, isolate the product via PCR-cleanup using Monarch PCR and DNA Clean-up Kit and elute the sample into 40 μL of nuclease-free water (pre-heated to 50 °C). If multiple products are present, excise the intended product from the gel and purify the DNA from the gel using a DNA Gel Extraction Kit (see Note 2).
4. Digest the PCR product and the *pyrE2*-marked plasmid pTA131 vector using the restriction enzymes *Bam*HI-HF and *Hin*dIII-HF in the following ratios in a 50 μL reaction. Incubated at 37 °C for 3 h.

Reagent	Volume
1 μg of PCR insert or pTA131 vector	40 μL
10X Cut Smart buffer	5 μL
*Bam*HI-HF	2.5 μL
*Hin*dIII-HF	2.5 μL

5. 30 min before the restriction enzyme digestion is completed, remove the pTA131 and add 6 μL of Antarctic Phosphatase Reaction Buffer (10X) and 1 μL of Antarctic Phosphatase enzyme. Incubate for 30 min at 37 °C. This step will dephosphorylate the ends of the plasmid DNA vector and minimize self-ligation.
6. Afterwards, separate the products by DNA electrophoresis (*e.g.*, 0.8 % agarose gel in 1X TAE buffer, 30 min, 90 V), excise the intended

product from the gel and purify the DNA from the gel into 10 μL of nuclease-free water (pre-heated to 55 °C) using a DNA Gel Extraction Kit ('gel extraction').

7. Maintain a 3:1 pmol concentration ratio of insert:vector for the ligation of the DNA to the pTA131 vector. Perform the ligation using Quick Ligase with 2X Quick Ligase Buffer using the following ratios.

Reagent	Volume
Nuclease-free water	0.5 μL
0.02 pmol of pTA131	3 μL
0.06 pmol of PCR insert	1 μL
2X Quick Ligase Buffer (see Note 3)	5 μL
Quick Ligase	0.5 μL

8. Incubate the ligation reaction at room temperature for 15 min and then chill on ice.
9. Transform *E. coli* TOP10 cells with the ligation mixture as follows. Thaw competent cells on ice. Add 2.5 μL of the ligation mixture per 100 μL of cells. Incubate the mixture on ice for 30 min. Heat shock the cells by immediately transferring the mixture to 37 °C for 1 min. Return the sample to ice for 1 min followed by transferring to room temperature. Allow the cells to recover by adding 1 mL of LB medium and incubating with aeration at 37 °C for 1 h. Harvest the cells by centrifugation at ~5000 × *g* for 5 min. Remove the spent culture broth by pouring off the liquid. Resuspend the cells with the excess broth remaining and spread plate on LB-Amp (100 μg/mL). Incubate the plates overnight at 37 °C (with the agar side on top and lid on bottom). Include IPTG and X-gal in the LB-Amp plates for blue/white selection.
   a. Deletion plasmids (pTA131) containing the *lacZ* fragment were screened by blue–white selection to distinguish recombinant clones from empty vectors. LB plates supplemented with 100 μg/mL ampicillin were surface-treated with 50 μL of X-gal (50 mg/mL prepared in 100 % N,N-dimethylformamide) and 100 μL of IPTG (23.8 mg/mL prepared in deionized water). The reagents were evenly distributed across the agar surface and plates were allowed to dry near a flame for approximately 30 min under sterile conditions.

Competent cells were plated according to standard transformation procedures and incubated overnight at 37 °C. White colonies were interpreted as containing plasmids with inserts disrupting *lacZ*, whereas blue colonies indicated absence of an insert.

10. Screen white colonies for a complete pre-KO plasmid by PCR using the pre-KO primers and separate the DNA by gel electrophoresis. Grow positive colonies overnight at 37 °C in 5 mL of LB-amp (100 μg/mL) with aeration. Isolate the plasmid using a PureLink Quick Plasmid Miniprep Kit and confirm the fidelity of the DNA sequence.

Preparing the KO plasmid:

1. Design knockout (KO) primers to delete the target gene from the preKO plasmid, while preserving the genomic region 500 bp upstream and downstream of the target gene and retaining all coding sequence from adjacent genes.

2. Use 2 ng of the pre-KO plasmid (isolated from TOP10 cells) as a template and perform "inverse" PCR using the KO primers.

3. Separate the PCR product by DNA electrophoresis and extract the PCR product from the gel into 20 μL of nuclease-free water.

4. Remove unwanted plasmid template that may co-purify by treating the sample in a 50 μL reaction with 5 μL of 10X rCutSmart buffer and 2.5 μL of *Dpn*I for 4 h at 37 °C, followed by PCR clean-up into 20 μL of nuclease-free water.

5. After removing the plasmid template with *Dpn*I, use a KLD Enzyme Mix to self-ligate the PCR product. Mix 1 μL of the PCR product with 3 μL of nuclease-free water, 5 μL of KLD reaction buffer (2X) and 1 μL of KLD enzyme mix (10X). Incubate at room temperature for 15 min and before using 2.5 μL of the mixture to transform *E. coli* TOP10 cells.

6. Perform colony PCR using the pre-KO primers to ensure that the target gene was removed from the plasmid.

7. Passage the KO-plasmid through *E. coli* GM2163 using the same protocol as described with TOP10. This step is important to demethylate the plasmid prior to transformation into *mrr+ H. volcanii* strains (*e.g.*, H26); however, it is unnecessary for *H. volcanii* Δ*mrr* strains (e.g., H1207) (Allers et al., 2010).

Generating pre-competent *H. volcanii* cells:

1. Streak the desired *H. volcanii* Δ*pyrE2* strain (*e.g.*, H26) from −80 °C stored glycerol stocks onto ATCC974 agar plates and incubate at 42 °C for 3–5 days.

2. Inoculate isolated colonies (3–5) into 50 mL of ATCC974 medium and incubate at 42 °C with shaking at 200 rpm until cultures reach an OD_{600} of 0.8–1.0. Dilute the culture 1:2 with fresh medium to obtain an accurate OD_{600} measurement.
3. Chill the cultures on ice for 10 min
4. Harvest the cells by centrifugation at 4000 × g for 10 min at 4 °C. Discard the supernatant, removing all excess with a pipettor.
5. Gently resuspend the cell pellet on ice with 2 mL of pre-chilled spheroplasting solution and continue to incubate on ice for 20 min
6. Keeping the samples on ice, distribute the pre-competent *H. volcanii* cells into 0.2 mL aliquots. Store at −80 °C until use.

H. volcanii mutation via pop-in/pop-out:

1. Transform the KO plasmid into the *H. volcanii* Δ*pyrE2* pre-competent cells overnight and plate on Hv-Ca^+ medium the following day to select for integration of the *pyrE2*+ KO plasmid at the target site. To transform *Hv*, use a rotating 15 mL conical tube. Add 10 μL of 0.5 M EDTA pH 8.0 and 200 μL of pre-competent cells to opposite sides of the inner wall of the tube. Mix the sample by allowing the tube to rotate for 10 min, turning the cells competent. Add 20 μL of the KO plasmid and rotate for an additional 10 min, followed by the addition of 230 μL of room temperature polyethylene glycol (PEG) solution for 10 min. Finally, add 10 mL of room-temperature regeneration buffer. Incubate the cells overnight at 42 °C. Plate 100 μL of cells directly and then centrifuge (5000 × *g*, 10 min, room temperature) and plate 100 μL and 10 μL of the cell pellet after removing the majority of spent culture broth.
2. Streak colonies for isolation on HvCa+ medium to ensure separation.
3. Perform colony PCR to check for integration of the KO plasmid using the pre-KO primers.
4. Inoculate positive colonies into 5 mL of ATCC974 supplemented with 5-FOA (50 ug/mL) and incubated at 42 °C until cells reach late-log phase (OD_{600} of 0.7–0.8 units/mL).
5. Perform serial dilutions of 10^{-1} through 10^{-4} of the liquid culture and plate on ATCC974 + 5-FOA. Incubate plates at 42 °C for 1 week.
6. Streak isolated colonies for isolation on ATCC974 + 5-FOA to ensure separation.
7. Incubate plates at 42 °C for 3–5 days.
8. Rescreen the isolated colonies by PCR using the pre-KO primer.

9. Repeat steps 6–8. Thus, conduct a total of three mutant screenings to ensure that the parental cells are absent.
10. For the final screen, design new primers to amplify 700 bp upstream and downstream of the gene of interest, and use these primers to confirm the mutation.
11. Grow the mutant strain in ATCC974 medium and store in 20 % (v/v) glycerol stocks at −80 °C (Bitan-Banin et al., 2003).

Complementation:

A similar strategy – to the generation of pre-KO plasmids – can be used to construct complementation plasmids. Notable differences include amplification of the coding sequence of the gene of interest from gDNA and ligation and digestion using different restriction endonucleases corresponding to the pJAM809 vector. This vector contains a non-native constitutive promoter as well as C-terminal *StrepII* tag to facilitate protein detection in *Hv* cells (Allers, 2010; Allers et al., 2010). These expression plasmids have an origin of replication distinct from the genome and can be transformed into the *H. volcanii* mutant cells to examine if the parental phenotype is restored (e.g., growth in the absence of thiamine).

These techniques, while seemingly basic and straightforward, provide the necessary skills to experimentally confirm the biosynthetic role of specific proteins involved in thiamine biosynthesis while also assessing the effects of polar mutations.

7. Site-directed mutagenesis

Site-directed mutagenesis (SDM) can be employed to investigate the roles of amino acids that may be critical for the function of proteins (enzymes or regulators) associated with thiamine biosynthesis and to define regulatory DNA elements controlling thiamine-dependent gene expression. This approach will provide insight into protein function and transcriptional regulation. One recommended SDM method to employ is reduce recycle PCR (rrPCR), a method that is optimized for high efficiency with large plasmids, requires only two sets of primers, and uses common enzymes (*Dpn*I and a proofreading DNA polymerase) (Jia et al., 2022). This type of strategy is exemplified by *Hv*Thi4 in which comparative sequence alignment and structural modeling with eukaryotic THI4-homologs from eukaryotic, bacterial, and archaeal systems. Past work targeted the conserved cysteine residue of *Hv*Thi4 corresponding to the sulfur-donating position in *Saccharomyces cerevisiae* THI4p for mutagenesis to assess its contribution to thiazole formation and found it was needed for growth in the absence of thiamine (Hwang et al., 2014).

Total time: 8 h (to implement the PCR-1 and PCR-2 reaction series).
Site-directed mutation generation:

The rrPCR methods according to Huiyong et al. (Jia et al., 2022) are described below and recommended for SDM due to their ease of implementation.

1. Design a forward primer and reverse anchor primer for SDM with the following guidelines in mind: (i) the forward primer should contain 12 unmodified nucleotides on both sides of the mutation, with the mutated bases in the center (see Note 4); (ii) the reverse anchor primer should be designed to amplify approximately 300 bp DNA fragment when paired with the forward primer; (iii) the primers should have a GC content of 40–70 %, T_m of 70–80 °C, at least one G or C at both the 5' and 3' ends, and minimal hairpin structures. Calculate primer T_m values; the NE Biolabs T_m calculator is useful for this calculation as it has settings for Phusion DNA polymerase in GC buffer and 500 nM primer concentration. Examine primer harpin structures; the IDT Technologies OligoAnalyzer can be used for this analysis with settings at: Oligo Conc 0.5 μM, Na^+ Conc 50 mM, Mg^{2+} Conc 1.5 mM, and dNTPs Conc 0.2 mM.
2. Using the plasmid designed to express the target gene in *H. volcanii* (see **Gene deletion and complementation**) as a DNA template (100 ng final concentration), perform a 100 μL PCR-1 reaction to generate a 'megaprimer' with the following reaction mixture and thermocycler conditions. Include a 'hot start' to facilitate product formation. Make sure the plasmid is purified from an *E. coli* strain that is *dam*+ *dcm*+ such as TOP10 to allow for later removal by *Dpn*I digestion. Include a 'no primer control' that has the DNA template mixed with all reagents except an additional 10 μL nuclease free water is used to replace the primers.

PCR-1 reaction mixture:

Reagent	1x Reaction
Nuclease free water	59 μL
5XGC buffer	20 μL
10 mM dNTP	2 μL

10 μM forward primer 1	5 μL
10 μM reverse anchor primer 2	5 μL
DNA template (20 ng/μL or 100 ng final)	5 μL
DMSO	3 μL
Phusion DNA polymerase (2 U/μL)	1 μL

PCR-1 Thermocycler conditions:

1 Cycle	98 °C	4 min	Initial denaturation
30 Cycles	98 °C	10 s	Denaturation
	75–80 °C; 5 °C above primer Tm	30 s	Annealing
	72 °C	15–30 s per kb	Extension
1 Cycle	72 °C	7 min	Final extension

1. Separate an aliquot of the PCR-1 reaction mixture (5 μL) by DNA electrophoresis to confirm the generation of the 'mega-primer'.
2. Purify the remaining PCR-1 reaction mixture using a PCR cleanup kit at a 2:1 ratio (see Note 5) into 40 μL of nuclease-free water.
3. Perform a PCR-2 reaction using the purified PCR-1 mixture as template with the following 50 μL reaction mixture and thermocycler conditions. Include a 'hot start' to facilitate product formation.

PCR-2 reaction mixture:

Reagent	1x Reaction
PCR-1 mixture after PCR cleanup	37 μL
5XGC buffer	10 μL
10 mM dNTP	1 μL
DMSO	1.5 μL
Phusion DNA polymerase (2 U/μL)	0.5 μL

PCR-2 Thermocycler conditions:

1 Cycle	98 °C	4 min	Initial denaturation
18 Cycles	98 °C	10 s	Denaturation
	72 °C	7 min 45 s	Annealing and extension
1 Cycle	72 °C	10 min	Final extension

1. After the PCR-2 reaction is completed, treat the sample with 1.5 μL of *Dpn*I (20 U/μL) for 3 h at 37°C and then purify the product using a PCR cleanup kit into 6 μL of nuclease free water.
2. Transform the PCR product into *E. coli* TOP10 cells following the same protocol previously described in **Gene deletion and complementation**.
3. Compare the number of colony forming units per selection plate (*e.g.*, LB Amp) of the SDM vs. no primer control samples. If the ratio is >2, then the likelihood of generating plasmids with the desired SDM is high. If so, the plasmids containing the SDM can be identified by DNA sequencing (Jia et al., 2022).

Impact of SDMs:

Point mutations were invaluable for researching the effects of specific catalytic residues in *Hv*THI4 (Cys165) and aided in our understanding of the role of the suicide-mediated cystine residue in sulfur-transfer and thiazole ring formation for thiamine biosynthesis (Hwang et al., 2014). Additionally, this technique was used for identified putative ThiN/ThiR palindromic and inverted-repeated operator motifs upstream of the *thi4* gene, thought to serve as repressor-protein binding sites (Hwang et al., 2014, 2017). Operator variants were also generated simultaneously and fused to a β-galactosidase reporter, allowing for additional understanding of thiamine-related regulation (Hwang et al., 2017). This technique serves as a valuable tool for allowing direct correlation between residue identity, protein expression, and enzymatic function.

Troubleshooting:

A less common issue can arise during PCR-2. Following PCR-1, separate the PCR product by DNA electrophoresis to verify successful megaprimer formation before proceeding to mutation plasmid generation. However, PCR-2 requires sufficient template DNA to remain in the reaction mixture. In some cases, despite successful megaprimer formation in PCR-1, an insufficient amount of template DNA may remain to support efficient amplification of the final product during PCR-2. To address this limitation, a modified PCR-2 protocol is described below.

1. Following PCR-1, separate the entire 100 μL reaction by DNA electrophoresis and extract the DNA from the gel into 40 μL of nuclease free water.
2. Perform PCR-2 with slight modifications in the volumes of reagents while maintaining the final concentrations. These modifications will allow for more DNA template (20 ng/μL or 100 ng total) to be added to increase efficiency. The rest of the protocol remains the same.

Modified PCR-2 reaction mixture:

Reagent	1x Reaction
PCR-1 mixture after gel extraction	37 μL
5XGC buffer	11 μL
10 mM dNTP	1.1 μL
DNA template (20 ng/μL or 100 ng final)	5 μL
DMSO	1.65 μL
Phusion DNA polymerase (2 U/μL)	0.55 μL

8. Growth assays and chemical complementation

Growth assays can be used to assess the functional consequences of gene deletions, SDMs, and regulatory mutations involved in thiamine biosynthesis (Hwang et al., 2014). Wild-type, deletion, complementation, empty-vector control, and SDM strains can be compared under defined growth conditions with and without chemical supplementation, as illustrated in Fig. 4.

Total time: Variable, several days for each successive growth curve. 5 days for initial culture and primary growth, four days for secondary growth curve.

1. Streak parent strain (H26), mutant, complementation strains, empty-vector controls, and SDM strains from 20% (v/v) glycerol stocks stored at −80 °C onto ATCC974 agar plates supplemented with novobiocin (0.2 μg/mL) when needed. Incubate plates at 42 °C for approximately three days in a plastic bag with moist Kimwipes to prevent drying out.

2. Inoculate single colonies into 4 mL of ATCC974 complex medium in 13 × 100 mm culture tubes and grow at 42 °C with aeration (e.g., rotary shaking at 200 rpm) until cultures reach log phase.
3. Harvest log-phase cultures by centrifugation (8600 × g, 2 min, room temperature). Discard the supernatant and resuspend the cells in 1 mL glycerol-minimal medium (GMM) lacking thiamine.
4. Recentrifuge the cells, remove the supernatant and resuspend in 1 mL of GMM lacking thiamine. Repeat this washing step (twice total) to remove residual thiamine.
5. Assess primary growth by direct sub-culture as follows. Transfer washed cells to a starting OD_{600} of 0.03 in 25 mL GMM supplemented with or without the thiamine-related supplements in 250-mL baffled Erlenmeyer flasks. Incubate cultures at 42 °C with shaking at 200 rpm until log phase is reached (OD_{600} of 0.3–0.6). The supplements include: thiamine (2.38 μM), the thiazole-precursor 4-methyl-5-(2-hydroxyethyl) thiazole (THZ; 2 μM), or the pyrimidine precursor 4-amino-5-hydroxymethyl-2-methylpyrmidine (HMP; 2 μM).
6. Sub-culture log-phase cells a second time to an OD_{600} of 0.03 in 25 mL fresh GMM containing either no supplement or the thiamine-related supplements. Incubate the cultures under the same conditions above using a microplate reader in standard 96-well-plates with water on the edge-wells to prevent desiccation.
7. Monitor growth by measuring optical density at 600 nm (OD_{600}) using a microplate reader. Record OD_{600} readings every 15 min for 99 h under continuous aeration via double-orbital shaking, with temperature maintained at 42 °C. Calculate generation times assuming that an OD_{600} of 1 corresponds to 1×10^9 CFU/mL. Include no-inoculum controls to evaluate background signal arising from the growth medium. Conduct all experiments at least twice with three biological replicates and three technical replicates per experiment, and report data mean ± standard deviation (S.D.) (Hwang et al., 2014).

9. Transcriptional reporter assays

Transcriptional reporter assays can be used to assess promoter activity upstream of thiamine biosynthetic genes and to determine the regulatory effects of ThiR ('inactive' ThiN domain HTH fusion) proteins under different metabolite conditions. One approach is to fuse the promoter regions of target genes to the β-galactosidase gene (*bgaH*) adapted from *Haloferax alicantei* (Hwang et al., 2017; Patenge et al., 2000).

Total time: 48–72 h (including overnight cultures).

Promoter region:

1. Predict putative operator motifs within promoter regions upstream of thiamine biosynthetic genes using MEME mast motif and FIMO scanning (Hwang et al., 2017).
2. Amplify the DNA regions upstream of the start codons (~500−1000 bp) by PCR using primers that include restriction endonuclease sites and promoter fragments to facilitate ligation into the pJAM809-derived plasmid containing the β-galactosidase reporter gene (*bgaH*), following a similar protocol to that discussed further in **gene deletion and complementation**.
3. Transform the resulting promoter-reporter plasmids into 'wild-type' parental and mutant *Hv* strains for functional analysis and promoter activity assessment, as previously described (Bitan-Banin et al., 2003).

β-galactosidase activity:

1. Grow the transformed strains in 4 mL GMM (42 °C, 200 rpm shaking) supplemented with or without thiamine or thiamine precursors (THZ, HMP).
2. Harvest cultures at log phase (OD_{600} 0.5−0.7) by centrifugation (8600 × *g*, 2 min, 4 °C).
3. Discard the supernatant and wash the cell pellet 2 times with buffer A (50 mM Tris-HCl pH 7.2, 2.5 M NaCl, and 10 μM $MnCl_2$). See Note 6 regarding pH assay conditions.
4. Resuspended the cell pellet in 300 μL of BgaH buffer (buffer A with 0.1 % w/v β-mercaptoethanol).
5. Lyse the cells using 100 μL of 0.1-mm Chemglass beads (3 times total, with 1 cycle of vortexing for 1 min with 2 min on ice).
6. Remove cell debris by centrifugation (8600 × *g* for 2 min at 4 °C).
7. Estimate the total protein concentration of the cell lysate in the supernatant using the Bradford assay according to manufacturer's protocols.
8. Measure the β-galactosidase activity in 100 μL assay mixture containing 5 μg of total protein from cell lysate (protein concentration can be increased if the reaction proceeds too slowly), up to 90 μL of *bgaH* buffer, and 10 μL (2.66 mM) of *o*- nitrophenyl-β-D--galactopyranoside (ONPG) in 100 mM potassium phosphate buffer, pH 7.2. Monitor the

reaction at 405 nm (A_{405}) using a spectrophotometer in consecutive 1-min intervals for 30 min at 25 °C. Background controls should include cell lysate alone and substrate alone, and the assay can be read for longer than 30 min if needed to capture slower reactions. See Note 7 for additional information.

9. Conduct all experiments at least twice with three biological replicates and three technical replicates per experiment, and report data mean ± standard deviation (S.D.) (Hwang et al., 2017).

Data Analysis:

Quantify the β-galactosidase activity using the ONPG-based assay and express the results as specific activity in milliunits of enzyme activity per milligram of total protein (mU mg^{-1}). One unit (U) of β-galactosidase activity is calculated as the amount of enzyme catalyzing the hydrolysis of 1 μmol of ONPG to o-nitrophenol (ONP) per min under the defined assay conditions (pH, temperature, buffer composition). Monitor ONP formation by measuring absorbance at 405 nm using a microplate reader for 100 μL of volume (path length ≈ 0.47 cm) and quantify the nanomoles of ONP produced based on a standard curve and the Beer-Lambert law (ε_{405}=3300 M^{-1} cm^{-1}). Perform the standard curve in the same *bgaH*-potassium phosphate buffer used for sample collection. Calculate the enzyme activity as nanomoles of ONPG hydrolyzed per minute, where 1 mU corresponds to 1 nmol min^{-1}, and normalized to the total protein content of each sample (determined using the Bradford assay) (Hwang et al., 2017; Rawls et al., 2010). Note that the ε_{405}=3300 M^{-1} cm^{-1} is sufficiently different from the standard molar extinction coefficients typically assayed for ONP formation (ε_{405}=4600 M^{-1} cm^{-1}), which can be attributed to the reduced path length and more neutral pH. See Note 6 for further guidance.

10. Conclusions and future directions

The genetic, biochemical, and computational approaches outlined in this chapter provide a comprehensive framework for investigating archaea, particularly the mechanisms of thiazole formation and transcriptional regulation. In past studies, the computational analysis and structural modeling strategies presented here guided the identification of archaeal THI4 homologs with conserved active-site residues, ThiR as a ThiN-domain transcriptional repressor, and operator motifs for ThiR binding and regulation of thiamine regulons, enabling SDM and transcriptional reporter assays to be developed. Gene deletion strategies, complementation with overexpression plasmids, and chemical rescue experiments facilitated understanding of the functional roles of thiamine-associated catalytic enzymes and regulators. These methods provided the framework for understanding archaeal thiamine biosynthesis, and its relationship to systems more readily characterized in eukaryotes and bacteria.

Despite these advances, several key questions remain. In *H. volcanii*, the designated model organism for cysteine-based conservation mechanisms, the in vivo contribution of exogenous vs enzyme-derived sulfur in HvTHI4 remains unresolved. The previously reported "suicide-mediated degradation pathway" mechanism regarding sulfur transfer for thiazole formation (Chatterjee et al., 2011) may be context-dependent, as it seems highly unlikely that the cell would waste a single THI4p enzyme for every thiazole-precursor formed. Future studies should directly monitor the fate of the active-site cysteine residue in HvTHI4 using mass spectrometry, particularly in the presence of exogenous sulfur sources like hydrogen sulfide. Previous studies in the thermophilic archaeon *Methanocaldococcus jannaschii* showed that THI4 can function as a histidine-mediated, iron-dependent enzyme capable of multiple turnovers using external sulfide (Eser et al., 2016). Research into the aerobic mesophilic bacterium *Thermovibrio ammonificans* has identified additional non-cysteine THI4 enzymes that also support multiple turnovers, even in low-sulfide environments (Joshi et al., 2021). Together, these findings suggest that non–suicide THI4 enzymes are widespread that alternative catalytic strategies may enable sustained thiamine biosynthesis across diverse metabolic and environmental contexts. Additionally, further research is needed on the ThiN/ThiR transcriptional repressor in *H. volcanii*, particularly to determine whether other regulators or cofactors influence the expression of thiamine-related genes. Future studies should address if archaea possess thiamine-specific activators, or if relief of repression alone is sufficient to upregulate these genes during thiamine starvation.

11. Notes

Note 1: The intensive mode of the Phyre2 web-based server can be used to predict protein structure, function, and mutations based on the target's homology modeling to similar enzymes, building a comprehensive

3D model. The user-friendly interface is amenable to modeling tasks and is related to other tools (i-TASSER, Swiss-Model, and HHpred) (Kelley et al., 2015). Newer models, such as AlphaFold3, are available, offering more comprehensive models of post-translational modifications and protein-ligand interactions (Abramson et al., 2024; David et al., 2022).

Note 2: While primers are designed and compared against the *Hv* DS2 genome NCBI accession NC_013967.1 (Hartman et al., 2010) in SnapGene to assess multiple-binding sites, situations arise where unintended products are generated during PCR, especially when amplifying from genomic DNA (gDNA). If this is the case, the entire PCR-reaction mixture can be separated by DNA electrophoresis and excised using gel extraction procedures.

Note 3: 2X Quick Ligase Buffer should be fully thawed at room temperature prior to use.

Note 4: When selecting the base-pair triplet sequence for the SDM primers, refer to the *H. volcanii* codon usage chart (https://www.kazusa.or.jp/codon/cgi-bin/showcodon.cgi?species=2246) and select the highest frequency per thousand percentage for the desired amino acid (Hartman et al., 2010).

Note 5: Although the optimal ratio for SDM PCR-cleanup is 2:1, it also works well at a 5:1 ratio (Jia et al., 2022).

Note 6: Although classical β-galactosidase assays are typically performed at acidic pH (4.5–6.0), the *bgaH* enzyme used in our transcriptional reporter assays originates from the halophilic and is adapted to the neutral-to-slightly alkaline cytoplasm of halophiles. Therefore, assays were performed at pH 7.2 to maintain enzyme stability and activity under physiologically relevant, high-salt conditions. Previous studies have shown that *bgaH* exhibits robust activity under these conditions (Gregor & Pfeifer, 2001; Holmes et al., 1997; Holmes & Dyall-Smith, 2000; Patenge et al., 2000).

Note 7: While we recommend using 5 μg of total protein from cell lysate, the amount of protein can be increased for slower reactions, as ultimately the β-galactosidase activity is normalized to the total protein content of each sample. Alternatively, for slower reactions the assay can be monitored for longer than 30 min

References

Abramson, J., Adler, J., Dunger, J., Evans, R., Green, T., Pritzel, A., ... Jumper, J. M. (2024). Accurate structure prediction of biomolecular interactions with AlphaFold 3. *Nature, 630*(8016), 493–500. https://doi.org/10.1038/s41586-024-07487-w.

Allers, T. (2010). Overexpression and purification of halophilic proteins in *Haloferax volcanii*. *Bioengineered Bugs, 1*(4), 290–292. https://doi.org/10.4161/bbug.1.4.11794.

Allers, T., Barak, S., Liddell, S., Wardell, K., & Mevarech, M. (2010). Improved strains and plasmid vectors for conditional overexpression of His-tagged proteins in *Haloferax volcanii*. *Applied and Environmental Microbiology, 76*(6), 1759–1769. https://doi.org/10.1128/AEM.02670-09.

Allers, T., Ngo, H.-P., Mevarech, M., & Lloyd, R. G. (2004). Development of additional selectable markers for the halophilic archaeon *Haloferax volcanii* based on the *leuB* and *trpA* genes. *Applied and Environmental Microbiology, 70*(2), 943–953. https://doi.org/10.1128/AEM.70.2.943-953.2004.

Altschul, S. F., Gish, W., Miller, W., Myers, E. W., & Lipman, D. J. (1990). Basic local alignment search tool. *Journal of Molecular Biology, 215*(3), 403–410. https://doi.org/10.1016/S0022-2836(05)80360-2.

Bailey, T. L., Johnson, J., Grant, C. E., & Noble, W. S. (2015). The MEME suite. *Nucleic Acids Research, 43*(W1), W39–W49. https://doi.org/10.1093/nar/gkv416.

Bitan-Banin, G., Ortenberg, R., & Mevarech, M. (2003). Development of a gene knockout system for the halophilic archaeon *Haloferax volcanii* by use of the *pyrE* gene. *Journal of Bacteriology, 185*(3), 772–778. https://doi.org/10.1128/JB.185.3.772-778.2003.

Butler, K. S., Rajput, A., & Chekan, J. R. (2025). ThiF-like enzyme chemistry in primary and secondary metabolism. *Chembiochem: a European Journal of Chemical Biology, 26*(21), e202500460. https://doi.org/10.1002/cbic.202500460.

Chatterjee, A., Abeydeera, N. D., Bale, S., Pai, P.-J., Dorrestein, P. C., Russell, D. H., ... Begley, T. P. (2011). *Saccharomyces cerevisiae* THI4p is a suicide thiamine thiazole synthase. *Nature, 478*(7370), 542–546. https://doi.org/10.1038/nature10503.

Cline, S. W., Lam, W. L., Charlebois, R. L., Schalkwyk, L. C., & Doolittle, W. F. (1989). Transformation methods for halophilic archaebacteria. *Canadian Journal of Microbiology, 35*(1), 148–152. https://doi.org/10.1139/m89-022.

Cline, S. W., Schalkwyk, L. C., & Doolittle, W. F. (1989). Transformation of the archaebacterium *Halobacterium volcanii* with genomic DNA. *Journal of Bacteriology, 171*(9), 4987–4991. https://doi.org/10.1128/jb.171.9.4987-4991.1989.

David, A., Islam, S., Tankhilevich, E., & Sternberg, M. J. E. (2022). The AlphaFold database of protein structures: A biologist's guide. *Journal of Molecular Biology, 434*(2), 167336. https://doi.org/10.1016/j.jmb.2021.167336.

Donovan, P. D., Holland, L. M., Lombardi, L., Coughlan, A. Y., Higgins, D. G., Wolfe, K. H., & Butler, G. (2018). TPP riboswitch-dependent regulation of an ancient thiamin transporter in Candida. *PLoS Genetics, 14*(5), e1007429. https://doi.org/10.1371/journal.pgen.1007429.

Dyall-Smith, M. (2009). The Halohandbook: protocols for haloarchaeal genetics. Version 7. 2. https://haloarchaea.com/halohandbook/.

Eser, B. E., Zhang, X., Chanani, P. K., Begley, T. P., & Ealick, S. E. (2016). From suicide enzyme to catalyst: The iron-dependent sulfide transfer in *Methanococcus jannaschii* thiamin thiazole biosynthesis. *Journal of the American Chemical Society, 138*(11), 3639–3642. https://doi.org/10.1021/jacs.6b00445.

Gregor, D., & Pfeifer, F. (2001). Use of a halobacterial *bgaH* reporter gene to analyse the regulation of gene expression in halophilic archaea. *Microbiology (Reading, England), 147*(7), 1745–1754. https://doi.org/10.1099/00221287-147-7-1745.

Hartman, A. L., Norais, C., Badger, J. H., Delmas, S., Haldenby, S., Madupu, R., ... Eisen, J. A. (2010). The complete genome sequence of *Haloferax volcanii* DS2, a model archaeon. *PLoS One, 5*(3), e9605. https://doi.org/10.1371/journal.pone.0009605.

Holmes, M. L., & Dyall-Smith, M. L. (2000). Sequence and expression of a halobacterial β-galactosidase gene. *Molecular Microbiology, 36*(1), 114–122. https://doi.org/10.1046/j.1365-2958.2000.01832.x.

Holmes, M., Pfeifer, F., & Dyall-Smith, M. (1994). Improved shuttle vectors for *Haloferax volcanii* including a dual-resistance plasmid. *Gene, 146*(1), 117–121. https://doi.org/10.1016/0378-1119(94)90844-3.

Holmes, M. L., Scopes, R. K., Moritz, R. L., Simpson, R. J., Englert, C., Pfeifer, F., & Dyall-Smith, M. L. (1997). Purification and analysis of an extremely halophilic β-galactosidase from *Haloferax alicantei*. *Biochimica et Biophysica Acta (BBA) – Protein Structure and Molecular Enzymology, 1337*(2), 276–286. https://doi.org/10.1016/S0167-4838(96)00174-4.

Hübenthal, A., Panchal, V., Brenk, R., & Mack, M. (2025). Functional analysis of thiamine pyrophosphate-responsive riboswitches in human bacterial pathogens of the ESKAPE group using a dual-luciferase reporter gene assay. *Journal of Bacteriology, 207*(11), e00308-25. https://doi.org/10.1128/jb.00308-25.

Hunter, S., Apweiler, R., Attwood, T. K., Bairoch, A., Bateman, A., Binns, D., ... Yeats, C. (2009). InterPro: The integrative protein signature database. *Nucleic Acids Research, 37*(Database issue), D211–D215. https://doi.org/10.1093/nar/gkn785.

Hwang, S., Cordova, B., Abdo, M., Pfeiffer, F., & Maupin-Furlow, J. A. (2017). ThiN as a versatile domain of transcriptional repressors and catalytic enzymes of thiamine biosynthesis. *Journal of Bacteriology, 199*(7), e00810–e00816. https://doi.org/10.1128/JB.00810-16.

Hwang, S., Cordova, B., Chavarria, N., Elbanna, D., McHugh, S., Rojas, J., ... Maupin-Furlow, J. A. (2014). Conserved active site cysteine residue of archaeal THI4 homolog is essential for thiamine biosynthesis in *Haloferax volcanii*. *BMC Microbiology, 14*(1), 260. https://doi.org/10.1186/s12866-014-0260-0.

Jia, H., Couto-Rodriguez, R., Johnson, S., Medina, S., Novillo, B., Huynh, P., ... Maupin-Furlow, J. A. (2022). Highly efficient and simple SSPER and rrPCR approaches for the accurate site-directed mutagenesis of large and small plasmids. *New Biotechnology, 72*, 22–28. https://doi.org/10.1016/j.nbt.2022.08.004.

Jones, P., Binns, D., Chang, H.-Y., Fraser, M., Li, W., McAnulla, C., ... Hunter, S. (2014). InterProScan 5: Genome-scale protein function classification. *Bioinformatics (Oxford, England), 30*(9), 1236–1240. https://doi.org/10.1093/bioinformatics/btu031.

Joshi, J., Li, Q., García-García, J. D., Leong, B. J., Hu, Y., Bruner, S. D., & Hanson, A. D. (2021). Structure and function of aerotolerant, multiple-turnover THI4 thiazole synthases. *Biochemical Journal, 478*(17), 3265–3279. https://doi.org/10.1042/BCJ20210565.

Jurgenson, C. T., Begley, T. P., & Ealick, S. E. (2009). The structural and biochemical foundations of thiamin biosynthesis. *Annual Review of Biochemistry, 78*(1), 569–603. https://doi.org/10.1146/annurev.biochem.78.072407.102340.

Kelley, L. A., Mezulis, S., Yates, C. M., Wass, M. N., & Sternberg, M. J. E. (2015). The Phyre2 web portal for protein modeling, prediction and analysis. *Nature Protocols, 10*(6), 845–858. https://doi.org/10.1038/nprot.2015.053.

Kelley, L. A., & Sternberg, M. J. E. (2009). Protein structure prediction on the web: A case study using the Phyre server. *Nature Protocols, 4*(3), 363–371. https://doi.org/10.1038/nprot.2009.2.

Lehmann, C., Begley, T. P., & Ealick, S. E. (2006). Structure of the *Escherichia coli* ThiS–ThiF complex, a key component of the sulfur transfer system in thiamin biosynthesis. *Biochemistry, 45*(1), 11–19. https://doi.org/10.1021/bi051502y.

Luo, Y., Liu, Y., Liu, J., Liu, X., Li, H., Wu, Y., ... Yan, J. (2025). Optimizing thiamine pyrophosphate metabolism enhances crop yield and quality. *Nature Communications, 17*(1), 165. https://doi.org/10.1038/s41467-025-66882-7.

Makarova, K. S., Wolf, Y. I., & Koonin, E. V. (2015). Archaeal clusters of orthologous genes (arCOGs): An update and application for analysis of shared features between *Thermococcales*, *Methanococcales*, and *Methanobacteriales*. *Life (Chicago, Ill.: 1978), 5*(1), 818–840. https://doi.org/10.3390/life5010818.

Maupin-Furlow, J. A. (2018). Vitamin B1 (thiamine) metabolism and regulation in archaea. In J. G. LeBlanc, & G. S. D. Giori (Eds.). *B group vitamins—Current uses and perspectives*. InTech. https://doi.org/10.5772/intechopen.77170.

Mendler, K., Chen, H., Parks, D. H., Lobb, B., Hug, L. A., & Doxey, A. C. (2019). AnnoTree: Visualization and exploration of a functionally annotated microbial tree of life. *Nucleic Acids Research, 47*(9), 4442–4448. https://doi.org/10.1093/nar/gkz246.

Miranda, H. V., Nembhard, N., Su, D., Hepowit, N., Krause, D. J., Pritz, J. R., ... Maupin-Furlow, J. A. (2011). E1- and ubiquitin-like proteins provide a direct link between protein conjugation and sulfur transfer in archaea. *Proceedings of the National Academy of Sciences, 108*(11), 4417–4422. https://doi.org/10.1073/pnas.1018151108.

Mistry, J., Chuguransky, S., Williams, L., Qureshi, M., Salazar, G. A., Sonnhammer, E. L. L., ... Bateman, A. (2021). Pfam: The protein families database in 2021. *Nucleic Acids Research, 49*(D1), D412–D419. https://doi.org/10.1093/nar/gkaa913.

Patenge, N., Haase, A., Bolhuis, H., & Oesterhelt, D. (2000). The gene for a halophilic β-galactosidase (*bga*H) of *Haloferax alicantei* as a reporter gene for promoter analyses in *Halobacterium salinarum*. *Molecular Microbiology, 36*(1), 105–113. https://doi.org/10.1046/j.1365-2958.2000.01831.x.

Pohl, M., Sprenger, G. A., & Müller, M. (2004). A new perspective on thiamine catalysis. *Current Opinion in Biotechnology, 15*(4), 335–342. https://doi.org/10.1016/j.copbio.2004.06.002.

Rawls, K. S., Yacovone, S. K., & Maupin-Furlow, J. A. (2010). GlpR represses fructose and glucose metabolic enzymes at the level of transcription in the haloarchaeon *Haloferax volcanii*. *Journal of Bacteriology, 192*(23), 6251–6260. https://doi.org/10.1128/JB.00827-10.

Robert, X., & Gouet, P. (2014). Deciphering key features in protein structures with the new ENDscript server. *Nucleic Acids Research, 42*(W1), W320–W324. https://doi.org/10.1093/nar/gku316.

Rodionov, D. A., Leyn, S. A., Li, X., & Rodionova, I. A. (2017). A novel transcriptional regulator related to thiamine phosphate synthase controls thiamine metabolism genes in archaea. *Journal of Bacteriology, 199*(4), e00743–16. https://doi.org/10.1128/JB.00743-16.

Sharma, V., Fedoseyenko, D., Joshi, S., Abdelwahed, S., & Begley, T. P. (2024). Phosphomethylpyrimidine synthase (ThiC): Trapping of five intermediates provides mechanistic insights on a complex radical cascade reaction in thiamin biosynthesis. *ACS Central Science, 10*(5), 988–1000. https://doi.org/10.1021/acscentsci.4c00125.

Sherwood, K. E., Cano, D. J., & Maupin-Furlow, J. A. (2009). Glycerol-mediated repression of glucose metabolism and glycerol kinase as the sole route of glycerol catabolism in the haloarchaeon *Haloferax volcanii*. *Journal of Bacteriology, 191*(13), 4307–4315. https://doi.org/10.1128/JB.00131-09.

Siddaramappa, S., Challacombe, J. F., DeCastro, R. E., Pfeiffer, F., Sastre, D. E., Giménez, M. I., ... Maupin-Furlow, J. A. (2012). A comparative genomics perspective on the genetic content of the alkaliphilic haloarchaeon *Natrialba magadii* ATCC 43099T. *BMC Genomics, 13*(1), 165. https://doi.org/10.1186/1471-2164-13-165.

Sudarsan, N., Barrick, J. E., & Breaker, R. R. (2003). Metabolite-binding RNA domains are present in the genes of eukaryotes. *RNA (New York, N. Y.), 9*(6), 644–647. https://doi.org/10.1261/rna.5090103.

Sun, J., Sigler, C. L., Beaudoin, G. A. W., Joshi, J., Patterson, J. A., Cho, K. H., ... Hanson, A. D. (2019). Parts-prospecting for a high-efficiency thiamin thiazole biosynthesis pathway. *Plant Physiology, 179*(3), 958–968. https://doi.org/10.1104/pp.18.01085.

Tazuya, K., Yamada, K., & Kumaoka, H. (1989). Incorporation of histidine into the pyrimidine moiety of thiamin in *Saccharomyces cerevisiae*. *Biochimica et Biophysica Acta (BBA) - General Subjects, 990*(1), 73–79. https://doi.org/10.1016/S0304-4165(89)80014-5.

Thompson, J. D., Higgins, D. G., & Gibson, T. J. (1994). CLUSTAL W: Improving the sensitivity of progressive multiple sequence alignment through sequence weighting, position-specific gap penalties and weight matrix choice. *Nucleic Acids Research, 22*(22), 4673–4680. https://doi.org/10.1093/nar/22.22.4673.

Zhang, X., Eser, B. E., Chanani, P. K., Begley, T. P., & Ealick, S. E. (2016). Structural basis for iron-mediated sulfur transfer in archael and yeast thiazole synthases. *Biochemistry, 55*(12), 1826–1838. https://doi.org/10.1021/acs.biochem.6b00030.

Zhou, G., Kowalczyk, D., Humbard, M. A., Rohatgi, S., & Maupin-Furlow, J. A. (2008). Proteasomal components required for cell growth and stress responses in the haloarchaeon *Haloferax volcanii*. *Journal of Bacteriology, 190*(24), 8096–8105. https://doi.org/10.1128/JB.01180-08.

CHAPTER SIX

Assessment of thiamine status in humans

Kerry S. Jones[a], Lucien Bettendorff[b], Elizabeth L. Frank[c], and Kyly C. Whitfield[d,*]

[a]Nutritional Biomarker Laboratory, Institute of Metabolic Science – Epidemiology, University of Cambridge, Cambridge, United Kingdom
[b]Laboratory of Neurophysiology, GIGA Institute, University of Liège, Liège, Belgium
[c]Department of Pathology, University of Utah Health, Salt Lake City, UT, United States
[d]Department of Applied Human Nutrition, Mount Saint Vincent University, Halifax, Nova Scotia, Canada
*Corresponding author. e-mail address: Kyly.Whitfield@msvu.ca

Contents

Abstract

Thiamine (vitamin B1) is an essential nutrient with established adverse health consequences associated with acute deficiency. Emerging evidence also suggests long-term neurocognitive deficits due to sub-clinical deficiency in early life, and there is a growing appreciation of a potential role for thiamine in Alzheimer's disease. Therefore, the assessment of thiamine biomarkers, particularly in large, population-based studies, is critical to grasp the true scope of this public health problem. The most reliable biomarkers of thiamine status are whole blood concentrations of thiamine diphosphate (ThDP) measured by high performance liquid chromatography (HPLC) with fluorescence detection or liquid chromatography tandem mass spectrometry (LC-MS/MS), and the functional assay of transketolase activity (erythrocyte transketolase activity coefficient, ETKAC). Plasma thiamine reflects recent intake, and is not a useful biomarker of

Methods in Enzymology, Volume 735
ISSN 0076-6879, https://doi.org/10.1016/bs.mie.2026.06.018

thiamine status. This manuscript describes the different biomarkers and methods of assessment, their advantages and disadvantages, and related cutoffs for deficiency. A table of 'performance characteristics' is provided to guide users on the quality and performance characteristics of each assay, and practicalities related to their cost, resource requirement, and maintenance. Provision of these insights will help inform decisions about inclusion of thiamine analysis in future work to better understand the extent of global thiamine deficiency.

1. Thiamine (vitamin B_1)

1.1 Introductory remarks

Thiamine (vitamin B1) is an essential water-soluble vitamin that has regained attention in recent years as a neglected nutrient of public health concern in low- and middle-income countries (LMIC) (Brown et al., 2025; Johnson et al., 2019). The diphosphorylated form of thiamine (thiamine diphosphate, ThDP) is a coenzyme in energy metabolism pathways for carbohydrates and branched-chain amino acids, as well as for neurotransmitter synthesis, so thiamine deficiency can cause reduced ATP synthesis, oxidative damage, and impaired neurological function (Whitfield et al., 2018). More recently, suboptimal thiamine intakes during infancy have been implicated in long-lasting neurocognitive delays, particularly in gross motor and language domains (Fattal-Valevski et al., 2005; Mimouni-Bloch et al., 2014). Given these vital roles in human health and development, accurate assessment of thiamine status is essential to spur clinical treatment among individuals, and to design public health interventions and policies at a population level.

Thiamine is required in relatively small quantities, with recommended nutrient intakes of 1.1 mg/day and 1.2 mg/day for women and men, respectively (World Health Organization and Food and Agriculture Organization of the United Nations, 2004). However, with limited body stores and a short half-life, routine intake of thiamine is necessary to prevent deficiency (Whitfield et al., 2018). Inadequate thiamine intakes are common among groups consuming limited, carbohydrate-rich diets, as much of the thiamine in cereals is removed in processing, and thiamine requirements increase with increasing carbohydrate intakes. Meat, particularly pork and organ meats, whole grains, and legumes are high in thiamine, while thiamine antagonists include betel nuts and tea, and thiaminase-containing raw fish, ferns, and African silkworm larvae (Whitfield et al., 2018). Excess thiamine intakes are not thought to be of concern because of rapid renal clearance of excess thiamine (World Health Organization and Food and Agriculture Organization of the United Nations, 2004).

1.2 Importance of assessing thiamine status

A recent modeling analysis estimated that 30% of the global population – or 2.2 billion people – have inadequate thiamine intakes (Passarelli et al., 2024). Thiamine deficiency has most commonly been associated with Southeast Asia, as beriberi was discovered in the Dutch East Indies (Indonesia) (Carpenter, 2000). While thiamine remains a nutrient of concern in the region, with recent reports of thiamine deficiency in Cambodia, Myanmar, and Laos PDR (Smith & Hess, 2021), there is growing evidence of thiamine deficiency beyond Southeast Asia, including reports from India (Hilal Ahmad et al., 2019; Koshy et al., 2018; Koshy et al., 2021), the Pacific Island nation of Kiribati (Green et al., 2021; Nilles et al., 2018), and throughout Africa (Bourassa et al., 2022; Johnson et al., 2019). Given the widespread concern about thiamine intakes throughout LMIC around the globe, there is a need for biological data to corroborate deficiency. At present, assessment of thiamine status is not often included in population-representative nutrition biomarker surveys, so there is a limited understanding of the global burden of thiamine deficiency (Brown et al., 2025). Exceptions are the United Kingdom National Diet and Nutrition Survey (NDNS) (Koulman et al., 2026), and recent surveys in Madagascar and Nigeria (Federal Government of Nigeria FGoN and the International Institute of Tropical Agriculture IITA, 2024; Rohner et al., 2026).

1.3 At-risk populations for thiamine deficiency

Thiamine deficiency manifests with several distinct clinical profiles (Smith et al., 2021). Historically, deficiency has been categorized by the type of beriberi: wet beriberi (cardiac presentation), dry beriberi (neurological symptoms), shoshin beriberi (rapid, acute clinical presentation), and infantile beriberi, or as Wernicke-Korsakoff syndrome with presentation among individuals with alcohol use disorder. Given the broad and overlapping symptoms of deficiency, 'thiamine deficiency disorders' (TDD) was coined to capture the broad spectrum of adverse outcomes (Whitfield et al., 2018).

Thiamine deficiency is generally uncommon in high-income settings, except among certain clinical populations (Marrs & Lonsdale, 2021; Rosewarne & Farina, 2026). Individuals with alcohol use disorders are the most common risk group for thiamine deficiency, which presents as

Wernicke-Korsakoff syndrome (Frank, 2015). Excessive alcohol consumption can impair thiamine absorption at a time of heightened thiamine requirements and often limited dietary thiamine intake (Gomes et al., 2021). Post-operative bariatric surgery patients are at risk due to limited nutrient intakes and/or vitamin malabsorption in cases of bypass or removal of the proximal small intestine (Frank, 2015). Finally, thiamine is vital in refeeding, including in cases of disordered eating as well as in the treatment of severe acute malnutrition (Hiffler et al., 2017; O'Brien et al., 2025). As such, the ability to measure thiamine status is important in clinical settings.

In LMIC settings, a monotonous diet reliant on a refined, non-fortified carbohydrate staple food is the greatest predictor of low thiamine status (Whitfield et al., 2018). Thiamine requirements increase with higher physical activity, and during periods of high metabolism, so groups such as pregnant/lactating women and labourers are at higher risk of deficiency. For instance, an outbreak of beriberi was reported among otherwise healthy young men working as illegal gold miners in French Guiana (Mosnier et al., 2017); here and in similar outbreaks, thiamine deficiency was explained by overall low dietary intake, a reliance on washed, polished rice, and heavy physical exertion (de Montmollin et al., 2002; Doung-ngern et al., 2007).

While thiamine deficiency is concerning at any lifestage, a major focus of recent research has been on infants (Gallant et al., 2021; Smith et al., 2024), given their comparatively high requirements relative to small body size and low food intakes (Fattal-Valevski, 2011; Whitfield et al., 2018). Infantile thiamine deficiency normally presents in the first six months, which is the exclusive breastfeeding period when the only source of thiamine is mother's milk (Luxemburger et al., 2003; Smith et al., 2024). The symptoms are fairly generic, including tachypnea, tachycardia, dysphonia, edema, cyanosis, colic, and vomiting (Smith et al., 2024), so infants may die before medical treatment is ever sought. Lactating mothers should be targeted to prevent infantile beriberi, as women with low thiamine intakes produce milk low in thiamine (Dror & Allen, 2018; Whitfield et al., 2018).

Infantile beriberi can be fatal, but even deficiency not severe enough to cause clinical presentation may have severe and long-lasting consequences. An unfortunate 'natural experiment' among Israeli infants resulted from a manufacturing error that erroneously excluded thiamine from infant formula (Fattal-Valevski et al., 2005). These otherwise well-nourished infants had low thiamine intakes in early life. Although many of these infants did not develop clinical beriberi symptoms, follow-ups through childhood highlighted persistent language and motor delays compared to their peers (Fattal et al., 2011; Harel et al., 2017; Mimouni-Bloch et al., 2014). This recent evidence suggests that sub-clinical thiamine deficiency may put millions of infants at risk of neurocognitive deficits, undercutting life-long well-being. Similarly, there is growing interest in the role of thiamine in dementia and Alzheimer's disease (Gibson et al., 1988; 2016).

Given the serious adverse outcomes of thiamine deficiency, and large geographic footprint of LMIC at risk of low dietary thiamine intakes, the assessment of thiamine biomarkers, particularly in large, population-based studies, is critical to grasp the true scope of this public health problem. Below we will highlight the most common thiamine assessment methods for blood, providing insights to help inform decisions about inclusion of thiamine analysis in future work.

2. Methods for thiamine assessment in human blood samples

2.1 Introductory remarks

Detection of thiamine and its vitamers in body fluids requires very sensitive methods. For instance, ThDP concentration in whole human blood is approximately 100 nM. Thiamine and thiamine monophosphate (ThMP) concentrations are approximately one order of magnitude lower. Typically, thiamine vitamers in biological samples can be separated by high performance liquid chromatography (HPLC) and detected either as fluorescent thiochrome derivatives or by mass spectrometry. Indeed, during the last two decades, powerful liquid chromatography tandem mass spectrometric (LC-MS/MS) methods were developed, allowing the detection of all thiamine derivatives and provitamins (Kim et al., 2019; Sambon et al., 2020). These methods, while more sophisticated than fluorescence-based HPLC methods, also require skilled technicians and the instrumentation is highly expensive.

A third common method exists, a functional assay for thiamine status assessment, the erythrocyte transketolase activity coefficient (ETKAC). Unlike HPLC and LC-MS/MS methods that report a direct assessment of thiamine concentrations in blood, this functional biomarker assesses the degree of saturation of the ThDP-dependent enzyme transketolase, providing insights into thiamine status of the individual (Jones et al., 2021).

Many variations of the general methods described here have been developed for a rapid, cost-effective, and routine determination of specific vitamers for the estimation of the thiamine status in humans (Vachali et al., 2025). Hence the choice of the method depends on factors such as the derivative to focus on, cost, and laboratory environment. Each of the methods is summarized below, alongside Table 1, where these methods are compared using numerous performance and logistical considerations.

2.2 Specimen requirements for HPLC and LC-MS/MS

Whole blood specimens collected using heparin or ethylenediaminetetraacetic acid (EDTA) as anticoagulants are suitable for direct measurement using chromatographic methods (HPLC, LC-MS/MS). Thiamine is photosensitive and specimens must be protected from light during storage and transport to the laboratory. Samples should be frozen at −70 °C to ensure complete lysis of erythrocytes. Samples frozen at −20 °C may yield falsely low thiamine concentrations. Plasma thiamine concentrations reflect recent intakes rather than thiamine nutritional status.

2.3 HPLC

Measurement of thiamine and its derivatives by HPLC involves separation by liquid chromatography typically achieved with a C18 chromatography column (Basiri et al., 2016; Lu & Frank, 2008; Zhang et al., 2017), followed by fluorescence detection.

2.3.1 Derivatization

UV detection of thiamine is neither sensitive nor selective enough, and is only useful in relatively pure pharmacological preparations. The most sensitive HPLC methods are based on derivatization of thiamine vitamers in fluorescent thiochrome derivatives (Fig. 1) followed by HPLC separation and fluorescence detection.

Various HPLC techniques have been described (Basiri et al., 2016; Bettendorff et al., 1991; Lu & Frank, 2008).

Typically, thiamine derivatives are oxidized by potassium ferricyanide in a strongly alkaline medium, hence the importance of neutralization to pH > 5 of the sample extract. Moreover, the formation of a thiochrome requires an intact thiazolium ring. Precolumn and post-column derivatization procedures have been described. According to our experience precolumn derivatization is easier, but in that case thiochrome derivatives

Table 1 Selected performance and use characteristics and minimal usage requirements for recommended assay methods for measuring thiamine status in human blood.

Performance and usage characteristics	HPLC	LC-MS/MS	ETK activity assay
1. **Assessment tool format and operator** *Minimal criteria:* Tool is operable by a laboratorian with specialist training. *Optimal criteria:* Tool is a benchtop analyzer that can be operated by a laboratorian with basic training on a laptop computer.	HPLC and fluorescence detector required in temperature-controlled lab. Instrument software required and substantial training	Instrumentation requires a temperature-controlled environment, electricity, gases, and a vacuum pump and trained laboratory technician	Requires a benchtop ELISA plate reader with integrated incubator, associated software and a trained laboratory technician. Not all plate readers are suitable due to uneven heating across the plate. A procedure to check plate reader suitability has been published (Jones et al., 2021)
2. **Software requirements** for assay data reduction and results management *Minimal criteria:* Results can be generated for multiple samples at a time, and the software does not have data security restrictions that would otherwise challenge exporting results in batch. *Optimal criteria:* As a minimum, and additionally, the software does not require the user to purchase a license.	HPLC software is required (often with a user license) to run samples and process data.	Software for instrument operation and data processing is required. A user license may be necessary.	Standard ELISA plate reader software is required to run the assay. All calculations can be performed with Microsoft Excel or other spreadsheet software

(continued)

Table 1 Selected performance and use characteristics and minimal usage requirements for recommended assay methods for measuring thiamine status in human blood. (*cont'd*)

Performance and usage characteristics	HPLC	LC-MS/MS	ETK activity assay
3. **Reagent stability** *Minimal criteria:* Long-term (6 months) storage, requiring refrigeration Optimal criteria: No refrigeration	Reagents require refrigeration and calibrator storage at −80 °C. A full assay kit can be purchased from Chromsystems	Reagents must be refrigerated. Calibrators, prepared controls, and internal standard solutions may be lyophilized and frozen at −80 °C.	Reagents require −20 °C freezer
4. Internal quality control *Minimal criteria:* Standard curves fit better than 0.95; sample controls; positive/negative controls required to monitor the quality of the kit, including low and high controls; calibration performed using appropriate reference method if available (e.g., monoplex immunoassay, HPLC, mass spectrometry). Optimal criteria: Same as minimum	In-house controls prepared from donor materials or Chromsystems controls	In-house controls prepared from donor materials or Chromsystems controls	In-house controls prepared from whole blood
a. External quality control *Minimal criteria:* Values pass performance targets set by external performance verification scheme or values for reference material are within 10% of target values. *Optimal criteria*: Same as minimum	INSTAND EQA scheme and RCPA quality assurance program have ThDP in whole blood. No standard reference material in whole blood is available	External proficiency testing is available from providers such as INSTAND and RCPA.	No external quality control, quality assessment scheme, or reference materials

b. Specimen type *Minimal criteria*: Venipuncture *Optimal criteria:* Capillary collection	Whole blood or red blood cells	Whole blood or red blood cells	Standard assay for which status cut-offs exist requires washed red blood cells. Whole blood and dried blood spots can be used but associated cut-offs have not been defined and agreement between sample types has not been established.
5. Analytical specificity *Minimal criteria:* 95% (tested compounds should not give a response of more than 5% of the value expected for the analyte of interest). *Optimal criteria*: Same as minimum	Good chromatographic separation of thiamine, ThMP, and ThDP. Well-maintained equipment allows confidence in analytes based on retention time and peak shape.	Good chromatographic separation of thiamine, ThMP and ThDP. Chromatographic elution without ion suppression. Good mass resolution.	Enzyme activity specific to ThDP-dependent transketolase
6. Precision (intra-assay variability) *Minimal criteria*: CV < 10% *Optimal criteria*: Same as minimum	<3% for ThDP, <2% for thiamine	<3% ThDP	Duplicate imprecision should not exceed 10%

(continued)

Table 1 Selected performance and use characteristics and minimal usage requirements for recommended assay methods for measuring thiamine status in human blood. (*cont'd*)

Performance and usage characteristics	HPLC	LC-MS/MS	ETK activity assay
7. Reproducibility (inter-assay variability) *Minimal criteria*: CV < 15% CV < 25% at upper and lower limits of quantification (LOQ) LOQ and reportable range are reported. *Optimal criteria*: CV < 10% LOQ and reportable range are reported.	ThDP and thiamine LOQ 2 nmol/L. Inter-assay variability: ThDP <8.5%; Thiamine <15%. Reporting range: ThDP 2 – 400 nmol/L; thiamine 2 – 20 nmol/L	ThDP < 10% if > 150 nmol/L; + /−15 nmol/L <150 nmol/L (RCPA criteria)	When monitoring the ETKac ratios of QC material CV should be <5%
8. Linearity (i.e., not of a standard curve, which is covered under internal QC, but of the analyte in the native matrix) *Minimal criteria:* r = 0.95 *Optimal criteria:* Same as minimum	Chromsystems lyophilized whole blood L2 diluted with Chromsystems lyophilized whole blood L1, $R^2 > 0.99$	Reporting range ThDP 20−1000 nmol/L	N/A
Minimal usage requirements	**HPLC**	**LC-MS/MS**	**ETK activity assay**
1. Power and water requirements Description: Does the instrument require uninterrupted daytime power or are power requirements minimal? Is access to de-ionized water required?	Uninterrupted electric power. Deionized water is required for reagent preparation unless purchasing kits from Chromsystems where all reagents are supplied.	Uninterrupted electric power. Deionized water. Gas supply (tanked or generator).	The plate reader requires uninterrupted power when in use. Access to deionized water is required for the preparation of the TRIS buffer.

2. Maintenance requirements and supply, services, and support Description: (1) How frequently does the instrument require maintenance (in relative terms, i.e., as compared with other assays for the same biomarker) and what is the cost of parts (in relative terms)? (2) What is the availability of servicing and support (i.e., can it be ordered and supplied directly by a manufacturer with global distribution, and is support offered via a network of global distributors)?	HPLCs require yearly preventive maintenance. Servicing and support are available from suppliers and third-party companies. $$	Annual or semi-annual preventive maintenance. Service and as-needed repairs require trained service providers. $$$$	Most plate readers will require a yearly service . Global distributors of plate readers are available $
3. Waste management (hazardous materials/chemicals) Description: What hazardous materials/chemicals are required? What waste management procedures are required as per WHO standards?	Methanol; Toxic by skin contact and inhalation. Highly flammable Orthophosphoric acid; Causes severe skin burns and eye damage. Trichloroacetic Acid; Causes severe skin burns and eye damage. May cause respiratory irritation. Very toxic to aquatic life. Sodium Hydroxide; Corrosive, Exothermic reaction when added to water to create 15% solution.	Organic reagents and solvents, including methanol and acids (formic, hydrochloric, phosphoric, trichloroacetic). Approved waste management and disposal system required.	Hydrochloric Acid. Corrosive. Should be disposed of by appropriate routes.

(continued)

Table 1 Selected performance and use characteristics and minimal usage requirements for recommended assay methods for measuring thiamine status in human blood. (*cont'd*)

Performance and usage characteristics	HPLC	LC-MS/MS	ETK activity assay
	All should be disposed of by appropriate routes, by a waste management company that removes chemical and solvent waste.		
4. Ancillary equipment and supplies Description: Determined by the method: e.g. plate washer, pipettes and tips, plate reader, and ancillary supplies necessary for the performance of a lab-based immunoassay. **Important note for both HPLC and LC-MS/MS methods:** derivatization in fluorescent thiochrome diphosphate derivative further requires sodium hydroxide and potassium ferricyanide as well as fluorescence spectrophotometer (more expensive than a spectrophotometer). It must be fluorescence spectrophotometer and	HPLC column: Agilent Poroshell 120 EC-C18 3.0 × 50 mm, 2.7 μm Centrifuge capable of being maintained at 4 °C Agilent Captiva 96 well lipid removal plates 2 mL polypropylene 96 well plates 12 × 75 mm borosilicate glass tubes 12 mL polypropylene tubes 2 mL cryovials Analytical balance PH meter Vortex mixer Multitube vortex mixer 200 μL Adjustable Pipette with tips 1 mL Adjustable Pipette with	Labware: Centrifuge, shaker, vortex mixer, pipettes and tips. Refrigerator, freezer. Supplies: Analytical column, 96-well plates and sealing mats, glass vials and caps	Multiskan™ FC microplate photometer with incubator Orbital microplate shaker with digital control Incubator with fan Refrigerated centrifuge Greiner UV-Star 96 well plates Microplate sealers Positive displacement pipette (50 μl – 200 μl) with disposable tips P20, P200, P1000 and P5000 Pipettes with tips 1 mL Pastettes Sarstedt 2 mL microtubes Falcon 5 mL, 15 mL and 50 mL tubes Roller mixer

not a simpler filter fluorometer which lacks selectivity.	filter tips 5 mL Adjustable Pipette with tips Eppendorf repeating Pipette with tips Volumetric flasks and measuring cylinders of various sizes Positive pressure processor or vacuum pump for 96 well plates Fluorescence spectrophotometer		
5. End-user price Description: In relative terms, i.e., as compared with other analytical methods for the same biomarker: (1) What is the capital equipment/ infrastructure costs? (2) What is the cost per assay (considering consumables as well as person time)?	Capital: $$$$ Per assay costs: $$	Capital: $$$$$$$$$$$$$$$ Per assay costs: $$$	Capital: $ Per assay cost: $

Footnote: The ETK assay requires less expensive equipment (microplate reader and photometer) than other methods, a small sample volume, and does not require calibration, but is only available in a small number of laboratories. Although not a direct measure of ThDP, similar functional assays are used for other nutrients, e.g. the erythrocyte glutathione reductase activity coefficient is used to assess vitamin B2 status (riboflavin). Commonly used cut-offs are based on measurements in washed red blood cells. Constant temperature over the whole plate is required as is an experienced analyst. HPLC is more expensive, no suitable internal standards exist, chromatographic separation is required as is a larger sample volume (>300 microliters), and high-throughput is challenging. A more robust laboratory infrastructure is also required. LC-MS/MS provides advantages in sensitivity and specificity and isotopically-labeled internal standards are available, but it is very expensive and methods can be difficult to set up. External proficiency testing schemes are available (e.g. RCPA and INSTAND), the use of which should be widely encouraged.

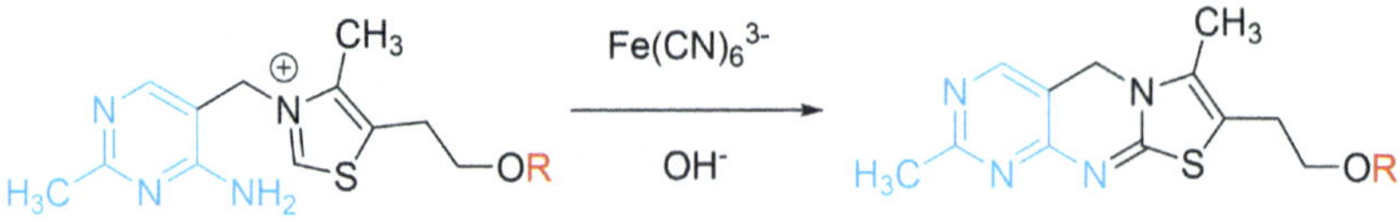

Fig. 1 Oxidation of thiamine to the fluorescent thiochrome in alkaline medium. The amino function of the pyrimidine ring and the thiazolium cycle, essential for cyclization, are drawn in blue. R = H, PO_3^{2-} (ThMP) or $P_2O_6^{3-}$ (ThDP).

are separated on the analytical column. As they are fluorescent only under alkaline conditions, an alkali-resistant column must be chosen.

For post-column derivatization, thiamine derivatives are separated on the column, but a further pump is required for injecting the oxidant into a mixing chamber with the column efflux. We found that post-column derivatization was more cumbersome and more subject to system failure than precolumn derivatization, mainly because of the high content of sodium hydroxide in the oxidant solution. A spectrofluorometer is to be preferred over a filter fluorometer because of a much better selectivity. Hence, some fluorescence-based HPLC methods are able to simultaneously determine thiamine, ThMP, ThDP, thiamine triphosphate, and adenosine triphosphate in human tissues (Gangolf et al., 2010).

As thiamine derivatives are present in sample solution at very low, sub-micromolar, concentrations, suitable standard solutions must be prepared in the same concentration range and particular attention must be paid for the material used for preparing the sample and further processing in order to avoid absorption of the material. According to our experience, polypropylene was neutral in that sense, but glass should be avoided.

Furthermore, recent data suggest that a non-negligible portion of ThDP can exist in an oxidized form in animal tissues (Pavlova et al., 2021), as well as in humans exposed to high doses of radiation. Indeed, one study found that ThDP was predominantly in an oxidized form in the blood of Chernobyl accident liquidators (Donchenko et al., 1997). Such forms, likely open-ring thiazole disulfide derivatives, require specific detection methods. Presently, both the biological role of these derivatives and their potential significance in assessing human thiamine status remain unknown.

2.3.2 Extraction

Many procedures are described in the literature for the extraction of thiamine and its vitamers from biological samples. As thiamine as well as all its natural vitamers are water-soluble, extraction can easily be realized by homogenization of the tissues in trichloroacetic acid (TCA) at a final concentration of 10 - 12% (with less, thiamine extraction may only be partial and with more acid, its elimination becomes challenging). TCA can easily be extracted with 3 times 3 volumes of water-saturated diethyl ether (preferably HPLC grade without stabilizer, which may interfere with fluorometric determination).

2.4 LC-MS/MS

From Fig. 1 it is evident that only thiamine derivatives with an intact thiazolium ring can be transformed into a thiochrome derivative. Therefore, several methods based on mass spectrometry detection have been developed (Kim et al., 2019; Verstraete et al., 2020). These techniques have the advantage of allowing the detection of derivatives with an open thiazolium ring that cannot be transformed into a fluorescent thiochrome (Sambon et al., 2019, 2020). However, mass spectrometric detection can also be used for the detection of conventional thiamine derivatives (i.e. thiamine, ThDP). Mass spectrometry methods are sensitive and selective. The complex sample preparation processes required for HPLC assays are not necessary, which shortens analysis time, and isotope-labelled compounds are available as internal standards to compensate for variability introduced during extraction and analysis.

2.4.1 Extraction

Frozen specimens are thawed at ambient temperature for two hours to ensure optimum recovery of ThDP. LC-MS/MS allows use of a stable-labelled internal standard, such as ThDP-d_3. Sample preparation can be as simple as protein precipitation with TCA followed by mixing and centrifugation prior to analysis (Vachali et al., 2025).

2.5 ETKAC

The cytosolic transketolase catalyzes two essential reactions of the non-oxidative (reversible) branch of the pentose phosphate pathway. It is essential for recycling ribulose 5-phosphate (precursor of ribose 5-phosphate) towards the glycolytic pathway (Li et al., 2019). Being central to metabolism, transketolase is present in all cells, including erythrocytes, providing an accessible tissue for the study of thiamine status.

Activity of the transketolase enzyme in erythrocytes has been used to assess thiamine status since the 1960s (Brin, 1962). Erythrocytes cannot synthesize proteins thus the transketolase levels represent protein production during precursor cell synthesis (Edwards, 2026). In the assay, transketolase activity is therefore largely determined by the availability of ThDP. The essence of the assay is simultaneous or consecutive measurements of transketolase activity in the absence of exogenous ThDP (basal activity rate) and with the addition of exogenous ThDP (activated rate). From the calculation of the 'activity coefficient' or the ratio of these two measurements, the erythrocyte transketolase activity coefficient (ETKAC) is derived. An ETKAC value near 1 indicates almost complete saturation of the transketolase enzyme and suggests functional thiamine sufficiency. A higher ratio indicates the degree of activation of the transketolase enzyme through the addition of exogenous ThDP. The greater the ratio, the greater the indicated deficit in ThDP. The assay is typically performed in erythrocytes washed in saline to remove white blood cells.

Early forms of the assay used colorimetric changes to determine transketolase activity whereas current assays monitor the rate of NADH oxidation by spectrophotometry (Edwards, 2026; Jones et al., 2021). Methods have evolved over the decades from test tube based, to automated clinical chemistry analyzers, with more recent methods utilizing a 96-well microplate format for convenience and high throughput (Jones et al., 2021). In the assay, transketolase catalyzes the metabolism of two reactions in the pentose phosphate pathway. The resultant glyceraldehyde 3-phosphate is metabolized to dihydroxyacetone phosphate, which in the final reaction is reduced to glycerol−3-phosphate by glycerol−3-phosphate dehydrogenase and NADH (Edwards, 2026; Jones et al., 2021).

A protocol published with the aim to standardize the transketolase activity assay has been used in a several studies, randomized controlled trials (Bourassa et al., 2022; Gallant et al., 2021; Hess et al., 2024; Koulman et al., 2026) and nutritional surveys of micronutrient status (Federal Government of Nigeria FGoN and the International Institute of Tropical Agriculture IITA, 2024; Rohner et al., 2026). Measurements are performed with a standard microplate reader available in many laboratories. This protocol has also been adopted and applied in other laboratories (Bandyopadhyay et al., 2025; Hauber et al., 2026; Therrien et al., 2026). The transketolase assay has also been applied to other tissues and species (Gibson et al., 1988; Giguère & Butterworth, 1987; Hauber et al., 2026; Therrien et al., 2026; Zhao et al., 2009) however, as recently discussed, consideration must be given to the assay conditions specific to each application beyond erythrocytes (Edwards, 2026).

The primary outcome of the assay is the ratio of activated to basal transketolase activity (ETKAC). Commonly, assay results are expressed as a percentage, typically known as the TPP effect. Results may also be expressed as the latency (the inactivated fraction of transketolase) (Edwards, 2026; Jones et al., 2021).

The activity coefficient, TPP effect, and latency all describe the relative change in enzyme activity with and without exogenous ThDP. Whilst this is an advantage in terms of the robustness of the assay and minimizes the impact of differences in the amount of enzyme it may become a limitation where differences in enzyme and/or tissue are systematically different between test groups. Where the apoenzyme is lower, as has been hypothesized to occur in long-term thiamine deficiency, less ThDP is required to stimulate the transketolase and therefore a ratio and TPP effect in the normal range may be observed. In this scenario, measurement of the absolute basal activity can provide additional information. However, further measurements and calculations are required to determine the absolute basal activity in units of enzyme activity, as well as measurements of either hemoglobin or protein in the tested sample. A protocol for the calculation of basal activity has been published (Jones et al., 2023), and a protocol for expressing brain basal transketolase activity normalized to total protein is included in this volume (Parkington et al., 2026).

2.6 Decision making tips for method selection

The primary aim of this manuscript is to compare the different assays to assess thiamine status and to guide users on the most appropriate assay for a given setting, whether primarily research or clinical. The elements in Table 1 provide guidance on the consumables and equipment for specific assays, their relative costs, the expertise associated with each, and other considerations such as the types of waste produced by each approach. Availability of equipment and staff, and accessibility of additional specialist support will clearly play a considerable role in determining the most feasible approach.

Three methods have been described, HPLC, LC-MS/MS and transketolase activity that align with current available and recommended assays (Brown et al., 2021). As alluded to above, there are primarily two types of micronutrient biomarker that provide information on nutritional status; those based on concentration and those related to metabolic function. This distinction is important and it is not the case that the two approaches necessarily give complete agreement. ThDP concentration measured by

HPLC or LC-MS/MS as described indicates the level of this form of thiamine circulating in the blood, whether metabolically available and active, or not. In contrast, the transketolase assays (both ETKAC and the basal transketolase activity) provide information on the functional capacity of the ThDP-dependent enzyme at the tissue level. Each status biomarker has been demonstrated to be responsive to thiamine depletion and supplementation (Baines & Davies, 1988; Brin, 1964; Gallant et al., 2021; Haro et al., 1966; Wood et al., 1980). Studies assessing agreement between them have reported mixed results, partly due to the challenges of analytical variability in both biomarkers and because measurements conducted in groups with little or no thiamine deficiency (Bailey et al., 1994; Baines & Davies, 1988). Whilst some studies observed no relationship between biomarkers (Bailey et al., 1994), others have found associations between ThDP and basal transketolase activity and/or ETKAC (Baines & Davies, 1988; Fidanza et al., 1989; Talwar et al., 2000). In terms of classification of thiamine deficiency, ThDP and basal transketolase activity performed better than between ThDP and ETKAC (Baines & Davies, 1988).

There are no-point-of-care or field-friendly methods to assess thiamine status, so in common to each of the three approaches are the requirement for laboratory facilities with reliable electricity and water supplies, standard computing resources, and support for waste management streams to manage biological and chemical waste. Several groups have explored the use of dried blood spots (and micro sampling devices) as an alternate sample collection method to venous blood draws to address some of the challenges such as cold chain management, which can be particularly challenging in LMIC (Huang et al., 2020; Mathew et al., 2019; Verstraete & Stove, 2021a, 2021b). However, dried blood spot methods still ultimately rely on HPLC or LC-MS/MS for analysis, so the considerations in Table 1 still apply.

The major difference between the methods are the capital outlay and the cost of servicing and maintenance, in particular for LC-MS/MS which would preclude this approach for many laboratories. In addition, liquid chromatography for either LC-MS/MS or HPLC requires reliable supplies of solvents. However, LC-MS/MS and to a lesser extent HPLC in general provide extremely versatile bioanalytical platforms. HPLC and LC-MS/MS platforms are relatively common in clinical settings in higher-income countries and therefore setting up of chromatography-based ThDP assays may be a more viable option in such settings. In contrast, the measurement of transketolase activity requires relatively little specialist equipment or servicing. However, this approach does require specialist chemicals, albeit in relatively small quantities.

As discussed above, ThDP is primarily found in erythrocytes rather than plasma, thus either whole blood or erythrocytes are required for analysis. Whilst for LC-MS/MS and HPLC, whole blood can be frozen for subsequent analysis, for the transketolase assays, washed erythrocytes are the preferred matrix to ensure the absence of white blood cells which, being very metabolically active, can contain high levels of transketolase enzyme (Buttery & Pannall, 1985; Smeets et al., 1971). This requirement for washed cells can be challenging in situations where blood samples are collected in the field, nevertheless it is possible as demonstrated by a number of recent studies in LMIC (Bourassa et al., 2022; Federal Government of Nigeria FGoN and the International Institute of Tropical Agriculture IITA, 2024; Gallant et al., 2021; Hess et al., 2024; Rohner et al., 2026). No reference method is available for the determination of ThDP in blood. Consequently, there are no accuracy-based external quality assessment (EQA) schemes. However, EQA schemes from the Royal College of Australian Pathologists Quality Assurance Programs (RCPAQAP) and INSTAND (Institut für Standardisierung und Dokumentation im medizinischen Laboratorium) are available, providing targets based on laboratory means for comparison of ThDP in whole blood. However, few laboratories measure or subscribe to these schemes and results can be dominated by commercially available assays. Such schemes are useful to understand between-laboratory variability and its causes (Hoad et al., 2013). Unfortunately, the situation is no better for transketolase assays where no EQA scheme exists. As more laboratories renew interest in thiamine assessment using this approach and published methods allow harmonization, opportunities may arise for assessment of interlaboratory comparisons and future standardisation. Lack of harmonization between methods and laboratories may also contribute to the difficultly and uncertainty surrounding cutoffs for thiamine deficiency as discussed in ***Section 3***.

As discussed at the start of this section, in most scenarios local expertise and resources will determine the feasibility of methodologies with a laboratory. More broadly, studies may rely upon international collaborators to measure thiamine status and then the decision of which assays to use may become more complex. In terms of cost, for the reasons described above, transketolase assays are likely to be less expensive than those using LC-MS/MS, and HPLC an intermediate option. Ideally, to build an evidence base, studies may choose to use both concentration and functional biomarkers. Previously published or locally available data may already be available for comparative purposes and may provide an incentive to use a particular method. ThDP concentration is a static measurement of thiamine status. On the other hand, and as has already been described, measurement of the transketolase activity provides a functional biomarker of thiamine status incorporating not only the availability of ThDP but also other factors such as enzyme availability and other cofactors (Edwards, 2026), some of which can be captured by the more detailed assessment of basal transketolase activity. The converse of this holistic functional biomarker is that many other factors, in addition to thiamine, may also impact the results, ranging from physiological, genetic and nutritional factors to methodological and data-related factors (Edwards, 2026), although these are somewhat negated by the presentation of relative (ETKAC) rather than absolute results.

This manuscript has focused on blood thiamine status as a marker of population status and in relation to clinical deficiency. However, thiamine is also measured in other tissues, particularly the brain. Both chromatographic methods and the transketolase assay can be applied to tissues (Gangolf et al., 2010; Gibson et al., 1988; Klooster et al., 2013; Parkington et al., 2026). Measurements by HPLC showed that the distribution and concentration of thiamine forms vary significantly between human tissues (Gangolf et al., 2010).

3. Interpretative criteria

Whilst it is well known that thiamine deficiency leads to the conditions described in ***Section 1*** above, there are other metabolic, physiologic and nutritional factors that have largely prevented the determination of cutoff values that can be used diagnostically. Of the methods shared here, ETKAC has the most agreed-upon cutoffs, with consortium groups commonly using ETKAC > 1.25 to define risk of thiamine deficiency, 1.15–1.25 to define marginal status, and ETKAC < 1.15 to define thiamine sufficiency (Institute of Medicine, 1998; Turck et al., 2016). But still, the evidence-base for these cutoffs is relatively small and largely derives from a handful of experimental studies and nutritional surveys conducted in the 1960s (Sauberlich, 1967).

There is no consensus around cutoffs for ThDP in blood. ThDP measurements by HPLC and LC-MS/MS can be made in erythrocytes (eThDP) or whole blood, the latter negating the requirement for blood processing, but the results may require correction for hemoglobin or hematocrit (Caballero et al., 2025; Evliyaoglu et al., 2019; Talwar et al., 2000). Using nationally-representative data from Cambodian Micronutrient Survey (linked with Cambodian Demographic and Health Survey 2014) where eThDP was measured on HPLC, Whitfield et al. reported status using numerous cutoffs from the literature (Whitfield et al., 2017). Depending on the cutoff employed, the prevalence of thiamine deficiency / low thiamine status among women of reproductive age ($n = 719$) ranged from 27 % (eThDP < 120 nmol/L) to 78 % (eThDP < 180 nmol/L) (Whitfield et al., 2017). The implications of the lack of cutoffs is obvious, as the urgency of action and subsequent policy solutions would differ whether one quarter versus three quarters of a population group were deficient.

Ideally, thiamine concentration cutoffs should be clinically meaningful, aligning with actual risk of clinical symptoms. A recent prospective cohort study in Lao PDR attempted to do exactly this: Hess et al. developed a clinical case definition for thiamine-responsive disorders among infants presenting to hospital, with blood samples taken before and after children were treated with supplemental thiamine (Hess et al., 2024). Those with favorable improvements in thiamine deficiency-related clinical symptoms were deemed thiamine-responsive, and hence would have been thiamine deficient upon hospital admission. Unfortunately, authors reported considerable overlap in ThDP concentrations between thiamine-responsive and non-responsive infants, so low sensitivity and specificity hindered their ability to set clinically predictive and meaningful cutoffs (Hess et al., 2024).

In the absence of cutoffs, reference ranges are often used to estimate risk of deficiency (Marrs & Lonsdale, 2021). However, as Wilkinson et al. wrote of their use of reference ranges, "the lower limit of normal, seen in a healthy population, cannot be assumed to be the upper limit of abnormal" (Wilkinson et al., 1997).

4. Summary and conclusions

The measurement of thiamine biomarkers is critical for assessing individual and population thiamine status, and for understanding links between thiamine and health and disease. In this manuscript we describe currently available biomarkers and summarize their advantages and disadvantages. Despite the well-established dietary recommendations for thiamine, the risk of deficiency persists in many populations across the world. Emerging evidence of long-term neurocognitive deficits due to sub-clinical deficiency and a growing appreciation of a potential role for thiamine in the prevention and treatment of Alzheimers's disease add to the need for reliable and well-characterized thiamine biomarkers.

The most reliable biomarkers of thiamine status are whole blood concentrations of ThDP measured by HPLC or LC-MS/MS, and the functional assay of transketolase activity. Plasma thiamine concentration is not a useful biomarker of thiamine status. The transketolase assay benefits from less expensive equipment and maintenance than other methods, and established cut-offs for deficiency and insufficiency. Efforts have been made to standardize the assay between laboratories, but cross-laboratory comparisons and materials for external quality control are lacking. HPLC is more expensive, a larger sample volume is required, and no suitable internal standards exist. LC-MS/MS provides advantages in sensitivity and specificity and isotopically-labeled internal standards are available, but equipment is expensive to purchase and maintain. External proficiency testing schemes are available and their use is encouraged to help understand differences between assays and laboratories. This is essential if progress is to be made on harmonizing cutoffs for whole blood ThDP concentration, which is currently a major barrier to use of ThDP as a meaningful biomarker for health outcomes. However, in practice, the availability of equipment and technical staff are likely to be strong determining factors in the choice of assay to assess thiamine status.

Acknowledgements

Authors wish to thank Warangkana Srichamnong, Sabri Bromage, Chawanphat Muangnoi, Pattamaporn Joompa, Sasiumphai Purttiponthanee, and other secretariat leaders of the Expert Consultation for 'Building capacity for joint measurement of a package of key biomarkers for routine surveillance of nutrient adequacy and nutrition-related non-communicable disease risk in low- and middle-income countries' (Institute of Nutrition, Mahidol University, Thailand; funded by Rockefeller Foundation), for convening these authors and drafting a template of Table 1.

KSJ - This work was supported by the National Institute of Health and Care Research (NIHR) Cambridge Biomedical Research Centre (NIHR2033312). The views expressed are those of the authors and not those of the NIHR or DHSC.

Author's Note:

References

Bailey, A. L., Finglas, P. M., Wright, A. J., & Southon, S. (1994). Thiamin intake, erythrocyte transketolase (EC 2.2.1.1) activity and total erythrocyte thiamin in adolescents. *The British Journal of Nutrition, 72*(1), 111–125. https://doi.org/10.1079/bjn19940014.

Baines, M., & Davies, G. (1988). The evaluation of erythrocyte thiamin diphosphate as an indicator of thiamin status in man, and its comparison with erythrocyte transketolase activity measurements. *Annals of Clinical Biochemistry, 25*(Pt 6), 698–705. https://doi.org/10.1177/000456328802500617.

Bandyopadhyay, S., Roy, A. K., Baker, S., Stephenson, K. K., Ge, X., Wang, Y., ... Schulze, K. J. (2025). Micronutrient biomarker selection and assay methods and performance in double-blind, randomized, controlled micronutrient dose response (MiNDR) trials among women of reproductive age and pregnant women in rural Bangladesh. *Current Developments in Nutrition, 9*(10), 107546. https://doi.org/10.1016/j.cdnut.2025.107546.

Basiri, B., Sutton, J. M., Hanberry, B. S., Zastre, J. A., & Bartlett, M. G. (2016). Ion pair liquid chromatography method for the determination of thiamine (vitamin B1) homeostasis. *Biomedical Chromatography, 30*(1), 35–41. https://doi.org/10.1002/bmc.3544.

Bettendorff, L., Peeters, M., Jouan, C., Wins, P., & Schoffeniels, E. (1991). Determination of thiamin and its phosphate esters in cultured neurons and astrocytes using an ion-pair reversed-phase high-performance liquid chromatographic method. *Analytical Biochemistry, 198*(1), 52–59.

Bourassa, M. W., Gomes, F., Jones, K. S., Koulman, A., Prentice, A. M., & Cerami, C. (2022). Thiamine deficiency in Gambian women of reproductive age. *Annals of the New York Academy of Sciences, 1507*(1), 162–170. https://doi.org/10.1111/nyas.14695.

Brin, M. (1962). Erythrocyte transketolase in early thiamine deficiency. *Annals of the New York Academy of Sciences, 98*, 528–541. https://doi.org/10.1111/j.1749-6632.1962.tb30574.x.

Brin, M. (1964). Erythrocyte as a biopsy tissue for functional evaluation of thiamine adequacy. *JAMA: the Journal of the American Medical Association, 187*, 762–766. https://doi.org/10.1001/jama.1964.03060230090022.

Brown, K. H., Moore, S. E., Hess, S. Y., McDonald, C. M., Jones, K. S., Meadows, S. R., ... Osendarp, S. J. M. (2021). Increasing the availability and utilization of reliable data on population micronutrient (MN) status globally: The MN data generation initiative. *The American Journal of Clinical Nutrition, 114*(3), 862–870.

Brown, K. H., Hess, S. Y., Moore, S. E., Combs, G. F., Jr, Cashman, K. D., McNulty, H., ... Osendarp, S. J. (2025). Neglected micronutrients-considering a broader set of vitamins and minerals in public health nutrition programs worldwide: A narrative review. *The American Journal of Clinical Nutrition, 122*(3), 680–694. https://doi.org/10.1016/j.ajcnut.2025.06.030.

Buttery, J. E., & Pannall, P. R. (1985). Kinetic transketolase assay: Use of whole-blood hemolysate as the sample. *Clinical Chemistry, 31*(6), 1086.

Caballero, A., Gonzalez-Silva, G., Gabriel-Medina, P., Cuadros, M., Ayora, A., Blanco-Grau, A., ... Villena, Y. (2025). Optimizing vitamin supplementation via reference interval update of vitamins A, E, B1, and B6 using HPLC. *Journal of Clinical Biochemistry and Nutrition, 76*(2), 148–155. https://doi.org/10.3164/jcbn.24-155.

Carpenter, K. J. (2000). *Beriberi, white rice, and vitamin B: A disease, a cause, a cure. berkley.* University of California Press.

de Montmollin, D., MacPhail, J., McMahon, J., & Coninx, R. (2002). Outbreak of beriberi in a prison in West Africa. *Tropical Doctor, 32*(4), 234–236. https://doi.org/10.1177/004947550203200419.

Donchenko, H. V., Chernukhina, L. O., Kuz'menko, I. V., & Parkhomenko, I. M. (1997). Blood vitamin levels in various population groups in the Ukraine suffering from sequelae to the accident at the Chernobyl power plant. *Ukrainskii Biokhimicheskii Zhurnal, 69*(3), 48–53.

Doung-ngern, P., Kesornsukhon, S., Kanlayanaphotporn, J., Wanadurongwan, S., & Songchitsomboon, S. (2007). Beriberi outbreak among commercial fishermen, Thailand 2005. *Southeast Asian Journal of Tropical Medicine and Public Health, 38*(1), 130–135.

Dror, D. K., & Allen, L. H. (2018). Overview of nutrients in human milk. *Advances in Nutrition, 9*(1), 278S–294S.

Edwards, K. A. (2026). Reassessing transketolase assays: Methodological considerations for detecting functional thiamine deficiency. *Annals of the New York Academy of Sciences, 1557*(1), e70245. https://doi.org/10.1111/nyas.70245.

Evliyaoglu, O., van Helden, J., Imöhl, M., & Weiskirchen, R. (2019). Vitamin B1 interpretation: Erroneous higher levels in non-anemic populations. *Nutrition (Burbank, Los Angeles County, Calif.), 60*, 25–29. https://doi.org/10.1016/j.nut.2018.09.002.

Fattal-Valevski, A. (2011). Thiamine (vitamin B1). *Journal of Evidence-Based Complementary & Alternative Medicine, 16*(1), 12–20.

Fattal-Valevski, A., Kesler, A., Sela, B. A., Nitzan-Kaluski, D., Rotstein, M., Mesterman, R., ... Eshel, G. (2005). Outbreak of life-threatening thiamine deficiency in infants in Israel caused by a defective soy-based formula. *Pediatrics, 115*, 233–238. https://doi.org/10.1542/peds.2004-1255.

Fattal, I., Friedmann, N., & Fattal-Valevski, A. (2011). The crucial role of thiamine in the development of syntax and lexical retrieval: A study of infantile thiamine deficiency. *Brain, 134*(6), 1720–1739.

Federal Government of Nigeria (FGoN) and the International Institute of Tropical Agriculture (IITA). 2024. National Food Consumption and Micronutrient Survey Final Report. Abuja and Ibadan, Nigeria: FGoN and IITA. 550 pp. (2021) https://www.unicef.org/nigeria/reports/national-food-consumption-and-micronutrient-survey-2021.

Fidanza, F., Simonetti, M. S., Floridi, A., Codini, M., & Fidanza, R. (1989). Comparison of methods for thiamin and riboflavin nutriture in man. *International Journal for Vitamin and Nutrition Research. Internationale Zeitschrift fur Vitamin- und Ernahrungsforschung. Journal International de Vitaminologie et de Nutrition, 59*(1), 40–47.

Frank, L. L. (2015). Thiamin in clinical practice. *Journal of Parenteral and Enteral Nutrition, 39*(5), 503–520.

Gallant, J., Chan, K., Green, T. J., Wieringa, F. T., Leemaqz, S., Ngik, R., ... Whitfield, K. C. (2021). Low-dose thiamine supplementation of lactating cambodian mothers improves human milk thiamine concentrations: A randomized controlled trial. *American Journal of Clinical Nutrition, 114*(1), 90–100. https://doi.org/10.1093/ajcn/nqab052.

Gangolf, M., Czerniecki, J., Radermecker, M., Detry, O., Nisolle, M., Jouan, C., ... Bettendorff, L. (2010). Thiamine status in humans and content of phosphorylated thiamine derivatives in biopsies and cultured cells. *PLoS One, 5*(10), e13616.

Gibson, G. E., Sheu, K. F., Blass, J. P., Baker, A., Carlson, K. C., Harding, B., & Perrino, P. (1988). Reduced activities of thiamine-dependent enzymes in the brains and peripheral tissues of patients with Alzheimer's disease. *Archives of Neurology, 45*(8), 836–840. https://doi.org/10.1001/archneur.1988.00520320022009.

Gibson, G. E., Hirsch, J. A., Fonzetti, P., Jordan, B. D., Cirio, R. T., & Elder, J. (2016). Vitamin B1 (thiamine) and dementia. *Annals of the New York Academy of Sciences, 1367*(1), 21–30. https://doi.org/10.1111/nyas.13031.

Giguère, J. F., & Butterworth, R. F. (1987). Activities of thiamine-dependent enzymes in two experimental models of thiamine deficiency encephalopathy: 3. Transketolase. *Neurochemical Research, 12*(3), 305–310. https://doi.org/10.1007/BF00972141.

Gomes, F., Bergeron, G., Bourassa, M. W., & Fischer, P. R. (2021). Thiamine deficiency unrelated to alcohol consumption in high-income countries: A literature review. *Annals of the New York Academy of Sciences, 1498*(1), 46–56.

Green, T. J., Whitfield, K. C., Daniels, L., Brown, R. C., & Houghton, L. A. (2021). Modeling thiamine fortification: A case study from kuria atoll, republic of kiribati. *Annals of the New York Academy of Sciences, 1498*(1), 108–115. https://doi.org/10.1111/nyas.14561.

Harel, Y., Zuk, L., Guindy, M., Nakar, O., Lotan, D., & Fattal-Valevski, A. (2017). The effect of subclinical infantile thiamine deficiency on motor function in preschool children. *Maternal and Child Nutrition, 13*(4), e12397. https://doi.org/10.1111/mcn.12397.
Haro, E. N., Brin, M., & Faloon, W. W. (1966). Fasting in obesity. Thiamine depletion as measured by erythrocyte transketolase changes. *Archives of Internal Medicine, 117*(2), 175–181. https://doi.org/10.1001/archinte.117.2.175.
Hauber, M. M., Todisco, V., Nordahl, O., Tibblin, P., Fridolfsson, E., Kärvegård, E., & Hylander, S. (2026). Thiamine allocation and deficiency status throughout the life cycle of cod. *Ecology and Evolution, 16*(1), e72828. https://doi.org/10.1002/ece3.72828.
Hess, S. Y., Smith, T. J., Arnold, C. D., Jones, K. S., Hampel, D., Hiffler, L., ... Kounnavong, S. (2024). Assessment of erythrocyte transketolase, whole blood thiamine diphosphate, and human milk thiamine concentrations to identify infants and young children responding favorably to therapeutic thiamine administration: Findings from the Lao Thiamine Study. *Current Developments in Nutrition, 8*, 103786. https://doi.org/10.1016/j.cdnut.2024.103786.
Hiffler, L., Adamolekun, B., Fischer, P. R., & Fattal-Vavleski, A. (2017). Thiamine content of F-75 therapeutic milk for complicated severe acute malnutrition: Time for a change? *Annals of the New York Academy of Sciences, 1404*(1), 20–26. https://doi.org/10.1111/nyas.13458.
Hilal Ahmad, S., Ahmad, S. B., & Jan, A. (2019). Thiamine deficiency related peripheral neuropathy in peripartum women of Kashmir, India: A hospital based study. *International Journal of Advances in Medicine, 6*(3), 662–667.
Hoad, K. E., Johnson, L. A., Woollard, G. A., Walmsley, T. A., Briscoe, S., Jolly, L. M., ... Greaves, R. F. (2013). Vitamin B1 and B6 method harmonization: Comparison of performance between laboratories enrolled in the RCPA Quality Assurance Program. *Clinical Biochemistry, 46*(9), 772–776. https://doi.org/10.1016/j.clinbiochem.2013.01.020.
Huang, Y., Gibson, R. A., & Green, T. J. (2020). Measuring thiamine status in dried blood spots. *Clinica Chimica Acta, 509*, 52–59. https://doi.org/10.1016/j.cca.2020.06.011.
Institute of Medicine. (1998). *Thiamin. Dietary reference intakes for thiamin, riboflavin, niacin, vitamin B6, folate, vitamin B12, pantothenic acid, biotin, and choline.* National Academy Press. 58–86.
Johnson, C. R., Fischer, P. R., Thacher, T. D., Topazian, M. D., Bourassa, M. W., & Combs, G. F., Jr. (2019). Thiamin deficiency in low- and middle-income countries: Disorders, prevalences, previous interventions and current recommendations. *Nutrition and Health, 25*(2), 127–151. https://doi.org/10.1177/0260106019830847.
Jones, K. S., Parkington, D. A., Cox, L. J., & Koulman, A. (2021). Erythrocyte transketolase activity coefficient (ETKAC) assay protocol for the assessment of thiamine status. *Annals of the New York Academy of Sciences, 1498*(1), 77–84. https://doi.org/10.1111/nyas.14547.
Kim, J., Jonus, H. C., Zastre, J. A., & Bartlett, M. G. (2019). Development of an IPRP-LC-MS/MS method to determine the fate of intracellular thiamine in cancer cells. *Journal of Chromatography. B, Analytical Technologies in the Biomedical and Life Sciences, 1124*, 247–255. https://doi.org/10.1016/j.jchromb.2019.05.037.
Klooster, A., Larkin, J. R., Wiersema-Buist, J., Gans, R. O. B., Thornalley, P. J., Navis, G., ... Bakker, S. J. L. (2013). Are brain and heart tissue prone to the development of thiamine deficiency? *Alcohol (Fayetteville, N. Y.), 47*(3), 215–221. https://doi.org/10.1016/j.alcohol.2012.12.014.
Koshy, R. M., Ismavel, V. A., Sharma, H., & Jacob, P. M. (2018). Suspected thiamine deficiency presenting as peripheral neuropathy among peripartum women in a hospital in rural Assam: A neglected public health problem. *CHRISMED Journal of Health and Research, 5*, 178–181.

Koshy, R. M., Thankaraj, S., Ismavel, V. A., & Solomi, C. V. (2021). The rediscovery of thiamine deficiency disorders at a secondary level mission hospital in northeast India. *Annals of the New York Academy of Sciences, 1498*(1), 96–107.

Koulman, A., Jones, K. S., Collins, D., Parkington, D., Amoutzopoulos, B., Page, P., & Forouhi, N. G. (2026). Ethnic differences in thiamine status persist after adjusting for diet; findings from the UK National Diet and Nutrition Survey. *Clinical Nutrition ESPEN, 72*, 102893. https://doi.org/10.1016/j.clnesp.2025.102893.

Li, M., Lu, Y., Li, Y., Tong, L., Gu, X.-C., Meng, J., ... Tong, X. (2019). Transketolase deficiency protects the liver from DNA damage by increasing levels of ribose 5-phosphate and nucleotides. *Cancer Research, 79*(14), 3689–3701. https://doi.org/10.1158/0008-5472.CAN-18-3776.

Lu, J., & Frank, E. L. (2008). Rapid HPLC measurement of thiamine and its phosphate esters in whole blood. *Clinical Chemistry, 54*(5), 901–906.

Luxemburger, C., White, N. J., ter Kuile, F., Singh, H. M., Allier-Frachon, I., Ohn, M., ... Nosten, F. (2003). Beri-Beri: The major cause of infant mortality in karen refugees. *Transactions of the Royal Society of Tropical Medicine and Hygiene, 97*(2), 251–255. https://doi.org/10.1016/s0035-9203(03)90134-9.

Marrs, C., & Lonsdale, D. (2021). Hiding in plain sight: Modern thiamine deficiency. *Cells, 10*, 2595. https://doi.org/10.3390/cells10102595.

Mathew, E. M., Sakore, P., Lewis, L., Manokaran, K., Rao, P., & Moorkoth, S. (2019). Development and validation of a dried blood spot test for thiamine deficiency among infants by HPLC-fluorimetry. *Biomedical Chromatography: BMC, 33*(11), e4668. https://doi.org/10.1002/bmc.4668.

Mimouni-Bloch, A., Goldberg-Stern, H., Strausberg, R., Brezner, A., Heyman, E., Inbar, D., ... Fattal-Valevski, A. (2014). Thiamine deficiency in infancy: Long-term follow-up. *Pediatric Neurology, 51*(3), 311–316. https://doi.org/10.1016/j.pediatrneurol.2014.05.010.

Mosnier, E., Niemetzky, F., Stroot, J., Pommier de Santi, V., Brousse, P., Guarmit, B., ... Nacher, M (2017). A large outbreak of thiamine deficiency among illegal gold miners in French Guiana. *American Journal of Tropical Medicine and Hygiene, 96*(5), 1248–1252. https://doi.org/10.4269/ajtmh.15-0906.

Nilles, E. J., Manaia, A., Ruaia, B., Huppatz, C., Ward, C., George, P., ... Tira, T. (2018). Re-emergence of thiamine deficiency disease in the Pacific Islands (2014 – 15): A case-control study. *PLoS One, 13*(6), e0198590. https://doi.org/10.1371/journal.pone.0198590.

O'Brien, M., Quirke, R., Gowan, R., & McNicholas, F. (2025). Prevalence of thiamine deficiency in anorexia nervosa: A systematic review and narrative synthesis. *European Child and Adolescent Psychiatry, 34*(4), 1281–1293.

Parkington, D. A., Koulman, A., Gibson, G. E., & Jones, K. S. (2026). A protocol for the measurement of brain transketolase activity. *Methods in Enzymology*.

Passarelli, S., Free, C. M., Shepon, A., Beal, T., Batis, C., & Golden, C. D. (2024). Global estimation of dietary micronutrient inadequacies: A modelling analysis. *The Lancet Global Health, 12*(10), e1590–e1599. https://doi.org/10.1016/S2214-109X(24)00276-6.

Pavlova, O., Stepanenko, S., Chehivska, L., Sambon, M., Bettendorff, L., & Yu, P. (2021). Thiamine deficiency in rats affects thiamine metabolism possibly through the formation of oxidized thiamine pyrophosphate. *Biochimica Et Biophysica Acta. General Subjects*, 129980. https://doi.org/10.1016/j.bbagen.2021.129980.

Rohner, F., Rakotovao, J. M., El-Mallah, C., Andrianarison, R., Randrenarizo, N., Wegmüller, R., ... Galetti, V. (2026). Burden and co-occurrence of anemia, micronutrient deficiencies, inflammation, and malaria among women, adolescents, and children in Madagascar: A national cross-sectional survey. *Journal of Nutrition*, 101610. https://doi.org/10.1016/j.tjnut.2026.101610.

Rosewarne, R. E., & Farina, N. (2026). Thiamine use in hospitalized patients: A clinical review. *Journal of Nutrition and Metabolism*, 7143046. https://doi.org/10.1155/jnme/7143046.
Sambon, M., Napp, A., Demelenne, A., Vignisse, J., Wins, P., Fillet, M., & Bettendorff, L. (2019). Thiamine and benfotiamine protect neuroblastoma cells against paraquat and β-amyloid toxicity by a coenzyme-independent mechanism. *Heliyon, 5*(5), e01710. https://doi.org/10.1016/j.heliyon.2019.e01710.
Sambon, M., Gorlova, A., Demelenne, A., Alhama-Riba, J., Coumans, B., Lakaye, B., ... Bettendorff, L. (2020). Dibenzoylthiamine has powerful antioxidant and anti-inflammatory properties in cultured cells and in mouse models of stress and neurodegeneration. *Biomedicines, 8*(9), 361. https://doi.org/10.3390/biomedicines8090361.
Sauberlich, H. E. (1967). Biochemical alternations in thiamine deficiency—their interpretation. *The American Journal of Clinical Nutrition, 20*, 528–542.
Smeets, E. H., Muller, H., & de Wael, J. (1971). A NADH-dependent transketolase assay in erythrocyte hemolysates. *Clinica Chimica Acta; International Journal of Clinical Chemistry, 33*(2), 379–386. https://doi.org/10.1016/0009-8981(71)90496-7.
Smith, T. J., & Hess, S. Y. (2021). Infantile thiamine deficiency in south and southeast Asia: An age-old problem needing new solutions. *Nutrition Bulletin, 46*(1), 12–25.
Smith, T. J., Arnold, C. D., Fischer, P. R., Trehan, I., Hiffler, L., Sitthideth, D., ... Hess, S. Y. (2024). A predictive model for thiamine responsive disorders among infants and young children: Results from a prospective cohort study in lao people's democratic republic. *Journal of Pediatrics, 268*, 113961. https://doi.org/10.1016/j.jpeds.2024.113961.
Smith, T. J., Johnson, C. R., Koshy, R., Hess, S. Y., Qureshi, U. A., Mynak, M. L., & Fischer, P. R. (2021). Thiamine deficiency disorders – A clinical perspective. *Annals of the New York Academy of Sciences, 1498*(1), 9–28.
Talwar, D., Davidson, H., Cooney, J., & St JO'Reilly, D. (2000). Vitamin B(1) status assessed by direct measurement of thiamin pyrophosphate in erythrocytes or whole blood by HPLC: Comparison with erythrocyte transketolase activation assay. *Clinical Chemistry, 46*(5), 704–710. https://doi.org/10.1093/clinchem/46.5.704.
Therrien, C. A., Baker, P. M., Garner, S., Swanson, H. K., & Neff, B. D. (2026). Dietary thiaminase alters morphology and decreases swimming performance of lake trout (Salvelinus namaycush). *Scientific Reports, 16*(1), 5493. https://doi.org/10.1038/s41598-026-35139-8.
Turck, D., Bresson, J., Burlingame, B., Dean, T., Fairweather-Tait, S., Heinonen, M., ... Neuhäuser-Berthold, M. (2016). Dietary reference values for thiamin. *EFSA Journal, 14*(12), 4653. https://doi.org/10.2903/j.efsa.2016.4653.
Vachali, P. P., Kish-Trier, E., Steckel, S., & Frank, E. L. (2025). Instrument multiplexing: Amplified throughput for measurement of vitamin B1 in whole blood using parallel liquid chromatography systems interfaced to a single mass spectrometer. *Clinical Chemistry, 71*(6), 713–722. https://doi.org/10.1093/clinchem/hvaf029.
Verstraete, J., Strobbe, S., Van Der Straeten, D., & Stove, C. (2020). The first comprehensive LC-MS/MS method allowing dissection of the thiamine pathway in plants. *Analytical Chemistry, 92*(5), 4073–4081. https://doi.org/10.1021/acs.analchem.9b05717.
Verstraete, J., & Stove, C. (2021a). Patient-centric assessment of thiamine status in dried blood volumetric absorptive microsamples using LC-MS/MS analysis. *Analytical Chemistry, 93*(4), 2660–2668. https://doi.org/10.1021/acs.analchem.0c05018.
Verstraete, J., & Stove, C. (2021b). Volumetric absorptive microsampling (VAMS) as a reliable tool to assess thiamine status in dried blood microsamples: A comparative study. *The American Journal of Clinical Nutrition, 114*(3), 1200–1207. https://doi.org/10.1093/ajcn/nqab146.
Whitfield, K. C., Bourassa, M. W., Adamolekun, B., Bergeron, G., Bettendorff, L., Brown, K. H., ... Combs, G. F., Jr. (2018). Thiamine deficiency disorders: Diagnosis, prevalence, and a roadmap for global control programs. *Annals of the New York Academy of Sciences, 1430*(1), 3–43. https://doi.org/10.1111/nyas.13919.

Whitfield, K. C., Smith, G., Chamnan, C., Karakochuk, C. D., Sophonneary, P., Kuong, K., ... Wieringa, F. T. (2017). High prevalence of thiamine (vitamin B1) deficiency in early childhood among a nationally representative sample of Cambodian women of childbearing age and their children. *PLoS Neglected Tropical Diseases, 11*(9), e0005814.

Wilkinson, T. J., Hanger, H. C., Elmslie, J., George, P. M., & Sainsbury, R. (1997). The response to treatment of subclinical thiamine deficiency in the elderly. *American Journal of Clinical Nutrition, 66*, 925–928.

Wood, B., Gijsbers, A., Goode, A., Davis, S., Mulholland, J., & Breen, K. (1980). A study of partial thiamin restriction in human volunteers. *The American Journal of Clinical Nutrition, 33*(4), 848–861. https://doi.org/10.1093/ajcn/33.4.848.

World Health Organization and Food and Agriculture Organization of the United Nations. (2004). Vitamin and mineral requirements in human nutrition (Second edition) https://www.who.int/publications/i/item/9241546123.

Zhao, Y., Pan, X., Zhao, J., Wang, Y., Peng, Y., & Zhong, C. (2009). Decreased transketolase activity contributes to impaired hippocampal neurogenesis induced by thiamine deficiency. *Journal of Neurochemistry, 111*(2), 537–546. https://doi.org/10.1111/j.1471-4159.2009.06341.x.

Zhang, X., Tang, X., Gibson, B., & Daly, T. M. (2017). Simple HPLC method with internal standard for evaluation of vitamin B1 status by use of whole blood. *J Appl Lab Med, 2*(3), 367–379. https://doi.org/10.1373/jalm.2017.024349.

CHAPTER SEVEN

Protocol for the measurement of human brain transketolase activity

Damon A. Parkington[a], Albert Koulman[a], Sarah A. Flowers[b], Gary E. Gibson[c], and Kerry S. Jones[a,*]

[a]Nutritional Biomarker Laboratory, Institute of Metabolic Science - Epidemiology, University of Cambridge, Cambridge, United Kingdom
[b]Department of Neuroscience, University of Virginia, Charlottesville, VA, United States
[c]Brain and Mind Research Institute, Weill Cornell Medicine, New York, NY, United States
*Corresponding author. e-mail address: kerry.jones@ims.cam.ac.uk

Contents

Methods in Enzymology, Volume 735
ISSN 0076-6879, https://doi.org/10.1016/bs.mie.2026.07.012

Abstract

Thiamine (vitamin B1) as thiamine diphosphate (or thiamine pyrophosphate) is a cofactor for enzymes involved in energy metabolism and is essential for brain function. Reliable biomarkers of thiamine status and function are necessary to understand thiamine's role in health and disease and provide mechanistic insights and to determine the efficacy of pharmacological interventions of thiamine analogues. The activity of the thiamine-dependent transketolase enzyme in erythrocytes is the established functional marker of whole-body thiamine status. In this chapter, we describe the analogous method optimized for the determination of transketolase activity in brain tissue and the calculation of the brain transketolase 'activity coefficient'. Provided are full method details for the preparation of brain material and instructions to perform the assay in 96-well plates using a microplate spectrophotometer. We provide details of suitable quality control materials and tips for troubleshooting. Instructions are also provided for the calculation of absolute basal transketolase activity. These protocols provide standardization for the determination of transketolase activity in human post mortem brain samples or from animal experiments and will be useful for researchers interested in the role of thiamine in brain-related conditions. Such protocols are essential to harmonize methods between laboratories for the benefit of comparisons between studies.

1. Introduction

Thiamine (vitamin B1) is an essential micronutrient. The thiamine diphosphate form (alternatively known as thiamine pyrophosphate (TPP)) is a cofactor for enzymes involved in energy metabolism including transketolase, pyruvate dehydrogenase and α-ketoglutarate dehydrogenase that are required for glucose metabolism. Glucose produced within the brain also provides the building blocks of carbon for the production of neurotransmitters, acetylcholine and glutamate (Gibson et al., 2016).

Symptoms of thiamine deficiency can be broad and non-specific affecting the cardiovascular, respiratory, gastrointestinal and nervous systems. Neurological symptoms are common in both infant and adult manifestations of thiamine deficiency and characterize dry beriberi, Wernicke encephalopathy and Korsakoff syndrome. Common clinical signs include peripheral neuropathies, nystagmus, ophthalmoplegia and ataxia; if left untreated these can progress to cognitive decline (Smith et al., 2021). The mechanisms of these conditions relate to the role of thiamine as a cofactor for the enzymes described above. Dry beriberi is associated with loss of myelin, whereas Wernicke encephalopathy is primarily a disorder of energy metabolism in tissues with high metabolic demand.

Thiamine deficiency has also been extensively associated with Alzheimer's disease (Gibson et al., 2016). Interest has grown in the potential therapeutic role of thiamine, and its analogues, as pharmacological agents in the treatment of Alzheimer's disease, underpinned by mechanistic studies (Gibson et al., 2022).

To further understand relationships between thiamine and health outcomes and disease, and to provide mechanistic insights in relation to thiamine and thiamine analogue pharmacological interventions, reliable biomarkers of thiamine concentration and function are required. Quantitative markers of thiamine forms, free thiamine, thiamine monophosphate, thiamine diphosphate and thiamine triphosphate can be measured using chromatography followed by UV or mass spectrometric detection (Jones et al., 2026). Functional assays were developed in the 1960s as a marker of thiamine status using measurement of the transketolase enzyme in red blood cells. Previous studies have also measured the activity of thiamine-dependent enzymes (ketoglutarate dehydrogenase and transketolase) in the brains of Alzherimer's disease patients (Gibson et al., 1988).

The transketolase activity assay, which assesses thiamine status and functional capacity, has undergone many developments (Edwards, 2026). More recent incarnations of the assay utilize a microplate format to determine erythrocyte transketolase 'activity coefficient' (ETKAC) where the ratio of basal activity to in vitro TPP-stimulated activity is calculated (Jones et al., 2021). This methodology can also be used to calculate the absolute basal activity per gram of hemoglobin (Jones et al., 2023). These assays are commonly applied in human studies to assess thiamine status in studies and nutritional surveys (Bourassa et al., 2022; Hess et al., 2020; Koulman et al., 2026; Ministere de la Sante Publique de Madagascar, 2025), but are not routinely available. However, standardized protocols (Jones et al., 2021, 2023) have been published and adopted by other labs (Bandyopadhyay et al., 2025) for erythrocyte transketolase activity.

To better understand thiamine functional capacity in other tissues and to underpin mechanistic understanding of the relationship between thiamine and disease, tissue biomarkers are required. Brain thiamine levels in particular are both lower and, in thiamine deficiency, are thought to decrease more slowly than other tissues (Bettendorff, 2023). Evidence has also shown that in Alzheimer's disease both brain thiamine transporters and brain transketolase activity are decreased (Gibson et al., 1988; Ramamoorthy et al., 2022). Transketolase activity has been measured in other tissues, in animals, particularly fish, (Hauber et al., 2026; Therrien et al., 2026), but also in human studies (Gibson et al., 1988; Giguère & Butterworth, 1987; Zhao et al., 2009). For tissue measurements of transketolase activity, it is essential to consider tissue specific assay parameters related to the availability of transketolase, thiamine and its metabolites, which all can vary between tissues (Edwards, 2026; Gangolf et al., 2010).

1.1 Rationale

Here, we describe a method to measure transketolase activity in brain tissue to calculate brain transketolase 'activity coefficient' (BTKAC) and basal activity per gram of protein in brain sample tissue taking advantage of 96-well plate format. The method is an adaptation of erythrocyte transketolase activity assay. Tissue measurements provide a functional and integrated biomarker of tissue-specific thiamine status that can be used to assess thiamine-related enzyme function in brain tissue, either from animal studies or human post-mortem brains to understand the significance and role of thiamine availability and/or thiamine-related clinical treatments. This protocol will help standardize measurements between laboratories to improve precision and allow comparison between studies.

1.2 Protocol

The following sections provide a step-by-step protocol for the measurement of transketolase activity in brain tissue. The protocol builds on previous methods measuring transketolase activity in erythrocytes (Jones et al., 2021, 2023) with adaptations for brain tissue measurements. The protocol includes instructions for the primary assay as well as techniques for the preparation of fresh brain tissue and quality controls and associated calculations, data analysis and troubleshooting.

1.3 Materials, equipment and reagents

1.3.1 Electrical equipment

- Analytical balance. Fisherbrand, product code: 15907500
- Digital microplate shaker, Thermo Scientific, product code: 88882006
- Digital multi-tube vortex mixer, DVX-2500, VWR, product code: 444-7061
- Fume cupboard, Monmouth Scientific Circulaire CT800
- Incubator, GALLENKAMP Economy Incubator with Fan Size 1

- Magnetic stirrer, Thermo Scientific RT Touch 22, with stirring bars, product code: 88880014
- Multiskan FC microplate photometer (with incubator), Thermo Scientific, product code: 51119100, if using an alternative plate reader see Precursor Techniques - Plate reader suitability
- pH meter, Fisherbrand, accumet AB150, product code: 12880633
- Refrigerated centrifuge, Eppendorf Centrifuge 5430 R with Rotor FA-45-48-11, product code: 5428000655
- Single tube vortex mixer, Vortex-Genie 2, Fisher Scientific, product code: 15557335

1.3.2 Consumables/other equipment

- 2 mL microtubes with lids, Sarstedt
- 5 mL, 15 mL, and 50 mL Falcon tubes
- 20 μL, 200 μL, 1000 μL and 5000 μL pipettes
- 96-well UV-transparent microplates, Fisher-Scientific, product code: 655801
- 500 mL and 1000 mL borosilicate glass bottle with lid
- 500 mL glass beaker
- Liquid nitrogen dewar (~3 L capacity) containing liquid nitrogen (LN_2)
- Microplate sealers
- Multichannel pipettes (for delivering 15 μL and 200 μL)
- Pestle and mortar (~300 mL) × 2
- Plastic trays (~40 × 30 cm internal diameter) × 2
- Polystyrene boxes (volume ~30 cm^3) × 2
- Sample rack for 2 mL microtubes
- Spatulas × 2
- Weighing boats

1.3.3 Chemicals

- 2-Mercaptoethanol, 99 %, pure, CAS: 60-24-2, Fisher Scientific
- α-Glycerophosphate dehydrogenase (GPDH)-triosephosphate isomerase (TPI) from rabbit muscle. GPDH 160 units.$mg_{protein}^{-1}$, TPI 1941 units.$mg_{protein}^{-1}$. 5 KU bottle, CAS: 9075-65-4, Merck
- β-Nicotinamide adenine dinucleotide reduced disodium salt, 97 %, CAS: 606-68-8, Fisher Scientific
- D-Ribose-5-phosphate disodium salt hydrate, 85%, [FC4.1] CAS: 18265-46-8, Fisher Scientific
- Ethylenediaminetetraacetic acid dipotassium salt dihydrate (K2-EDTA), 99 %, CAS: 25102-12-9, Fisher Scientific

- Hydrochloric acid solution, 2 M, Fisher Chemical, CAS: 7647-01-0, Fisher Scientific
- Potassium phosphate monobasic (white crystals), Fisher BioReagents, CAS: 7778-77-0, Fisher Scientific
- Sodium hydroxide, 10 % (w/v) (100 $g.L^{-1}$), CAS: 1310-73-2, Ricca Chemical Company
- Sodium hypochlorite 4–6 % (NaOCl), CAS: 7681-52-9, Fisher Scientific
- Thiamine pyrophosphate chloride, CAS: 154-87-0, Fisher Scientific
- Triton X-100, CAS: 9036-19-5, Fisher Scientific
- Tris(hydroxymethyl)aminomethane base (Tris base), CAS: 77-86-1, Fisher Scientific
- Tween 20, CAS: 9005-64-5, Fisher Scientific

2. Safety considerations and standards

The use of standard laboratory PPE (lab coat, eye protection and gloves) is required. Use of a safety cabinet is required when handling neat 2-mercaptoethanol. It is important to check the safety data sheets to ensure compliance with local rules. Appropriate ethical permission procedures and local rules should be followed for obtaining, storing and handling of human and animal tissue samples. In the UK, human brain tissue for research should be obtained and handled in accordance with the legal framework of the Human Tissue Act 2004 and researchers must comply with regulations set by the Human Tissue Authority. Outside of the UK local guidelines should be followed.

3. Preparation of ground frozen brain tissue

1. Quarter fill 2 polystyrene boxes with LN_2 from the dewar. NB. Boxes will require regular refilling to maintain this level throughout the procedure
2. Transfer brain fragments (in their containers) from −80 °C storage into one of the LN_2 containing polystyrene boxes, place the sample rack into the second polystyrene boxes and fill with 2 mL microtubes required for storage of the ground samples
3. Sterilize tools (pestle and mortar, spatulas and tweezers) in 1 % sodium hypochlorite and rinse with distilled water and place in a plastic tray

labelled "sterile". After each use place tools in a plastic tray labelled "dirty". The tools must be cleaned and sterilized and returned to the "sterile" plastic tray before each use to prevent cross contamination

4. Half fill the 500 mL borosilicate glass beaker with LN_2 and place in the spatula, tweezers and pestle to cool. Place the mortar into the second LN_2 containing polystyrene box to cool
5. Remove the cooled mortar from the polystyrene box. Use the tweezers to transfer brain fragments to the mortar. Use the cooled pestle to grind the tissue to a fine powder
6. Use the spatula to transfer the ground tissue immediately to one of the cooled 2 mL microtubes. Label the tubes with the sample ID
7. Immediately transfer the tubes to −80 °C storage (or temporarily store on dry ice before transferring to −80 °C storage if available and more convenient)

NB. The method described in this protocol is for ~50 mg of ground tissue. It is recommended to weigh the microtube before and after adding the tissue and calculating the exact weight.

4. Preparation of control material

It is important to prepare a stock of control material which can be analyzed with each batch to monitor between batch variation. This material could also be shared with other laboratories to assess inter-laboratory agreement.

To prepare quality control (QC) material a bulk of brain tissue is required. This does not need to be human brain; ovine and murine brain samples are both suitable for use with this assay. Brain tissue should be frozen as soon as possible after collection.

Using the instructions in "Preparation of ground frozen brain tissue" grind the frozen brain tissue to a powder. Take three 50 mL Falcon tubes and label as low, medium and high control. Add 3, 5 and 8 g of ground brain tissue respectively. Add 45 mL of cell lysis buffer to each tube.

1. Vortex mix on multi-tube vortex mixer at 1450 rpm (use intermittent setting) for 20 min
2. Centrifuge at 10,000 rpm for 10 min at + 4 °C
3. The supernatant is the lysate which can be stored in labelled 0.5 mL aliquot tubes at −80 °C

5. Reagent preparation

Prepare the cell lysis buffer and Tris buffer prior to the day of analysis. Once prepared both can be stored at + 4 °C for up to 1 month.

Thiamine pyrophosphate (TPP), D-ribose-5-phosphate (D-R5P) and reduced nicotinamide adenine dinucleotide (NADH) must be prepared on the day of analysis.

5.1 Preparation of cell lysis buffer reagent

The cell lysis buffer (potassium phosphate 50 mM; potassium EDTA 1 mM, 2-mercaptoethanol 1 mM, Triton X-100 0.2 %, pH 7.8) is required to lyse the samples of ground tissue in preparation for the analysis. Table 1 shows the components required to prepare cell lysis buffer following the steps below.

Table 1 Reagents required for the preparation of the cell lysis buffer.

Reagent	CAS	Final concentration	Amount (for 500 mL)
Potassium Phosphate Monobasic	7778-77-0	50 mM	3.402 g
2-Mercaptoethanol density 1.11	60-24-2	1 mM	35 μL
K2-EDTA	25102-12-9	1 mM	0.202 g
Triton X-100	9002-93-1	0.2%	1 mL
10 % w/v NaOH	—	n/a	As required

1. Dissolve 3.402 g of potassium phosphate in 450 mL deionized H_2O in a 500 mL volumetric flask
2. In a fume hood, add 35 μL 2-mercaptoethanol, cap the bottle, and mix by inverting 10 times
3. Add 1 mL Triton X-100 and mix by inverting 10 times. Place on a magnetic stitrrer and use a transfer pipette to slowly add 10 % w/v NaOH to adjust the pH to 7.8. Make up to 500 mL with deionized H_2O

Store at + 4 °C for up to 1 month.

5.2 Preparation of 100 mM Tris buffer

100 mM Tris buffer is required for the preparation of assay reagents and to dilute the lysates for analysis of transketolase activity. Table 2 shows the components required to prepare Tris buffer following the steps below.

Table 2 Reagents required for the preparation of Tris buffer.

Reagent	CAS	Final Concentration	Amount (for 500 mL)
Tris base	77-86-1	0.1 mM	12.1 g
2 M hydrochloric acid	7647-01-0	n/a	As required

1. Dissolve 12.1 g of Tris base in 750 mL deionized H_2O in a 1000 mL borosilicate glass bottle mix using a magnetic stirrer
2. Whilst on the magentic stirrer, check the pH using the pH meter and use a transfer pipette to add 2 M hydrochloric acid to adjust the pH to 7.6
3. Make up to 1000 mL with deionized H_2O

Store at + 4 °C for up to 1 month.

5.3 Preparation of 1 % Tween 20

A solution of 1 % Tween 20 is required for the preparation of the assay reagent. Table 3 shows the components required to prepare 1% Tween 20. First, make 10 % Tween 20 solution following the steps below (NB. Do not attempt to pipette neat Tween 20 quantitatively).

1. Measure 9 mL deionized H_2O into a 10 mL measuring cylinder
2. Add 1 mL Tween 20 below the meniscus using a transfer pipette. Use the water meniscus and graduations to determine the volume
3. Transfer to a 15 mL Falcon tube and mix by inverting 10 times.

Store at + 4 °C for up to 6 months.

Use a 10-fold dilution of the 10 % Tween 20 to produce a 1 % solution following the steps below:

4. Add 1 mL of 10 % Tween 20 into a 10 mL measuring cylinder
5. Make up to 10 mL with deionized H_2O
6. Transfer to a 15 mL Falcon tube and mix by inverting 10 times

Store at + 4 °C for up to 6 months.

5.4 Preparation of 7.813 mM thiamine pyrophosphate solution

In the assay, transketolase activity is measured both with the addition of TPP (activated rate) and without TPP addition (basal rate). Use the components listed in Table 4 to prepare a 7.813 mM TPP solution following the steps below:

Table 3 Reagents required for the preparation of 1 % Tween 20.

Reagent	Final concentration	Amount (for 10 mL)
Tween 20	10 % Tween 20	1 mL
Deionized H_2O	—	9 mL

Table 4 Reagents required for the preparation of thiamine pyrophosphate solution.

Reagent	CAS	Final Concentration	Amount (sufficient for one microplate)
Thiamine pyrophosphate (TPP)	154-87-0	7.813 mM	4.5–5.0 mg
Tris buffer	n/a	n/a	Calculate as per instructions

1. Prepare a 7.813 mM TPP solution in a 5 mL Falcon tube (MW = 460.77 Da, 3.6 mg.mL^{-1}) in Tris buffer. 1 mL is required per microplate for analysis.
2. Mix on a roller mixer for 10 min
3. Store at + 4 °C for up to 1 day

5.5 Preparation of 0.358 M D-ribose-5-phosphate solution

he D-R5P solution is required as a component of the assay reagent. In the assay D-R5P is required as a substrate for transketolase. Table 5 shows the components required to prepare D-R5P solution following the steps below.

Table 5 Reagents required for the preparation of ribose-5-phosphate solution.

Reagent	CAS	Final concentration	Amount (sufficient for one microplate)
D-Ribose-5-phosphate (D-R5P)	18265-46-8	0.358 M	4.5–5.0 mg
Tris buffer	n/a	n/a	Calculate as per instructions

Table 6 Reagents required for the preparation of nicotinamide adenine dinucleotide solution.

Reagent	CAS	Final concentration	Amount (sufficient for one plate)
Nicotinamide adenine dinucleotide (NADH)	606-68-8	5.29 mM	4.5–5.0 mg
Tris buffer	n/a	n/a	Calculate as per instructions

1. Prepare a 0.358 M D-R5P solution in 100 mM Tris buffer in a 5 mL Falcon tube (MW = 330.10 Da, 82.5 $mg.mL^{-1}$). 1 mL is required per microplate for analysis
2. Mix on a roller mixer for 10 min
3. Store at + 4 °C for up to 1 day

5.6 Preparation of 5.29 mM (NADH)

The NADH solution is a required component of the assay reagent (see 2.7). In the assay reaction the oxidation of NADH to NAD^+ is measured spectrophotometrically at 340 nm. Table 6 shows the components required to prepare NADH solution following the steps below.

1. Prepare a 5.29 mM NADH solution in a 5 mL Falcon tube (MW = 709.41 Da, 3.75 $mg.mL^{-1}$) in 100 mM Tris buffer (1 mL is required per microplate for analysis)
2. Mix on a roller mixer for 10 min
3. Store at + 4 °C for up to 1 day

5.7 Preparation of assay reagent

The assay reagent (Tris 100 mM, NADH 0.24 mM, D-R5P 16.2 mM, Tween 20 0.004%, GPDH- TPI 1.5 $U.mL^{-1}$) is added to the microplate wells containing samples, QCs and blanks prior to reading the plate spectrophotometrically at 340 nm (see Section 7). Preparation of the constituents of the assay reagent are described in the steps above except for GPDH-TPI which is supplied in solution ready to use. The combination of enzymes in the GPDH-TPI solution are required to maintain the reactions in the assay which allow for the conversion of NADH to NAD^+. Table 7 shows the components required to prepare the assay reagent following the steps below.

Table 7 Assay reagent preparation component list.

Reagent	Amount (sufficient for one plate)
Tris buffer	20 mL
0.358 M D-Ribose-5-phosphate solution	1 mL
5.29 mM NADH solution	1 mL
α-Glycerophosphate dehydrogenase (GPDH)-triosephosphate isomerase (TPI) from rabbit muscle	34 units
1 % Tween 20	80 μL

1. Using a measuring cylinder transfer 20 mL Tris buffer into a 50 mL Falcon tube.
2. Add 1 mL 0.358 M D-R5P solution
3. Add 1 mL 5.29 mM NADH solution
4. Add 34 units GPDH-TPI
5. Add 80 μL 1 % Tween 20
6. Mix on a roller mixer for 10 min
7. Store at + 4 °C for up to 1 day

Note: 34 units of GPDH-TPI are required for each plate. The bottles are supplied containing 5 KU activity but the volume of solution changes from batch to batch and is shown on the bottle. To calculate the volume in mL required to give 34 units use the following equation:

$$\frac{34 * \textit{starting volume of } 5 \textit{ KU bottle}}{5000} = \textit{volume of solution containing } 34 \textit{ units } (ml)$$

6. Preparation of brain tissue lysates for analysis

These steps should usually be completed at the beginning of the day of analysis; when removing ground brain tissue samples from the freezer only remove what can be analyzed in one day and based on 20 samples per plate. Ideally a minimum of 50 mg of weighed brain tissue is required.

1. Pre-chill the centrifuge to + 4 °C

2. Turn on the incubator and allow it to reach + 37 °C
3. Turn on the plate reader and allow it to reach + 37 °C
4. Remove controls from −80 °C storage and mix on a roller mixer until thawed
5. Remove tubes containing 50 mg ground brain samples from freezer and allow to equilibrate to room temperature
6. Add 450 μL of cell lysis buffer (prepared in step 5.1) to each sample tube. Control samples are stored as lysates so do not require this step – see Section 4. If the weight of the tissue in the tube is not 50 mg, multiply the weight of tissue in mg by 9 to calculate the μL of cell lysis buffer required
7. Mix on a multi-tube vortex mixer at 1450 rpm (use intermittent setting) for 20 min
8. Centrifuge at 11,000*g* for 10 min at + 4 °C

Store sample lysates and thawed controls at + 4 °C until ready to commence the assay procedure. Lysates must be used on the same day or stored at −80 °C.

7. Measurement of brain transketolase activity

In this step the transketolase activity of the samples and quality control samples (QC) is measured with the addition of exogenous TPP (activated) and basally, both in duplicate.
NB. For instructions on how to prepare QC material see (Section 4).

1. Produce 1 in 2 dilutions of QC and tissue lysates in Tris (150 μL lysate + 150 μL 100 mM Tris)
2. Pipette 30 μL quadruplicates of each sample, QC and blank as per the plate map layout in Fig. 1
3. Using a multichannel pipette, add 15 μL 100 mM Tris buffer to rows A, B, E, F (basal). Add 15 μL TPP preparation to rows C, D, G, H (activated)
4. Cover the plate with a self-adhesive seal and mix the plate on the plate shaker at 600 rpm for 60 s
5. Carefully place the plate in the incubator at + 37 °C for 15 min
6. Remove plate from the incubator. Remove the plate seal and, using a multichannel pipette, add 200 μL of the 'assay reagent' per well. Re-seal with a new plate-seal
7. Mix the plate on the plate shaker at 600 rpm for 30 s
8. Remove the plate seal. Load plate onto the plate reader

	1	2	3	4	5	6	7	8	9	10	11	12
A	BLANK	1	3	5	7	9	QC	12	14	16	18	20
B	BLANK	1	3	5	7	9	QC	12	14	16	18	20
C	BLANK	1	3	5	7	9	QC	12	14	16	18	20
D	BLANK	1	3	5	7	9	QC	12	14	16	18	20
E	QC	2	4	6	8	10	11	13	15	17	19	QC
F	QC	2	4	6	8	10	11	13	15	17	19	QC
G	QC	2	4	6	8	10	11	13	15	17	19	QC
H	QC	2	4	6	8	10	11	13	15	17	19	QC

Fig. 1 Recommended plate map layout for the measurement of transketolase activity in brain lysates. Shaded wells are activated with exogenous TPP, unshaded wells are the basal wells without the addition of exogenous TPP.

7.1 Read plate

In this step the absorbance at 340 nm is measured 60 times in each of the wells at 1-minute intervals at a controlled temperature of + 37 °C.

In the first step of the reaction, transketolase catalyzes the reversible transfer of a two-carbon keto group to an aldose sugar in the pentose phosphate pathway. The D-glyceraldehyde-3-phosphate produced is isomerized to dihydroxyacetone phosphate (DHAP) via the enzyme triose phosphate isomerase. DHAP is then reduced to glycerol -3-phosphate by the GPDH enzyme coupled with oxidation of NADH to NAD^+. Oxidation of NADH to NAD^+ is observable spectrophotometrically as a reduction in absorbance at 340 nm.

$$\text{Ribose} - 5 - \text{Phosphate} + \text{Xylulose} - 5 - \text{Phosphate} \xrightarrow{\text{Transketolase}} \text{D} - \text{glyceraldehyde} - 3 - \text{Phosphate}$$

$$\text{Glyceraldehyde} - 3 - \text{Phosphate} \xrightarrow{\text{TPI}} \text{Dihydroxyacetone Phosphate (DHAP)}$$

$$\text{DHAP} + \text{NADH} + \text{H}^+ \xrightarrow{\text{GDH}} \text{Glycerol} - 3 - \text{Phosphate} + \text{NAD}^+$$

7.2 Calculation of brain transketolase 'activity coefficient' (BTKAC)

The BTKAC is the ratio of transketolase activity in the activated sample compared to the basal sample. A ratio of 1 demonstrates saturation of the transketolase in the sample with endogenous TPP, indicative of thiamine sufficiency. Ratios >1 indicate that enhanced transketolase activity was measured with the addition of exogenous TPP in the activated compared to the basal sample. This indicates that transketolase in the sample was not saturated with endogenous TPP suggesting potential deficit in thiamine related enzyme activity. There is currently no BTKAC ratio cut-off to indicate thiamine insufficiency.

1. The first 40 readings, where the temperature of the plate equilibrates with that of the reader, are not included in the calculations
2. Calculate the mean absorbance change per minute measured in each well (including the blank) during the last 20 readings
3. Calculate the mean absorbance change per minute in the four blank wells and subtract this mean value from the measurements for each individual well
   Note: All basal and activated activities are measured in duplicate. Duplicate measurements allow for an assessment of analytical imprecision. Use the following formula to calculate the percent difference between two values:
   $$\frac{(V_1 - V_2)}{((V_1 + V_2)/2)} \times 100$$
   V_1 = mean change in absorbance duplicate "1"
   V_2 = mean change in absorbance duplicate "2"
   A percent difference of up to +/−10% is acceptable. Analysis for samples with a percent difference of >10% should be repeated.
4. Calculate the mean rate for each activated and basal duplicate pair and divide the mean rate of the activated duplicate samples by the mean rate of the basal duplicate samples to calculate the transketolase 'activity coefficient'

For QC samples, keep a record of values for each microplate assayed. Calculate the mean, standard deviation and %CV for each QC sample. After the first 10 microplates have been assayed the results can be reviewed and providing a %CV < 4% is achieved (see Note below). A ± 2 standard deviation range can be calculated and applied using Westgard (Westgard Rules, 2023) or local rules to assess the validity of results of subsequent assays. Usually, if QC results on a microplate fall outside the defined 2 standard deviation range the results should be considered unreliable and the assay should be repeated.

NB.: For this assay %CV > 4 indicates suboptimal precision and suggests either imprecise pipetting or equipment failure.

8. Measurement of basal transketolase activity per gram of protein

In order to calculate absolute basal transketolase activity, measurement of the protein concentration of the sample is required.

8.1 Protein concentration measurement

Measurement of the protein concentration of the Tris diluted lysate allows for calculation of transketolase activity per gram of protein. This is useful for interpretation of samples with a transketolase 'activity coefficient' close to 1 (indicating basal TPP saturation of available enzyme) as saturation can coincide with inadequate levels of TPP if there is a reduced availability of enzyme in the tissue. Expressing the basal activity corrected for protein may improve the reliability of the result by normalizing activity for protein in the sample.

Use of the Pierce™ Bradford Plus Protein Assay Reagent (catalogue number 23288) following the microplate protocol (Pierce™ Bradford Plus Protein Assay Reagent) is recommended for calculation of the protein concentration of the Tris diluted lysates.

8.2 Calculation of basal transketolase activity per gram of protein

Use the Beer-Lambert law to calculate the basal activity, as follows:

$$\Delta A \times \frac{245 \times 1000}{\textit{Response Coefficient} \times 30} = \textit{basal activity, in U}$$

- ΔA is the change in abosrbance per minute in the last 20 readings of the basal sample
- 245 is the total volume (μL) of sample and reagents per well
- 30 is the volume (μL) of diluted brain tissue lysate per well
- The response coefficient represents the slope of an NADH calibration curve measured using the specific conditions for this assay. Calculation of

	1	2	3	4	5	6	7	8	9	10	11	12
A	0 μg/mL standard	500 μg/mL standard	2000 μg/mL standard	1	5	9	13	17	QC 1			
B	0 μg/mL standard	500 μg/mL standard	2000 μg/mL standard	1	5	9	13	17	QC 1			
C	25 μg/mL standard	750 μg/mL standard	QC 1	2	6	10	14	18	QC 2			
D	25 μg/mL standard	750 μg/mL standard	QC 1	2	6	10	14	18	QC 2			
E	125 μg/mL standard	1000 μg/mL standard	QC 2	3	7	11	15	19	QC 3			
F	125 μg/mL standard	1000 μg/mL standard	QC 2	3	7	11	15	19	QC 3			
G	250 μg/mL standard	1500 μg/mL standard	QC 3	4	8	12	16	20				
H	250 μg/mL standard	1500 μg/mL standard	QC 3	4	8	12	16	20				

Fig. 2 Recommended plate map layout for the determination of total protein concentration in brain lysates.

the response coefficient is explained in the Section 9 (Precursor Techniques) and will need to be calculated separately for each laboratory as it can vary depending on specific laboratory conditions.
- U is enzyme activity in μmol/min

Dividing the *basal activity, in U* by the protein concentration of the assay dilution of the brain tissue lysate gives basal activity in U.mg $_{\text{protein}}{}^{-1}$. (Fig. 2).

9. Precursor techniques

9.1 Plate reader suitability

The plate reader must have the ability to maintain a controlled precise temperature in all 96 wells of the plate to produce accurate and precise measurements. A protocol for assessing the reader suitability has been published previously (Jones et al., 2023).

9.2 Calculation of response coefficient

To ensure accuracy it is recommended that the NADH extinction coefficient is calculated for the assay and laboratory conditions rather than using a generic published value. Also, the path length (which for a microplate method is the precise depth of the sample + reagent mixture in each well). These values can be calculated in combination as a response coefficient: First produce a serial dilution of NADH which can be used to produce an NADH standard curve in a microplate. A stock of assay reagent made without addition of NADH should be used as diluent. Use a 1 in

20 dilution (e.g., 0.1 mL + 1.9 mL) of 5.29 mM NADH in the NADH-free assay diluent to produce a 0.265 mM solution, then use serial 1 in 2 dilutions (e.g., 1 mL + 1 mL) of the 0.265 mM solution to produce 0.132 mM and 0.066 mM solutions and use the NADH-free assay diluent as a blank. Add 245 μL of the 0.265, 0.132, 0.066 mM NADH solution and the blank to a microplate in duplicate. Measure the absorbance of each well at 340 nm and calculate the mean absorbance of the duplicates. Plot the absorbance against the mM NADH concentration and calculate the slope. The slope is the response coefficient.

10. Analysis and statistics

Plotting absorbance against time allows the analyst to visualize the reaction and check that the reaction is maintained for the full 60 min. Fig. 3 is an example of an absorbance curve if the reaction has worked as expected. In the case where there is insufficient reagent to maintain the reaction the plots of the absorbance curve may plateau at baseline before the end of the analysis time (Fig. 4). If reagent depletion is observed such as in Fig. 4, it is recommended that the analysis is repeated using a further dilution of the sample. For example, a 2-fold dilution of the brain tissue lysate in cell lysis buffer and analyze according to instructions.

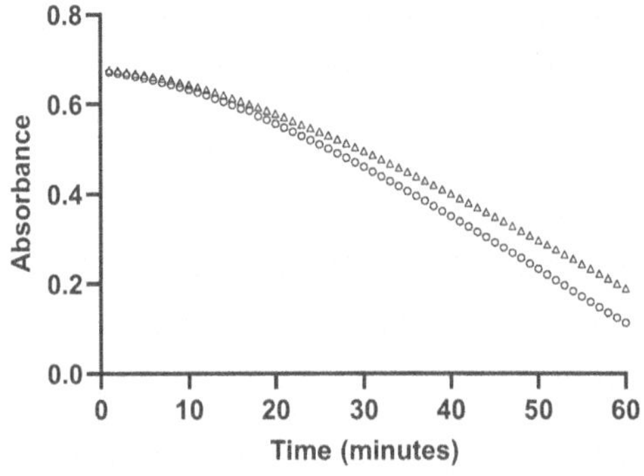

Fig. 3 Basal (open triangles) and activated (open circles) rates for a human brain tissue sample. In this example, the graph shows a steady and constant decrease in absorbance as the NADH is oxidized to NAD^+ during the final 20 min of the reaction (readings for the first 40 min are not used whilst the plate reaches temperature).

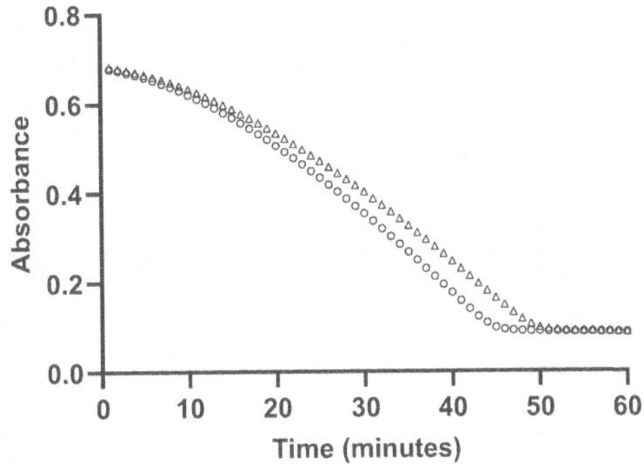

Fig. 4 Basal (open triangles) and activated (open circles) rates for human brain tissue sample where there is insufficient reagent. In this example the graph does not show a steady and constant decrease. The reaction slowing for final 15 min suggests that the enzyme in the sample has depleted the assay reagents.

11. Troubleshooting & optimization

Table 8 presents problems that may be encountered during analysis and possible solutions.

Table 8 Problems and solutions encountered in the brain transketolase activity assay.

Problem	Solution
Unacceptable imprecision (between batch or within batch)	• Determine suitability of reader (see precursor techniques) • Check reaction traces for abnormalities • Repeat using more concentrated lysate dilution
Low rates of reaction	• Repeat using a more concentrated lysate dilution
Reagent depletion (Fig. 4)	• Repeat with a less concentrated lysate dilution

12. Conclusions

The assessment of transketolase enzyme activity, measured via the depletion of NADH, as a biomarker of the functional capacity of tissue thiamine has a long history. Its primary application has been in the assessment of thiamine status in humans through measurement of erythrocyte transketolase activity for which detailed protocols have been published. In this chapter, we have described a protocol to measure transketolase activity in 96-well plate format using a spectrophotometer and optimized for brain tissue samples. The protocol is applicable to studies of thiamine-related enzyme activity in brains obtained from animal studies and human brain samples post-mortem. The described techniques provide the tools to further investigate the potential role of thiamine, and its analogues, such as Benfotiamine, in the treatment of Alzheimer's disease and other neuropathies.

Acknowledgements

This research was supported by the National Institute of Health and Care Research (NIHR) Cambridge Biomedical Research Centre (NIHR2033312). The views expressed are those of the authors and not those of the NIHR or Department of Health and Social Care (DHSC). Research was also supported by NIH grants R01AG072505, R01AG082228 and RO1AGO76634.

References

Bandyopadhyay, S., Roy, A. K., Baker, S., Stephenson, K. K., Ge, X., Wang, Y., et al. (2025). Micronutrient biomarker selection and assay methods and performance in double-blind, randomized, controlled micronutrient dose response (MiNDR) trials among women of reproductive age and pregnant women in rural Bangladesh. *Current Developments in Nutrition, 9*(10), 107546. https://doi.org/10.1016/j.cdnut.2025.107546.

Bettendorff, L. (2023). Synthetic thioesters of thiamine: Promising tools for slowing progression of neurodegenerative diseases. *International Journal of Molecular Sciences, 24*(14), https://doi.org/10.3390/ijms241411296.

Bourassa, M. W., Gomes, F., Jones, K. S., Koulman, A., Prentice, A. M., & Cerami, C. (2022). Thiamine deficiency in Gambian women of reproductive age. *Annals of the New York Academy of Sciences, 1507*(1), 162–170. https://doi.org/10.1111/nyas.14695.

Edwards, K. A. (2026). Reassessing transketolase assays: Methodological considerations for detecting functional Thiamine deficiency. *Annals of the New York Academy of Sciences, 1557*(1), e70245. https://doi.org/10.1111/nyas.70245.

Gangolf, M., Czerniecki, J., Radermecker, M., Detry, O., Nisolle, M., Jouan, C., et al. (2010). Thiamine status in humans and content of phosphorylated thiamine derivatives in biopsies and cultured cells. *PLoS One, 5*(10), e13616. https://doi.org/10.1371/journal.pone.0013616.

Gibson, G. E., Feldman, H. H., Zhang, S., Flowers, S. A., & Luchsinger, J. A. (2022). Pharmacological thiamine levels as a therapeutic approach in Alzheimer's disease. *Frontiers in Medicine, 9*, 1033272. https://doi.org/10.3389/fmed.2022.1033272.

Gibson, G. E., Hirsch, J. A., Fonzetti, P., Jordan, B. D., Cirio, R. T., & Elder, J. (2016). Vitamin B1 (thiamine) and dementia. *Annals of the New York Academy of Sciences, 1367*(1), 21–30. https://doi.org/10.1111/nyas.13031.

Gibson, G. E., Sheu, K. F., Blass, J. P., Baker, A., Carlson, K. C., Harding, B., et al. (1988). Reduced activities of thiamine-dependent enzymes in the brains and peripheral tissues of patients with Alzheimer's disease. *Archives of Neurology, 45*(8), 836–840. https://doi.org/10.1001/archneur.1988.00520320022009.

Giguère, J. F., & Butterworth, R. F. (1987). Activities of thiamine-dependent enzymes in two experimental models of thiamine deficiency encephalopathy: 3. Transketolase. *Neurochemical Research, 12*(3), 305–310. https://doi.org/10.1007/BF00972141.

Hauber, M. M., Todisco, V., Nordahl, O., Tibblin, P., Fridolfsson, E., Kärvegård, E., & Hylander, S. (2026). Thiamine allocation and deficiency status throughout the life cycle of cod. *Ecology and Evolution, 16*(1), e72828. https://doi.org/10.1002/ece3.72828.

Hess, S. Y., Smith, T. J., Fischer, P. R., Trehan, I., Hiffler, L., Arnold, C. D., et al. (2020). Establishing a case definition of thiamine responsive disorders among infants and young children in Lao PDR: protocol for a prospective cohort study. *BMJ Open, 10*(2), e036539. https://doi.org/10.1136/bmjopen-2019-036539.

Jones, K. S., Parkington, D. A., Bourassa, M. W., Cerami, C., & Koulman, A. (2023). Protocol and application of basal erythrocyte transketolase activity to improve assessment of thiamine status. *Annals of the New York Academy of Sciences, 1521*(1), 104–111. https://doi.org/10.1111/nyas.14962.

Jones, K. S., Parkington, D. A., Cox, L. J., & Koulman, A. (2021). Erythrocyte transketolase activity coefficient (ETKAC) assay protocol for the assessment of thiamine status. *Annals of the New York Academy of Sciences, 1498*(1), 77–84. https://doi.org/10.1111/nyas.14547.

Jones, K. S., Bettendorff, L., Frank, E. L., & Whitfield, K. C. (2026). Assessment of thiamine status in humans. *Methods in Enzymology*. https://doi.org/10.1016/bs.mie.2026.06.018 in press.

Koulman, A., Jones, K. S., Collins, D., Parkington, D., Amoutzopoulos, B., Page, P., et al. (2026). Ethnic differences in thiamine status persist after adjusting for diet; findings from the UK national diet and nutrition survey. *Clinical Nutrition ESPEN, 72*, 102893.

Ministère de la Santé Publique de Madagascar; & Office National de Nutrition; UNICEF; CAETIC Développement; Institut de Recherche pour le Développement; GroundWork. (2025). Madagascar National Micronutrient Survey Report 2024.

PierceTM Bradford Plus Protein Assay Reagent. (n.d.) Retrieved https://www.thermofisher.com/order/catalog/product/23238.

Ramamoorthy, K., Yoshimura, R., Al-Juburi, S., Anandam, K. Y., Kapadia, R., Alachkar, A., et al. (2022). Alzheimer's disease is associated with disruption in thiamin transport physiology: A potential role for neuroinflammation. *Neurobiology of Disease, 171*, 105799.

Smith, T. J., Johnson, C. R., Koshy, R., Hess, S. Y., Qureshi, U. A., Mynak, M. L., et al. (2021). Thiamine deficiency disorders: A clinical perspective. *Annals of the New York Academy of Sciences, 1498*(1), 9–28. https://doi.org/10.1111/nyas.14536.

Therrien, C. A., Baker, P. M., Garner, S., Swanson, H. K., & Neff, B. D. (2026). Dietary thiaminase alters morphology and decreases swimming performance of lake trout (*Salvelinus namaycush*). *Scientific Reports, 16*(1), 5493.

Westgard Rules. Retrieved 22 September 2023 from https://www.westgard.com/mLtirule.htm.

Zhao, Y., Pan, X., Zhao, J., Wang, Y., Peng, Y., & Zhong, C. (2009). Decreased transketolase activity contributes to impaired hippocampal neurogenesis induced by thiamine deficiency. *Journal of Neurochemistry, 111*(2), 537–546.

Further reading

Whitfield, K. C., Bourassa, M. W., Adamolekun, B., Bergeron, G., Bettendorff, L., Brown, K. H., et al. (2018). Thiamine deficiency disorders: Diagnosis, prevalence, and a roadmap for global control programs. *Annals of the New York Academy of Sciences, 1430*(1), 3–43.

CHAPTER EIGHT

A general method for the assay of thiamine pyrophosphokinase activity

Lucien Bettendorff*
Laboratory of Neurophysiology, GIGA Institute, University of Liège, Liège, Belgium
*Corresponding author. e-mail address: L.Bettendorff@uliege.be

Contents

Abstract

Thiamine diphosphate (ThDP) is a coenzyme for several essential enzymes in cell metabolism (mainly transketolase, pyruvate and 2-oxoglutarate dehydrogenases). In eukaryotic cells ThDP is synthesized by pyrophosphorylation of thiamine (vitamin B1) by ATP or UTP and catalyzed by thiamine pyrophosphokinase isoform 1 (TPK1). Complete inactivation of TPK1 is lethal, but recently several hereditary mutations in the corresponding gene were described in human families. The phenotype is generally severe with acute encephalopathy and fatal outcome when not treated with excess thiamine. This suggests that the mutated enzyme still retains a fraction of its native activity. In order to estimate the effect of the mutations on enzyme activity it is essential to benefit from a reliable assay of TPK1 activity. This enzyme has a low K_M for thiamine ($K_M < 0.1$ μM) but a low turnover ($k_{cat} = 0.03\ s^{-1}$) requiring very sensitive methods for the detection of the product ThDP by HPLC followed by fluorometric

Methods in Enzymology, Volume 735
ISSN 0076-6879, https://doi.org/10.1016/bs.mie.2026.06.026

(after derivatization into fluorescent thiochrome derivatives) or mass spectrometric detection. Moreover, ThDP is a noncompetitive inhibitor of TPK1 with a $K_I = 0.4$ μM. Hence, when measuring TPK1 activity care must be taken that the reaction is stopped before inhibition becomes significant. Here, we present a powerful HPLC method coupled to fluorometric detection allowing TPK1 assay either in the forward or the reverse direction. The importance of such a method is further highlighted by recent data showing decreased expression of TPK1 in blood and brain of Alzheimer patients making TPK1 assay an interesting alternative to the estimation of TPK1 expression.

1. Introduction

Thiamine (vitamin B1) is an essential nutrient for animals, which are unable to synthesize this molecule. After intestinal absorption thiamine enters the bloodstream and it is rapidly captured by the cells by means of specific transporters. The best known of these transporters are ThTr1 and ThTr2, which are the products of genes SLC19A2 and SLC19A3 belonging to the folate transporter family (Makarchikov et al., 2025; Marcé-Grau et al., 2019). Inside the cells thiamine is pyrophosphorylated to thiamine diphosphate (ThDP) by a specific enzyme, thiamine pyro- or diphosphokinase (TPK1, EC 2.7.6.2) (Fig. 1):

$$Thiamine + ATP \rightleftarrows ThDP + AMP \quad (1)$$

While the co-substrate is generally indicated as ATP, recent data suggest that, at least in mammals, UTP could be a better substrate (Sahu et al., 2024). Another isoform, TPK2, is present only in plants (Luo et al., 2025).

ThDP is an essential coenzyme for over 40 enzymes and enzyme complexes, among which 6 are found in animals (Prajapati et al., 2022). They include such important enzymes such as transketolase in the pentose phosphate pathway and the mitochondrial enzyme complexes pyruvate, 2-oxoglutarate and branched chain 2-oxoacid dehydrogenases. These enzymes occupy key strategic positions in cellular catabolism and ThDP deficiency impairs their activity with severe consequences on cell survival. In humans, thiamine deficiency mainly affects the nervous and the cardiovascular systems resulting in beriberi or Wernicke-Korsakoff syndrome (a complication of chronic alcoholism) (Smith et al., 2021). In addition to nutritional thiamine deficiency, mutations leading to partial inactivation of ThDP-dependent enzymes, thiamine and ThDP transporters, and TPK1 leads to thiamine deficiency disorders with severe consequences for the nervous system (Makarchikov et al., 2025). In some cases, the patients

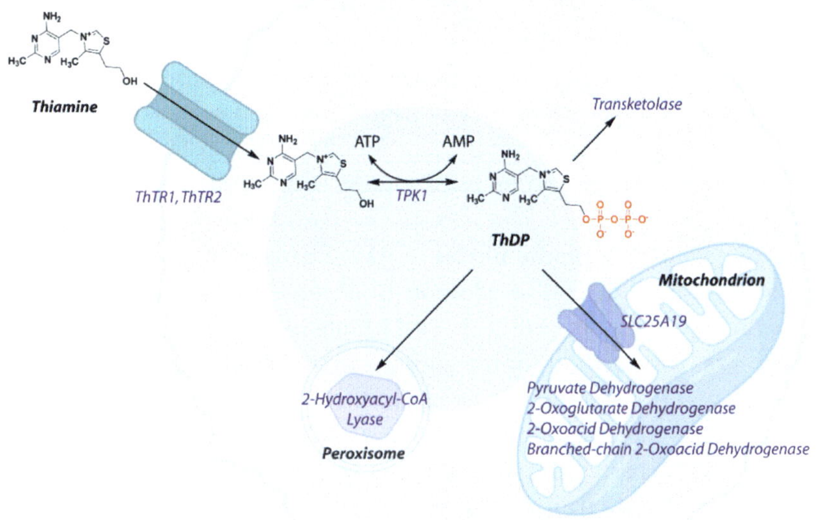

Fig. 1 Thiamine is transported into eukaryotic cells by the specific plasma membrane transporters ThTR1 and ThTR2. Inside the cytoplasm thiamine is pyrophosphorylated to ThDP by TPK1. ThDP is a coenzyme for the cytosolic transketolase. The bulk of ThDP is transported into mitochondria by the specific inner mitochondrial membrane ThDP transporter (TM) SLC25A19 (Jiang, 2024) where it associates with at least four 2-oxoacid dehydrogenases. ThDP is also coenzyme for the peroxisomal 2-hydroxyacyl-CoA-lyase.

respond to the administration of high doses of thiamine, highlighting the importance of identifying the underlying cause and start a therapeutic strategy before irreversible lesions occur.

With the advent of cheap and high throughput DNA sequence facilities, several mutations in the human *tpk1* gene were identified, with widespread and severe consequences: decreased blood ThDP levels, decreased pyruvate utilization and decreased amounts of the transketolase product D-sedoheptulose-7-phosphate (Sahu et al., 2024). While extremely rare, mutations in *tpk1* are often underdiagnosed. TPK1 deficiency is known as thiamine dysfunction syndrome 5 (OMIM 614458) and, as of beginning 2026, only about 28 patients worldwide are known suffering from this disease (Dallan et al., 2025; Zhao et al., 2023). Clinical manifestations can be extremely severe, with acute encephalopathy, seizures and elevated lactate plasma concentrations with often fatal outcome. Improvements can be observed in response to thiamine treatment (Thompson et al., 2023). It must be emphasized that if TPK1 deficiency is very rare, because most probably only patients with a residual enzyme activity are viable. It can be expected that total inactivation of *tpk1* will results in embryos dying at early stages of development.

Very recently, an interesting observation has shown that ThDP levels are decreased in blood and brain of patients with Alzheimer's disease and that astrocytic TPK1 deficiency exacerbates Aβ burden leading to spatial memory deficits in a mouse model of AD (Zhang et al., 2026).

For these reasons and in addition to a purely academic interest, it is important to evaluate the effect of TPK1 mutations on enzyme activity and to have a reliable method for measuring TPK1 activity. This is particularly important as TPK1 is inhibited by its product ThDP, potentially leading to an important underestimation of enzyme activities. For instance, one of the first enzymatic characterizations of this enzyme concluded that the equilibrium (eq. 1) was strongly displaced to the left ($K_{eq} = 10^{-5}$) (Peterson et al., 1975), though the reaction is coupled to ATP consumption. This result is of course erroneous and due to product inhibition by ThDP. For that reason, it is important not only to measure the forward reaction but also the reverse reaction (Sambon et al., 2022). Here we report a reliable and high sensitivity HPLC-based method for the determination of TPK1 activity in both directions (forward and reverse).

2. Assay of TPK1 activity

2.1 Rationale

TPK1 has a very high apparent affinity for its substrate thiamine ($K_M < 0.1$ μM), moreover the ThDP product is a non-competitive inhibitor of the enzyme with a $K_I \sim 0.4$ μM (for a litterature review see Sambon et al., 2022). Hence, for an accurate determination of the enzyme activity, thiamine concentrations must be ≤0.1 μM and the reaction must be stopped before significant ThDP concentrations accumulate. In order to overcome this difficulty, we show that the determination of the reverse reaction is also possible by detecting the formation of thiamine. Both methods require a very sensitive method for the determination of either ThDP in the forward direction or thiamine for the reverse reaction. A sufficiently high sensitivity can be achieved by a HPLC separation using fluorescent detection after derivatization of thiamine derivatives in fluorescent thiochrome (Fig. 2).

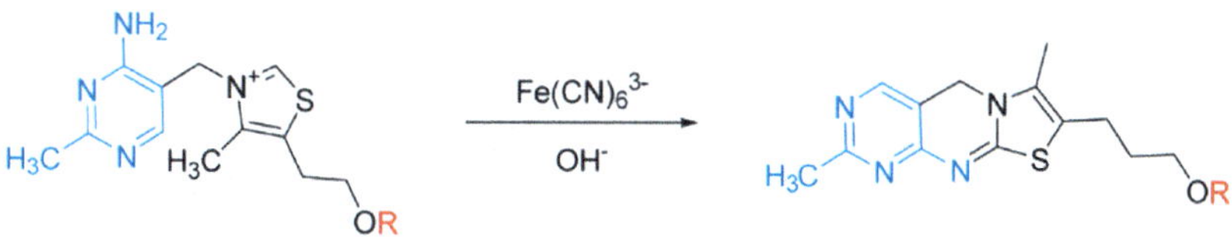

Fig. 2 Oxidation of thiamine derivatives to thiochrome derivatives in strongly alkaline medium, typically 5% NaOH and 0.2 mM potassium ferricyanide with λ_{ex} = 365–370 nm and λ_{em} = 433–435 nm (Bettendorff et al., 1991; Lu & Frank, 2008) (Thiamine, R = H; ThMP, R = $-PO_3^{2-}$; ThDP, R = $-PO_2^{-}-O-PO_3^{2-}$).

While the TPK protein can be easily quantified using antibody-based techniques, there is no simple (colorimetric) method available for the assay of TPK activity. This is due in large part because of the low specific activity 1 $nmol.mg^{-1}.s^{-1}$, corresponding to $k_{cat} = 0.03\ s^{-1}$ and the low K_M of the enzyme for thiamine (low nanomolar range) (Sambon et al., 2022). This requires that any TPK assay must be able to reliably detect nanomolar concentrations of substrate. This is highly improbable with absorbance-based techniques in the absence of specific chromophores. In addition, without chromatographic separation discrimination between thiamine and ThDP would be impossible.

2.2 Materials, equipment and reagents

- HPLC equipment for isocratic or gradient elution equipped with a flow-through fluorescence spectrometer. A fluorescence spectrometer is required rather than a filter fluorometer as it provides a much better selectivity. Injection loop 20 μL.
- Hamilton PRP-1 HPLC column, 5 μm, 150 × 4.1 mm, pore size 100 Å (N° 79444). Alternatively, a 250 × 4.1 mm column (N° 79820) from the same manufacturer can be used.
- ATP (A2383 Sigma-Aldrich)
- AMP (01970 Fluka Biochemika)
- Thiamine (T4625 Sigma-Aldrich)
- Thiamine monophosphate (T8637 Sigma-Aldrich)
- ThDP (Sigma-Aldrich C8754)
- $MgSO_4$ (5886 Merck)
- Tetrahydrofuran (HPLC grade, without stabilizer)
- Diethyl ether (analytical grade without stabilizer)
- Tris-HCl buffer (0.1 M, pH 7.5), analytical grade
- NaOH (prepare an 15% solution in water, 3.75 M solution), analytical grade

- Potassium ferricyanide, analytical grade
- Trichloroacetic acid, analytical grade
- Na_2HPO_4, analytical grade
- NaH_2PO_4, analytical grade
- Tetra-n-butylammonium hydrogen sulfate (TBA, 8.18858 Sigma-Aldrich)

2.3 Protocols

The protocol is described for recombinant mouse TPK1 (Sambon et al., 2022) and can be adapted for any enzyme source. All solutions are prepared with milli-Q water. The preparation of a suitable enzyme source is a prerequisite. Because of the low enzymatic activity care must be taken to avoid unnecessary dilution of the sample during preparation. All steps should be carried out at 4 °C.

2.3.1 Forward reaction: Thiamine + ATP → ThDP + AMP

1. Prepare an ATP solution (20 mM) adjusted to pH 7 with NaOH 15 %.
2. Prepare a $MgSO_4$ solution (25 mM).
3. Prepare an enzyme solution from the stock solution (in this case 12.9 $mg.mL^{-1}$), of approximately 1.29 $\mu g.mL^{-1}$ by dilution in Tris-HCl buffer (180 mM, pH 7.5).
4. For the forward reaction, the assay mixture (300 μL) contained Tris-HCl (0.084 M, pH 7.5), 2.0 mM ATP, 2.5 mM MgSO4 and 0.5 μM thiamine (Table 1). Note that under Mg^{2+} and ATP concentrations close to physiological ones, an optimum enzyme activity is obtained with Mg^{+2}/ATP ratio of 1.2 (Sambon et al., 2022). The specific activity of the preparation (pure recombinant protein, see 2.4.1) was 60 $nmol.mg^{-1}.min^{-1}$ (k_{cat} = 0.03 s^{-1},

Table 1 Incubation medium for the forward reaction (alternatively, a 10 mM ATP – 12.5 mM $MgSO_4$ can be prepared and 60 μL are added to the reaction mixture).

	Initial concentration	Volume	Final concentration
Thiamine	3 μM	50 μL	0.5 μM
ATP	20 mM	30 μL	2 mM
$MgSO_4$	25 mM	30 μL	2.5 mM
Tris-HCl (pH 7.5)	180 mM	140 μL	84 mM
TPK	1.14 $\mu g.mL^{-1}$	50 μL	0.19 $\mu g.mL^{-1}$
Total		300 μL	

considering a molecular mass of 29, 473 Da for the recombinant His-tagged protein). In the incubation medium (300 μL), the enzyme concentration was 0.19 μg.mL^{-1} corresponding to an activity of 11.10^{-6} μmol.mL^{-1}.min^{-1} or 3.3.10^{-6} μmol.min^{-1} (U, enzyme units).

All solution should be at 4 °C and the incubation medium prepared in an ice bath. The incubation is carried out for 20 min at 37 °C with the controls being incubated in the absence of enzyme. The reaction is stopped by addition of 60 μL trichloroacetic acid (TCA, 60 %). After centrifugation (10,000 g, 15 min), the TCA is extracted with 1.5 mL diethyl ether by vigorous shaking for 10 s using a table vortex mixer. The ethereal phase is siphoned with a glass pipette connected to a vacuum pump. The procedure is repeated twice. Bubble nitrogen through the remaining non-ether layer to thoroughly remove any dissolved ether. ThDP and thiamine are determined by HPLC in the aqueous phase (Bettendorff et al., 1991).

2.3.2 Reverse reaction: ThDP + AMP → Thiamine + ATP

1. Prepare an AMP solution (500 mM) adjusted to pH 7 with NaOH 15 %.
2. Prepare a $MgSO_4$ solution (30 mM).
3. For the reverse reaction, which proceeds at a slower rate than the forward reaction, the assay mixture (300 μL) contained Tris-HCl (0.1 M, pH 7.5), AMP (50 mM), $MgSO_4$ (30 mM), ThDP (0.3 μM) and an enzyme concentration of 2.4 μg.mL^{-1} (Table 2). As the reverse reaction is much slower than the forward reaction, a much higher enzyme concentrations must be used than for the forward reaction.
4. The incubation is carried out for 20 min at 37 °C with the controls being incubated in the absence of enzyme. The reaction is stopped by addition of 60 μL TCA (60 %). The samples are treated exactly as for the forward reaction (2.3.1.). The samples can be stored at −20 °C until use.

Table 2 Incubation medium for the forward reaction.

	Initial concentration	Volume	Final concentration
ThDP	1.8 μM	50 μL	0.3 μM
AMP	500 mM	30 μL	50 mM
$MgSO_4$	300 mM	30 μL	30 mM
Tris-HCl (pH 7.5)	180 mM	140 μL	84 mM
TPK	14.4 μg.mL^{-1}	50 μL	2.4 μg.mL^{-1}
Total		300 μL	

2.3.3 HPLC procedure

1. *Derivatization.* Thiamine is derivatized into fluorescent thiochromes (Fig. 2) by addition of 80 μL alkaline ferricyanide solution (4.3×10^{-3} M potassium ferricyanide in 15 % sodium hydroxide) to 50 μL of the extracted sample. Inject the sample onto the HPLC column within 1 min of derivatization. Longer delays may lead to side reactions leading to decreased fluorescence.
2. The chromatographic system is equipped with a fluorescence spectrometer. The excitation wavelength is set at 365 nm and the emission wavelength at 433 nm. A PRP-1 reversed-phase analytical column (150 or 250 mm) is used, protected by a PRP-1 guard column cartridge (25–2.3 mm I.D.). A 20-μL injection loop was used. An isocratic system is sufficient for the analysis, but a system with two pumps for gradient elution allows automatic washing of the column after use.
3. The mobile phase is composed of sodium phosphate buffer (50 mM), tetrabutylammonium hydrogen sulfate (TBA, 25 mM), and 4 % tetrahydrofuran adjusted at pH 9.0 with NaOH. Elution is isocratic at a flow rate of 0.5 $mL.min^{-1}$. A stock solution of phosphate buffer and TBA can be prepared and filtered on 0.45 μm pore ultrafilters. Before use, add 4 % THF and homogenize.
4. Prepare standard solutions of thiamine, thiamine monophosphate, ThDP at 10 mM. The standard solutions are diluted and combined to one solution containing 0.25 μM of each derivative. Thiamine monophosphate is included to detect any side reaction catalyzed by phosphatases and leading to hydrolysis of ThDP in whole tissue samples. Thiamine (reverse reaction) and ThDP (forward reaction) formed are estimated by comparing the peak areas to those of the standard solution.
5. When starting a day's analysis, run a 10-minute gradient from 90 % methanol to water. Wash the column with water for 10 more minutes before connecting the column to the elution buffer. Allow a further 45 min for the column to equilibrate in the elution buffer. Indeed, equilibration of the column with the relatively low concentration of THF requires at least 10 column volumes of buffer. At the end of the day put the column in water and make a 2-hour wash with water followed by a 1-hour gradient to methanol 90 % and a 2-hour wash in 90 % methanol, all at a flowrate of 0.25 $mL.min^{-1}$.
6. Any other HPLC separation method allowing a good resolution between the thiamine and ThDP peaks can be used (Jonus et al., 2018; Lu & Frank, 2008), provided the detection limit is at least 10 $nmol.L^{-1}$.

Fig. 3 shows a chromatogram obtained for a typical reaction carried out with the recombinant TPK.

2.3.4 Determination of kinetic parameters of TPK

TPK1 activity was measured as a function of thiamine concentration in the presence of an ATP concentrations close to physiological (2 mM) and 2.5 mM Mg^{2+} (Fig. 4). The relationship was hyperbolic and non-linear regression to the Michaelis-Menten equation gave a K_M value of 0.072 μM (R^2 = 0.96). The V_{max} was calculated to be 60 $nmol.mg^{-1}.min^{-1}$, corresponding to a k_{cat} of 0.03 s^{-1} for the recombinant enzyme (MW = 29,473). When the same data were drawn as a Lineweaver-Burk, a linear relationship between $1/v_i$ and 1/[thiamine] was obtained (Fig. 4, insert).

As we previously discussed at length, it was impossible to obtain an equilibrium constant for the forward reaction because of the retro-inhibition of TPK1 by the product ThDP (Sambon et al., 2022). Therefore, we

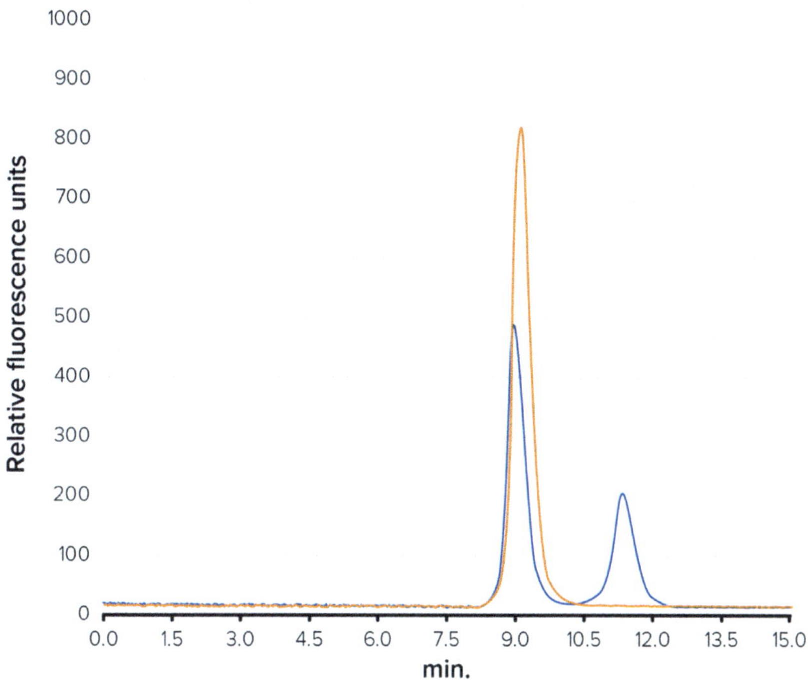

Fig. 3 Superposition of two chromatograms (relative fluorescence at 433 nm as a function of elution time) illustrating the synthesis of ThDP by purified recombinant TPK1. In orange, a control obtained in the absence of enzyme and in blue in the presence of enzyme (0.19 μg. mL^{-1}). The incubation was carried out as described in Table 1. The first peak (9 min) corresponds to the substrate thiamine and the second peak (11.5 min) corresponds to the product ThDP.

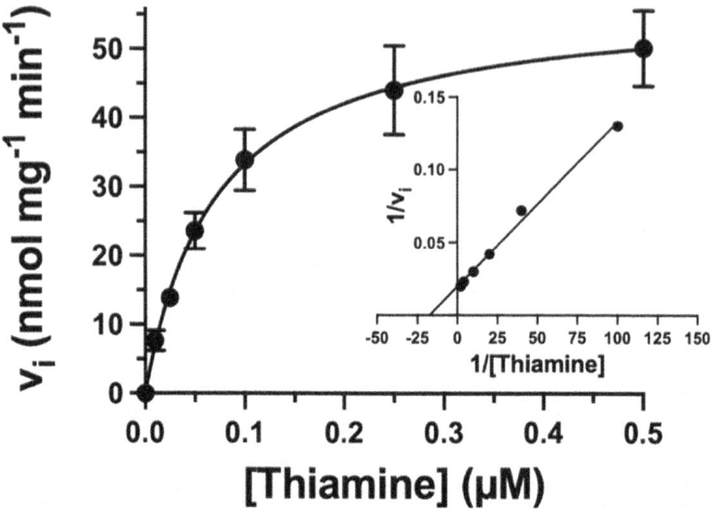

Fig. 4 Kinetic properties of mouse recombinant TPK1. Dependence of ThDP synthesis on thiamine concentration (2 mM ATP, 2.5 mM Mg^{2+}). The data were fitted to the Michaelis-Menten equation using non-linear regression. The inset shows a Lineweaver-Burk plot of the same data fitted to a linear regression line. *From Sambon et al. (2022).*

used the reverse reaction which can proceed at a significant, though very low rate, provided there is a large excess of AMP (≥5 mM) and a relatively high amount of enzyme (Fig. 5). First, the activity was estimated by measuring the appearance of thiamine in the assay medium at two different concentrations of ThDP, 0.3 and 3.0 μM and a large excess of AMP (50 mM) (Fig. 5A).

The initial rate is strongly dependent on AMP concentration between 5 and 50 μM (Fig. 5B), suggesting a low affinity (K_M ~20 mM). In contrast, the affinity for ThDP seems to be very high: apparent K_M values of 0.043 ± 0.017 μM (5 mM AMP) and 0.053 ± 0.011 μM (50 mM AMP) were calculated (Fig. 5B). A maximum specific activity of 5.3 nmol.mg^{-1}.min^{-1} (V_{max} in Fig. 5B) was calculated, corresponding to a k_{cat} of 0.0026 s^{-1}, a value 10–20 times lower than for the forward reaction.

As for the forward reaction, the reverse reaction absolutely required Mg^{2+} ions with an EC_{50} (apparent K_a) of 7.1 ± 1.7 mM and 12 ± 5 mM in the presence of 5 or 50 mM AMP, respectively (Fig. 5C).

For the determination of K_{eq}, we repeated the experiment under different conditions (1 μM ThDP, 5 mM AMP and 15 mM Mg^{2+}) and

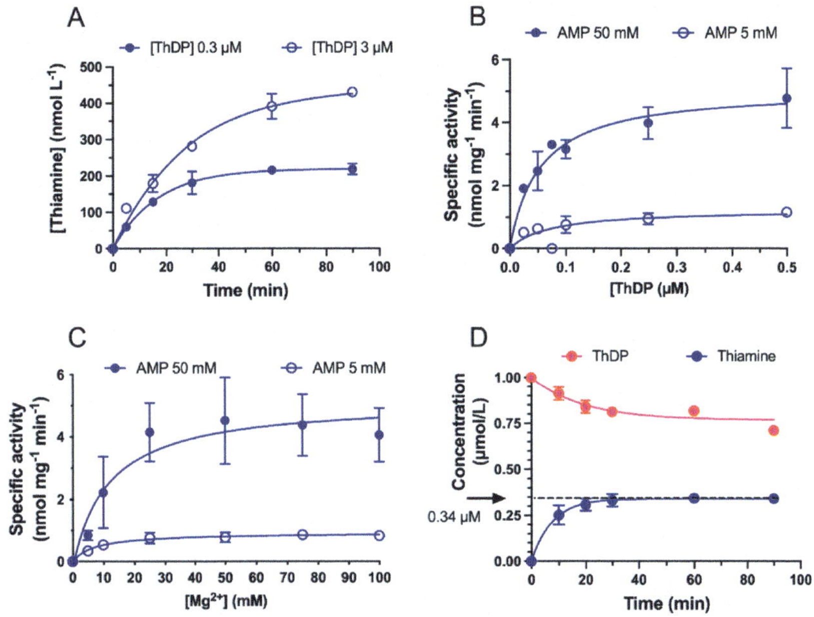

Fig. 5 Kinetic parameters for the mouse TPK reverse reaction ThDP + AMP → Thiamine + ATP at 37 °C. (A) Kinetics of thiamine synthesis at two concentrations of ThDP (0.3 and 3 µM). The AMP concentration was 50 mM and Mg^{2+} concentration was 30 mM. (B) Dependence of the initial rate of the TPK reverse reaction on ThDP concentration at two different concentrations of AMP and at $[Mg^{2+}] = 30$ mM. (C) Dependence of the TPK reverse reaction on Mg^{2+} concentration at two different concentrations of AMP and at [ThDP] = 3 µM. (D) Kinetics of the reverse reaction catalyzed by TPK with ThDP (1 µM), AMP (5 mM), Mg^{2+} (15 mM). The enzyme concentrations were 2.42 µg.mL^{-1} in A, B and C and 86 µg.mL^{-1} in D. Data were analyzed and regression curves were generated using GraphPad Prism 9. In panels A and D, the curves were obtained by a non-linear fit for one phase association: $Y = Y_{max} \times (1 - e^{-kt})$. In panel C, we used a agonist concentrtaion vs response model: $Y = Y_{max} \times x/(EC_{50} + x)$, while a Michaelis-Menten model was used in panel B. (Mean ± SD, n = 3–6). *From Sambon et al. (2022).*

measured the disappearance of ThDP and the appearance of thiamine as a function if time (Fig. 5D). In order to rapidly reach a plateau and thus minimize enzyme denaturation, we used a significantly higher enzyme concentration (86 µg.mL^{-1} instead of 2.42 µg.mL^{-1}). The plateau was reached after 30 min and maintained over a significant time with $[\text{Thiamine}]_{eq} = 0.34$ µM (Fig. 5D). We calculated the apparent equilibrium constant using the simplified equation (Eq. (I)) used by Peterson et al., 1975 (Peterson et al., 1975; Sambon et al., 2022). It is assumed that at equilibrium the thiamine and ATP concentrations are equal and the AMP concentration, largely in excess with respect to the ThDP concentration, is considered equal to its initial value:

$$K_{eq} = \frac{[ATP]_{eq} \times [Th]_{eq}}{[AMP]_{eq} \times [ThDP]_{eq}} = \frac{([Th]_{eq})^2}{[AMP]_{init} \times ([ThDP]_{init} - [Th]_{eq})} \quad (2)$$

$$K_{eq} = \frac{(0.34 \cdot 10^{-6})^2}{5000 \cdot 10^{-6} \times (1 \cdot 10^{-6} - 0.34 \cdot 10^{-6})} = 3.5 \cdot 10^{-5}$$

With $K_{eq} = 3.5\ 10^{-5}$ for the reverse reaction (ThDP + AMP → Thiamine + ATP), the reaction is largely in favor of the physiological direction, i.e., the synthesis of ThDP from ATP. The order of magnitude of the K_{eq} was confirmed under several different conditions (Sambon et al., 2022).

2.4 Precursor techniques

2.4.1 Preparation of recombinant human or mouse TPK1

TPK1 from any species can be prepared as described with well-established techniques (Nosaka et al., 1993, 1999, 2001; Onozuka & Nosaka, 2003; Sahu et al., 2024; Sambon et al., 2022). Preparation of recombinant precursor techniques are now most often handled by contract with specialized technical platforms. In our case, mouse TPK1 was produced by the Protein Factory of the Center for Protein Engineering at the University of Liège (Belgium) as a His-tagged protein and purified to homogeneity as described earlier (Sambon et al., 2022). In this study, we used a batch at a concentration of 12.9 mg.mL^{-1}, as determined by UV absorption at 280 nm.

2.4.2 Preparation of TPK extracts from biological sources

This can be done from animal (Artsukevich et al., 1977; Peterson et al., 1975), plant (Mitsuda et al., 1975, 1979) or bacterial (Sanemori & Kawasaki, 1980) sources. Note that enterobacteria such as *E. coli* lack TPK. Considering the importance of thiamine metabolism in brain tissue, we will briefly describe the method for the isolation of TPK from human brain samples (Bettendorff et al., 1996).

Typically, a minimum of 25 mg of brain tissue is required. Here we describe the procedure for 100 mg of brain tissue all volumes can be adapted according to the amount of starting material.

1. 100 mg of human brain tissue are rinsed in 0.32 M sucrose containing 10 mM Hepes-Tris (pH 7.4) and homogenized in a Teflon—glass homogenizer in 1 mL of Tris-HC1 buffer (20 mM, pH 7.4) containing 2 mM 2-mercaptoethanol and 1 mM EDTA.
2. Centrifuge (15,000 × *g*, 40 mm) and use the supernatant for determination of TPK activity as described above.

2.5 Safety considerations and standards

2.5.1 General remark

All procedures must be approved by the local ethical committees, in particular the use of animal tissues, human biopsies or post-mortem tissues.

2.5.2 Specific safety considerations for some chemicals

- *Diethyl ether:* Keep away from heat, hot surfaces, sparks, open flames and another ignition sources. Wear protective gloves/eye protection. Store in a well-ventilated place. Keep cool.
- *Methanol:* flammable liquids, highly toxic (even vapors) in oral, dermal contact, even in small amounts. Work in well-ventilated area. Do not leave open to the air. Check with your local environmental office to dispose of excess methanol.
- *NaOH:* highly irritating, caustic, corrosive, Wear chemical-resistant gloves, goggles, and face shields, ensuring proper ventilation. Always add NaOH to water—never water to NaOH—to avoid violent boiling. Store in sealed containers, away from acids and metals. Do not store in glass vessels
- *Tetrahydrofuran (THF):* flammable: very low flash point (−14 °C), highly flammable vapors, Wear solvent-resistant gloves (nitrile thickness ≥0.4 mm or butyl). Preferentially work under an extractor hood at all times.
- *Trichloroacetic acid:* Causes severe skin burns and eye damage. May cause respiratory irritation. Very toxic to aquatic life with long lasting effects. Wear protective gloves/eye protection. In general, never dispose of it down the drain. Collect in labelled canisters for halogenated/non-halogenated solvent waste according to your establishment's recommendations.

2.6 Analysis and statistics

The main point with this enzyme assay is to be sure to work as close as possible to conditions of initial velocity so as to prevent inhibition by excess of the product ThDP. Therefore, the enzyme concentration and the time of incubation should be set so as that less than 10 % of thiamine is consumed for the forward reaction. The HPLC separation allows the simultaneous visualization of the thiamine and ThDP peaks on the same chromatogram facilitating the estimation of the ThDP/thiamine ratio.

The enzyme has a very low catalytic power in the reverse direction. It must however be kept in mind that at 2.4 $\mu g.mL^{-1}$ and assuming a molecular mass of 29.5 kDa for the recombinant enzyme, the enzyme concentration is 0.08 μM, the same order of magnitude as the substrate concentration. Hence, an essential condition for the validity of the Michaelis-Menten equation ($[E] \ll [S]$) is not verified and, the reverse reaction should only be used when the forward reaction is not feasible, for instance for the determination of equilibrium constants (Sambon et al., 2022).

2.7 Related techniques

Estimation of TPK expression either by estimation of mRNA or protein abundance can be an alternative (Jonus et al., 2018; Xia et al., 2024; Zhang et al., 2026). However, the abundance of the protein is not necessarily related to the enzyme activity as it does not consider regulations of the enzyme activity by allosteric regulators, covalent modification. Nevertheless, these techniques represent an interesting alternative when kinetic properties are the main issue of the study. For estimation of the residual activity in mutated human TPK1 and its correlation with the severity of the disease, there is no alternative to an enzymatic technique.

2.8 Pros and cons

Pros	Cons
Direct method High sensitivity for detection of ThDP and thiamine. Allows assay of the forward (physiological) and reversed direction with the same equipment, but the equilibrium constant can only be measured in the reverse reaction	Requires sophisticated equipment While it was generally assumed that the phosphoryl donor is ATP, some reports suggest that the more physiological substrate could be UTP (Sahu et al., 2024).

2.9 Alternative methods/procedures

According to our knowledge there are no existing valuable alternative to HPLC coupled either to fluorescence or mass-spectrometric detection.

Sometimes yeast pyruvate decarboxylase is used for enzymatic detection of ThDP (Parkhomenko et al., 2011). However, this method is indirect, with questionable reliability and does not allow assay of the reverse reaction.

One could imagine mass spectrometric determination of ThDP formed directly in the assay medium without prior HPLC separation, but no such method has been published to our knowledge.

Another option could be the quantification of AMP for the forward reaction (Mondal et al., 2017) or ATP for the reverse reaction. The limiting factor is the sensitivity of these methods as the assay of TPK requires the detection of product concentrations as low as 10 nM. Furthermore, in biological samples, where AMP is already present at relatively high concentrations, it would be difficult to detect the increment of AMP produced by TPK. On the other hand, the use of AMP (luciferase/luciferin) assay kits has the advantage of being rapid and allowing the assay of many samples simultaneously using microtiter plates.

2.10 Troubleshooting & optimization

Problem	Solution
The enzyme (recombinant or native) is soluble and relatively resistant towards denaturation.	Nevertheless, care must be taken to keep the preparation on ice.
Too low enzyme activity, to be detected	Use a preparation with a higher enzyme content. This should not be an issue with recombinant enzyme. For native enzyme, avoid unnecessary dilution during preparation. Use a more sensitive fluorescence detector
Thiamine and ThDP peaks are poorly separated (low resolution)	Change column or use a PRP-1column with a length of 25 cm rather than 15 cm.
When recombinant enzymes are used, the total protein concentrations are very low and posing a risk of adsorption tube walls.	Use polypropylene tubes for enzyme assays. Alternatively, bovine serum albumin can (0.1 $mg.mL^{-1}$) be added in the reaction medium.
ThDP and thiamine (to a lesser extent) is subject to adsorption on glass surfaces	Use polypropylene tubes.
The use of a high AMP concentration interferes with the elution on the HPLC column by considerably shifting the retention times	When higher AMP concentrations (>5 mM) are used, the sample can be diluted 10 times in water before injection on the HPLC system.

2.11 Summary

The TPK assay described here is an extremely powerful method allowing the assay of the enzyme activity from nearly every source (microorganisms, plants, animal tissues or recombinant sources). The method allows the determination of the enzyme activity in the forward (physiological) and reverse direction. Concerning the fact that the enzyme is subject to product inhibition by ThDP, testing the activity in the reverse direction can be a valuable control. However, the estimation of TPK activity is arduous as the enzyme is inhibited by excess product ThDP in the forward direction and must be stopped before a significant inhibition occurs. This also make it impossible to measure an equilibrium constant in the forward direction (Sambon et al., 2022). The reverse reaction is so slow that enzyme concentrations must be used in the same order of magnitude as the substrate ThDP, not fulfilling Michaelis-Menten conditions.

3. Conclusion

TPK1 is an essential enzyme for most living organisms. The enzyme has a very high affinity for its substrates, thiamine for the forward reaction and ThDP for the reverse reaction and the k_{cat} is low ($0.03\,s^{-1}$ for the forward reaction and 10 times less for the reverse reaction, Sambon et al., 2022). This requires not only a very sensitive method but also a method able to separate the different thiamine derivatives in a single step before detection. In view of the recent developments implicating reduced TPK1 activity in Alzheimer's disease patients, determination of TPK1 activity in human tissues will be a challenging issue. In addition, several mutations in the human *tpk1* gene have been described (Makarchikov et al., 2025; Marcé-Grau et al., 2019). The characterization of these mutant enzymes and the implication of the mutations on enzyme activity requires a reliable method for the assay of TPK1 activity. In view of the competitive inhibition of TPK1 by the product ThDP the assay is challenging and requires carefully executed control experiments. In summary, we present here a sensitive reliable method for the assay of TPK1 activity in animal tissues or recombinant proteins produced in the laboratory.

Acknowledgements

LB is Research Director of the Fonds de la Recherche Scientifique – FNRS.

References

Artsukevich, I. M., Voskoboev, A. I., & Ostrovskii, I. u M. (1977). Purification and several properties of thiamine pyrophosphokinase from rat liver. *Voprosy Meditsinskoi Khimii, 23*(2), 203–210.

Bettendorff, L., Mastrogiacomo, F., Kish, S. J., & Grisar, T. (1996). Thiamine, thiamine phosphates, and their metabolizing enzymes in human brain. *Journal of Neurochemistry, 66*(1), 250–258.

Bettendorff, L., Peeters, M., Jouan, C., Wins, P., & Schoffeniels, E. (1991). Determination of thiamin and its phosphate esters in cultured neurons and astrocytes using an ion-pair reversed-phase high-performance liquid chromatographic method. *Analytical Biochemistry, 198*(1), 52–59.

Dallan, A., Reynolds, G., Canavese, C., Carli, D., Luca, M., Gazzin, A., ... Mussa, A. (2025). The wide phenotypic spectrum of thiamine metabolism dysfunction syndrome 5 and its treatment. *Orphanet Journal of Rare Diseases, 20*(1), 160. https://doi.org/10.1186/s13023-025-03665-9.

Jiang, Z. (2024). SLC25A19 is required for NADH homeostasis and mitochondrial respiration. *Free Radical Biology & Medicine, 222*, 317–330. https://doi.org/10.1016/j.freeradbiomed.2024.06.019.

Jonus, H. C., Hanberry, B. S., Khatu, S., Kim, J., Luesch, H., Dang, L. H., ... Zastre, J. A. (2018). The adaptive regulation of thiamine pyrophosphokinase-1 facilitates malignant growth during supplemental thiamine conditions. *Oncotarget, 9*(83), 35422–35438. https://doi.org/10.18632/oncotarget.26259.

Lu, J., & Frank, E. L. (2008). Rapid HPLC measurement of thiamine and its phosphate esters in whole blood. *Clinical Chemistry, 54*(5), 901–906.

Luo, Y., Liu, Y., Liu, J., Liu, X., Li, H., Wu, Y., ... Yan, J. (2025). Optimizing thiamine pyrophosphate metabolism enhances crop yield and quality. *Nature Communications, 17*(1), 165. https://doi.org/10.1038/s41467-025-66882-7.

Makarchikov, A. F., Wins, P., & Bettendorff, L. (2025). Biochemical and medical aspects of vitamin B1 research. *Neurochemistry International, 185*, 105962. https://doi.org/10.1016/j.neuint.2025.105962.

Marcé-Grau, A., Martí-Sánchez, L., Baide-Mairena, H., Ortigoza-Escobar, J. D., & Pérez-Dueñas, B. (2019). Genetic defects of thiamine transport and metabolism: A review of clinical phenotypes, genetics, and functional studies. *Journal of Inherited Metabolic Disease, 42*(4), 581–597. https://doi.org/10.1002/jimd.12125.

Mitsuda, H., Takii, Y., Iwami, K., & Yasumoto, K. (1975). Purification and properties of thiamine pyrophosphokinase from parsley leaf. *Journal of Nutritional Science and Vitaminology, 21*(2), 103–115. https://doi.org/10.3177/jnsv.21.103.

Mitsuda, H., Takii, Y., Iwami, K., Yasumoto, K., & Nakajima, K. (1979). Enzymatic formation of thiamine pyrophosphate in plants. *Methods in Enzymology, 62*, 107–111. https://doi.org/10.1016/0076-6879(79)62202-4.

Mondal, S., Hsiao, K., & Goueli, S. A. (2017). Utility of adenosine monophosphate detection system for monitoring the activities of diverse enzyme reactions. *Assay and Drug Development Technologies, 15*(7), 330–341. https://doi.org/10.1089/adt.2017.815.

Nosaka, K., Kaneko, Y., Nishimura, H., & Iwashima, A. (1993). Isolation and characterization of a thiamin pyrophosphokinase gene, THI80, from Saccharomyces cerevisiae. *Journal of Biological Chemistry, 268*(23), 17440–17447.

Nosaka, K., Onozuka, M., Kakazu, N., Hibi, S., Nishimura, H., Nishino, H., & Abe, T. (2001). Isolation and characterization of a human thiamine pyrophosphokinase cDNA. *Biochimica et Biophysica Acta, 1517*(2), 293–297.

Nosaka, K., Onozuka, M., Nishino, H., Nishimura, H., Kawasaki, Y., & Ueyama, H. (1999). Molecular cloning and expression of a mouse thiamin pyrophosphokinase cDNA. *Journal of Biological Chemistry, 274*(48), 34129–34133.

Onozuka, M., & Nosaka, K. (2003). Steady-state kinetics and mutational studies of recombinant human thiamin pyrophosphokinase. *Journal of Nutritional Science and Vitaminology (Tokyo), 49*(3), 156–162.

Parkhomenko, Y. M., Kudryavtsev, P. A., Pylypchuk, S. Y., Chekhivska, L. I., Stepanenko, S. P., Sergiichuk, A. A., & Bunik, V. I. (2011). Chronic alcoholism in rats induces a compensatory response, preserving brain thiamine diphosphate, but the brain 2-oxo acid dehydrogenases are inactivated despite unchanged coenzyme levels. *Journal of Neurochemistry, 117*(6), 1055–1065.

Peterson, J. W., Gubler, C. J., & Kuby, S. A. (1975). Partial purification and properties of thiamine pyrophosphokinase from pig brain. *Biochimica et Biophysica Acta, 397*(2), 377–394.

Prajapati, S., Rabe von Pappenheim, F., & Tittmann, K. (2022). Frontiers in the enzymology of thiamin diphosphate-dependent enzymes. *Current Opinion in Structural Biology, 76*, 102441. https://doi.org/10.1016/j.sbi.2022.102441.

Sahu, U., Villa, E., Reczek, C. R., Zhao, Z., O'Hara, B. P., Torno, M. D., ... Ben-Sahra, I. (2024). Pyrimidines maintain mitochondrial pyruvate oxidation to support de novo lipogenesis. *Science, 383*(6690), 1484–1492. https://doi.org/10.1126/science.adh2771.

Sambon, M., Pavlova, O., Alhama-Riba, J., Wins, P., Brans, A., & Bettendorff, L. (2022). Product inhibition of mammalian thiamine pyrophosphokinase is an important mechanism for maintaining thiamine diphosphate homeostasis. *Biochimica Et Biophysica Acta. General Subjects, 1866*(3), 130071. https://doi.org/10.1016/j.bbagen.2021.130071.

Sanemori, H., & Kawasaki, T. (1980). Purification and properties of thiamine pyrophosphokinase in Paracoccus denitrificans. *Journal of Biochemistry (Tokyo), 88*(1), 223–230.

Smith, T. J., Johnson, C. R., Koshy, R., Hess, S. Y., Qureshi, U. A., Mynak, M. L., & Fischer, P. R. (2021). Thiamine deficiency disorders: A clinical perspective. *Annals of the New York Academy of Sciences, 1498*(1), 9–28. https://doi.org/10.1111/nyas.14536.

Thompson, Z. E., Boyd, N. K., Khoshnood, M. M., & Santoro, J. D. (2023). Thiamine metabolism dysfunction syndrome 5 (THMD5) mimicking acute disseminated encephalomyelitis. *American Journal of Medical Genetics. Part A, 191*(12), 2868–2872. https://doi.org/10.1002/ajmg.a.63376.

Xia, Y., Qian, T., Fei, G., Cheng, X., Zhao, L., Sang, S., & Zhong, C. (2024). Low expression of thiamine pyrophosphokinase-1 contributes to brain susceptibility to thiamine deficiency. *Neuroreport*. https://doi.org/10.1097/WNR.0000000000002094.

Zhang, S.-Z., Ma, Y., Ding, Y., Yin, Y.-Q., Wang, B.-W., Hu, G., & Zhou, J.-W. (2026). Astrocytic TPK1 mitigates amyloid pathology via TFEB-mediated endocytosis. *Experimental Neurology*, 115640. https://doi.org/10.1016/j.expneurol.2026.115640.

Zhao, D., Liu, M., Jiang, H., Song, T., Xu, C., Duan, X., ... Fang, F. (2023). Thiamine pyrophosphokinase deficiency: Report of two Chinese cases and a literature review. *Frontiers in Pediatrics, 11*, 1173787. https://doi.org/10.3389/fped.2023.1173787.

CHAPTER NINE

Measurement of thiaminase I activity in biological samples: 4-nitrothiophenol (4-NTP) assays, optimization, and interpretation

Katie A. Edwards[a,b,*], Esther R. Angert[b], Clifford E. Kraft[c], Eileen A. Randall[c], and Patricia C. Wolfe[a]

[a]Department of Pharmaceutical Sciences, Binghamton University, Binghamton, NY, United States of America
[b]Department of Microbiology, Cornell University, Ithaca, NY, United States of America
[c]Department of Natural Resources and the Environment, Cornell University, Ithaca, NY, United States
*Corresponding author. e-mail address: kedwards@binghamton.edu

Contents

Methods in Enzymology, Volume 735
ISSN 0076-6879, https://doi.org/10.1016/bs.mie.2026.06.028

Abstract

Thiaminase I enzymes catalyze the cleavage of thiamine (vitamin B1) and have been implicated in dietary and ecological thiamine deficiency syndromes affecting fish, wildlife, livestock, and other organisms. This chapter focuses on the colorimetric 4-nitrothiophenol (4-NTP) assay for thiaminase I activity and discusses alternative techniques. Protocols for sample collection, tissue homogenization, extraction, enzyme preservation, assay optimization, and data analysis are presented in a stepwise format. The application of the assay to bacterial culture supernatants, which offer valuable mechanistic insights at the microbial level, and to more complex tissue homogenates to better understand thiamine deficiency, is discussed. Factors influencing assay performance and interpretation, including endogenous thiamine carryover, endpoint versus kinetic measurements, and physiological relevance of measured in vitro activity, are considered. Emphasis is placed on data processing and normalization. Under the conditions employed, the assay has a limit of detection of 50 pmol 4-NTP degraded per minute, corresponding to 1 ng of thiaminase from *Paenibacillus thiaminolyticus*. The assay measures maximal catalytic capacity, though we recommend caution when translating thiaminase activities measured under substrate-saturating, elevated-temperature in vitro conditions using homogenized and de-compartmentalized samples to physiological thiamine degradation in vivo. In the latter, thiaminase-mediated thiamine degradation is constrained by substrate availability, an organism's body temperature, biochemical degradation processes, inhibitory factors, and physiological compartmentalization. Collectively, the approach and considerations presented here provide a framework for measuring and interpreting thiaminase I activity in complex biological matrices.

1. Introduction

Thiamine supplementation has widespread beneficial effects, including activation of disease resistance pathways in plants (Ahn et al., 2005; Boubakri et al., 2013); prevention of polioencephalomacia in livestock (McGuirk, 1987); support of reproductive viability in aquatic organisms (Brown, Brown, et al., 2005; Reed et al., 2023); and mitigation of metabolic dysfunction associated with diabetes in humans (Obrenovich & Monnier, 2003; Thornalley, 2005). In mammalian systems, thiamine in its diphosphate form [thiamine diphosphate (TDP)] functions as an essential cofactor for enzymes central to carbohydrate metabolism, branched-chain amino acid catabolism, and fatty acid oxidation, as well as for pathways supporting nucleotide and amino acid biosynthesis and cellular redox balance (Agyei-Owusu & Leeper, 2009; Frank et al., 2007). Disruption of thiamine availability through both enzymatic degradation by thiaminases and production of toxic thiamine antimetabolites can therefore have broad metabolic consequences that are directly relevant to the biochemical context in which thiaminase activity is measured (Jurgenson et al., 2009; Kraft & Angert, 2017; Shannon et al., 2025; Yazdani et al., 2025; Zilles et al., 2000).

Because total body thiamine stores in humans are limited (~30–50 mg), deficiency can develop rapidly (within ~2–3 weeks) (Mancinelli & Ceccanti, 2009). Thiamine deficiency may arise from inadequate intake, impaired absorption, degradation during food processing or storage, or exposure to thiamine-degrading agents such as sulfites and thiaminases (Ahoua et al., 2007; Tomasulo et al., 1968; Vimokesant et al., 1975; Zoltewicz et al., 1984). Among these factors, thiaminases have attracted particular attention due to their occurrence in various microbes, plants, and fish. The association of thiamine depletion with a fish-based diet was first described in the early 1940s (Green et al., 1941, 1942; Green & Evans, 1940; Krampitz & Woolley, 1944; Lee, 1948; Neilands, 1947; Reddy et al., 1948; Sealock & Goodland, 1944; Sealock et al., 1943; Smith & Proutt, 1944; Spitzer et al., 1941; Woolley, 1941), with early recognition of its possible enzymatic nature (Deutsch & Hasler, 1943; Sealock et al., 1943; Spitzer et al., 1941; Woolley, 1941).

Thiaminase enzymes catalyze the cleavage of thiamine into its pyrimidine and thiazole moieties via nucleophilic substitution. Thiaminase I (EC 2.5.1.2) uses a range of exogenous nucleophiles, including heterocyclic amines and sulfhydryl compounds, whereas thiaminase II (EC 3.5.99.2) uses water as the nucleophile (Fig. 1).

Thiaminase II enzymes, recognized as thiaminase enzyme A (TenA)-family proteins involved in thiamine salvage (Jenkins et al., 2007), can catalyze the hydrolytic cleavage of thiamine. Genes encoding TenA-family proteins with thiaminase II-like activity are widely distributed across taxa and are primarily associated with thiamine salvage pathways (Jenkins et al., 2007; Stentz et al., 2023; Zallot et al., 2014) rather than documented dietary or ecological thiamine deficiency syndromes. In contrast, thiaminase I, through nucleophile-dependent exchange reactions, has been implicated with thiamine depletion in ecological and dietary contexts (Edwin & Jackman, 1970; Honeyfield et al., 2005; Melnick et al., 1945).

Bacterial thiaminase I is a secreted enzyme first described in the 1950s (Kimura, 1965). Decades later, mirroring previous recognition of thiaminase II

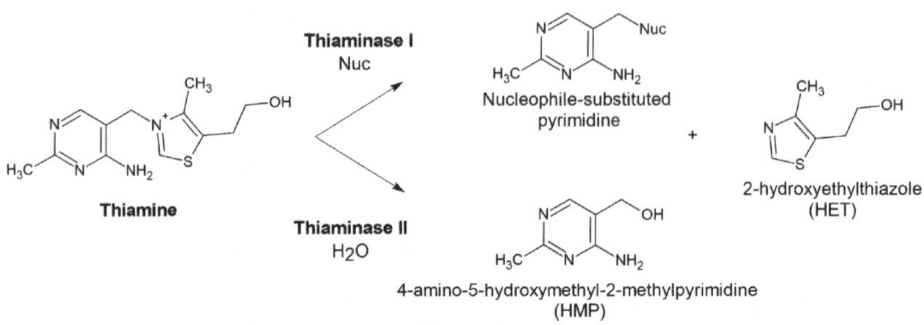

Fig. 1 Cleavage of thiamine into pyrimidine and thiazole fragments by (top) thiaminase I in the presence of an external nucleophile, and (bottom) thiaminase II in the presence of water. Thiaminase I produces a nucleophile-substituted pyrimidine derivative and 2-hydroxyethylthiazole (HET), whereas thiaminase II produces 4-amino-5-hydroxymethyl-2-methylpyrimidine (HMP) and HET. In the 4-NTP assay, 4-nitrothiophenol (4-NTP) serves as the nucleophile for thiaminase I.

function, Sannino et al. described the role of thiaminase I as an enzyme that salvages thiamine precursors from degraded forms of thiamine for reuse in thiamine synthesis (Sannino et al., 2018). The existence of a TenA-like protein exhibiting thiaminase activity has been reported in zebrafish and other aquatic organisms, and gene homologs have been identified in several other fish species that are associated with thiamine deficiency (Richter et al., 2023; Richter et al., 2009; Stentz et al., 2023). However, whether these fish proteins evolved primarily for thiamine degradation, salvage/remodeling functions, detoxification of thiamine analogues, or other aspects of thiamine homeostasis remains unresolved. The physiological relevance of observed in vitro thiaminase activity under endogenous substrate and cosubstrate conditions remains incompletely defined. In human and rat urine, thiazole and pyrimidine fragments have been identified, indicating thiamine catabolism (Balaghi & Pearson, 1967; Ziporin et al., 1965). This was observed also in germ-free rats, suggesting the existence of an endogenous mammalian breakdown pathway (Suzuoki et al., 1968). These mammalian studies did not elucidate a specific enzyme or pathway, which could include hydrolytic (TenA/thiaminase II-like), nucleophile transfer (thiaminase I-like), or chemical (e.g., oxidative) processes.

Thiaminase I enzymes are relatively rare in nature and have been identified in a limited number of bacterial species, including the human pathogens responsible for post-infectious hydrocephalus (*Paenibacillus thiaminolyticus),* melioidosis (*Burkholderia pseudomallei),* and botulism (*Clostridium botulinum)* (Gassiep et al., 2020; Paulson et al., 2020; Ringe et al., 2014). Sannino & Angert showed that other bacterial pathogens have similar gene sequences, indicating that pathogenicity might be associated with thiaminase I (Sannino & Angert, 2017). Thiaminase I has also been found in non-human pathogens such as *Paenibacillus apiarius* and *Burkholderia thailandensis* (Kraft et al., 2014), although the latter has been associated with isolated opportunistic infections (Hall et al., 2023). Thiaminase I activity has additionally been found in some plants, including bracken fern (*Pteridium aquilinum*), horsetail (*Equisetum arvense*), and nardoo (*Marsilea drummondii*) (Gassiep et al., 2020; McCleary & Chick, 1977; Paulson et al., 2020; Ringe et al., 2014). Although thiaminase-associated thiamine deficiency is uncommon in humans, these enzymes are considered responsible for thiamine depletion in ecological and veterinary contexts, particularly in fish and ruminants (Honeyfield et al., 2002; Richter et al., 2009; Zajicek et al., 2009). Notably, *P. thiaminolyticus*, a microbe associated with fish thiamine deficiency (Honeyfield et al., 2002; Richter et al., 2009; Zajicek et al., 2009), has been implicated in rare human infections (Di Micco et al., 2021) and more recently been identified as a contributor to post-infectious hydrocephalus in infants (Paulson et al., 2020), highlighting the emerging clinical relevance of thiaminase-producing organisms.

Mortality of commercially valuable salmonine fisheries from thiamine deficiency was first reported in the Laurentian Great Lakes region of the U.S. and in the Baltic Sea, and more recently was observed along the Pacific coast of North America (Brown, Fitzsimons, et al., 2005; Mantua et al., 2025). In all of these geographical regions, fish mortality from thiamine deficiency was associated with predation upon forage fish from which thiaminase I has been measured, including smelt, dace, alewife, and Baltic herring (Hirn & Pekkanen, 1975). Prior to mortality, the predator fish exhibit common neurological deficits and poor fecundity attributed to thiamine deficiency. However, thiaminase I is heat-labile, and in fish harvested for pet foods or human consumption, residual thiaminase I enzyme activity depends on the processing steps employed (Gnaedinger & Krzeczkowski, 1966; Wolfe et al., 2023).

The influence of thiaminase I on a wide range of animals has fostered the need for robust analytical methods to quantify thiaminase activity, including spectrophotometric assays based on nucleophilic exchange reactions, such as the 4-nitrothiophenol (4-NTP) assay (Hanes et al., 2007). Nucleophiles are electron-rich species that can donate electron density to form covalent bonds, and commonly contain lone pairs, π bonds, or other electron-rich functional groups. In thiaminase I-catalyzed exchange reactions, a wide variety of nucleophilic cosubstrates have been reported to include sulfhydryl compounds such as β-mercaptoethanol and dithiothreitol (DTT); nicotinic acid; amino acids and related compounds, including lysine, cysteine, and taurine; and heterocyclic or aromatic organic bases such as pyridoxine, pyridine, aniline, *m*-nitroaniline, and quinoline (Edwards et al., 2023; Fujita et al., 1952; Honeyfield et al., 2010; Murata, 1982).

Despite decades of work on the biochemistry of thiaminase I, a comprehensive understanding of its impact on animal and environmental health is still lacking. This provides an incentive to develop improved protocols for assaying and interpreting thiaminase I activity.

1.1 Definition

Measurement of thiaminase I activity in biological and environmental samples using a 4-Nitrothiophenol (4-NTP) assay.

1.2 Rationale

The 4-NTP assay utilizes 4-nitrothiophenolate (Fig. 2) as a nucleophilic cosubstrate for thiaminase I (Hanes et al., 2007). 4-NTP exhibits a high molar extinction coefficient (13,650 M^{-1} cm^{-1}) with a maximum absorbance at 411 nm, enabling sensitive spectrophotometric detection. As the thiaminase-mediated degradation of thiamine proceeds, 4-NTP is consumed, thereby decreasing absorbance at 411 nm.

A reducing agent [tris(2-carboxyethyl)phosphine (TCEP)] is included to prevent non-enzymatic oxidation of 4-NTP, which would otherwise decrease absorbance (Hanes et al., 2007). TCEP may additionally help maintain enzymatic activity under assay conditions. Under conditions where endogenous nucleophiles are limited, the reaction rate is driven primarily by 4-NTP, which outcompetes native nucleophiles present in crude preparations.

4-NTP was identified as an effective cosubstrate based on favorable kinetic parameters and higher activity relative to other nucleophiles, including aniline and thiophenol (Hanes et al., 2007; Honeyfield et al., 2010). Assay conditions have been optimized to include saturating concentrations of 4-NTP to ensure maximal enzyme activity and avoid substrate limitation (Honeyfield et al., 2010), sample centrifugation through 30 μm pore membranes to minimize light scattering interferences by particles (Honeyfield et al., 2010), adaptation to a microplate format to improve throughput and replication (Hanes et al., 2007; Kraft et al., 2014), matching the extraction buffer to the assay buffer (Boggs et al., 2019), and inclusion of a protease inhibitor to minimize enzyme degradation (Wolfe et al., 2023). This assay has been demonstrated using recombinant thiaminase I expressed in *Escherichia coli* and tested with environmental and

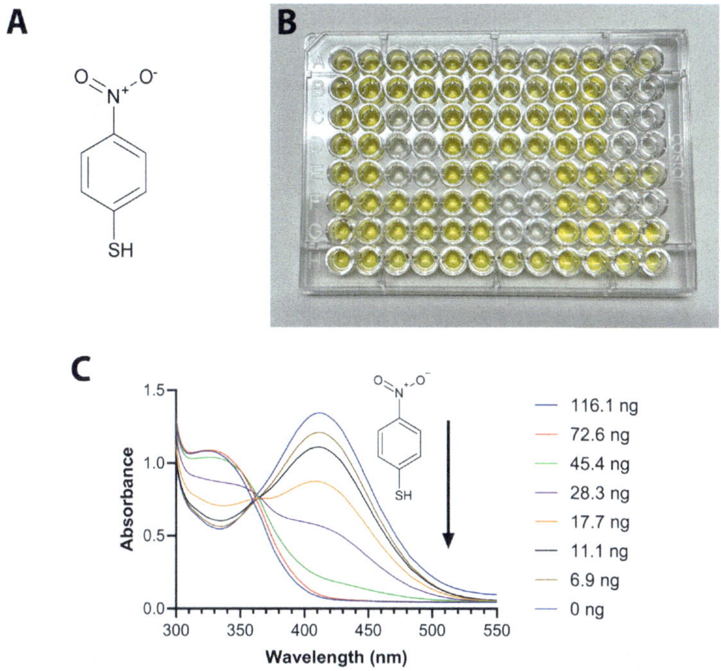

Fig. 2 (A) Structure of 4-nitrothiophenol, (B) 4-NTP assay with samples in duplicate columns, without and with added thiamine in the presence of 4-NTP. Wells with low thiaminase activity remain yellow in color (e.g., columns 1 & 2), while those with high thiaminase activity become clear as 4-NTP is depleted (e.g., wells C3, C4, D3, D4, E3, E4), (C) Absorbance spectra of 4-NTP depletion in the presence of increasing amounts of recombinant *P. thiaminolyticus* thiaminase in a background of whole anchovy extract (0.8 mg tissue/well).

biological samples, including microbial samples, forage fish, plant tissues, and fish processed for food (Boggs et al., 2019; Hanes et al., 2007; Kraft et al., 2014; Sannino et al., 2018; Wolfe et al., 2023).

As with other enzymes, the goal of thiaminase extraction from tissues is to maximize yield while preserving enzymatic activity. Extraction from fish presents additional considerations, as fish tissues exhibit high protease activity, particularly in muscle and viscera. Moreover, fish proteases may be somewhat cold-adapted (Khiari, 2022), requiring rapid sample processing in as cold an environment as possible. To manage these challenges, we adapted methods using dry ice and a consumer-grade coffee grinder to homogenize fish tissue quickly and efficiently (Kraft et al., 2014; Zajicek et al., 2005) and storage at −80 °C for later thiaminase extraction.

2. Materials, equipment, and reagents

Unless otherwise indicated, chemicals and disposable plasticware should be purchased from reliable sources such as Thermo Fisher Scientific (Waltham, MA), VWR Scientific (Radnor, PA), or Sigma-Aldrich (St. Louis, MO; Burlington, MA).

2.1 For bacterial positive control for the 4-NTP assay

1. Frozen 20 % glycerol stocks of the bacterium *P. thiaminolyticus*, stored at −80 °C (ATCC 13023)
2. Microcentrifuge tubes 1.5 mL (e.g., 89126−724, VWR)
3. Sterile petri dishes
4. BactoTryptic Soy Broth (TSB) medium (211825, BD). For plates, per liter, add 15 g granulated agar (214530, BD)
5. Bunsen burner and spark lighter, gas
6. Stick applicator (autoclaved)
7. Autoclaved glass culture tubes with metal caps (6−10 tubes)
8. 5 mL sterile glass or plastic pipettes and pipettor
9. Reusable glass bottle with screw cap lid for liquid culture medium 1 L (06−414−1D, Corning)
10. Erlenmeyer flask, 1 L capacity (e.g., CLS49801L, Sigma)
11. Aluminum foil
12. Autoclave
13. Water bath (e.g., New Brunswick Waterbath Shaker C76) or shaker incubator (e.g., Innova 44 incubator shaker)
14. Microcentrifuge that can centrifuge 1.5 mL tubes at 10,000 × *g* or higher (e.g., Eppendorf models 5425, 5415)
15. Ultralow freezer for storing samples at −80 °C.

2.2 For processing of plant and animal tissues

1. Cold room at 4 °C (preferable)
2. Dry ice pellets, approximately equal in volume to available samples
3. Coffee grinder with removable stainless-steel cup (preferable) (e.g., Krups model GX332)
4. Resealable plastic bags, quart size
5. 15 mL centrifuge tubes for −80 °C storage
6. Sample spatula
7. Tongs
8. Timer

2.3 For running the 4-NTP assay

1. Microplate reader with thermal control, shaking, and kinetic read capabilities at 411, 900, and 975 nm (e.g., Varioskan LUX)
2. Biological samples frozen in −80 °C storage.
3. Clear microtiter plates (e.g., Corning Costar 3795)
4. Low protein-binding microcentrifuge tubes 2 mL (e.g., EP022600044, Eppendorf)
5. Microcentrifuge tubes 1.5 mL (e.g., 89126−724, VWR)
6. 50 mL Falcon conical bottom centrifuge tubes (e.g., 352070, Corning Life Sciences)
7. 15 mL Falcon conical bottom centrifuge tubes (e.g., 352096, Corning Life Sciences)
8. 30 μm centrifugal filters (e.g., Pierce 89868)
9. Tris(2-carboxyethyl)phosphine (TCEP) hydrochloride
10. Potassium phosphate monobasic/dibasic
11. Sodium chloride
12. Sodium hydroxide
13. Protease Inhibitor (e.g., HALT, Thermo Scientific)
14. Thiamine hydrochloride
15. 4-nitrothiophenol (4-NTP)
16. Dimethyl sulfoxide (DMSO), anhydrous
17. Thiaminase positive control (e.g., cell-free culture supernatant from *P. thiaminolyticus* or recombinant *C. botulinum* thiaminase I)
18. HPLC grade water (JT Baker, Phillipsburg, NJ).
19. Parafilm (Amcor)
20. Nitrogen gas (preferred) or vacuum pump for deoxygenation

3. Protocols

3.1 Bacterial supernatant preparation

Cell-free culture supernatant from *P. thiaminolyticus* is a convenient positive control for the 4-NTP assay. The preparation of bacterial samples requires a clean work environment and aseptic techniques. Before starting any procedure, clean the workbench with 70 % alcohol. Be sure to use new disposable gloves while working with bacterial samples.

3.2 Prepare culture media

For liquid cultures, prepare 500 mL of TSB in a 1 L reusable glass medium bottle, following the manufacturer's recommendations. For pouring plates, prepare 500 mL of TSB with agar (7.5 g agar per 500 mL of medium) in a 1 L Erlenmeyer flask. Cover the flask with an aluminum foil lid. Sterilize the media using an autoclave. Cap and autoclave empty culture tubes. Autoclave applicator sticks. We recommend placing them in a reusable glass jar, then covering it with aluminum foil.

3.3 Bacterial culture and supernatant preparation

Prepare TSA plates using standard microbiological technique. When the medium in the petri plates (TSB+Agar, TSA) has solidified and dried overnight, start a fresh culture of *P. thiaminolyticus*. Retrieve frozen bacterial glycerol stock from –80 °C freezer. Using a sterile applicator, scrape the surface of the frozen bacterial stock and streak it onto a TSA plate to produce isolated colonies. Incubate the streaked plate at 37 °C overnight (~16 h). If growth is too light to see individual colonies, incubate the plate for a few additional hours.

Prepare several culture tubes by pipetting 5 mL of prepared TSB into each test tube using aseptic technique. Select a colony from the streaked plate and inoculate a culture tube containing TSB. Incubate the culture tubes in a 37 °C water bath with agitation overnight.

The next day, the inoculated liquid media should be turbid. Pipet 1.5 mL of culture into individual 1.5 mL microcentrifuge tubes. Centrifuge at 13,000 × *g* for 5 min. Use a pipet to transfer the supernatant into a clean microcentrifuge tube, then aliquot into single-use tubes. We recommend making aliquots of 100 μL per tube to minimize freezing and thawing of the samples. Store the cell-free supernatant frozen at −80 °C until needed.

3.4 Homogenization of tissue samples

Accurate and reproducible downstream analyses assume that samples were collected and stored appropriately. Biological samples (e.g., fish and plant tissues) should be stored at −80 °C after collection. To minimize sample warming and possible enzymatic degradation during processing, homogenization should ideally take place in a cold room at 4 °C using prechilled equipment.

1. On the day prior to the analysis, place the coffee grinder, spatulas, and disposable sample bags in a cold room at 4 °C overnight so that all tools and supplies are prechilled before use.
2. Transfer approximately 10−20 g of fish tissue into a clean, prechilled removable cup of a coffee grinder. Using tongs, add approximately the same amount of dry ice to the cup.

   a. We prefer using a coffee grinder with a removable stainless-steel cup (e.g., Krups model GX332) rather than an integrated cup/mechanical grinder device.
   b. A removable cup simplifies washing between samples, allows the grinder itself to remain at 4 °C to minimize condensation within the motor housing and internal mechanics, and permits replacement of the cup/blade assembly without replacement of the entire grinder if the blade becomes damaged.
   c. Dry ice helps to maintain the sample in a frozen, brittle state during homogenization, improving pulverization efficiency while minimizing thermal degradation.
3. Grind the sample with dry ice using repeated pulses at the highest speed until a fine, homogeneous powder is obtained.
   a. The pulsing helps to ensure material accumulating in the lid and walls of the grinder cup returns to contact with the blade while also limiting the possibility of the grinder motor exceeding its duty cycle or overheating.
   b. Although different sample types may require more pulses or total grinding time, maintaining consistent homogenization parameters within a given sample type will improve reproducibility between samples. Fine pulverization is needed for consistency in activity measurements (Zajicek et al., 2005).
4. Transfer the ground sample and remaining dry ice powder into a resealable plastic bag.
5. Store the bag in a −20 °C freezer overnight, with the bag left open to allow the CO_2 to sublimate and vent.
6. After visible dry ice is no longer present, transfer the samples to 15 mL centrifuge tubes for long-term storage at −80 °C. This transfer should be performed in the 4 °C cold room as quickly as possible to minimize sample warming.
   a. During this transfer, also weigh ~150 mg as a subsample into a 2 mL microcentrifuge tube for ease of analysis. Be sure to record the weights of the samples accurately.

3.5 Preparation of soluble protein extracts from tissue samples

The homogenized tissue needs to be extracted in an aqueous medium that maintains enzyme activity and uses sufficient buffer volume relative to sample mass to ensure complete and consistent extraction. The extracts are then centrifuged, clarified by filtration, and either stored or, preferably, immediately analyzed.

1. Remove the 2 mL centrifuge tubes containing ground biological samples from −80 °C storage and allow them to thaw on ice.
2. Add 5 mL ice-cold tissue extraction buffer per g sample to each tube.
   a. Our current extraction protocol uses TCEP in PBS and includes the HALT protease inhibitor (Wolfe et al., 2023), as described below. This formulation was adopted as a precautionary measure to preserve enzyme activity during extraction.
   b. Previous reports have used an extraction solution composed of 100 mM phosphate buffer at pH 6.5 (Hanes et al., 2007; Kraft et al., 2014; Zajicek et al., 2009; Zajicek et al., 2005).
   c. Some sample types, depending on their inherent hydration status and fluid absorption characteristics, may require more extraction buffer to achieve complete and consistent extraction (Wolfe et al., 2023).
3. Vortex samples for 10 s on the highest speed, followed by a 1-minute rest on ice. This process is repeated three times.
4. Centrifuge sample tubes at 17,000 × *g* for 20 min 4 °C.
5. Transfer supernatants to 0.8 mL centrifuge columns with 30 μm filters and centrifuge at 8000 × g for 5 min at 4 °C to remove particulates.
   a. This step is optional and increases the expense of consumables, but is highly recommended, especially for low activity samples.
   b. A smaller pore filter may be desirable for some sample types.
   c. We strongly recommend the use of low-protein binding tubes for filtrate collection.
6. The clarified filtrates should be stored in low protein-binding tubes at −80 °C for subsequent thiaminase and total protein concentration analyses.
   a. Our preference is to assay for thiaminase activity immediately. We observed a loss of activity after samples were frozen and thawed.
7. Determine the protein concentration using the microBCA assay following the manufacturer's instructions.
   a. Based on our experience, we suggest the following starting dilutions for the microBCA assay to ensure protein concentrations are within the linear range of the assay ($0.5-20\ \mu g_{protein} \cdot mL^{-1}$):
      i. Raw fish tissues (e.g., liver, muscle) 1:100 dilutions of extracts at 1 g per 5 mL (Wolfe et al., 2023)
      ii. Cooked/processed fish tissues: 1:100 dilutions of extracts at 1 g per 5 mL (Wolfe et al., 2023)
      iii. Formulated pet food/treats: 1:1000 dilutions of extracts at 1 g per 5 mL
      iv. Plant extracts: 1:200 dilutions of extracts at 1 g per 5 mL
      v. Bacterial culture supernatants: 1:500

4. 4-NTP assay

The 4-NTP assay consists of adding a working reagent containing 4-NTP both without thiamine (thiamine −) and with thiamine (thiamine +) to identical sample wells. The loss of 4-NTP, a strong chromophore with a maximum absorbance at 411 nm, is monitored kinetically. The differential rate between thiamine + and thiamine - wells is attributed to thiaminase I activity.

1. Before initiating the assay, consider the plate layout, number of samples, availability of controls, and a priori knowledge of the expected level of activity.
   a. The concentration should be empirically determined for a given sample type and activity level, although we suggest some starting places based on our experience and prior reports below:
      i. Raw fish tissues (e.g., liver, muscle)
         - 1.2 mg per well (3 μL extracted at 1 g per 2.5 mL) (Kraft et al., 2014)
         - 2.4 mg per well (6 μL extracted at 1 g per 2.5 mL) (Boggs et al., 2019)
      ii. Cooked/processed fish tissues
         - 3.2 mg per well (16 μL extracted at 1 g per 5 mL) (Wolfe et al., 2023)
      iii. Bacterial culture supernatants
         - 3 μL, with serial dilutions detectable to 1:256 (Kraft et al., 2014)
      iv. Affinity-purified recombinant thiaminase
         - 1:250 – 1:1000, depending on protein concentration
   b. In the assay as described below, we use 8 μL diluted supernatant per well.
      i. Example:

$$8\ \mu\text{L} \times \left(\frac{1\ g}{5\ mL}\right) \times \left(\frac{1\ mL}{10^3\ \mu L}\right) \times \left(\frac{10^3\ mg}{g}\right) = 1.6\ \text{mg tissue}$$

      ii. This volume can be adjusted to accommodate more or less tissue depending on activity levels. However, we caution against overloading the assay by increasing tissue mass to detect a signal, which risks increasing the background.
   c. It is highly recommended that the plate be set up with 6 adjacent wells per sample, with triplicate wells for the thiamine - solution containing 4-NTP only and triplicate wells for the thiamine + solution containing both thiamine and 4-NTP.

Although duplicate wells are often sufficient for samples with expected high activity, triplicate or quadruplicate wells are necessary to differentiate low activity from the background signal reliably.

   d. Room must be left on the plate to accommodate controls: a minimum of a blank (extraction buffer only) and positive control (recombinant thiaminase or thiaminase positive culture supernatant).

2. Prepare a 20x PBS solution (1 M potassium phosphate, 2 M sodium chloride) at the desired pH for extraction and working reagent solutions.
3. Turn on the plate reader and allow it to equilibrate at 37 °C for at least 30 min while other steps are carried out.
4. Ensure that a kinetic protocol is loaded and ready for use.
   a. Temperature: 37 °C
   b. Plate shaking: 5 s on low speed (360 rpm)
   c. Kinetic read at 411 nm every minute for 30 min (optionally, 1 h)
   d. Pathlength correction (automatic or include 900 and 977 nm as the final read)
5. Remove samples from freezer storage and allow to thaw on ice.
6. Once thawed, centrifuge samples at 4 °C for 10 min at 17,000 × *g* and return to ice until needed.
   a. This step is not needed if samples are assayed immediately after extraction
7. While samples are thawing and centrifuged, deoxygenate HPLC-grade water by vigorously bubbling with nitrogen gas or stirring under vacuum for at least 15 min. It is important to deaerate thoroughly.
   a. For convenience, we use a 50 mL Falcon tube with three 1.5 mm holes drilled in the lid and a long, blunt needle (Fig. 3). The extra holes beside that for the needle are for venting.
8. Prepare 500 mM TCEP hydrochloride:
   a. Weigh ~65 mg TCEP HCl into a 1.5 mL microcentrifuge tube.
   b. Dilute to 500 mM with HPLC-grade water.
   c. The 500 mM TCEP solution may be aliquoted and frozen for future use.
9. Prepare 20 mL TCEP buffer just before use:
   a. Add 1 mL 20x PBS pH 9.0 and 400 μL 500 mM TCEP to deoxygenated HPLC-grade water in a 50 mL Falcon tube, mix, then dilute to a final 20 mL volume. Adjust pH to 6.7 with 1 N NaOH or 1 N HCl if needed.
   b. Include protease inhibitor in the TCEP buffer if desired. This is strongly recommended if samples are not to be assayed immediately.

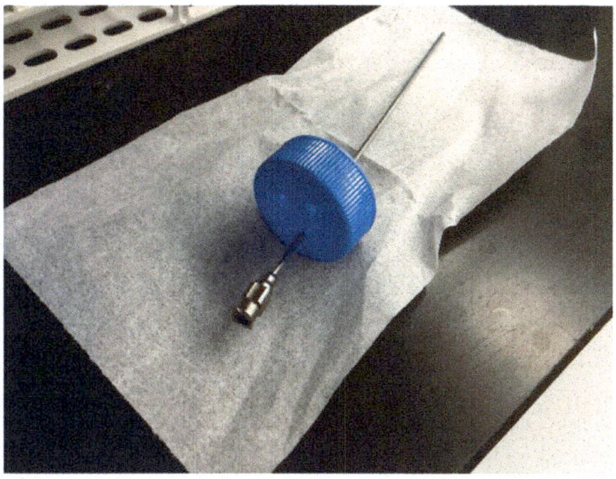

Fig. 3 Cap a 50 mL centrifuge tube with a blunt needle and holes for venting during nitrogen purging.

   c. TCEP solutions in phosphate buffer should be prepared fresh to minimize oxidation and degradation over time.

10. Prepare 100 mM thiamine hydrochloride:
    a. Weigh ~10 mg thiamine hydrochloride into a 1.5 mL microcentrifuge tube.
    b. Dilute to 100 mM with the requisite volume of HPLC-grade water.
    c. The 100 mM thiamine HCl solution may be aliquoted and frozen for future use.
11. Prepare 200 mM 4-NTP in DMSO:
    a. Weigh ~ 10 mg 4-NTP in a 1.5 mL microcentrifuge tube.
    b. Dilute 4-NTP with DMSO to 200 mM (31.0 $mg.mL^{-1}$).
    c. Vortex vigorously to dissolve.
    d. This solution needs to be prepared fresh. It may not be stored aliquoted as moisture in the headspace of the tubes will destroy it; thus, weigh as little as possible to minimize waste.
    e. Store in a dark drawer until use.
12. Prepare 13 mL of 4-NTP at 280 μM in TCEP buffer in a 15 mL centrifuge tube:
    a. This corresponds to 18.2 μL 200 mM 4-NTP in DMSO in a final volume of 13 mL TCEP buffer.
13. Divide the 280 μM 4-NTP into separate 6.5 mL volumes in two 15 mL Falcon tubes.
    a. One tube will be used for the 4-NTP control only (thiamine −), the other will contain both 4-NTP and thiamine (thiamine +) to assay thiaminase activity.

   b. Dividing the same 4-NTP solution into two tubes ensures that both the thiamine + and thiamine - solutions have identical starting concentrations of 4-NTP.

14. Prepare working assay solutions (thiamine – and thiamine +):
    a. Add 28.9 μL of 100 mM thiamine to the one tube containing 6.5 mL 280 μM 4-NTP.
    b. Add 28.9 μL HPLC-grade water to the other tube containing 6.5 mL 280 μM 4-NTP.
    c. Reseal promptly and vortex solutions thoroughly.
15. Pipette 32 μL of TCEP buffer into the wells of a Corning 3795 clear 96-well microplate.
    a. This buffer volume assumes that 8 μL of sample is added; however, we routinely will adjust the sample volume up or down according to expected activity and correspondingly compensate with buffer volume such that the overall volume is 40 μL.
16. Pipette 8 μL of sample supernatant into each well (six wells for triplicate analyses, four wells for duplicate analyses). Pipet up and down to ensure complete transfer and thorough mixing.
    a. Pipet 8 μL of either the *P. thiaminolyticus* culture supernatant or the recombinant thiaminase enzyme as a positive control.
       i. Ensure that the culture used as a positive control has activity.
       ii. As an example, a 1:1000 dilution of the affinity-purified enzyme originally at a 3.3 mg.mL^{-1} protein concentration yields a strong positive signal.
    b. Pipet 8 μL of the TCEP sample diluent similarly as a negative control.
17. Transfer the working assay solutions to two separate disposable microplate reagent trays.
    a. Using an 8-channel pipettor, add 100 μL of the thiamine - solution to one set of wells.
    b. Then add 100 μL of the thiamine + solution to the other set of wells.
    c. The same set of tips may be used for first the thiamine - wells followed by the thiamine + wells, provided that no contact between the sample solution and the reagents is made. This reduces waste and critical time during reagent addition for this kinetic assay.
    d. Ensure the addition is done rapidly and forcefully to enhance mixing upon reagent addition.

   e. The wells may then be either manually pipetted to mix, use a plate shaker, or immediately inserted into the plate reader, where the first assay step is plate shaking. Our preference is the latter for efficiency and consistency.

18. Read the assay plate immediately using the pre-loaded program.

Notes:

1. Thiaminases from different sources have different pH optima, with many in the mildly acidic range for maximal activity (pH 4.6–6.6) (Edwards et al., 2023; Porter & Pinger, 2025; Sealock et al., 1943; Wittliff & Airth, 1968; Zajicek et al., 2009; Zajicek et al., 2005).

2. pH 6.9 is commonly used in the literature for the 4-NTP assay. The molar extinction coefficient of 4-NTP is pH-dependent in phosphate buffer, with decreases at pH values < 7.5, however this pH dependence was attenuated in solutions containing TCEP (Fig. 6). Therefore, activity calculations should use an extinction coefficient determined in the same final buffer matrix whenever possible. In the present assay buffer, consisting of 50 mM potassium phosphate, 100 mM NaCl, and 10 mM TCEP, the final measured pH was 6.7, and the experimentally-determined extinction coefficient was 13,418 $M^{-1}.cm^{-1}$, not significantly different from the reported 13,650 $M^{-1}.cm^{-1}$ at pH 7.2 (Hanes et al., 2007). We prepare our 20x PBS stock at an alkaline pH so that the final pH upon dilution into the working solution and the addition of TCEP HCl matches the desired pH for analyses, facilitating routine use.

3. Non-specific degradation of thiamine increases above pH 7.0.

4. We are aware that the article by Honeyfield et al. (Honeyfield et al., 2010) was retracted; however, it is our understanding that the reason was a unit calculation problem that affected the reporting of 4-NTP rates, not the methodology or the valuable insights it contains.

5. Safety considerations and standards

Dry ice should be handled only with tongs or insulated gloves. Only small amounts should be brought into the cold room to reduce the possibility of excess CO_2 levels building up in the enclosed space.

This method has been used with fish muscle, swim bladder, liver, processed dog foods and treats, and pin bones. However, consumer-grade coffee grinders are not designed for larger or especially hard bones. In many models, the stainless-steel blade is attached to the grinder housing via a plastic component that can shear if the sample is too hard or if excessive sample volume creates too much resistance within the confined grinding chamber.

Solutions containing 4-NTP should be disposed of as hazardous waste in accordance with institutional policies.

6. Analysis and statistics

6.1 Procedure for analysis

The analysis of 4-NTP data is complicated by low activity levels in most fish tissue matrices. We illustrate below the analysis procedure in bacterial culture media, where activity is straightforward to analyze. Following a consistent process is even more critical when activity levels are low.

1. Export the kinetic data from the instrument software to Excel, Google Sheets, or another suitable spreadsheet
2. Determine the pathlength for each well using measurements at 977 nm and 900 nm

   The pathlength (b) is calculated as follows:

$$\frac{(A_{977} - A_{900})}{0.173}$$

3. Correct the absorbance data by the pathlength
4. Copy data into GraphPad Prism in an X-Y plot with the X-axis as time and a single absorbance (Y-value) for each point (Fig. 4A). The analyses can also be completed in Excel or Google Sheets; however, GraphPad Prism is recommended for this purpose

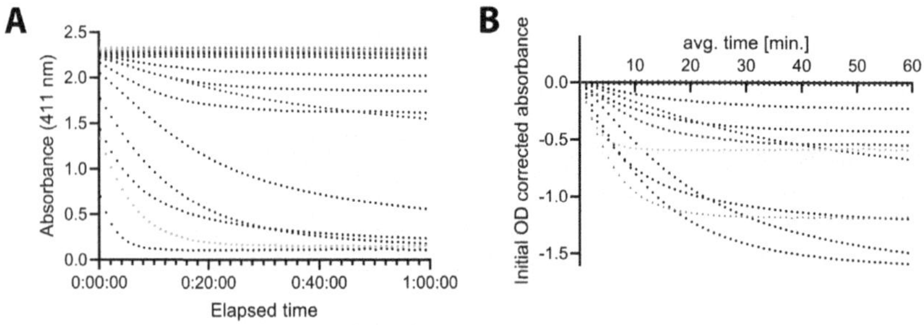

Fig. 4 (A) Kinetic loss of 4-NTP as measured at 411 nm, (B) Initial OD-corrected loss of 4-NTP measured at 411 nm.

5. Deduct the initial OD. This is not necessary for slope determination, but does improve visual clarity (Fig. 4B)
6. Convert the time to minutes if not exported as such
7. Using the straight-line equation under non-linear regression, determine the slopes over defined periods (e.g., 0–4 min, 4–9 min, 9–16 min)
   a. The straight-line option under non-linear regression allows the time range to be set, whereas in linear regression, the entire range is set. Ensure that only the points in the range of interest are included in the fit (Fig. 5A)
   b. It also provides for the detection and removal of individual time point outliers, which can improve the reliability of the slope.
8. Review the plots for each individual curve visually to detect aberrations.
9. Adjust the time range accordingly to capture the regions with the greatest slopes:
   a. In our hands, for microbial culture supernatants, we typically select the range from 1–4 min (Fig. 5B). Beyond this range, slopes from some samples decrease.
   b. However, for more complicated matrices with very low activity, such as fish tissues, individual time periods must be evaluated due to differences in the lag time across samples.
      i. Low levels of activity will also be more impacted by the initial thermal equilibration to 37 °C, hence later time points or pre-equilibrating the working solution to 37 °C may be preferable.
      ii. The use of pre-determined, defined static endpoints or kinetic reads without reviewing the individual curves runs the risks of missing the maximal window or including more points and diluting the slope (e.g., choosing a defined endpoint of 10 min or a 2–5 min fixed window).

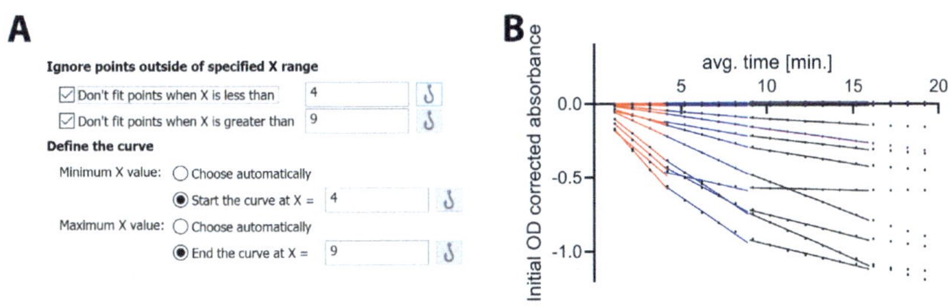

Fig. 5 (A) Straight line fitting parameters in GraphPad Prism, (B) Slope determination using linear fits over 1–4 min, 4–9 min, and 9–16 min.

10. Confirm the ranges with the greatest slopes and best fits by plotting the slopes and R^2 values for each time range. Visually review slopes with fits below an R^2 of 0.95.

Notes:

- It is very important not to rely on automatic data processing and instead to visually assess each curve before and after applying curve fitting.
- Although the assay is typically run in triplicate with and without 4-NTP, each replicate should be interrogated separately.
- It is not uncommon to have sporadic spikes or inconsistent time periods, especially for samples with high background and low activity. By manually observing each curve, data that would contribute to high replicate variability can often be salvaged by selecting a different time range for the slope.
- We strongly discourage reliance on only the initial value and predefined endpoint (e.g., reading only after 10 min) instead of determining the slope from kinetic readings. Sporadic absorbance aberrations at the start and end points, along with suboptimal time selection, can yield erroneous data with this practice.

11. The slopes in OD/min. (A) are then converted to $pmol.min^{-1}$ using the Beer-Lambert Law (A = ε x b x [C]).

a. The molar extinction coefficient for 4-NTP is pH-dependent, decreasing from 13,666 $M^{-1}.cm^{-1}$ at pH 8.0–11,762 $M^{-1}.cm^{-1}$ at pH 6.5 (Fig. 6A) However, this pH-dependence was largely mitigated after dilution in TCEP buffer. Although an extinction coefficient of 11,925 $M^{-1}.cm^{-1}$ would be predicted for pH 6.7 phosphate buffer alone, an experimental value of 13,418 $M^{-1}.cm^{-1}$ was obtained in the same buffer including TCEP (Fig. 6B).

b. Fish tissue had a minimal effect on 4-NTP absorbance. In concentrated anchovy extracts (1.2 μg per well dry anchovy, equivalent to ~4.0 mg wet tissue) prepared in TCEP buffer, the extinction coefficient did not differ significantly from TCEP buffer alone and neither differed significantly from the published value of 13,650 M^{-1} cm^{-1}, determined at pH 7.2 (Hanes et al., 2007). Therefore, we recommend retaining the published extinction coefficient (13,650 M^{-1} cm^{-1}) for consistency with previous studies unless substantially different buffer systems or pH conditions are used.

c. Rearrange the Beer-Lambert law:

i. $[C] = \left(\frac{\frac{A}{b}}{\varepsilon}\right)$

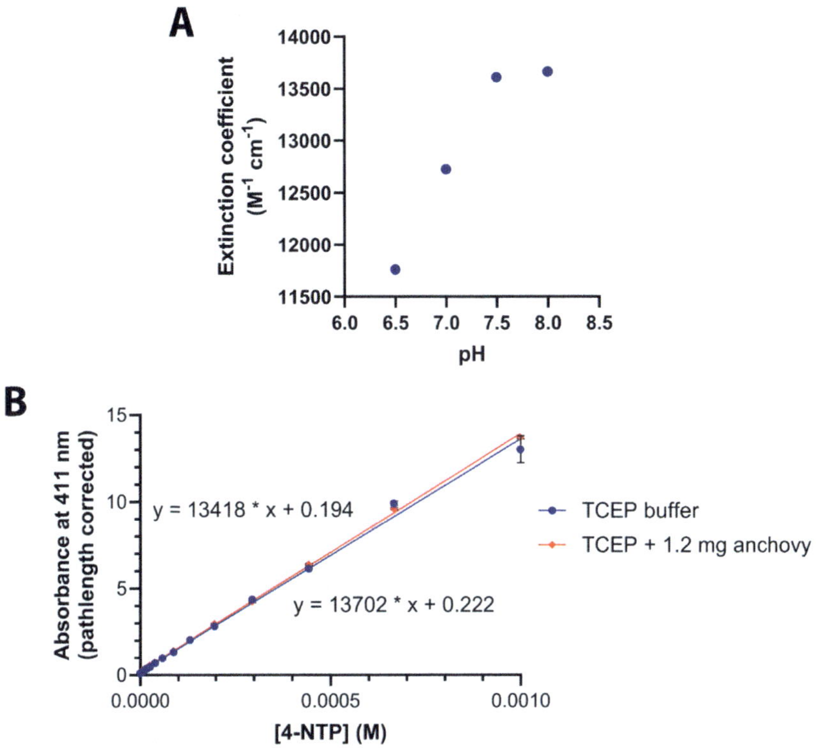

Fig. 6 (A) pH dependence of the molar extinction coefficient of 4-NTP in 20x PBS at pH 6.5 to pH 8.0 (before addition of TCEP). (B) Calibration curve for 4-NTP in TCEP buffer, pH 6.7, with and without 1.2 mg anchovy tissue per well (dry, estimated 4 mg wet tissue), with linear equations shown. Each point represents the average of triplicate measurements.

ii. The data were already pathlength corrected in step 1 (A/b = Abs.min^{-1}.cm^{-1}).

iii. $[C] = \frac{\frac{A}{b}}{13{,}650 M^{-1} cm^{-1}}$

iv. [C] is in units of mol.L^{-1}

d. The total assay volume is 140 μL.

i. 100 μL 4-NTP solution, 32 μL TCEP buffer 8 μL sample

e. To convert OD.min^{-1}.cm^{-1} to pmol.min^{-1}, multiply by 10,256.4.

$$\frac{X \frac{OD}{\min}}{cm} \times \frac{\frac{mol}{L} * cm}{13,650} \times \frac{10^{12}\, pmol}{mol} \times \frac{L}{10^{6} \mu L} \times 140\, \mu L$$

$$= \frac{X \frac{OD}{\min}}{cm} \times 10,256.4 \quad pmol * cm = X\, \frac{pmol}{\min}$$

12. Average replicates and report values for both thiamine + and thiamine – wells.
13. Deduct the $pmol.min^{-1}$ value for thiamine - wells from the thiamine + wells.
14. To report values on a mass normalized basis ($pmol.\ min^{-1}.mg_{tissue}{}^{-1}$), use the following equation:

$$\frac{X\frac{OD}{\min}}{cm} \times \frac{10,\ 256.4\,pmol * \mathrm{cm}}{X\ mg\ tissue}$$

15. Assay-specific values for the mg tissue per well, measured $OD.min^{-1}$ values with pathlength adjustment, and pH-dependent extinction coefficient (if substantially different than published values) should be used.
16. To report values as specific activity ($pmol.min^{-1}.\mu g_{protein}{}^{-1}$), divide the $pmol.min^{-1}$ value by the μg protein in the sample

6.2 Normalization and reporting results

We recommend reporting both the + thiamine and - thiamine rates, along with the conventionally reported difference between the two, which is typically interpreted as thiamine-specific activity. As discussed further in Sections 5.3 and 5.4, differential activity is derived from the subtraction of two independently measured rates, and uncertainty in the thiamine - background can become a substantial contributor to the overall variance at low activity levels. Presentation of the individual reaction rates, therefore, provides additional transparency and allows assessment of the magnitude of the thiamine-responsive signal relative to matrix-dependent background activity.

We recommend reporting results based on both the rate per unit of sample (e.g., $pmol.min^{-1}$ per mg tissue, μL culture supernatant, etc.) and per unit of protein ($pmol.min^{-1}.\mu g_{protein}{}^{-1}$). The former is often more biologically relevant from an exposure perspective. For example, a predator or consumer ingests a given mass of tissue or food, and mass-based normalization therefore provides an estimate of the potential capacity for thiamine degradation associated with that mass.

However, mass-based normalization alone does not account for differences in process extraction efficiency or compositional differences between samples or matrices. Hence, we strongly recommend determining the protein concentration and reporting the thiamine degradation on a specific activity basis ($pmol.min^{-1}.\mu g_{protein}^{-1}$). A sample may exhibit high activity on a mass basis simply because protein extraction from that matrix is more efficient, while the enzyme activity relative to the extracted protein is comparatively low. Wet-weight normalization reflects the potential exposure burden per unit of consumed material mass, whereas protein normalization better reflects enzyme abundance/activity relative to the extracted biological material.

This distinction becomes particularly important when comparing tissues or processed products with markedly different water contents. In our silver carp studies, freeze-dried samples exhibited approximately three-fold greater thiaminase activity on a mass-normalized basis relative to raw tissue, despite exhibiting lower activity when normalized to extracted protein content (Wolfe et al., 2023). This reflects the substantial concentration of solids and protein accompanying dehydration rather than a true increase in catalytic efficiency. Thus, expression solely on a wet-weight basis may overestimate the apparent enhancement of thiaminase activity in dehydrated or compositionally altered products.

We observed this in our work on silver carp prepared using different processing methods (e.g., raw, baked, microwaved, freeze-dried). Freeze-dried tissue exhibited substantially lower loss upon drying (0.6 %) relative to raw fish (76.2 %), indicating a lower water content. Correspondingly, protein concentrations obtained following otherwise identical sample extraction procedures were $23.33\ mg.mL^{-1}$ and $5.1\ mg.mL^{-1}$, respectively. These differences likely reflect both variation in water content and differences in the ease of protein release from the matrices during extraction. More broadly, we have observed substantial differences in water content and extractable protein concentrations across tissues and processing methods.

Quantitatively, this variation in tissue water content may substantially affect interpretation. In lake trout livers, we observed losses upon drying ranging from 63.4 to 78.5 % (Edwards et al., 2026). Hypothetically, this range would equate to substantially more thiaminase activity per dry tissue mass in the sample with greater loss upon drying.

Example 1: 63.4 % loss upon drying:

100 g wet tissue − 63.4 g loss = 36.6 g dry tissue.

$10\ nmol.min^{-1}.g_{\text{wet tissue}}^{-1} \times 100\ g\ \text{wet tissue}/36.6\ g\ \text{dry tissue} = .$

$27.3\ nmol.min^{-1} g_{\text{dry tissue}}^{-1}.$

Example 2: 78.5 % loss upon drying:

100 g wet tissue − 78.5 g loss = 21.5 g dry tissue.
10 $nmol.min^{-1}.g_{wet\ tissue}{}^{-1}$ × 100 g wet tissue/21.5 g dry tissue = .
46.5 $nmol.min^{-1}.g_{dry\ tissue}{}^{-1}$.

Thus, identical activities on a wet-mass basis would differ by approximately 1.7-fold when expressed on a dry-mass basis solely because of differences in hydration.

6.3 Method performance

Fig. 7 shows the response of spiking increasing amounts of affinity-purified recombinant *P. thiaminolyticus* into a homogenized whole anchovy sample. The commercially available anchovy marketed as a dog treat was air-dried, so the 0.8 mg dry tissue mass in the assay, assuming a 70 % moisture loss upon drying, represented approximately 2.67 mg wet tissue mass. Fig. 7A illustrates the pmol degraded over time while Fig. 7B shows the concentration-dependent response in units of $pmol.min^{-1}$ in the presence and absence of thiamine. The anchovy sample itself did not show detectable thiaminase activity. Across all no-thiamine wells (Fig. 7B), the matrix background averaged 39.1 ± 2.8 $pmol.min^{-1}$ ($n = 24$). After subtracting paired no-thiamine wells, the response was linear over the tested concentration range. Using the background-corrected tissue-only controls (0 ng added thiaminase), the analytical limit of detection, defined as the mean blank response plus three times its standard deviation, was 49.9 $pmol.min^{-1}$. Based on the calibration curve, this corresponded to 1.0 ng of recombinant thiaminase (Fig. 7C).

Importantly, this analytical limit of detection is a property of the assay and is expressed in terms of 4-NTP loss ($pmol.min^{-1}$). When activity is normalized to tissue mass, the corresponding detection limit becomes dependent on the sample load. For example, an analytical detection limit of 50 $pmol.min^{-1}$. would correspond to 50 $pmol.min^{-1}.mg^{-1}$ when 1 mg of tissue was assayed, versus 25 $pmol.min^{-1}.mg^{-1}$ when 2 mg of tissue was assayed. Similarly, the minimum detectable amount of thiaminase of 1 ng would equate to 1 $ng_{thiaminase}.mg_{tissue}{}^{-1}$ at a 1 mg tissue load, but 0.5 $ng_{thiaminase}.mg_{tissue}{}^{-1}$ at a 2 mg tissue load. Consequently, mass normalized rates are not fixed assay characteristics but vary according to sample loading and the basis for normalization (e.g., dry versus wet mass of tissue.).

These observations do not invalidate prior reports of low thiaminase activity. Rather, they highlight several analytical considerations that become increasingly important as measured activities approach background

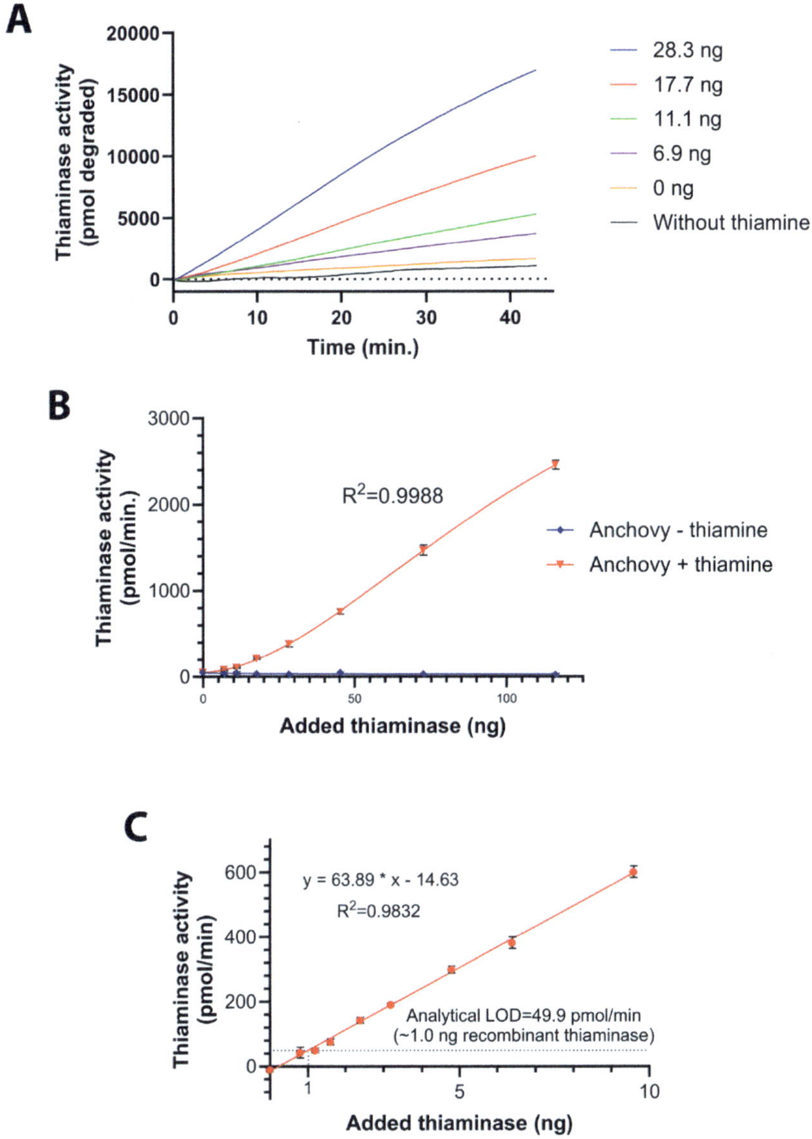

Fig. 7 Thiaminase activity in the presence of increasing amounts of recombinant *P. thiaminolyticus* thiaminase in a background of air-dried, whole anchovy extract (0.8 mg per well). (A) pmol 4-NTP degraded as a function of time (min.) and (B) pmol.-$min^{-1}.mg_{dry\ tissue}^{-1}$ as a function of thiaminase amount (μg). A four-parameter logistic equation was used to fit the calibration data, with each point assayed in triplicate for both thiamine + and thiamine - conditions. (C) Linear response of thiaminase activity ($pmol.min^{-1}$) relative to added recombinant thiaminase showing the limit of detection. 0.4–1.2 mg tissue (dry mass, estimated 1.3–4.0 mg wet mass) was used to generate these data, including thiaminase spikes at 2, 4, and 8 ng $thiaminase.\mu g_{tissue}^{-1}$ (dry mass).

variability. In the present study, low-magnitude activities were disproportionately influenced by the subtraction of paired no-thiamine controls, resulting in proportionally greater uncertainty than observed for higher-activity samples. Consequently, low differential activities should be interpreted cautiously unless supported by matrix-specific validation, adequate replication, and clear separation from background variability.

An additional consideration is that thiaminase activity is typically reported after normalization to tissue mass, whereas assay sensitivity is fundamentally determined by the amount of 4-NTP depleted and detected per well. Under the conditions examined here, the analytical LOD expressed in $pmol.min^{-1}$ was independent of tissue loading, whereas the equivalent detection limit expressed as $pmol.min^{-1}.mg_{tissue}{}^{-1}$ varied according to the amount of tissue analyzed and the basis for normalization (e.g., wet versus dry mass). Absolute rates of 4-NTP depletion, background variability, and tissue loading are not routinely reported, thus direct comparison of very low activities across studies can be challenging.

6.4 Avoiding overinterpretation

It is important to avoid overinterpreting the presence and impact of thiaminase activity when measured rates are difficult to distinguish from background signal or when low levels determined under in vitro, substrate-saturating conditions are assumed to be independently causative of thiamine deficiency in vivo.

Native biological systems differ substantially from the optimized conditions employed in analytical assays. In tissues and gastrointestinal contents, the local availability of thiamine, compatible cosubstrates, and pH conditions may vary considerably, while endogenous inhibitory compounds may further constrain catalytic efficiency (Edwards et al., 2023). Thus, despite tissue thiamine levels typically in the single to low double-digit $nmol.g^{-1}$ range, thiaminase-positive animals are rarely fully thiamine-depleted. This likely reflects biological constraints on thiamine accessibility and degradation that are not captured by in vitro assays. Most tissue stores of thiamine are intracellular as thiamine diphosphate (TDP), which is largely protein-bound as a coenzyme. This inherently imposes access constraints on thiaminase-mediated degradation in intact tissues, as intracellular TDP pools are protected by cellular compartmentalization and membrane barriers. The relatively large size of thiaminase I (~42 – 47 kDa in bacteria) further limits access to these intracellular stores. Consequently, in vivo thiamine degradation is constrained by cellular compartmentalization and by the rate at which thiamine equilibrates extracellularly between enzyme-bound and intracellular stores. Likewise, gastrointestinal thiaminase activity would be expected to reduce dietary thiamine availability but cause a more gradual reduction in systemic tissue reserves. Organ distribution of thiaminase activity has also been demonstrated (Wolfe et al., 2023; Zajicek et al., 2005). Whole-tissue and whole-fish homogenates are often used in vitro for thiaminase measurements, thereby placing thiaminase in direct contact with the tissue milieu and bypassing natural barriers.

Further, the enzyme activity at human body temperature (37 °C), as used in this assay, is markedly higher than that observed experimentally at ambient temperature, let alone the 1–24 °C body temperatures expected in fish associated with thiamine deficiency in the environment. Thiaminase activity measured in alewife was found to approximately double with every 10 °C difference in temperature (Zajicek et al., 2005).

Consistent with Michaelis-Menten kinetics, measured enzyme activity depends strongly on substrate availability. Physiologically, although thiaminase I enzymes are often described as having promiscuous substrate specificity, recent work suggests that substrate utilization among naturally occurring compounds may be more restricted than previously assumed (Edwards et al., 2023). Further, endogenous cosubstrates such as nicotinic acid or pyridoxine may be concentration- and proximity-limited for efficient in vivo thiamine degradation. Consequently, assays performed under substrate-saturating conditions and with a highly effective substrate measure the maximum catalytic potential of the enzyme under the assay conditions employed, rather than rates necessarily achieved in vivo. Likewise, inhibitory substances that may meaningfully suppress activity under physiological conditions may be partially or fully outcompeted in vitro by high substrate concentrations.

Beyond substrate availability and inhibitor presence, consideration of the enzyme's fate in a biological system after ingestion is also important. For example, if a prey fish contained high reported levels of this enzyme and was consumed by a predator fish, the enzyme would be subject to acidic gastric conditions, proteolytic degradation, dilution, and kinetic constraints within the GI tract. This would likely be the case whether the thiaminase originated from microbial, plant, or fish sources. Many thiaminases have mildly acidic pH optima (pH 4.6–6.6) (Edwards et al., 2023; Porter & Pinger, 2025; Sealock et al., 1943; Wittliff & Airth, 1968; Zajicek et al., 2009; Zajicek et al., 2005). Zajicek et al. hypothesized that diets with an elevated fat content may attenuate gastric acidity, potentially allowing some thiaminases to retain activity (Zajicek et al., 2009). If the source was microbial, however, it is conceivable that thiaminase-producing microorganisms could colonize the gut microbiota and provide a continual source of enzyme production within the gut. A precedent for gastrointestinal production of thiaminase exists in ruminants, where proliferation of thiaminase-producing microorganisms within the rumen has long been implicated in the development of polioencephalomalacia and thiamine deficiency (Edwin & Jackman, 1970). Whether analogous processes occur in fish gastrointestinal microbiota remains unclear. Further, fish gastrointestinal tracts have been reported to exhibit uptake of intact proteins through endocytotic mechanisms, suggesting that the gastrointestinal tract may not present an absolute barrier to dietary or microbiota-produced enzymes (Debnath & Saikia, 2021; McLean & Ash, 1987; McLean & Donaldson, 1990; Rombout et al., 1985). Whether such processes influence the persistence or physiological impact of dietary thiaminases remains unclear. Overall, the extent to which measured thiaminase activity contributes to physiologically meaningful thiamine depletion under native biological conditions remains an important unresolved question in the literature. However, given the varied physicochemical and physiological constraints, high in vitro activity does not necessarily translate directly into equivalent thiamine-degrading capacity in vivo.

While assessing thiaminase activity in relatively dilute, simple matrices such as cell culture media is straightforward, interpreting these analyses in complex tissue matrices is more challenging. The premise of the 4-NTP assay is to compare the rate of 4-NTP disappearance in the presence of exogenous thiamine versus its absence, under conditions where high concentrations of 4-NTP are available as a non-endogenous cosubstrate. The differential rate is attributed to thiamine-dependent thiaminase activity, whereas the rate observed in the absence of added thiamine is attributed to non-thiamine-dependent degradation of 4-NTP. This approach inherently assumes that endogenous thiamine concentrations in the sample are low. Otherwise, endogenous thiamine could also support thiaminase-mediated consumption of 4-NTP in the 'thiamine -' condition, increasing the apparent background rate and underestimating thiaminase activity.

To evaluate the magnitude of this potential effect, we performed the assay using a range of added thiamine concentrations rather than a single saturating concentration (Fig. 9). In recombinant thiaminase preparations, substantial enhancement of 4-NTP depletion was observed at added thiamine concentrations greater than 10 μM (Fig. 8A). However, some biological matrices exhibited substantial background degradation and measurable responses at low thiamine concentrations, potentially reflecting endogenous thiamine, matrix-dependent effects, or differences in apparent substrate availability. In an experiment using freeze-dried silver carp, we observed a measurable enhancement of 4-NTP depletion at 0.6 μM thiamine in the 100 μL working reagent, corresponding to 0.06 nmol thiamine per well (Fig. 8B). With 3.2 mg tissue per well, this is equivalent to an added 18.75 $nmol_{thiamine}{\cdot}g_{tissue}^{-1}$, an amount physiologically plausible in fish tissue samples. However, we acknowledge that freeze-drying may increase the contribution of endogenous thiamine by concentrating tissue constituents and disrupting physiological compartmentalization relative to unprocessed tissues.

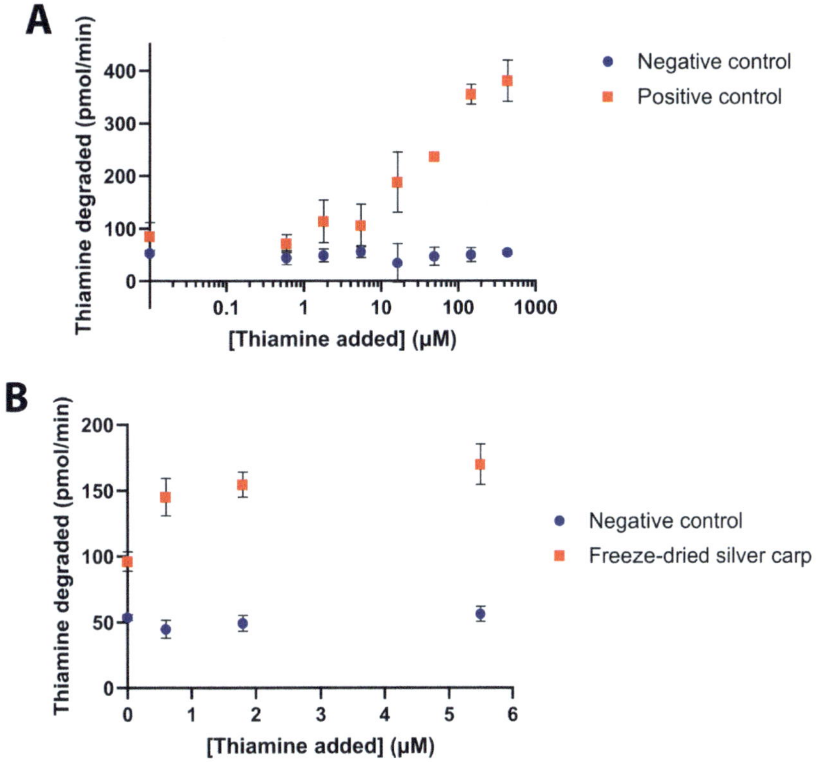

Fig. 8 Amount of thiamine degraded ($pmol.min^{-1}$) as a function of added thiamine concentration in (A) recombinant *Clostridium botulinum* thiaminase I and (B) freeze-dried silver carp relative to negative control.

Endogenous thiamine may contribute to the apparent background rate observed in the absence of added thiamine, particularly in samples with higher native thiamine concentrations. While unlikely to fully account for elevated background degradation, endogenous thiamine may contribute measurably in certain matrices and should be considered when interpreting low differential activities.

Additional matrix-dependent chemical or enzymatic effects cannot be excluded. Although substantial matrix suppression was not observed in the tissue range examined (0.4–1.2 mg dry tissue), tissue loading remains an important analytical consideration as it influences carryover of endogenous thiamine and other matrix constituents. Increasing tissue loads may also introduce greater amounts of inhibitory substances, such as Cu^{2+} and ascorbic acid (Edwards et al., 2023), potentially complicating analyses.

Further, while normalization to tissue mass facilitates biological comparison among samples, the underlying analytical signal is measured as 4-NTP depletion per well. Consequently, identical tissue-normalized activities may arise from substantially different absolute rates of 4-NTP depletion depending on the amount of tissue analyzed. This, in turn, affects the degree of separation between the measured activity and background and may influence sensitivity, particularly near the limit of detection.

7. Alternative techniques for evaluation

7.1 Precursor techniques

Some earlier thiaminase I assays included using colorimetric or fluorescent product formation to monitor the remaining intact thiamine (Melnick & Field, 1939b; Nishimune et al., 2000; Sealock & White, 1949). Reactions with diazotized p-aminoacetophenone were carried out following selective adsorption of thiamine from complex matrices (Melnick & Field, 1939b). Additional spectrophotometric approaches exploited nucleophile-dependent base-exchange reactions, leading to the formation of heteropyrithiamine (Kenten, 1957) and anilothiamine (Wittliff & Airth, 1968), which exhibit different spectral characteristics. A fluorescence assay using a novel substrate (TanA) was further developed, allowing for kinetic measurement of the thiaminase I-induced cessation of fluorescence quenching and release of a cleaved pyrimidine-based fluorescent product (Zhu et al., 2013).

Other assays for thiaminase activity monitored the amount of intact thiamine remaining by measuring its oxidation to fluorescent thiochrome

(Fig. 9) (Edwards et al., 2023; Nishimune et al., 2000; Porter & Pinger, 2025; Sealock & White, 1949). For example, we have routinely done this when screening the efficiency of natural cosubstrates of microbial thiaminases in culture media (Edwards et al., 2023).

However, direct detection of thiochrome in complex matrices presents several challenges, including its formation as a natural oxidation product of thiamine, competing reactions that form non-fluorescent products, and sensitivity to the conditions used for thiochrome formation (Edwards et al., 2017; Melnick & Field, 1939a; Soliman, 1981). Hence, thiochrome-based techniques require careful validation in complex matrices. The matrix must not contribute substantial intrinsic fluorescence, quench thiochrome fluorescence, form fluorophores that share the spectral properties of thiochrome, or interfere with the alkaline ferricyanide oxidation step required for thiochrome formation.

We initially evaluated the conversion of residual thiamine to thiochrome as a measure of thiaminase activity in fish tissues. Subsequently, we adapted the method to instead quantify endogenous thiamine by adding recombinant thiaminase and a non-interfering cosubstrate. However, strong tissue-dependent matrix effects complicated quantitative measurement. As shown in Fig. 10A, calibration curves generated in neutralized TCA extracts of fish liver and muscle exhibited substantially reduced slopes relative to buffer controls, indicating suppression of thiochrome formation, fluorescence, or both. These effects can be differentiated by comparing externally generated thiochrome added to the matrix with thiochrome formed in situ from thiamine. Using this approach (Fig. 10B), liver extracts exhibited minimal fluorescence quenching, but impaired thiochrome formation, whereas muscle extracts strongly suppressed thiochrome fluorescence. Similar matrix-dependent effects are likely relevant to pre-column thiochrome derivatization methods used for HPLC-based thiamine analysis.

These limitations can be mitigated with appropriate controls, including thiamine-depleted matrix blanks, matrix-matched positive controls, and sample dilution to reduce matrix and background interference. However, a thorough validation is warranted in any new matrix.

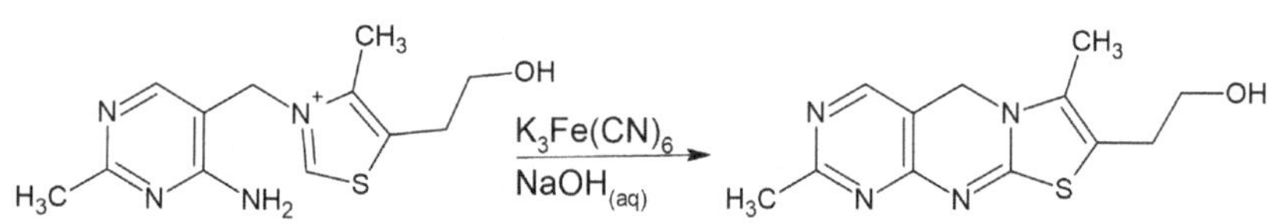

Fig. 9 Oxidation of thiamine (non-fluorescent) to thiochrome (fluorescent) using potassium ferricyanide in alkaline solution. *Reproduced from Edwards et al. 2024, Scientific Reports 14:10269, licensed under CC by 4.0.*

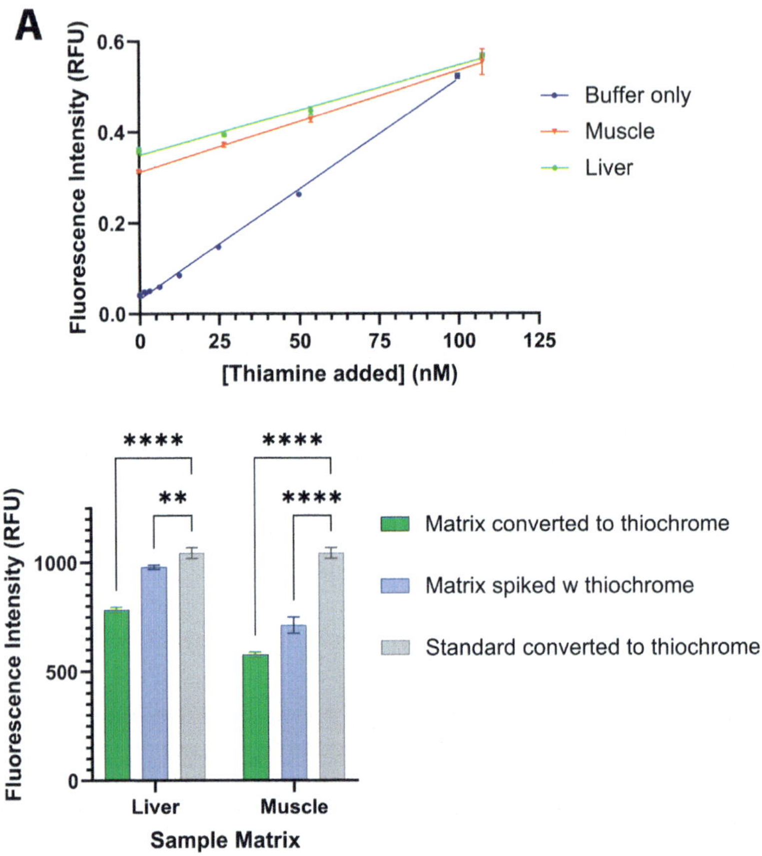

Fig. 10 (A) Thiamine calibration curves in neutralized TCA extracts of lake trout muscle and liver (3.56 and 4.44 $mg_{tissue}.mL^{-1}$, respectively) relative to neutralized TCA alone (buffer only). A substantially milder slope was observed in the liver and muscle sample extracts than in buffer alone, indicating suppression in thiochrome formation or fluorescence in these matrices. (B) Thiamine converted to thiochrome in situ or spiked with externally formed thiochrome in neutralized TCA extracts of thiamine-depleted liver and muscle matrices versus buffer alone.

Nevertheless, because any thiochrome assay relies on endpoint determination of residual thiamine rather than continuous kinetic monitoring, interpreting low activity levels remains challenging. Small errors or variability in the initial or final thiamine measurements can disproportionately affect the calculated apparent rate of thiamine degradation. Furthermore, the useful temporal window for detection may vary substantially between matrices and activity levels. Excessively short incubations may fail to resolve small differences in thiamine concentration from analytical variability, whereas prolonged incubations may compress differences as substrate depletion approaches completion or background degradation accumulates. Consequently, end-point-based approaches that do not directly observe the kinetic window may be more susceptible to spurious estimates of thiaminase activity.

Optimization of thiaminase assays has emphasized the use of high concentrations of preferred cosubstrates (e.g., pyridine) to ensure maximal enzyme activity, as subsaturating substrate conditions can underestimate thiaminase activity (Edwin, 1979; Edwin & Jackman, 1974; Honeyfield et al., 2010; McCleary & Chick, 1977). Assays relying solely on endogenous nucleophiles have also been reported. However, reaction rates may be limited by the availability and composition within the sample matrix (Nishimune et al., 2000; Zajicek et al., 2005). Thiamine itself can inhibit thiaminase activity in the absence of sufficient cosubstrates (Agee & Airth, 1973; Nishimune et al., 2000; Zajicek et al., 2005) and because nucleophile availability is not standardized among samples, such approaches may introduce additional variability when comparing tissues or species.

The identity of the cosubstrate is a critical parameter in any thiaminase I assay. While some assays use nicotinic acid as a cosubstrate, some thiaminase I enzymes have a greater affinity towards pyridoxine or pyridine than nicotinic acid (Edwards et al., 2023; McCleary & Chick, 1977). In fact, some thiaminases cannot utilize the latter (Bos & Kozik, 2000; Fujita et al., 1952). Nicotinic acid reportedly serves as a functional cosubstrate in bacterial, but not fish sources of thiamine (Bos & Kozik, 2000; Fujita et al., 1952). In our hands, pyridoxine was a far more efficient cosubstrate than nicotinic acid for recombinant thiaminase from *C. botulinum*, though 4-NTP as a non-natural substrate was slightly more efficient (Fig. 11) (Edwards et al., 2023). We note that nicotinic acid appeared to inhibit the enzymatic reaction at high concentrations. However, we cannot rule out the possibility that it may form a nucleophilic-substituted product with thiaminase that fluoresces at wavelengths identical to those of thiochrome.

Radiometric assays using thiamine labelled with ^{14}C on its thiazole ring can quantify enzymatic cleavage by release of labelled thiazole products following organic extraction (Edwin & Jackman, 1974; Edwin & Jackman, 1970). The released thiazole compound is then extracted into an immiscible organic solvent prior to measurement. The polarity differences permit resolution of ^{14}C-labeled thiazole from the more hydrophilic intact ^{14}C-labelled thiamine. This is an elegant approach that also allows thiaminase determinations to be extended to thiaminase II measurements, since detection does not depend on a separate cosubstrate. However, ^{14}C-thiazole-labeled

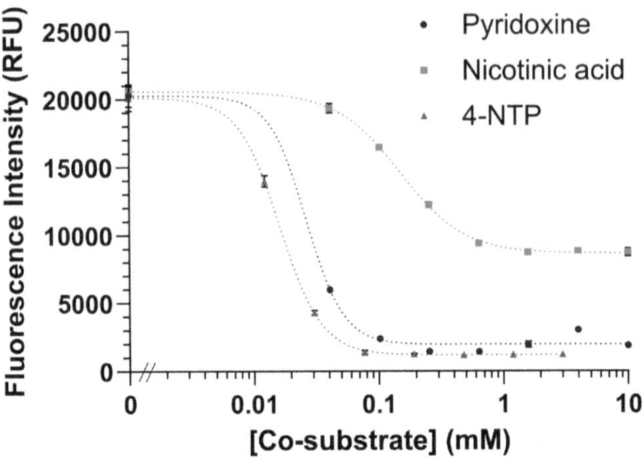

Fig. 11 Cosubstrates for thiaminase I from *C. botulinum.* 10 μM thiamine, 10 nM–10 mM 4-NTP, nicotinic acid, and pyridoxine in 20 mM MES/20 mM NaCl, pH 6.5 were incubated in the presence of 0.5 μg.mL^{-1} thiaminase I at 37 °C for 24 h. 5 μL of the reaction mixture was diluted with 45 μL water, then 100 μL 0.05 % potassium ferricyanide in 15% (w/v) sodium hydroxide was added prior to measurement using 360/40 nm excitation, 450/50 nm emission. *Reproduced from Edwards et al. 2023, Scientific Reports 13:7008, licensed under CC by 4.0.*

thiamine no longer appears to be commercially available. ^{3}H-labelled options available in radiochemical manufacturers' catalogs are not specifically labeled on the thiazole ring. These would not provide a viable alternative in this assay. While ^{14}C-thiazole-labelled thiamine can be custom-synthesized, the high cost and additional regulatory constraints of radioassays may be prohibitive for many labs.

7.2 Related techniques

We have evaluated the specificity of thiaminase I using UPLC-MS/MS to identify products formed during thiaminase-catalyzed nucleophilic substitution. MS readily detects the released 2-hydroxyethylthiazole (Fig. 1), and the corresponding pyrimidine-nucleophile adduct can be inferred by HPLC UV detection based on altered polarity/retention behavior. However, in our hands, direct MS identification of the pyrimidine conjugate has been challenging. We suspect that this adduct may be labile under ionization conditions, leading to poor recovery of the intact molecular ion or to fragmentation during analysis. We note broader concerns about thiamine recovery due to adsorption onto glass and other materials, as well as poor TDP recovery by MS detection (Edwards et al., 2024; Edwards et al., 2017).

7.3 Pros and cons

Pros	Cons
Colorimetric measurements do not carry the regulatory constraints, expense, and potential health risks of radiometric assays.	Radiometric markers are more readily detectable in complex matrices than a colorimetric or fluorescent output.
	4-NTP can undergo non-specific degradation, generating a high non-specific background signal.
	Radiometric assays measure the release of ^{14}C-labeled thiazole, which is highly specific to thiaminase-cleavage
	Radiometric assays are not subject to the same pH constraints as 4-NTP.
4-NTP analyses use widely available UV/Vis plate readers	Fluorescence-based analyses require fluorescence microtiter plate readers, while radiometric analyses require scintillation counters. Both are accessible but are less widely available than UV/Vis plate readers.
4-NTP is commercially available	^{14}C-thiazole thiamine requires custom synthesis
The microplate format allows samples and controls to be run with replication and with moderate throughput	A required organic solvent extraction step in radiometric assays or protein precipitation in fluorescence assays based on thiochrome detection increases manual labor and time. Difficult to carry out numerous replicates to assess low levels of activity.

Although thiaminase I does not require a metal cofactor for catalysis, its activity is sensitive to divalent cations, with many metals causing inhibition. These effects vary depending on the enzyme source and metals present. For example, Zn^{2+} has been reported to cause complete inhibition of thiaminase I isolated from freshwater prawn and variegated grasshoppers (*Zonocerus variegatus*), whereas Ni^{2+} enhanced activity in these preparations (Ehigie et al., 2013; Ona et al., 2015). Other divalent cations, such as Ca^{2+}, Mg^{2+}, Mn^{2+}, and Sn^{2+}, substantially inhibited the freshwater prawn enzyme (Ona et al., 2015). We found that Cu^{2+} was a potent and irreversible inhibitor of the *C. botulinum* thiaminase, though other metal ions had an overall limited effect (Edwards et al., 2023). Ascorbic acid was also inhibitory, but reversible (Edwards et al., 2023).

Chelation by EDTA can also reduce activity (Ona et al., 2015), likely through disruption of enzyme structure or the removal of stabilizing ions, rather than the elimination of an essential catalytic metal cofactor. We found that EDTA both increased the background rate of 4-NTP degradation and decreased the thiamine-specific rate. Hanes et al. also evaluated the impact of EDTA during the development of the 4-NTP assay (Hanes et al., 2007).

Additional inhibitors include small molecules such as 2-mercaptoethanol and 6-aminohexanoic acid (Ehigie et al., 2013). Notably, 4-methyl-5-β-hydroxyethylthiazole, the thiazole product released during thiaminase-catalyzed cleavage, has been reported to inactivate thiaminase from *Bacillus thiaminolyticus* (reclassified as *Paenibacillus thiaminolyticus* in 1997 (Shida et al., 1997)) and other biological sources (Fujita et al., 1952). As this molecule is cleaved from thiamine during thiaminase-catalyzed breakdown, this may result in product inhibition under experimental assay conditions if too much thiamine is available as a substrate.

8. Troubleshooting and optimization

Problem	Solution
High rate of 4-NTP degradation in the absence of thiamine	Ensure that water used for buffer preparation is thoroughly deoxygenated
Loss of signal from recombinant positive control	Add a dynamic blocking reagent to minimize losses from nonspecific interactions with the well or tube surfaces.
Negative or highly variable difference or thiamine - signal	Repeat the assay at a higher tissue concentration and with more replicates

9. Summary

In this chapter, we describe the application of the 4-NTP assay to measure thiaminase I activity across a range of biological samples, including microbial cultures, plant tissues, fish tissues, and processed pet foods. The assay uses commercially available reagents, standard microplate instrumentation, and kinetic monitoring of 4-NTP depletion to provide a sensitive measure of thiaminase I activity, with a limit of detection of ~50 pmol 4-NTP depleted per minute, corresponding to 1 ng thiaminase, under the conditions employed.

Particular attention is warranted when evaluating low activity samples. Since thiaminase activity is typically calculated as the difference between paired measurements obtained in the presence and absence of added thiamine, uncertainty in background rates can become a major contributor to overall variability. The analytical sensitivity of the assay is fundamentally determined by absolute 4-NTP depletion per well, whereas results are commonly reported after normalization to tissue mass. Consequently, interpretation and comparison of very low reported activities may be influenced by background variability, sample loading, and normalization approach. Reporting individual reaction rates with sufficient replication and considering analytical sensitivity in terms of absolute 4-NTP depletion can improve interpretation. We strongly recommend collecting activity data over a time course rather than relying on static endpoint measurements. The time corresponding to the maximal reaction rate may vary among samples, and isolated aberrant measurements can disproportionately influence the results, particularly for low-activity samples approaching the background signal.

While thiaminase activity is commonly used as a proxy for thiamine-degrading potential, assay-derived activities should not be interpreted as in vivo thiamine degradation rates. Thiaminase activities determined under elevated temperatures, substrate saturation with a uniquely high affinity cosubstrate, and homogenization of whole samples, which overcome physiological barriers, do not reflect physiological conditions in vivo. Consequently, reported activities are more appropriately viewed as indicators of maximal catalytic capacity, rather than actual rates of thiamine depletion. Consideration of endogenous substrate availability, compartmentalization, tissue organization, temperature, and inhibitory factors remains essential when relating measured thiaminase activity to biological outcomes.

References

Agee, C. C., & Airth, R. L. (1973). Reversible inactivation of thiaminase I of *Bacillus thiaminolyticus* by its primary substrate, thiamine. *Journal of Bacteriology, 115*(3), 957–965.

Agyei-Owusu, K., & Leeper, F. J. (2009). Thiamin diphosphate in biological chemistry: analogues of thiamin diphosphate in studies of enzymes and riboswitches. *The FEBS Journal, 276*(11), 2905–2916. https://doi.org/10.1111/j.1742-4658.2009.07018.x.

Ahn, I.-P., Kim, S., & Lee, Y.-H. (2005). Vitamin B(1) functions as an activator of plant disease resistance. *Plant Physiology, 138*(3), 1505–1515. http://www.ncbi.nlm.nih.gov/pmc/articles/PMC1176421/.

Ahoua, L., Etienne, W., Fermon, F., Godain, G., Brown, V., Kadjo, K., ... Guerin, P. J. (2007). Outbreak of beriberi in a prison in Cote d'Ivoire. *Food and Nutrition Bulletin, 28*(3), 283–290. http://www.ncbi.nlm.nih.gov/pubmed/17974361.

Balaghi, M., & Pearson, W. N. (1967). Comparative studies of the metabolism of 14C-pyrimidine-labeled thiamine, 14C-thiazole-labeled thiamine and 35S-labeled thiamine in the rat. *Journal of Nutrition, 91*(1), 9–19. https://doi.org/10.1093/jn/91.1.9.

Boggs, K., Spooner, D. E., Honeyfield, D. C., Shull, D., Wertz, T., & Sweet, S. (2019). Phylogeny and foraging mode correspond with thiaminase activity in freshwater fishes: potential links to environmental factors. *Freshwater Science, 38*(3), 605–615. https://doi.org/10.1086/704927.

Bos, M., & Kozik, A. (2000). Some molecular and enzymatic properties of a homogeneous preparation of thiaminase I purified from carp liver. *Journal of Protein Chemistry, 19*(2), 75–84.

Boubakri, H., Poutaraud, A., Wahab, M. A., Clayeux, C., Baltenweck-Guyot, R., Steyer, D., ... Soustre-Gacougnolle, I. (2013). Thiamine modulates metabolism of the phenylpropanoid pathway leading to enhanced resistance to *Plasmopara viticola* in grapevine. *BMC Plant Biology, 13*, 31. http://www.ncbi.nlm.nih.gov/entrez/query.fcgi?cmd=Retrieve&db=PubMed&dopt=Citation&list_uids=23442597.

Brown, S. B., Brown, L. R., Brown, M., Moore, K., Villella, M., Fitzsimons, J. D., ... Wolgamood, M. (2005). Effectiveness of egg immersion in aqueous solutions of thiamine and thiamine analogs for reducing early mortality syndrome. *Journal of Aquatic Animal Health, 17*(1), 106–112. https://doi.org/10.1577/H03-075.1.

Brown, S. B., Fitzsimons, J. D., Honeyfield, D. C., & Tillitt, D. E. (2005). Implications of thiamine deficiency in great lakes salmonines. *Journal of Aquatic Animal Health, 17*(1), 113–124. https://doi.org/10.1577/H04-015.1.

Debnath, S., & Saikia, S. K. (2021). Absorption of protein in teleosts: A review. *Fish Physiology and Biochemistry, 47*(2), 313–326. https://doi.org/10.1007/s10695-020-00913-6.

Deutsch, H. F., & Hasler, A. D. (1943). Distribution of a Vitamin B1 destructive enzyme in fish. *Proceedings of the Society for Experimental Biology and Medicine, 53*(1), 63–65. https://doi.org/10.3181/00379727-53-14186.

Di Micco, R., Schneider, M., & Nüesch, R. (2021). Postoperative *Paenibacillus thiaminolyticus* wound infection, Switzerland. *Emerging infectious Diseases, 27*(7), 1984–1986. https://doi.org/10.3201/eid2707.203348.

Edwards, K. A., Kleiner, D., Mangal, B. M., & Sonare, B. S. (2026). An updated transketolase activity assay to probe thiamine utilization in fish. *Comparative Biochemistry and Physiology Part C: Toxicology & Pharmacology, 300*, 110418. https://doi.org/10.1016/j.cbpc.2025.110418.

Edwards, K. A., Randall, E. A., Wolfe, P. C., Angert, E. R., & Kraft, C. E. (2023). Dietary factors potentially impacting thiaminase I-mediated thiamine deficiency. *Scientific Reports, 13*(1), 7008. https://doi.org/10.1038/s41598-023-34063-5.

Edwards, K. A., Randall, E. A., Wolfe, P. C., Kraft, C. E., & Angert, E. R. (2024). Pre-analytical challenges from adsorptive losses associated with thiamine analysis. *Scientific Reports, 14*(1), 10269. https://doi.org/10.1038/s41598-024-60910-0.

Edwards, K. A., Tu-Maung, N., Cheng, K., Wang, B., Baeumner, A. J., & Kraft, C. E. (2017). Thiamine Assays—Advances, challenges, and caveats. *ChemistryOpen, 6*(2), 178–191. https://doi.org/10.1002/open.201600160.

Edwin, E. E. (1979). [24] Determination of thiaminase activity using thiazole-labeled thiamine. *Methods in enzymology*. Academic Press. 113–117. https://doi.org/10.1016/0076-6879(79)62204-8.

Edwin, E. E., & Jackman, R. (1974). A rapid radioactive method for determination of thiaminase activity and its use in the diagnosis of cerebrocortical necrosis in sheep and cattle. *Journal of the Science of Food and Agriculture, 25*(4), 357–368. http://www.ncbi.nlm.nih.gov/pubmed/4857780.

Edwin, E. E., & Jackman, R. O. Y. (1970). Thiaminase I in the development of cerebrocortical necrosis in sheep and cattle. *Nature, 228*(5273), 772–774. https://doi.org/10.1038/228772a0.

Ehigie, L. O., Okonji, R. E., & Ehigie, F. A. (2013). Biochemical properties of thiaminase, a toxic enzyme in the gut of grasshoppers (*Zonocerus variegatus* Linn). *Cameroon Journal of Experimental Biology, 9*(1), 9–16.

Frank, R. A. W., Leeper, F. J., & Luisi, B. F. (2007). Structure, mechanism and catalytic duality of thiamine-dependent enzymes. *Cellular and Molecular Life Sciences, 64*(7), 892. https://doi.org/10.1007/s00018-007-6423-5.

Fujita, A., Nose, Y., Kozuka, S., Tashiro, T., Ueda, K., & Sakamoto, S. (1952). Studies on thiaminase. I. Activation of thiamine breakdown by organic bases. *Journal of Biological Chemistry, 196*(1), 289–295.

Gassiep, I., Armstrong, M., & Norton, R. (2020). Human melioidosis. *Clinical Microbiology Reviews, 33*(2), https://doi.org/10.1128/cmr.00006-19.

Gnaedinger, R. H., & Krzeczkowski, R. A. (1966). Heat inactivation of thiaminase in whole fish. *Commercial Fisheries Review, 28*(8), 11–14. https://spo.nmfs.noaa.gov/sites/default/files/pdf-content/mfr2883.pdf.

Green, R. G., Carlson, W. E., & Evans, C. A. (1941). A deficiency disease of foxes produced by feeding fish: B1 avitaminosis analogous to wernicke's disease of man. *Journal of Nutrition, 21*(3), 243–256. https://doi.org/10.1093/jn/21.3.243.

Green, R. G., Carlson, W. E., & Evans, C. A. (1942). The inactivation of vitamin B1 in diets containing whole fish: Two figures. *Journal of Nutrition, 23*(2), 165–174. https://doi.org/10.1093/jn/23.2.165.

Green, R. G., & Evans, C. A. (1940). A deficiency disease of foxes. *Science, 92*(2381), 154–155. https://doi.org/10.1126/science.92.2381.154.

Hall, C. M., Stone, N. E., Martz, M., Hutton, S. M., Santana-Propper, E., Versluis, L., ... Wagner, D. M. (2023). *Burkholderia thailandensis* isolated from the environment, United States. *Emerging Infectious Diseases, 29*(3), 618–621. https://doi.org/10.3201/eid2903.221245.

Hanes, J. W., Kraft, C. E., & Begley, T. P. (2007). An assay for thiaminase I in complex biological samples. *Anal Biochem, 368*(1), 33–38. https://doi.org/10.1016/j.ab.2007.06.001.

Hirn, J., & Pekkanen, T. J. (1975). Quantitative analysis of thiaminase activity in certain fish species. *Nord Veterinary Medicine, 27*(12), 646–648.

Honeyfield, D., Hinterkopf, J., Fitzsimons, J., Tillitt, D., Zajicek, J., & Brown, S. (2005). Development of thiamine deficiencies and early mortality syndrome in lake trout by feeding experimental and feral fish diets containing thiaminase. *Journal of Aquatic Animal Health, 17*, 4–12. https://doi.org/10.1577/H03-078.1.

Honeyfield, D. C., Hanes, J. W., Brown, L., Kraft, C. E., & Begley, T. P. (2010). Comparison of thiaminase activity in fish using the radiometric and 4-nitrothiophenol colorimetric methods. *Journal of Great Lakes Research, 36*(4), 641–645. https://doi.org/10.1016/j.jglr.2010.07.005.

Honeyfield, D. C., Hinterkopf, J. P., & Brown, S. B. (2002). Isolation of thiaminase-positive bacteria from alewife. *Transactions of the American Fisheries Society, 131*(1), 171–175. https://doi.org/10.1577/1548-8659(2002)1312.0.CO;2.

Jenkins, A. H., Schyns, G., Potot, S., Sun, G., & Begley, T. P. (2007). A new thiamin salvage pathway. *Nature Chemical Biology, 3*(8), 492–497. https://doi.org/10.1038/nchembio.2007.13.

Jurgenson, C. T., Begley, T. P., & Ealick, S. E. (2009). The structural and biochemical foundations of thiamin biosynthesis. *Annual Review of Biochemistry, 78*, 569–603. https://doi.org/10.1146/annurev.biochem.78.072407.102340.

Kenten, R. H. (1957). The partial purification of properties of a thiaminase from bracken [Pteridium aquilinum (L.) Kuhn]. *Biochemical Journal, 67*(1), 25–33. https://doi.org/10.1042/bj0670025.

Khiari, Z. (2022). Sustainable upcycling of fisheries and aquaculture wastes using fish-derived cold-adapted proteases. *Frontiers in Nutrition, 9*, 875697. https://doi.org/10.3389/fnut.2022.875697.

Kimura, R. (1965). Thiamine decomposing bacteria. In K. Shimazono, & E. Katsura (Eds.). *Review of Japanese literature on beriberi and thiamine* (pp. 255–274). Vitamin B Research Committee of Japan.

Kraft, C., & Angert, E. (2017). Competition for vitamin B1 (thiamin) structures numerous ecological interactions. *Quarterly Review of Biology, 92*, 151–168.

Kraft, C. E., Gordon, E. R., & Angert, E. R. (2014). A rapid method for assaying thiaminase I activity in diverse biological samples. *PLoS One, 9*(3), e92688. https://doi.org/10.1371/journal.pone.0092688.

Krampitz, L. O., & Woolley, D. W. (1944). The manner of inactivation of thiamine by fish tissue. *Journal of Biological Chemistry, 152*(1), 9–17. https://doi.org/10.1016/S0021-9258(18)72084-X.

Lee, C. F. (1948). Thiaminase in fishery products: A review. *Commercial Fisheries Review, 10*(4), 7–17. https://spo.nmfs.noaa.gov/sites/default/files/pdf-content/mfr1042.pdf.

Mancinelli, R., & Ceccanti, M. (2009). Biomarkers in alcohol misuse: Their role in the prevention and detection of thiamine deficiency. *Alcohol and Alcoholism, 44*(2), 177–182. https://doi.org/10.1093/alcalc/agn117.

Mantua, N. J., Bell, H., Todgham, A. E., Daniels, M. E., Rinchard, J., Ludwig, J. M., ... Johnson, R. C. (2025). Widespread thiamine deficiency in California salmon linked to an anchovy-dominated marine prey base. *Proceedings of the National Academy of Sciences, 122*(26), e2426011122. https://doi.org/10.1073/pnas.2426011122.

McCleary, B. V., & Chick, B. F. (1977). The purification and properties of a thiaminase I enzyme from nardoo (*Marsilea drummondii*). *Phytochemistry, 16*(2), 207–213. https://doi.org/10.1016/S0031-9422(00)86787-4.

McGuirk, S. M. (1987). Polioencephalomalacia. *Veterinary Clinics of North America: Food Animal Practice, 3*(1), 107–117. https://doi.org/10.1016/S0749-0720(15)31183-X.

McLean, E., & Ash, R. J. (1987). Intact protein (antigen) absorption in fishes: Mechanism and physiological significance. *Journal of Fish Biology, 31*, 219–223.

McLean, E., & Donaldson, E. M. (1990). Absorption of bioactive proteins by the gastrointestinal tract of fish: A review. *Journal of Aquatic Animal Health, 2*(1), 1–11. https://doi.org/10.1577/1548-8667(1990)0022.3.CO;2.

Melnick, D., & Field, H. (1939a). Chemical determination of vitamin B1: I. Reaction between thiamine in pure aqueous solution and diazotized p-aminoacetophenone. *Journal of Biological Chemistry, 127*(2), 505–514. https://doi.org/10.1016/S0021-9258(18)73803-9.

Melnick, D., & Field, H. (1939b). Chemical determination of vitamin B1: II. Method for estimation of the thiamine content of biological materials with the diazotized p-aminoacetophenone reagent. *Journal of Biological Chemistry, 127*, 515. http://www.jbc.org/content/127/2/515.full.pdf?sid=b2d71702-471c-44bf-b506-bd706e75b313.

Melnick, D., Hochberg, M., & Oser, B. L. (1945). Physiological availability of the vitamins: II. The effect of dietary thiaminase in fish products. *Journal of Nutrition, 30*(2), 81–88. https://doi.org/10.1093/jn/30.2.81.

Murata, K. (1982). Actions of two types of thiaminase on thiamin and its analogues. *Annals of the New York Academy of Sciences, 378*(1), 146–156. https://doi.org/10.1111/j.1749-6632.1982.tb31193.x.

Neilands, J. B. (1947). Thiaminase in aquatic animals of nova scotia. *Journal of the Fisheries Research Board of Canada, 7a*(2), 94–99. https://doi.org/10.1139/f47-011.

Nishimune, T., Watanabe, Y., Okazaki, H., & Akai, H. (2000). Thiamin is decomposed due to *Anaphe* spp. entomophagy in seasonal ataxia patients in Nigeria. *Journal of Nutrition, 130*(6), 1625–1628. https://doi.org/10.1093/jn/130.6.1625.

Obrenovich, M. E., & Monnier, V. M. (2003). Vitamin B1 blocks damage caused by hyperglycemia. *Science of Aging Knowledge Environment, 2003*(10), PE6. http://www.ncbi.nlm.nih.gov/entrez/query.fcgi?cmd=Retrieve&db=PubMed&dopt=Citation&list_uids=12844520.

Ona, E., Emuebie, O., Adeola, F., Oluwaseun, E., & Adeolu, F. (2015). Properties of thiaminase in freshwater prawn *Machrobrachium rosenbergii*. *International Journal of Research in Applied, Natural, and Social Sciences, 3*(4), 37–48.
Paulson, J. N., Williams, B. L., Hehnly, C., Mishra, N., Sinnar, S. A., Zhang, L., ... Schiff, S. J. (2020). *Paenibacillus* infection with frequent viral coinfection contributes to post-infectious hydrocephalus in Ugandan infants. *Science Translational Medicine, 12*(563), https://doi.org/10.1126/scitranslmed.aba0565.
Porter, D., & Pinger, C. (2025). Measuring thiaminase activity in fish extracts using fluorescence spectrophotometry. *ACS Measurement Science Au, 5*(6), 833–841. https://doi.org/10.1021/acsmeasuresciau.5c00091.
Reddy, K. K., Giri, K. V., & Das, R. (1948). Thiaminase system in fresh water mussel. *Enzymologia, 12*, 238–245.
Reed, A. N., Rowland, F. E., Krajcik, J. A., & Tillitt, D. E. (2023). Thiamine supplementation improves survival and body condition of hatchery-reared steelhead (*Oncorhynchus mykiss*) in Oregon. *Veterinary science, 10*(2), https://doi.org/10.3390/vetsci10020156.
Richter, C. A., Evans, A. N., Heppell, S. A., Zajicek, J. L., & Tillitt, D. E. (2023). Genetic basis of thiaminase I activity in a vertebrate, zebrafish *Danio rerio*. *Scientific Reports, 13*(1), 698. https://doi.org/10.1038/s41598-023-27612-5.
Richter, C. A., Wright-Osment, M. K., Zajicek, J. L., Honeyfield, D. C., & Tillitt, D. E. (2009). Quantitative polymerase chain reaction (PCR) assays for a bacterial thiaminase I gene and the thiaminase-producing bacterium *Paenibacillus thiaminolyticus*. *Journal of Aquatic Animal Health, 21*(4), 229–238. https://doi.org/10.1577/h07-054.1.
Ringe, H., Schuelke, M., Weber, S., Dorner, B. G., Kirchner, S., & Dorner, M. B. (2014). Infant botulism: Is there an association with thiamine deficiency? *Pediatrics, 134*(5), e1436–e1440. https://doi.org/10.1542/peds.2013-3378.
Rombout, J. H., Lamers, C. H., Helfrich, M. H., Dekker, A., & Taverne-Thiele, J. J. (1985). Uptake and transport of intact macromolecules in the intestinal epithelium of carp (*Cyprinus carpio L.*) and the possible immunological implications. *Cell and Tissue Research, 239*(3), 519–530. https://doi.org/10.1007/bf00219230.
Sannino, D., & Angert, E. R. (2017). Genomic insights into the thiamin metabolism of *Paenibacillus thiaminolyticus* NRRL B-4156 and P. apiarius NRRL B-23460. *Standards in Genomic Sciences, 12*, 59. https://doi.org/10.1186/s40793-017-0276-9.
Sannino, D. R., Kraft, C. E., Edwards, K. A., & Angert, E. R. (2018). Thiaminase I provides a growth advantage by salvaging precursors from environmental thiamine and its analogs in *Burkholderia thailandensis*. *Applied and Environmental Microbiology, 84*(18), https://doi.org/10.1128/aem.01268-18.
Sealock, R. R., & Goodland, R. L. (1944). Thiamine inactivation by the Chastek-Paralysis factor. Inhibition of thiamine inactivation. *Journal of the American Chemical Society, 66*(4), 507–510. https://doi.org/10.1021/ja01232a001.
Sealock, R. R., Livermore, A. H., & Evans, C. A. (1943). Thiamine inactivation by the fresh-fish or Chastek-Paralysis factor. *Journal of the American Chemical Society, 65*(5), 935–940. https://doi.org/10.1021/ja01245a046.
Sealock, R. R., & White, H. S. (1949). Neopyrithiamine and the thiaminase of fish tissues. *Journal of Biological Chemistry, 181*, 393. http://www.jbc.org/content/181/1/393.full.pdf.
Shannon, K. C., Colwell, F. S., Crump, B. C., Brennan, E., St. John, G., Gould, R., ... Suffridge, C. P. (2025). Bacimethrin, an allelopathic vitamin B1 antagonist, is linked with microbial gene expression patterns in a hypereutrophic watershed. *PLOS ONE, 20*(11), e0335861. https://doi.org/10.1371/journal.pone.0335861.

Shida, O., Takagi, H., Kadowaki, K., Nakamura, L. K., & Komagata, K. (1997). Transfer of *Bacillus alginolyticus, Bacillus chondroitinus, Bacillus curdlanolyticus, Bacillus glucanolyticus, Bacillus kobensis*, and *Bacillus thiaminolyticus* to the genus *Paenibacillus* and emended description of the genus *Paenibacillus*. *International Journal of Systematic Bacteriology, 47*(2), 289–298. https://doi.org/10.1099/00207713-47-2-289.

Smith, D. C., & Proutt, L. M. (1944). Development of thiamine deficiency in the cat on a diet of raw fish. *Proceedings of the Society for Experimental Biology and Medicine, 56*(1), 1–3. https://doi.org/10.3181/00379727-56-14565.

Soliman, A. G. (1981). Comparison of manual and benzenesulfonyl chloride-semi-automated thiochrome methods for determination of thiamine in foods. *Journal - Association of Official Analytical Chemists, 64*(3), 616–622.

Spitzer, E. H., Coombes, A. I., Elvehjem, C. A., & Wisnicky, W. (1941). Inactivation of vitamin B1 by raw fish. *Experimental Biology and Medicine, 48*(1), 376–379. https://doi.org/10.3181/00379727-48-13326.

Stentz, R., Cheema, J., Philo, M., & Carding, S. R. (2023). A possible aquatic origin of the thiaminase TenA of the human gut symbiont *Bacteroides thetaiotaomicron*. *Journal of Molecular Evolution, 91*(4), 482–491. https://doi.org/10.1007/s00239-023-10101-8.

Suzuoki, Z., Tominaga, F., Matsuo, T., Sumi, Y., & Miyakawa, M. (1968). Metabolism of thiamine and thiamine tetrahydro-furfuryl disulfide to 4-methylthiazole-5-acetic acid in conventional and germfree rats under various dosing conditions. *Journal of Nutrition, 96*(4), 433–444. https://doi.org/10.1093/jn/96.4.433.

Thornalley, P. J. (2005). The potential role of thiamine (vitamin B1) in diabetic complications. *Current Diabetes Reviews, 1*(3), 287–298. http://www.ncbi.nlm.nih.gov/entrez/query.fcgi?cmd=Retrieve&db=PubMed&dopt=Citation&list_uids=18220605.

Tomasulo, P. A., Kater, R. M., & Iber, F. L. (1968). Impairment of thiamine absorption in alcoholism. *The American Journal of Clinical Nutrition, 21*(11), 1341–1344. http://www.ncbi.nlm.nih.gov/pubmed/5699730.

Vimokesant, S. L., Hilker, D. M., Nakornchai, S., Rungruangsak, K., & Dhanamitta, S. (1975). Effects of betel nut and fermented fish on the thiamin status of northeastern Thais. *The American Journal of Clinical Nutrition, 28*(12), 1458–1463. http://www.ncbi.nlm.nih.gov/pubmed/803009.

Wittliff, J. L., & Airth, R. L. (1968). Extracellular thiaminase I of *Bacillus thiaminolyticus*. I. Purification and physicochemical properties. *Biochemistry, 7*(2), 736–744. https://doi.org/10.1021/bi00842a032.

Wolfe, P. C., Tuske, A. M., Tillitt, D. E., Allen, F., & Edwards, K. A. (2023). Understanding and mitigating thiaminase activity in silver carp. *Current Research in Food Science, 6*, 100502. https://doi.org/10.1016/j.crfs.2023.100502.

Woolley, D. (1941). Destruction of thiamine by a substance in certain fish. *Journal of Biological Chemistry, 141*, 997–998.

Yazdani, M., Suffridge Christopher, P., Liu, F., Costello Cait, M., Zhou, Z., St. John, G., ... Ahner Beth, A. (2025). Harmful algal bloom species *Microcystis aeruginosa* releases thiamin antivitamins to suppress competitors. *mBio, 16*(8), e01608–01625. https://doi.org/10.1128/mbio.01608-25.

Zajicek, J., Brown, L., Brown, S., Honeyfield, D., Fitzsimons, J., & Tillitt, D. (2009). Variations of thiaminase I activity pH dependencies among typical great lakes forage fish and *Paenibacillus thiaminolyticus*. *Journal of Aquatic Animal Health, 21*, 207–216. https://doi.org/10.1577/H07-052.1.

Zajicek, J. L., Tillitt, D. E., Honeyfield, D. C., Brown, S. B., & Fitzsimons, J. D. (2005). A method for measuring total thiaminase activity in fish tissues. *Journal of Aquatic Animal Health, 17*(1), 82–94. https://doi.org/10.1577/H03-083.1.

Zallot, R., Yazdani, M., Goyer, A., Ziemak, M. J., Guan, J. C., McCarty, D. R., ... Hanson, A. D. (2014). Salvage of the thiamin pyrimidine moiety by plant TenA proteins lacking an active-site cysteine. *The Biochemical Journal, 463*(1), 145–155. https://doi.org/10.1042/bj20140522.

Zhu, W., Zajicek, J. L., Tillitt, D. E., & Glass, T. E. (2013). TanA: A fluorogenic probe for thiaminase activity. *Analytical Methods, 5*(2), 446–448. https://doi.org/10.1039/C2AY26047F.

Zilles, J. L., Croal, L. R., & Downs, D. M. (2000). Action of the thiamine antagonist bacimethrin on thiamine biosynthesis. *Journal of Bacteriology, 182*(19), 5606–5610. https://doi.org/10.1128/jb.182.19.5606-5610.2000.

Ziporin, Z. Z., Nunes, W. T., Powell, R. C., Waring, P. P., & Sauberlich, H. E. (1965). Thiamine requirement in the adult human as measured by urinary excretion of thiamine metabolites. *Journal of Nutrition, 85*(3), 297–304. https://doi.org/10.1093/jn/85.3.297.

Zoltewicz, J. A., Kauffman, G. M., & Uray, G. (1984). A mechanism for sulphite ion reacting with vitamin B1 and its analogues. *Food Chemistry, 15*(2), 75–91. https://doi.org/10.1016/0308-8146(84)90101-8.

CHAPTER TEN

Engineered nanopore biosensor for thiamine quantification: Experimental design

Jan P. Dobert[a,1], Sofia B. Relvas[a,1], Gerard Carrera i Cardona[b], Florian L.R. Lucas[a,*], and Carsten Wloka[a,*]
[a]BIOS Lab on a Chip Group, MESA+ Institute for Nanotechnology, Technical Medical Centre, University of Twente, Enschede, The Netherlands
[b]Lab for Nanobiology, Department of Chemistry, KU Leuven, Leuven, Belgium
*Corresponding authors. e-mail address: florian.lucas@utwente.nl; c.wloka@utwente.nl

Contents

Abstract

Biosensing using protein nanopores has become a powerful tool for the analysis of single molecules, enabling for example portable devices for long-read DNA sequencing. By inserting a single pore into a lipid bilayer and applying an electric field, one can create a single-molecule sensor using the ionic current flowing through the pore. Thiamine can be quantified with such a technique using the Cytolysin A pore and a thiamine-binding protein TbpA, which imparts its affinity to the nanopore. When thiamine passes through the pore, it gets captured by TbpA, causing a conformational change that alters the flow of ions through the pore and thus causes a change in current. The result is a biosensor pair with high selectivity that allows quantification of thiamine even in complex solutions. Here, we provide instructions

[1] These authors contributed equally to this work.

Methods in Enzymology, Volume 735
ISSN 0076-6879, https://doi.org/10.1016/bs.mie.2026.07.014

for the implementation of this measurement technique, which includes the production of the required recombinant proteins as well as construction and operation of the electrochemical measuring setup.

1. Introduction

Biological nanopores are proteins that can insert themselves into electrically isolating lipid membranes as well as synthetic polymers (Li et al., 2024; Vreeker et al., 2025). These transmembrane structures then enable the transport of ions and molecules across the membrane depending on their physical properties. When an electrical potential is applied, an ionic current flows through the pore which can be measured. A wide abundance of nanopores exist with different inner dimensions, ranging from just a few nanometres (alpha-haemolysin) to tens of nanometres (pneumolysin) (Clarke et al., 2009; Tilley et al., 2005). Whereas DNA can be readily isolated from complex matrices due to the uniform charge of the phosphate groups, it becomes more complicated to measure other molecules, especially in complex samples. Nanopores can detect small peptides and folded proteins at the single-molecule level. However, these molecules are always surrounded by similar species, making discrimination difficult if they are in low abundance. Intriguingly, proteins that bind very specifically to molecules can also be trapped inside a nanopore. For example, substrate-binding proteins (SBPs) bind to ligands in complex environments and display a conformational change that alters the steric hindrance they present within the nanopore lumen. SBPs may even be bioengineered to bind molecules for which no natural binding protein exists in the future.

One of the most suitable nanopores for such a nanopore/SBP system is the Cytolysin A (ClyA) nanopore, with an inner diameter that generally allows the trapping of small, folded proteins (~20–50 kDa, depending on the shape) (Galenkamp et al., 2018; Soskine et al., 2013). The thiamine-binding protein TbpA binds thiamine and its phosphorylated derivatives, undergoing a conformational change leading to a measurable change in ionic current flow across the pore. The time TbpA spends in bound and unbound states is dependent on the thiamine concentration. Thus, by trapping TbpA within the ClyA nanopore, it is possible to determine the concentration of thiamine in the surrounding solution with high precision (Fig. 1) (Lucas et al., 2021, 2023). In this protocol, we describe the

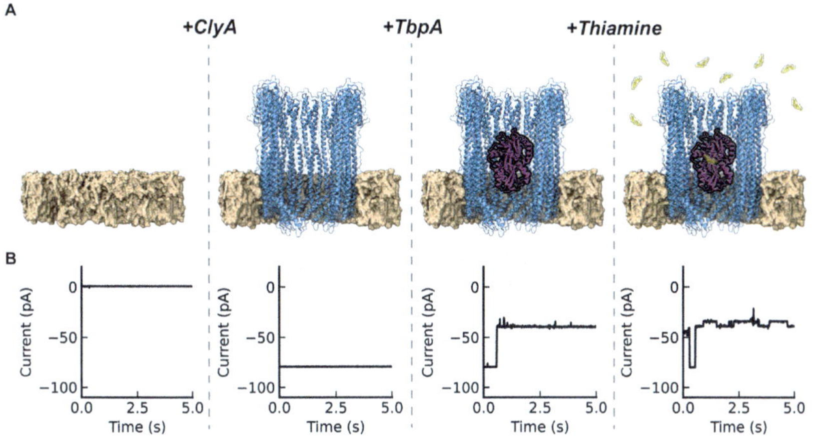

Fig. 1 (A) Illustration of thiamine quantification using a biosensor pair consisting of thiamine binding protein TbpA (purple) and the protein nanopore cytolysin A (ClyA, blue). A single ClyA pore is inserted into a lipid bilayer (beige) and captures TbpA. When thiamine (yellow) passes through the pore, it can bind to TbpA, allowing detection and quantification on a single-molecule basis. (B) Traces corresponding to biosensor state shown in A. *(A) Figure created using ChimeraX (Pettersen et al., 2021). Protein structures are obtained from the PDB (TbpA: 2QRY (Soriano et al., 2008), ClyA pore: 6MRT (Peng et al., 2019)). Lipid bilayer taken from CHARMM-GUI Archive (Jo et al., 2007, 2008).*

full workflow that allows such measurements. We explain the steps required to obtain both ClyA and TbpA, how to set up the measurement apparatus, and measure its conductance. The methodology to properly analyze the resulting data is presented in the chapter 11 *Engineered nanopore biosensor for thiamine quantification: Data analysis* of this *Methods in Enzymology* volume.

2. Rationale and theory

For our experiments we employ the engineered *Salmonella typhi* ClyA-AS variant, also used by several laboratories in the last decade (Li, Fahie et al., 2022; Shorkey et al., 2025; Soskine et al., 2013). We recombinantly express and purify ClyA-AS in its soluble, monomeric, form. To obtain the oligomeric form, we use the detergent *n*-dodecyl-β-D-maltoside (DDM) at a higher than critical micellar concentration (CMC) to form compact micelles in which ClyA-AS monomers can insert and assemble into a full pore (Galenkamp et al., 2021). Once oligomerized, a mixture of mainly 12, 13, and 14-meric assemblies is present. However, only the 12-mer (dodecamer) can be used to capture TbpA for longer times. To separate the different oligomeric forms, we perform blue native page polyacrylamide gel electrophoresis (BN-PAGE, Section 3.2.5). This is crucial because we need to preserve the oligomeric state and the structure of the pores. ClyA-AS monomers are approximately 34 kDa in size, thus, the oligomeric assemblies are between 400–500 kDa. We use an acrylamide gel with a gradient (4–20 %), which provides good separation of larger proteins. The acrylamide gel also functions as a very stable long-term storage matrix for the ClyA oligomers. We isolate the different oligomeric forms by physically cutting the bands out of the gel with a sharp scalpel, separate them into tiny pieces, and store them embedded inside the acrylamide matrix at −80 °C for years.

TbpA displays a high affinity towards thiamine. While such high-affinity binding is advantageous for detecting ultra-low concentrations, it hinders effective analysis in a nanopore setup. To facilitate more frequent unbinding events and accelerate data acquisition, we engineered (decreased) the binding affinity by introducing a single amino acid substitution (Y27A) in the binding pocket, resulting in more dynamic binding kinetics and enabling a more robust analysis (Lucas et al., 2021). Purification of TbpA from *E. coli* requires the use of periplasmic expression. Expression into the periplasm is beneficial for proteins that form disulfide bonds because the periplasm provides an oxidative environment in which these bonds can form (contrary to the cytoplasm, which is a reducing environment). To target proteins to the periplasm, they can be modified with an N-terminal periplasmic signal sequence. TbpA is natively a periplasmic protein and contains an N-terminal signal sequence (LKKCLPLLLLCTAPVFA) that is cleaved after transport to the periplasm. Importantly, we added the N-terminal HisTag on TbpA after this signal sequence to preserve periplasmic expression. Expression of TbpA is done similarly to ClyA, but the purification from the periplasm requires a distinct protocol (Section 3.2.6, Fig. 2). In contrast to cytoplasmic purification, we use for ClyA, where the whole cell is lysed, extraction of periplasmic proteins is carried out using osmotic shock to selectively break the outer bacterial membrane and release only the periplasm, leaving the inner membrane containing the cytoplasm intact.

We add purified ClyA-AS 12mers and TbpA into a recording chamber that hosts the lipid membrane, analysis buffers, and the electrodes for

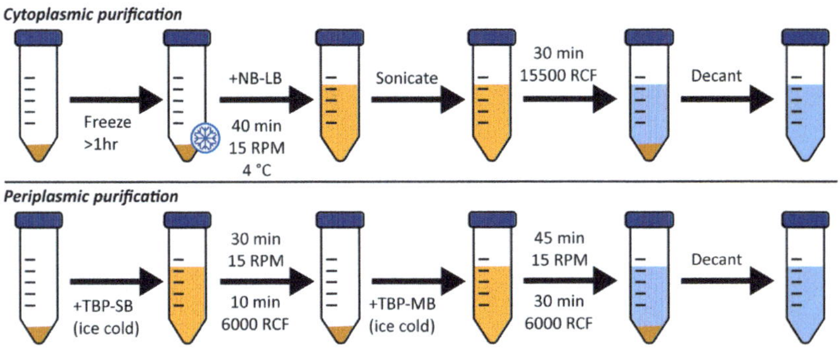

Fig. 2 Workflow overview of the critical steps described within this method, show casing the differences between cytoplasmic and periplasmic purification. NP-LB is the nanopore lysis buffer, TBP-SB is the thiamine binding protein sucrose buffer, TBP-MB is the thiamine binding protein magnesium buffer.

readout. Naturally, the recording chamber must be constructed from non-conductive material. We describe a simple manufacturing process in Section 4. However, recording chambers are also available commercially (*e.g.*, Elements SRL). Between the two half-chambers, we place a membrane with a ~100 μm hole (Fig. 3). The membrane is a thin polytetrafluoroethylene (PTFE) sheet with a hole created *via* dielectric breakdown using a spark generator. Alternatively, polyimide can be used for this membrane. While the PTFE membrane from Section 4.2.2 can be replaced with a polyimide membrane instead, it should be noted that these membranes are generally more rigid.

The method to form lipid bilayers on the PTFE aperture is based on the Montal-Mueller method (Montal & Mueller, 1972), which exploits the amphipathic nature of lipids and their tendency to self-assemble into Langmuir-Blodgett monolayers at polar-apolar interfaces. This lipid bilayer is a soft-condensed matter that self-assembles and remains semi-fluid, allowing reformation with the same protocol after disruption. The bilayer behaves as a dielectric and acts as a capacitor that, upon application of an electrical potential, separates the electrical charges across the membrane.

The electrical equivalent circuit is best described as a capacitor in parallel with a (very large) resistance (in the order of 10 GΩ). When a nanopore inserts into the lipid membrane, it acts as a resistor in parallel with the bilayer. The measurement setup also contains access resistance (R_a) in series with the bilayer, which we consider negligible ($R_a \approx 0$) (Fig. 4A).

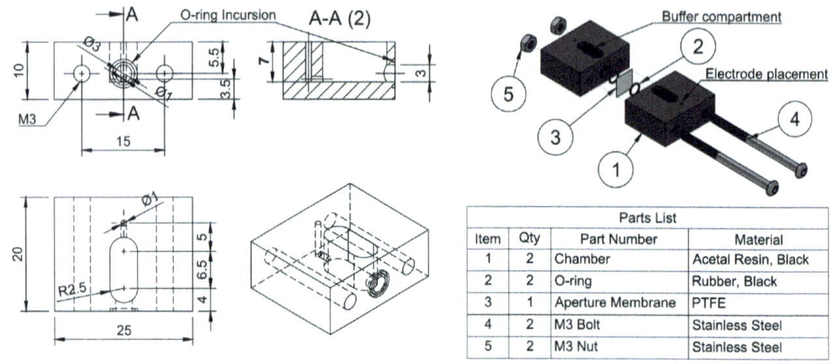

Parts List			
Item	Qty	Part Number	Material
1	2	Chamber	Acetal Resin, Black
2	2	O-ring	Rubber, Black
3	1	Aperture Membrane	PTFE
4	2	M3 Bolt	Stainless Steel
5	2	M3 Nut	Stainless Steel

Fig. 3 Recording chamber for lipid bilayer formation using the Montal-Mueller method.

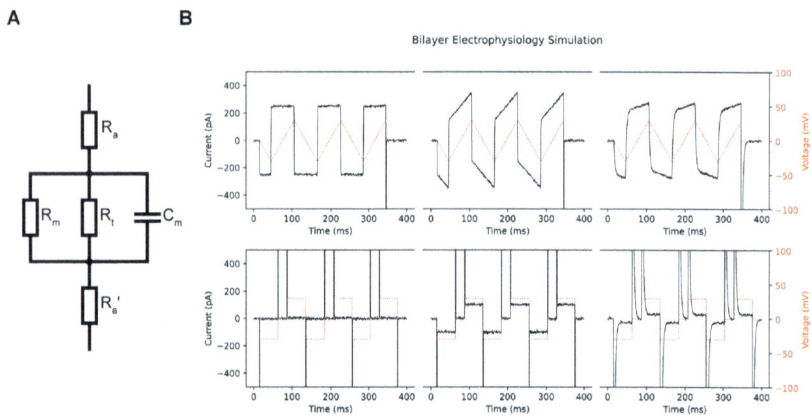

Fig. 4 Electrical equivalent circuit and expected signal of the lipid bilayer. (A) Electrical equivalent circuit of a nanopore embedded within a lipid membrane. (B) Predicted electrophysiology signal upon stimulation with a triangular wave (top) and under a gap-free potential (bottom), from left to right: An "ideal" bilayer, a bilayer with a nanopore embedded, and a "leaky" bilayer.

Solving this equivalent circuit yields a first-order linear ordinary differential equation relating the measured current to the applied voltage (Eq. 1, where R_a equals $R_a + R_a$' in Fig. 4A).

In practice, during bilayer formation, we apply a triangular wave with a constant ramp of $\pm 1\ V.s^{-1}$; since dV/dt is constant, the output current is directly proportional to the capacitance, allowing it to be read out directly to assess the properties of the formed bilayer. Conversely, under a constant holding potential (gap-free recording), the capacitive term vanishes, and the same circuit reduces to a direct proportionality between current and conductance; this is the regime used for single-molecule recordings.

$$R_a C_m \frac{dI(t)}{dt} + \frac{R_a + R_{mt}}{R_{mt}} I(t) = C_m \frac{dV(t)}{dt} + \frac{V(t)}{R_{mt}} \overset{R_a \approx 0}{\Rightarrow} I(t) = C_m \frac{dV(t)}{dt} + \frac{V(t)}{R_{mt}}, \; with \; C_m = \varepsilon \frac{A}{d} \quad (1)$$

where R_a is the access resistance (in Ohm, Ω), C_m is the membrane capacitance (in Farad, F), $I(t)$ is the observed current as a function of time (in Ampere, A), R_{mt} is the reciprocal sum of the reciprocal resistances of the lipid bilayer (R_m) and the nanopore (R_t) $\left(R_{mt} = \frac{R_m * R_t}{R_m + R_t}\right)$ in the equivalent circuit (in Ohm, Ω), $V(t)$ is the voltage potential as a function of time (in Volt, V), ε is the absolute permittivity of the bilayer, A is the surface area of bilayer (in m^2), and d is the bilayer thickness (in m).

Bilayers with larger surface area embed nanopores more readily but are noisier; experimentally, bilayers permitting ClyA-AS insertions show a capacitance of ~200–400 pF. To confirm bilayer formation, we apply a larger potential of 1 V, which breaks bilayers but leaves thicker, more "oily" membranes intact. Because the permittivity is intrinsic and the aperture area is fixed, a recurring capacitance value reliably indicates bilayer formation. It is good practice to characterize the surface area of each recording chamber based on the capacitance, as shown previously (Maglia et al., 2010; Wloka et al., 2021).

The output waveform reveals two non-idealities (Fig. 4B). A leaky bilayer produces a non-zero slope at the plateaus of the square wave, the signature of a finite, rather than infinite, R_{mt} (the leak term V/R_{mt} grows linearly with voltage). Such bilayers are unsuitable for nanopore insertion and should be ruptured and reformed. The second non-ideality is a significant access resistance ($R_a \gg 0$), which appears instead as a slow rise rather than an instantaneous jump. When seen, the electrodes should be cleaned, sanded, and re-bleached because a high access resistance can distort the nanopore measurements.

Having discussed the scientific background of our methodology, in the following we will go through the whole experimental process step-by-step, starting with the preparation of the recombinant proteins (Section 3), followed by the assembly of the measurement setup (Section 4) and finally data acquisition (Section 5).

3. Preparation of cytolysin A and TbpA-Y27A variant thiamine binding protein

3.1 Materials, equipment and reagents

3.1.1 Transformation

- Competent *E. coli* BL21(DE3) or similar strain with IPTG-inducible T7-polymerase expression
- Plasmid DNA (75–150 ng.μL^{-1})
  - Y27A-TbpA plasmid
    - Vector: pT7-derived (ampicillin resistance)
    - Insert: TbpA (Uniprot: P31550)
    - Mutations: Y27A
    - Modifications: N-terminal HisTag after signal peptide
  - ClyA-AS plasmid
    - Vector: pT7-derived (ampicillin resistance)
    - Insert: Cytolysin A (Uniprot: Q8Z727)
    - Mutations: C87A, L99Q, E103G, F166Y, I203V, C285S, K294R, H307Y (Soskine et al., 2013)
    - Modifications: C-terminal HisTag
  - *Note: We advise to design and commercially obtain fully assembled expression plasmids from a gene synthesis service.*
- SOC medium
- LB-agar plates supplemented with 100 μg.mL^{-1} ampicillin
- Drigalski spatula (T-shaped spreader)
- Ultrasonication bath (Branson M2800)
- Incubator at 37 °C

3.1.2 Expression

- Erlenmeyer flask (100 mL and 1 L)
- LB-Lennox (ClyA) or TB (TbpA) medium
- Ampicillin 100 mg.mL^{-1} (1000 × stock)
- Shaking incubator capable of maintaining 37 °C, 25 °C and 200 rpm
- IPTG 0.5 M (1000 × stock)
- spectrophotometer & plastic cuvettes (1 mL)

3.1.3 Gravity flow HisTag purification

- Ni-NTA agarose beads (Qiagen)
- Centrifuge for microtubes
- Gravity flow column

3.1.4 FPLC HisTag purification

- FPLC machine (ÄKTA go/pure or equivalent)
- HisTrap HP 1 mL chromatography column (Cytiva, #29051021)
- Superloop 50 mL (Cytiva, #18111382)

3.1.5 Thiamine binding protein (TBP) purification

- Sucrose Buffer (TBP-SB):
  - o 50 mM Tris-HCl pH 7.5
  - o 20 % w/v Sucrose
  - o 1 mM EDTA
  - o *Store 4 °C*
- TBP Wash Buffer (TBP-WB):
  - o 50 mM Tris-HCl pH 7.5
  - o 150 mM NaCl
  - o 20 mM imidazole
    - ▪ *Always check/adjust pH after adding imidazole.*
- TBP Magnesium Buffer (TBP-MB):
  - o 5 mM $MgCl_2$
  - o *Store 4 °C*
    - ▪ *Hypotonic shock, just magnesium chloride, temperature is very important.*
- TBP Elution Buffer (TBP-EB):
  - o 50 mM Tris-HCl pH 7.5
  - o 150 mM NaCl
  - o 300 mM imidazole
    - ▪ *Always check/adjust pH after adding imidazole.*

3.1.6 Cytolysin A (ClyA) purification

- Sonicator
- NP Lysis Buffer (NP-LB):
  - o 15 mM Tris-HCl pH 7.5
  - o 150 mM NaCl
  - o 20 mM imidazole
    - ▪ *Always check pH after adding imidazole.*
  - o 1 mM $MgCl_2$
    - ▪ *Required for DNAseI function.*
  - o 0.2 $U.mL^{-1}$ DN*Ase*I
    - ▪ *Add fresh, breaks down residual DNA.*
  - o 0.2 $mg.mL^{-1}$ Lysozyme
    - ▪ *Add fresh, breaks down peptidoglycan in the cell wall.*

- NP Wash Buffer (NP-WB):
  - o 15 mM Tris-HCl, pH 7.5
  - o 150 mM NaCl
  - o 20 mM imidazole
    - *Always check pH after adding imidazole.*
- NP Elution Buffer (NP-EB):
  - o 15 mM Tris-HCl, pH 7.5
  - o 150 mM NaCl
  - o 300 mM imidazole
    - *Always check pH after adding imidazole.*

3.1.7 Nanopore oligomerization

- DDM (*n*-dodecyl-β-D-maltoside, 10 % stock)
- Heating block at 37 °C
- Native page gel (4–20 %)
- Native page gel running system (Criterion, Bio-Rad)
- 1x Native Running Buffer
- 20x Cathode Buffer Additive (Blue solution)

3.2 Step-by-step protocol

3.2.1 Purification of cytolysin A (ClyA) from E. coli and generation of nanopore

3.2.1.1 Transformation

1. Pre-warm SOC medium at 37 °C.
2. Take an aliquot of *E. coli* BL21(DE3) competent cells from the −80 °C freezer and thaw the cells on ice.
3. Add 1 μL of ClyA-AS plasmid DNA (75–150 $ng.\mu L^{-1}$) to 50 μL *E. coli* BL21(DE3) competent cells (from step 2).
4. Keep on ice for 15 min.
5. Sonicate for 12 s in an ultrasonication bath (room temperature).
6. Keep on ice for 2 min.
7. Add 250 μL pre-warmed SOC medium.
8. Incubate at 37 °C for 20 min.
9. Spread the mixture with the Drigalski spatula on a LB-agar plate with ampicillin.
10. Grow overnight at 37 °C.
    a. *Place the dish upside down, with the agar being on top in order to prevent condensation.*

Notes:

- *Other E. coli expression strains can be used in place of BL21(DE3), but the yield may be different.*
- *Instead of sonication, suitable bacteria can also be transformed via heat shock (45 s at 42 °C).*

3.2.1.2 Expression

1. Use culture plate grown overnight (See 3.2.1.1) to prepare a liquid bacteria suspension using either of the options below.
   a. *Option 1: create a pre-culture from a single colony (standard procedure).*
      i. *Inoculate 20 mL of LB medium containing 100 μg.mL^{-1} ampicillin in a 100 mL culture flask. Using a pipette tip, carefully scrape off a single bacteria colony from the plate and suspend it in the culture medium by pipetting up and down.*
      ii. *Grow pre-culture overnight at 37 °C and 200 rpm (25 mm orbit radius).*
   b. *Option 2: Skip pre-culture by harvesting whole plate (time-saving trick).*
      i. *Suspend all colonies on the culture plate in 1 mL of LB medium by repeatedly pipetting the medium over the agar until the colonies are suspended and the medium is cloudy.*
      ii. *This suspension replaces the pre-culture from option 1.*
   c. *Note: Option 1 yields a clonal, genetically homogeneous culture derived from a single bacterium. Option 2 theoretically allows for genetic heterogeneity within the culture as it is derived from multiple colonies. However, we assume that all colonies on the plate contain the identical expression plasmid since they are transformed from a pure plasmid stock and selected via the encoded ampicillin resistance.*
2. Inoculate 200 mL of LB medium containing 100 μg.mL^{-1} ampicillin (in 1 L culture flasks) with 1 mL of pre-culture (option 1) or 1 mL of solubilized colonies (option 2).
   a. *The culture volume should not exceed 20 % of the maximum capacity of the Erlenmeyer flask (e.g., 200 mL culture in a 1000 mL flask). This ensures sufficient aeration of the culture and is crucial for growth and protein expression.*
3. Incubate the expression culture on an orbital shaking incubator for 2–4 h at 37 °C and 200 rpm (25 mm orbit radius) while monitoring the optical density (OD_{600}) every 30 min.
   a. *Take a 1 mL sample of the culture and add it to a plastic cuvette. Make sure to mix the culture before taking the sample.*
   b. *Measure absorption at 600 nm (OD_{600}) using a* spectrophotometer.

4. Proceed to the next step when OD_{600} reaches 0.4–0.6.
   a. *Note: At an OD_{600} of 0.2–1.0, the bacteria are in exponential growth (log phase), and induction of protein expression will be most efficient.*
5. Add at a final concentration of 0.5 mM IPTG (*e.g.* 200 µL IPTG of 0.5 M stock) to induce protein expression.
6. Incubate the expression culture in an orbital shaking incubator overnight at 21 °C and 200 rpm (25 mm orbit radius).
   a. *Generally, expression at a lower temperature can improve protein folding and provide larger yields with longer expression duration.*

3.2.1.3 Pellet storage

1. Harvest the cells by centrifugation for 15–30 min at 4600 g and 4 °C.
   a. *Centrifuge until pellet is solid and separate from supernatant. "Sludge"-like cells may not clump and can be removed with step 2.*
   b. *Balance tubes precisely.*
2. Discard the supernatant and freeze the cell pellets for at least 1 h at −20 °C.
   a. *Freezing and thawing the pellet can improve lysis efficiency.*
   b. *Frozen pellets can be stored long-term.*

3.2.2 Preparation of cell lysate

1. Carefully resuspend frozen cell pellets in 20 mL NP-LB using a pipette.
   a. *Important: avoid foam formation.*
   b. *Mixture should look smooth without chunks.*
2. Incubate the lysate for 40 min at 4 °C under constant slow rotation (~15 rpm) on a tube rotator.
3. Sonicate the cells on ice (at 40 % output power, duty cycle 70 %) for 120 s to break the cells open.
   a. *Make sure to pack the ice around the sample tube. This prevents the tube from moving during sonication.*
   b. *Check that the tip is fully submersed (sample will foam up significantly otherwise) and be careful to not touch the sides of the tube.*
   c. *Keep an eye on the machine and immediately stop sonication if sound changes. This can indicate foaming of the sample or the tip touching the tube. This can damage your sample and sample tube and will prevent proper lysis.*
4. Pellet the cell debris by centrifugation for 30 min at 15,500 g and 4 °C.
   a. *Balance tubes precisely.*
   b. *The protein is in supernatant.*
   c. *If not completely clear, centrifuge longer (saves time during the next steps).*

i. *The supernatant should look as clear as fresh broth.*

3.2.3 Gravity flow HisTag purification

1. Wash Ni-NTA beads with NP-WB (3 times).
   a. *100 μL beads per 50 mL of culture (200 μL of Ni-NTA suspension), adjust according to yield.*
2. Incubate the supernatant from 3.2.2 with the washed Ni-NTA beads for 15–25 min at 4 °C on a rolling mixer or tube rotator.
   a. *Long incubation results in aggregation.*
3. Equilibrate a gravity flow column with approximately 2 column volumes (2 mL) of NP-WB, to remove potential contaminants and to check the flow of the column.
4. Add the incubated Ni-NTA beads/solution to the column.
5. Wash the column with at least 10 column volumes (20 mL) of NP-WB.
   a. *20 mL is bare minimum.*
6. Overflow the column with wash buffer and pipette off the "foam".
   a. *This step removes the lysozyme.*
7. Let the column flow until nearly dry.

Next 2 steps are to be performed in very rapid succession:

8. Blow air through the beads with a syringe (dry beads).
9. Add 200 μL elution buffer per 100 μL beads directly on the Ni-NTA beads.
10. Incubate the Ni-NTA beads 2–5 min.
11. Place the column on a fresh 1.5 mL tube.

Next 3 steps are to be performed in very rapid succession:

12. Blow air through the beads with a syringe (dry beads).
    a. *The protein is in the elute, so catch it in the 1.5 mL tube (monomers).*
13. Remove column from the 1.5 mL tube.
14. Add 20 % ethanol (storage).
15. Close the caps and place in fridge for later recovery.
16. Check the protein concentration of the monomers with haemolytic activity (Lucas et al., 2023), Nanodrop, or Bradford assay.
17. *Store monomers at 4 °C, do not freeze the ClyA monomers.*

3.2.4 Alternative HisTag purification using fast protein liquid chromatography (FPLC)

The following protocol is designed for a HisTrap HP, 1 mL chromatography column (Cytiva, #29051021) with sample application using a 50 mL

Superloop (Cytiva, #18111382). Optionally, buffers can be pre-cooled and kept on ice throughout the purification for added protein stability.

Make sure to read the manufacturer's instructions for operating the chosen chromatography column (max. flow rates/pressures, handling information).

1. Attach and prepare HisTrap column according to the manufacturer's instructions.
   a. *Columns are stored in 20 % EtOH. When changing from ethanol to buffer, always perform a washing step with water in between to avoid precipitation of buffer ingredients on the column.*
2. Equilibrate with 5 column volumes of wash buffer at a flow rate of 1 $mL.min^{-1}$.
3. Fill lysate into Superloop and attach to injection valve.
   a. *Make sure Superloop is installed in the correct direction.*
4. Inject sample onto column with a flow rate of 0.5 $mL.min^{-1}$ until loop is empty.
   a. *Optional: collect analytical sample of flow-through to check binding efficiency.*
5. Wash column with wash buffer until a stable UV baseline is achieved, (10–20 column volumes) at a flow rate of 1 $mL.min^{-1}$.
6. Elute using a linear gradient from 0–100 % elution buffer over 5 mL at 0.5 $mL.min^{-1}$ flow rate and gather eluate in fractions of 250/500 μL. Add an additional 5 mL elution step at 100 % concentration after the gradient to make sure all proteins elute from the column.
   a. *Bound protein should elute as a single peak.*
   b. *Reverse flow elution can improve sharpness of the peak (not all chromatography systems support this).*
7. Check the protein concentration of peak fractions.
   a. *Optional: Run analytical samples (10 μL) of lysate, flow through and elution fractions on an SDS-PAGE and stain the gel to assess binding efficiency and purity of fractions.*
8. Storage of HisTrap:
   a. *Wash with 5 column volumes of water at 1 $mL.min^{-1}$.*
   b. *Equilibrate with 5 column volumes of 20 % ethanol at 1 $mL.min^{-1}$.*
   c. *Store column at 4 °C.*
9. Store purified monomers at 4 °C, **do not freeze** the ClyA monomers.

3.2.5 Oligomerization (ClyA)

1. To oligomerize ClyA monomers, dilute the protein concentration to 1 $mg.mL^{-1}$.

  a. *1 mg.mL^{-1} is the absolute maximum concentration.*
  b. *Higher concentrations will aggregate before assembling.*
2. Add 0.2 % w/v DDM to the ClyA monomers and incubate for 30 min at 37 °C.
  a. *If the concentration of ClyA monomers is too high, the protein will precipitate (irreversible).*
  b. *This step is time sensitive.*
    i. *Incubation can be extended to up to 2 h, no longer.*
  c. *This step is temperature sensitive.*
3. Add 5 × BN-PAGE SB to the protein sample with a final concentration of 1 × BN-PAGE SB.
4. Prepare the BN-PAGE Native gel (4–20 %) by removing the packaging and placing the gel in the Criterion gel system.
5. Fill the Criterion tank and the top compartment with 1 × Native Running Buffer and remove the comb, wash the wells of the gel by pipetting buffer into the wells.
6. Load the protein samples into each well (maximum of 45 μL per well, can be different per gel type) and add 2 mL of 20 × Cathode Buffer Additive to the top compartment.
  a. *The sample contains glycerol and therefore stays in the wells. Mix the top compartment until it has a uniform color (do not pipette near the wells).*
7. Allow the Native gel to run at 120 V for 10 min, after 10 min discard the blue solution in the top compartment and wash several times with water, including the wells.
8. Fill the top compartment again with 1 × Native Running Buffer and continue running the gel at 120 V for approximately 1 h, until the protein bands are nicely separated.
  a. *The gradient prevents the protein from running off the gel*
9. Cut out the protein bands corresponding to the desired oligomeric state (Fig. 5) and store the gel bands in microtubes at −80 °C.

Fig. 5 Single lane of a blue native (4–20 %) gel run with oligomerized ClyA. The 14, 13, and 12 -meric oligomers are indicated. The gel is running from left to right.

a. *Although optional, we advise to cut the gel bands into very small pieces to make extraction faster and waste less protein. A minuscule band piece will yield sufficient pores, even in 0.5 mL volumes. It is not necessary to measure protein concentration after extraction.*
   i. *Protein concentration after extraction of the gel band is not informative. Only active protein nanopores can insert into the bilayer, and this gradually reduces after extraction from the gel band. Quantification of the active protein concentration should be performed instead by using electrophysiology protocol 4.2.2 after each extraction, updating during experiments.*
b. *Gel pieces containing protein can be stored for years at −80 °C.*

3.2.6 Purification of thiamine binding protein (TbpA) from E. coli

1. Transform and express cells according to protocol 3.1.1 and 3.1.2 using:
   a. Plasmid: Y27A-TbpA
   b. Medium: TB-medium + ampicillin
   Harvest the cells by centrifugation for 15 min at 4600 g and 4 °C.
   c. *Balance tubes very carefully.*
2. Discard the supernatant and immediately start with the periplasmic purification.
   a. *Important:* ***do not freeze*** *the cell pellet at this point.*
3. Resuspend the cell pellets (from 100 mL culture) in 50 mL ice cold TBP-SB.
4. Incubate the cells for 30 min at room temperature (~22 °C) on a tube shaker or rotator.
   a. *The sucrose creates osmotic pressure while the EDTA weakens the outer bacterial membrane.*
5. Pellet the cells by centrifugation for 10 min at 6000 g and 4 °C.
   a. *Cells are still intact. Target protein is in pellet.*
6. Resuspend the cell pellets (from 100 mL culture) in 50 mL ice cold TBP-MB.
7. Incubate the cells for at least 45 min (maximum of 2 h) at room temperature (~22 °C).
   a. *The sudden change in osmotic pressure ruptures the weakened outer membrane, releasing the periplasm.*
8. Pellet the cells by centrifugation for 30 min at 6000 g and 4 °C.
   a. *Target protein was released from periplasm and is now in the supernatant.*
9. Purify target protein from supernatant using protocol 3.2.3. or alternative protocol 3.2.4.

4. Measurement setup

4.1 Materials, equipment, and reagents

4.1.1 Recording chamber (Fig. 3)

- 2 pieces of polyoxymethylene POM (25 mm × 20 mm × 10 mm)
- 1 mm drill bit
- 5 mm flat-face mill
- 3 mm drill bit
- 5 mm hole saw
  - *Can also use a face or ball mill.*
  - Dimensions *can be altered depending on the O-ring and additional clearance.*
- Rubber O-ring (5 mm)
  - *Hole saw and O-ring should match.*
- M3 screw 45 mm or longer
- M3 nut

4.1.2 PTFE membrane

- Polytetrafluoroethylene (PTFE) sheet (Goodfellow FP30-FM-000200, 0.025 mm thickness)
- *Option 1:* Spark generator (able to output several thousand of Volts)
- *Option 2:* A small sewing needle

4.2 Step-by-step protocol

4.2.1 Recording chamber

1. Place a piece of POM (25 mm × 20 mm × 10 mm (XYZ)) facing XY.
   a. *We manufacture the recording chamber out of POM due to its low conductivity and dielectric polarization.*

      Cautionary note: While polyvinylchloride (PVC) seems like an equally suitable material, it tends to exhibit measurable dielectric polarization that results in a buildup of capacitance across the chamber.
2. Drill a 1 mm hole (7 mm depth) for placement of the electrode.
3. Flat-face mill a 5 mm × 12 mm × 7 mm slit.
4. Place the chamber facing XZ.
5. Drill two 3 mm through-holes on the sides.
6. Drill one 3 mm hole such that the bottom is flat with the slit (step 3).
7. Drill a 1 mm hole through the slit until it protrudes from the electrode hole (step 2).

8. Add a 5 mm incursion around the opening (step 6) to host a rubber O-ring.
   a. *Adjust the available O-ring, it is to seal the edges of the membrane.*

4.2.2 PTFE membrane

1. Place a PTFE sheet into a rigid holder with an opening on either side.
   a. *The holder can be made from any non-conductive material, including cardboard.*
2. Puncture a hole in the membrane of approximately 100 μm using either option below.
   Option 1: using dielectric breakdown
3. Place the anode and cathode on either side of the membrane.
4. Spark a hole with a 200 ms pulse.
   a. *Do not touch the electrodes when sparking.*
   Option 2: using physical puncture
5. Using a small needle, puncture a small hole in the membrane of ~100 μm.
   a. *Make sure that the membrane does not protrude and stays flat.*
      i. *For this reason, dielectric breakdown is preferred.*

Note: The protocol to generate this membrane does not form perfectly round holes on purpose, as the attachment of the *n*-hexadecane and lipids is enhanced with surface roughness.

4.2.3 Recording chamber assembly

1. Place one half-chamber (4.2.1) facing the O-ring upwards.
   a. Cut the PTFE membrane (4.2.2) such that it is larger than the half chamber.
2. Place the PTFE membrane such that the aperture is centered.
   a. *Tape one side with a piece of scotch tape.*
   b. *Extend the PTFE membrane across such that it is tight and tape this side.*
3. Carefully attach the second half-chamber.
   a. *The PTFE membrane can bend easily.*
   b. *Do not fold the PTFE membrane* accidentally.
4. Screw the two half-chambers tight using the M3 screws.
5. Carefully remove the Scotch tape.
6. Cut away excess PTFE.
   a. *The chamber is now assembled similar to* Fig. 3.

5. Nanopore recording

5.1 Materials, equipment, and reagents

- Ag|AgCl electrodes (Sigma-Aldrich, CAS number 7440-22-4, diameter of 1.0 mm, ≥99.99 % trace metals basis)
- Microcapillary tubes (Sigma-Aldrich, Drummond Microcaps®, 10 μL, SKU number P1924)
- Clear glass vials (Sigma-Aldrich, volume 2 mL, SKU number 27078)
- Screw cap, solid top with PTFE liner (Sigma-Aldrich, for use with 2 mL vial, SKU number 27091-U)
- *n*-Hexadecane (Sigma-Aldrich, CAS number 544-76-3)
  - *Hexadecane has a melting point of approximately 18 °C, therefore, it should be stored at room temperature as it solidifies at lower temperatures. However, when hexadecane is dissolved in pentane at 2.5 % (v/v), the melting point decreases to that of pentane (approximately −130 °C). Therefore, the oil solution may be stored either at −20 °C or at room temperature.*
- 1,2-diphytanoyl-*sn*-glycero-3-phosphocholine (DPhPC, Avanti Polar Lipids, CAS number 207131-40-6)
  - *DPhPC lipids must be dissolved in oleophilic solvents like pentane due to their hydrophobic nature. Pentane is highly volatile, evaporating rapidly at room temperature. Glass vials with a PTFE cap are recommended for storage, and the vial cap may also be sealed with Parafilm to minimize evaporation. The lipid solution is typically stored at −20 °C or −80 °C. Storage at room temperature is also possible, provided that exposure to sunlight is avoided. However, shelf life is then decreased.*
- Pentane (Sigma-Aldrich, CAS number 109-66-0)
- NaCl (Sigma-Aldrich, CAS number 7647-14-5)
- Tris-HCl (Roche, CAS number 1185-53-1)
- *n*-dodecyl-β-D-maltoside (DDM, Glycon, product number D97002)
- Ethylenediaminetetraacetic acid (EDTA, Sigma-Aldrich, CAS number 60-00-4)
- ClyA Extraction Buffer (CEB):
  - 150 mM NaCl
  - 15 mM Tris-HCl at pH 7.5
    - *The pH of a Tris-based buffer is temperature-sensitive.*
  - 0.02 % DDM
    - *DDM supplementation in the ClyA-AS extraction buffer must be at a concentration above its critical micelle concentration (approximately 0.17 mM in water) to enable micelle formation and effective solubilization of the nanopores.*

- o 10 mM EDTA
    - ■ *EDTA is also added to the CEB as it chelates metal ions, preventing them from affecting protein conformation.*
- o Store 4 °C, or frozen.

- Recording Buffer (RB):
    - o 150 mM NaCl
    - o 15 mM Tris-HCl at pH 7.5 (25 °C)
        - ■ *The pH of a Tris-based buffer is temperature-sensitive.*

5.2 Step-by-step protocol

5.2.1 Preparation of the recording chamber and formation of the bilayer lipid membrane (BLM)

After assembling the recording device with a PTFE membrane containing an aperture of approximately 100 μm in between the two compartments, the following steps are required to initiate nanopore experiments.

1. Place the chamber (Fig. 3) such that the ZY axis is horizontal (one can see the PTFE membrane).
2. Apply 5 μL of 2.5 % (v/v) *n*-hexadecane in pentane onto both sides of the PTFE membrane.
    - **a.** *Use a microcapillary. Pentane is volatile, so regular pipettes do not work.*
        - **i.** *Assemble the microcapillary into the holder (bulb).*
        - **ii.** *Fill the microcapillary by placing it into the hexadecane-pentane mixture. In principle the capillary forces are sufficient to draw 10 μL. If needed, apply slight negative pressure (squeeze the bulb).*
        - **iii.** *Place the end of the microcapillary above the PTFE and cover the aperture.*
            - **1.** *Gently squeeze the bulb.*
        - **iv.** *The chamber can be turned before all pentane evaporates to apply on the other side.*
    - **b.** *Hexadecane is required to provide a strong oleophilic binding of the lipid layer in subsequent steps. Pentane is "just" a carrier medium that allows the uniform application of n-hexadecane.*
    - **c.** *One can also choose to coat only one side of the PTFE, using 5−10 μL of oil solution.*
3. Wait 1 min.
    - **a.** *Allows the pentane to evaporate.*
    - **b.** *Gently tap the chamber against the bench from several sides to promote more even spreading.*

4. Place the chamber such that the XY axis is horizontal (flat standing on a surface).
5. Add 500 μL of recording buffer to each compartment.
6. Add about 20 μL of lipid solution (8 mg.mL^{-1}) to each compartment by using a microcapillary tube.
   a. *Over time, pentane will evaporate, hence, the actual concentration of lipids will increase over time.*
      i. *When in doubt, be conservative, more lipids can be added if BLM formation fails.*
   b. *This step deposits a thin phospholipid monolayer via solvent evaporation.*
      i. *The phospholipid tails will point towards the liquid-gas (air) interface, while the phosphate headgroups face the aqueous phase.*
7. Wait 2 min.
   a. *Allows the pentane to evaporate.*
8. Mix the contents of each compartment by pipetting up and down using a pipette set to 500 μL.
   a. *Be gentle, as the container is open and spills will affect the recording.*
   b. *Avoid pipetting air/bubbles into the chamber.*
9. Insert the Ag|AgCl electrodes into the small openings adjacent to each compartment.
   a. *Connect one electrode to ground and the other one to the working electrode of the amplifier.*
   b. *The compartment connected to ground is designated the cis compartment, whereas the compartment connected to the working electrode is hereafter referred to as the trans compartment.*
10. Gently pipette up and down the volume in the compartments.
    a. *Pipette up until there is no electrical connection.*
    b. *Slowly pipette down such that the electrical connection is restored.*
       i. *Repeat if observing a closed circuit (full conductance, infinite current)* (Maglia et al., 2010).
11. The capacitance over the compartments will change to the specific capacitance of DPhPC multiplied by the aperture area.
    a. *Measure at 10 kHz with a Bessel-filter set to 2 kHz.*
12. *Recommended:* Perform a Fourier transform to determine if the electrical grid results in increased noise.
    a. *There is a power or grounding issue in the setup if you observe a large spike at 50 Hz (or 60 Hz).*

Additional notes:

The *cis/trans* nomenclature refers to the *cis* and *trans* sides of ClyA-AS. The nanopore is added to the compartment connected to the ground. Thus, nanopore insertion into the membrane separating the two compartments occurs preferentially with the *cis* side of ClyA-AS facing the *cis* compartment.

After each nanopore experiment, the chamber must be rinsed with a gentle stream of warm water, followed by 96 % ethanol. If more thorough cleaning is required, pure pentane can be used. Then, the chamber can be blow-dried with nitrogen and gently heated on a hot plate at 40 °C for about 10–15 min to ensure complete drying.

5.2.2 Single-nanopore insertion into the BLM

The insertion of ClyA-AS nanopore into the lipid bilayer is a probabilistic process. However, several practical considerations can increase the likelihood of successful nanopore insertion into the membrane:

1. Extract oligomerized dodecameric ClyA-AS from blue-native (BN) polyacrylamide gel.
   a. *BN gel preparation is described in Section 3.2.5.*
   b. *Extract using 20 µL CEB until the gel becomes transparent and the solution blue.*
      i. *Overnight extraction is sufficient to extract the nanopores from the gel into the solution.*
   c. *Monomers can also form pores; however, they yield 12, 13, and 14-mers.*
   d. *Only dodecameric ClyA-AS (12-mer) can capture Y27A-TbpA.*
   e. *Extract can be stored for 1–3 weeks at 4 °C.*
   f. *The exact concentration of ClyA-AS in the gel band is difficult to quantify. Working dilutions can be determined empirically during the next step.*
2. Add ClyA-AS into the *cis* compartment of the recording chamber.
   a. *Dip and swirl a small pipette tip into the ClyA-AS solution, then rinse the outside of the tip by briefly dipping it into clean recording buffer (RB). This last step ensures that excess pores are removed. Finally, dip the tip into the cis compartment solution.*
   b. *The final concentration of ClyA pores in the chamber needs to be very low. Only a single pore should insert. The concentration required for this is too low to be quantified. If no pore insertion is observed, add more sample. If the bilayer ruptures, the pore concentration was likely too high and multiple pores inserted at once. In that case, try again with a diluted pore sample.*

   i. *ClyA-AS is unstable in RB. It will stop inserting after some time (~10 min). If too many pores were added, wait before adding new pore sample.*
   ii. *The buffer solution in the chamber can alternatively be replaced carefully with fresh RB.*
3. Apply a triangular voltage with a constant voltage ramp of ± 1 Volt per second.
   a. *Measure at 10 kHz with a Bessel-filter set to 2 kHz.*
   b. *Repeatedly apply between ± 15 mV over the duration of 30 ms.*
   c. *Repeatedly rupturing and reforming the bilayer increases the probability of nanopore insertion.*
      i. *Application of 1 V or more ruptures the bilayer (use the zapping command).*
      ii. *Reform the bilayer by gently pipetting up and down (repeat protocol 5.2.1 step 10).*
4. Wait until ClyA-AS inserts into the bilayer.
   a. *If ClyA-AS does not insert within ~10 min, repeat from step 2.*
5. Correct the offset potential.
   a. *Apply a constant 0 mV and change the offset such that the current is 0 pA.*
      i. *The offset provides a counter potential, hence, 0 mV applied is not actually 0 mV.*
6. Check the current output when ± 35 mV is applied.
   a. *In RB, ClyA-AS will exhibit about −60 pA at −35 mV, and about + 65 pA at + 35 mV in the correct configuration* (Wloka et al., 2021).
   b. *Measure at 10 kHz with a Bessel-filter set to 2 kHz.*
   c. *Notice that the pore is not perfectly symmetrical, as its current should always be slightly higher at positive potential. If the current is higher at the negative potential, then ClyA-AS was inserted "upside-down", meaning the cis opening is directed towards the trans compartment of the chamber.*
      i. *In this case, the nanopore can still be used for the experiment however, the sensor components (TBP and thiamine) must be added to the trans compartment, and the polarity of the applied voltage must be reversed for the measurement.*
      ii. *If multiple pores insert the output current is the sum of all the individual nanopore currents.*
   d. *It is also important to check if the current is fluctuating. If fluctuations are observed, leave the system at 0 mV potential for approximately 5 min. Applying higher potentials, such as ± 180 mV, can also help stabilize the pore, as the fluctuations may result from transient clogging by impurities.*

5.2.3 Electrical recordings of thiamine using ClyA-AS and TBP in a BLM

Note that during the measurement, the bilayer may rupture, the nanopore may become unstable, or additional nanopores may insert stochastically.

1. Add 15 nM Y27A-TbpA (protocol 3.2.6) to the *cis* compartment.
   a. *Measure by Bradford assay the protein concentration (preferred) or Nanodrop.*
   b. *Cis is now referring to the orientation of ClyA-AS, not the electrodes.*
   c. *Mixing within the compartment should be performed carefully, as abrupt fluid movement may lead to membrane rupture or nanopore exiting from the bilayer.*
   d. *Stochastic interactions from Y27A-TbpA entrapment into ClyA-AS can be observed.*
2. Start recording.
   a. *You may wish to start recording earlier and segregate the Y27A-TbpA traces in pre-processing.*
   b. *Measure at 10 kHz with a Bessel-filter set to 2 kHz.*
3. Apply a continuous potential of −45 mV.
4. Add sample containing thiamine to the *cis* compartment.
   a. *A new calibration curve (2, 4, 8, 16, 32, 64, 128, 256, 512, 1024, 2048, and 4096 nM thiamine) should be* obtained at least once *to calibrate the fully assembled setup.*
   b. *Mixing within the compartment is essential but must be performed carefully.*
   c. *Sample includes:*
      i. *Calibration samples.*
      ii. *Biological fluids such as urine. Blood is less likely to work because the large numbers of blood proteins are easily captured from cis.*

6. Conclusions

Biological nanopores offer exquisite reproducibility but lack an inherent affinity towards an analyte, apart from its dimensions. If one wishes to measure less abundant molecules in complex samples, it is thus necessary to equip a nanopore with an adapter with a high affinity towards a molecule. Many proteins have been the product of natural evolution to bind small molecules in highly complex samples. There are many substrate-binding proteins, such as TbpA, which bind diverse small molecules and can potentially be used in nanopore sensing. Combined with high-throughput platforms, point-of-care devices using biological nanopores can become a reality.

Acknowledgements

S.B.R., J.P.D., and C.W. were supported by sectorplan 2.0. G.C.C. was supported by the Research Foundation–Flanders (11Q4E24N). F.L.R.L. was supported by the Dutch Research Council (NWO) file number VI.Veni.242.400 of the research programme VENI under the grant [grant ID https://doi.org/10.61686/GKPYM97413].

Molecular graphics and analyses were performed with UCSF ChimeraX, developed by the Resource for Biocomputing, Visualization, and Informatics at the University of California, San Francisco, with support from National Institutes of Health R01-GM129325 and the Office of Cyber Infrastructure and Computational Biology, National Institute of Allergy and Infectious Diseases.

References

Clarke, J., Wu, H.-C., Jayasinghe, L., Patel, A., Reid, S., & Bayley, H. (2009). Continuous base identification for single-molecule nanopore DNA sequencing. *Nature Nanotechnology, 4*(4), 265–270. https://doi.org/10.1038/nnano.2009.12.

Galenkamp, N. S., Soskine, M., Hermans, J., Wloka, C., & Maglia, G. (2018). Direct electrical quantification of glucose and asparagine from bodily fluids using nanopores. *Nature Communications, 9*(1), 4085. https://doi.org/10.1038/s41467-018-06534-1.

Galenkamp, N. S., Van Meervelt, V., Mutter, N. L., van der Heide, N. J., Wloka, C., & Maglia, G. (2021). Preparation of cytolysin A (ClyA) nanopores. *Methods in Molecular Biology,* 11–18. https://doi.org/10.1007/978-1-0716-0806-7_2.

Jo, S., Kim, T., & Im, W. (2007). Automated builder and database of protein/membrane complexes for molecular dynamics simulations. *PLoS ONE, 2*(9), e880. https://doi.org/10.1371/journal.pone.0000880.

Jo, S., Kim, T., Iyer, V. G., & Im, W. (2008). CHARMM-GUI: A web-based graphical user interface for CHARMM. *Journal of Computational Chemistry, 29*(11), 1859–1865. https://doi.org/10.1002/jcc.20945.

Li, F., Fahie, M. A., Gilliam, K. M., Pham, R., & Chen, M. (2022). Mapping the conformational energy landscape of Abl kinase using ClyA nanopore tweezers. *Nature Communications, 13*(1), 3541. https://doi.org/10.1038/s41467-022-31215-5.

Li, P., Liang, D., Yang, E., Zeb, M., Huang, H., Sun, H., ... Ma, W. (2024). Bio-nanopore technology for biomolecules detection. *Advanced Biotechnology, 2*(4), 45. https://doi.org/10.1007/s44307-024-00051-7.

Lucas, F. L. R., Finol-Urdaneta, R. K., Van Thillo, T., McArthur, J. R., van der Heide, N. J., Maglia, G., ... Wloka, C. (2023). Evidence of cytolysin A nanopore incorporation in mammalian cells assessed by a graphical user interface. *Nanoscale, 15*(42), 16914–16923. https://doi.org/10.1039/D3NR01977B.

Lucas, F. L. R., Piso, T. R. C., van der Heide, N. J., Galenkamp, N. S., Hermans, J., Wloka, C., & Maglia, G. (2021). Automated electrical quantification of Vitamin B1 in a bodily fluid using an engineered nanopore sensor. *Angewandte Chemie International Edition, 60*(42), 22849–22855. https://doi.org/10.1002/anie.202107807.

Maglia, G., Heron, A. J., Stoddart, D., Japrung, D., & Bayley, H. (2010). Analysis of single nucleic acid molecules with protein nanopores. *Methods in Enzymology,* 591–623. https://doi.org/10.1016/S0076-6879(10)75022-9.

Montal, M., & Mueller, P. (1972). Formation of bimolecular membranes from lipid monolayers and a study of their electrical properties. *Proceedings of the National Academy of Sciences, 69*(12), 3561–3566. https://doi.org/10.1073/pnas.69.12.3561.

Peng, W., de Souza Santos, M., Li, Y., Tomchick, D. R., & Orth, K. (2019). High-resolution cryo-EM structures of the E. coli hemolysin ClyA oligomers. *PLOS ONE, 14*(5), e0213423. https://doi.org/10.1371/journal.pone.0213423.

Pettersen, E. F., Goddard, T. D., Huang, C. C., Meng, E. C., Couch, G. S., Croll, T. I., ... Ferrin, T. E. (2021). UCSF ChimeraX: Structure visualization for researchers, educators, and developers. *Protein Science, 30*(1), 70–82. https://doi.org/10.1002/pro.3943.

Shorkey, S. A., Zhang, Y., Sharp, J., Clingman, S., Nguyen, L., Chen, J., & Chen, M. (2025). Tracking flaviviral protease conformational dynamics by tuning single-molecule nanopore tweezers. *Biophysical Journal, 124*(1), 145–157. https://doi.org/10.1016/j.bpj.2024.11.017.

Soriano, E. V., Rajashankar, K. R., Hanes, J. W., Bale, S., Begley, T. P., & Ealick, S. E. (2008). Structural similarities between thiamin-binding protein and thiaminase-I suggest a common ancestor. *Biochemistry, 47*(5), 1346–1357. https://doi.org/10.1021/bi7018282.

Soskine, M., Biesemans, A., De Maeyer, M., & Maglia, G. (2013). Tuning the size and properties of ClyA nanopores assisted by directed evolution. *Journal of the American Chemical Society, 135*(36), 13456–13463. https://doi.org/10.1021/ja4053398.

Tilley, S. J., Orlova, E. V., Gilbert, R. J. C., Andrew, P. W., & Saibil, H. R. (2005). Structural basis of pore formation by the bacterial toxin pneumolysin. *Cell, 121*(2), 247–256. https://doi.org/10.1016/j.cell.2005.02.033.

Vreeker, E., Grünewald, F., van der Heide, N. J., Bonini, A., Marrink, S. J., Tych, K. (Kasia), & Maglia, G. (2025). Nanopore-functionalized hybrid lipid-block copolymer membranes allow efficient single-molecule sampling and stable sensing of human serum. *Advanced Materials, 37*(15), https://doi.org/10.1002/adma.202418462.

Wloka, C., Galenkamp, N. S., van der Heide, N. J., Lucas, F. L. R., & Maglia, G. (2021). Strategies for enzymological studies and measurements of biological molecules with the cytolysin A nanopore. *Methods in Enzymology,* 567–585. https://doi.org/10.1016/bs.mie.2021.01.007.

CHAPTER ELEVEN

Engineered nanopore biosensor for thiamine quantification: Data analysis

Gerard Carrera i Cardona[a], Sofia B. Relvas[b], Jan P. Dobert[b], Carsten Wloka[b], and Florian L.R. Lucas[b,*]
[a]Lab for Nanobiology, Department of Chemistry, KU Leuven, Leuven, Belgium
[b]BIOS, Lab on a Chip group, MESA+ Institute for Nanotechnology, Technical Medical Centre, University of Twente, Enschede, The Netherlands
*Corresponding author. e-mail address: florian.lucas@utwente.nl

Contents

Abstract

Biological nanopores equipped with substrate-binding proteins are powerful biosensor pairs capable of measuring low-abundance molecules in complex samples. One example of such a system pairs the thiamine-binding protein (TbpA) and the Cytolysin A (ClyA) nanopore. TbpA has been evolved to selectively bind thiamine in even the most complex samples. When a ClyA nanopore traps TbpA, even a single experiment can generate a vast amount of data in a very short time. While at the research level, even a single nanopore can accurately quantify thiamine concentrations, the setup time prohibits a practical application, such as in point-of-care devices. Such devices would likely house hundreds to millions of pores, rendering manual data analysis impossible. To bridge the gap between biosensor development and practical devices, it is therefore necessary to develop algorithms which can analyze and interpret data automatically. Here we outline our approach to nanopore data analysis to tackle the challenges in automatic analysis.

Methods in Enzymology, Volume 735
ISSN 0076-6879, https://doi.org/10.1016/bs.mie.2026.07.020

1. Introduction

Biological nanopores enable the electrical detection of individual molecules. These nanopores are inserted into an electrically isolating lipid bilayer (membrane) (Li et al., 2024). Applying an electrical potential across the membrane drives ions through the pore, generating an ionic current. By recording the ionic current over time, we can measure the steric exclusion that molecular species cause when they are inside the nanopore. The internal dimensions of nanopores determine which molecular species exhibit a measurable signal. The Cytolysin A (ClyA) nanopore has internal dimensions that are ideal for the capture of the Y27A mutant of thiamine-binding periplasmic protein (Y27A-TbpA) (Galenkamp et al., 2018; Soskine et al., 2013). This protein exhibits conformational changes when it binds to thiamine, resulting in a change in the steric exclusion and therefore the measurable ionic current (Lucas et al., 2021). The Y27A-TbpA is highly specific to thiamine, allowing it to be utilized directly in complex samples such as urine. Since the Y27A mutations reduce the dissociation constant of TbpA, we can observe rapid binding and unbinding (Fig. 1). The nanopore itself acts as a transducer of the Y27A-TbpA, sporadically allowing its insertion into its lumen. Consequently, the signal we observe from the experiment is complex, since it includes active protein but also any contaminants and requires robust analysis, which we provide with the protocols described in this chapter.

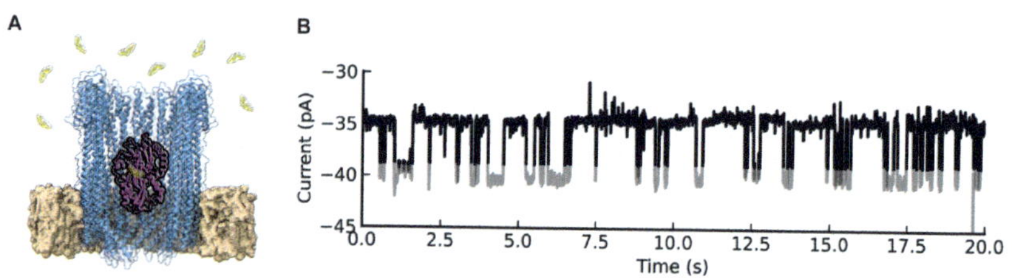

Fig. 1 (A) Illustration of thiamine quantification using the biosensor pair consisting of thiamine binding protein TbpA (purple) and the protein nanopore cytolysin A (ClyA, blue). A single ClyA pore is inserted into a lipid bilayer (beige) and captures TbpA. When thiamine (yellow) passes through the pore, it can bind to TbpA, which allows detection and quantification on a single-molecule basis. Figure created using ChimeraX (Pettersen et al., 2021). Protein structures obtained from the PDB (TbpA: 2QRY (Soriano et al., 2008), ClyA pore: 6MRT (Peng et al., 2019)). Lipid bilayer taken from CHARMM-GUI Archive (Jo et al., 2007, 2008). (B) Representative trace of Y27A-TbpA inside ClyA-AS in 150 mM NaCl under an applied potential of −45 mV recorded at 10 kHz and filtered down to 100 Hz. The grey areas represent the unbound Y27A-TbpA and the black areas represent the bound Y27A-TbpA.

2. Rationale and theory

The ionic current recorded over time (trace) is obtained from a setup in which nanopores are inserted into lipid bilayers, using, for instance, the Montal-Mueller methods (Montal & Mueller, 1972). However, the ionic current is in theory independent of the method of creating the lipid bilayer and the analysis is therefore (most likely) general. Since the signals observed from Y27A-TbpA in ClyA are relatively long lived (high milliseconds to seconds), we only need to consider the steady-state regime. That is, the signals appear to be instantaneous. The membrane with an inserted nanopore behaves as a resistor in parallel with the membrane capacitance. Under a constant holding potential, the capacitive term vanishes, and the current reduces to Eq. 1.

$$I(t) = V \cdot G_p \tag{1}$$

where I is the current over time, V is the applied potential, and G_p is the conductance of the pore. Because the voltage is held constant, the current is proportional to the conductance, allowing us to use the measured current as a direct proxy for conductance.

The recorded traces contain data from pores that include active protein, non-active protein, impurities, nothing trapped within ClyA, as well as noise. We segment the data to ensure that we only utilize active protein for the quantification of the thiamine concentration. First, we separate the Y27A-TbpA blockades from the rest of the signal. For this, the Rapid Event Detection (RED) algorithm is used since the signal of the Y27A-TbpA entering the nanopore is relatively big (tens of pA) and long lived. The RED algorithm consists of establishing a baseline and applying a threshold-crossing method. If the data is higher than this threshold, the data is flagged as meaningful. This threshold can be defined as 6 times the standard deviation.

Active Y27A-TbpA contains the conductance levels of the pore with binding protein in two states, bound to thiamine and not bound to thiamine. To understand how to extract concentrations from this data we first need to understand the underlying mathematics. The occupancy fraction of the bound state (θ) as a function of ligand concentration can be described by the Hill equation describing a binding event (Eq. 2). In this case the binding of thiamine to Y27A-TbpA follows a simple 1:1 stoichiometry:

each Y27A-TbpA binds a single thiamine molecule with no cooperativity, so the Hill coefficient $n_H = 1$. The occupancy is therefore directly dependent on thiamine concentration.

$$Y27A - TbpA + Thiamine \underset{k_{off}}{\overset{\mathrm{k_{on}}}{\rightleftarrows}} Y27A - TbpA - Thiamine$$

$$\theta = \frac{[L]^{n_H}}{K_d + [L]^{n_H}} \tag{2}$$

where $[L]$ is the ligand concentration, K_d is the apparent dissociation constant, and n_H is the Hill coefficient which quantifies cooperativity: $n_H > 1$ indicates positive cooperativity, $n_H < 1$ negative cooperativity, and $n_H = 1$ non-cooperative (independent) binding. The dissociation constant is the ratio of the off-rate to the on-rate, $K_d = k_{off}/k_{on}$ (with k_{off} in s^{-1} and k_{on} in $M^{-1}.s^{-1}$). For a two-state system, the mean dwell time in the bound state (τ_{bound}) is the inverse of the rate leaving that state ($1/k_{off}$), and the mean dwell time in the unbound state ($\tau_{unbound}$) is the inverse of the rate leaving it ($1/([L]\cdot k_{on})$). Substituting these allows the Hill equation to be rewritten in terms of the mean dwell times in each state (Eq. 3).

$$\theta = \frac{[L]\cdot k_{on}}{k_{off} + [L]\cdot k_{on}}; \quad \theta = \frac{\tau_{bound}}{\tau_{bound} + \tau_{unbound}} \tag{3}$$

We could now determine the bound fraction based on these average dwell times and use the variance between thousands of events to determine the error in our observed concentrations. In statistical terms, we would determine some distribution around the data and ask *how likely the data is given our hypothesis*, which consolidates into an error estimation. However, it would require a (very) long time as we would need to gather thousands of events before we can extract a statistically meaningful concentration.

Fortunately, we have a trick up our sleeve to turn our time-dependent signal into a state-dependent signal. We change the statistical question to *how likely our hypothesis is given our data*. To do so we first must rewrite our bound fraction in terms of the probability of finding TbpA bound and not bound to thiamine, given the data D (Eq. 4). This rewriting is justified because, under the assumption that the time average of a single trajectory equals the ensemble average, so the fraction of time spent in a state equals its equilibrium probability, the mean dwell time in each state is proportional to its steady-state occupancy; the bound fraction can therefore be expressed equivalently in terms of the posterior state probabilities.

$$\theta = \frac{P(bound|\mathcal{D})}{P(bound|\mathcal{D}) + P(unbound|\mathcal{D})} \tag{4}$$

Where θ is the bound fraction, $P(bound|\mathcal{D})$ is the probability of the bound state given the data, $P(unbound|\mathcal{D})$ is the probability of the unbound state given the data. When we measure the thiamine concentration, we are not given the probability of states given the data. However, we can describe the likelihood of finding Y27A-TbpA in the bound and unbound state given the data as the likelihood of the data given the bound and unbound state of Y27A-TbpA (Eqs. 5a and 5b) using Bayes Theorem (Eq. 6) (Ramachandran & Tsokos, 2021). In other words, we can look at the data and determine how likely it is that it is resulting from the bound or unbound state. With these likelihoods, we can then determine the bound fraction directly (Eq. 7). This method is known as empirical Bayesian inference. It is empirical as we base it solely on observation. It is Bayesian as we assume that the probability of our states can be described in terms of our prior belief. It is inference as we draw a conclusion on something we do not directly observe.

$$P(bound|\mathcal{D}) = \frac{P(\mathcal{D}|bound) \cdot P(bound)}{P(\mathcal{D})} \tag{5a}$$

$$P(unbound|\mathcal{D}) = \frac{P(\mathcal{D}|unbound) \cdot P(unbound)}{P(\mathcal{D})} \tag{5b}$$

$$P(A|B) = \frac{P(B|A) \times P(A)}{P(B)} \tag{6}$$

$$\theta = \frac{P(\mathcal{D}|bound) \cdot P(bound)}{P(\mathcal{D}|bound) \cdot P(bound) + P(\mathcal{D}|unbound) \cdot P(unbound)} \tag{7}$$

where θ is the bound fraction, $P(\mathcal{D} \mid bound)$ is the posterior probability of the bound state given the data, $P(unbound|\mathcal{D})$ is the posterior probability of the unbound state given the data, $P(bound)$ is the prior probability of the bound state, $P(unbound)$ is the prior probability of the unbound state, $P(\mathcal{D}|bound)$ is the probability of the data given the bound state, $P(\mathcal{D}|unbound)$ is the probability of the data given the unbound state, $P(\mathcal{D})$

is the prior probability of the data. $P(A \mid B)$ is the posterior probability of A given B, $P(B \mid A)$ is the probability of B given A, $P(A)$ is the prior probability of A, and $P(B)$ is the prior probability of B.

These terms may be confusing but are quite simple from a practical perspective. Let's start by talking about the prior probabilities of finding the thiamine in the bound and unbound states, respectively: P(unbound) and $P(bound)$. To determine the P(unbound), we first need to measure the occupancy of the protein without added ligand. We determine P(bound) when the protein is bound maximally to the ligand, by oversaturating the sensor. The signal is normally distributed around the mean, hence can fit a normal distribution around the signal. Fortunately, the mean is the average of the data, and the variance is the standard deviation squared; hence it can be directly interpreted from the data.

The bound fraction can be calculated from each (part of) the signal using the prior probabilities, which we use to determine the concentration of ligand based on the calculated bound fraction we must use a calibration curve. We can modify the Hill equation to have an offsetting term a and a scaling term b, and rewrite it as the concentration of ligand as a function of bound fraction (Eq. 8) with these correcting terms. The dissociation constant does not change between batches, yet a calibration curve will also yield this constant and can be used as a check. The intrinsic closing may differ for each measurement setup as it also considers off-target binding that may occur due to minor contaminants in the detection buffer (corrected for by terms a and b).

$$\theta = a + b\left(\frac{[L]^{n_H}}{K_d + [L]^{n_H}}\right) \Rightarrow [L](\theta) = \left(\frac{(\theta - a)^* K_d}{a + b - \theta}\right)^{\frac{1}{n_H}} \tag{8}$$

where a is the offset, b is the scaling factor, $[L]$ is the ligand concentration, K_d is the apparent dissociation constant, and n_H is the Hill coefficient, and θ is the bound fraction. We fit this equation to a calibration curve, which is the measured bound fraction with respect to the added concentration of ligand. The concentrations to be measured should lay between the zero-concentration (blank) and saturation. As can be seen from the Hill equation, the concentrations in the calibration should not be linearly spaced, as it would make the calibration too dependent on its tails, which we call "leveraged". For Y27A-TbpA we find a calibration curve with concentrations of 2, 4, 8, 16, 32, 64, 128, 256, 512, 1024, 2048, and 4096 nM highly suitable.

To determine the distributions, we bin our data into a histogram. The choice of bin width matters: too wide, and structure is lost, too narrow, and the histogram becomes sparse and noisy. We use the Freedman-Diaconis rule, which is roughly optimal for normally distributed data.

$$\text{Bin Width} = 2\frac{IQR\,(x)}{\sqrt[3]{n_p}} \tag{9}$$

where n_p is the number of datapoints and IQR(x) is the difference between the 0.75 and 0.25 quantiles. This rule of thumb reduces the leverage of outliers and generally keeps bins neither too small nor too large. It remains an approximation, so the result should be checked: because the underlying signal is distributed around its mean and variance, a well-binned histogram is continuous, and a spike-like histogram with empty intermediate bins indicates the bin width is too small.

Having obtained a concentration from the bound fraction via the calibration curve, we need an estimate of its error. Here we have two options, one: measure once again thousands of events and calculate its spread, or two: estimate the error based on the variance we expected from our model. The derivative of our model (Eq. 10) clearly shows that the most unlikely state for our protein to be is in the pure bound or pure unbound state. This is logical as pure unbound state is under the detection limit, and pure bound state is outside the dynamic range. We use the derivative of our expected concentration (Eq. 10) as a weight to skew the results towards our most trusted outcome. Note that this derivative is negative over the working range ($a < \theta < a + b$), so its absolute value is used as the weight.

$$\frac{d}{d\theta}[L](\theta) = \frac{b * \left(\frac{(\theta - a) * K_d}{a + b - \theta}\right)^{\frac{1}{nH}}}{n_H\,(a - \theta)\,(a + b - \theta)} \tag{10}$$

where a is the offset, b is the scaling factor, $[L]$ is the ligand concentration, K_d is the apparent dissociation constant, and n_H is the Hill coefficient, and θ is the bound fraction.

The error has two sources. Within a single pore, repeated measurements scatter around that pore's mean; across different pores, the pore means themselves scatter. Assuming both are normally distributed, the variance in the concentration [L] decomposes by the law of total variance into a within-pore and a between-pore term (Eq. 11).

$$Var\,(X) = E\,[Var\,(X|\,Pore)] + Var\,[E\,(X|\,Pore)] \tag{11}$$

Where $Var\,(X)$ is the total variance, $E\,[Var\,(X|\,Pore)]$ is the expected value of the variance within a single nanopore (Eq. 12), $Var\,[E\,(X|\,Pore)]$ is the variance of the expected value between different nanopores (Eq. 13).

$$E\,[Var\,(X|\,Pore)] = \frac{\sum_{j=1}^{M} W_j \cdot \sigma_{aj}^2}{\sum_{j=1}^{M} W_j} \tag{12}$$

where $E\,[Var\,(X|\,Pore)]$ is the expected value of the variance within a single nanopore, M is the total number of measurements within a single nanopore, w are the associated weights for each measurement, and σ^2 is the variance of each measurement within a single nanopore.

The second term, $Var\,[E\,(X|\,Pore)]$ is the between-sample variance — the variance expected between the mean concentrations of different pores, where μ_{aj} is the mean for pore j and μ_{total} the overall mean:

$$Var\,[E\,(X|\,Pore)] = \frac{\sum_{j=1}^{M} W_j \cdot (\mu_{aj} - \mu_{total})^2}{(M-1)\frac{\sum_{j=1}^{M} W_j}{M}} \tag{13}$$

where $Var\,[E\,(X|\,Pore)]$ is the variance of the expected value between different nanopores, μ_{aj} is the mean for pore j, μ_{total} the overall mean, M is the total number of measurements within a single nanopore and w are the associated weights for each measurement.

The total standard deviation is the square root of the total variance:

$$\sigma_{total} = \sqrt{Var\,(X)} \tag{14}$$

where σ_{total} is the error expected when combining measurements across multiple pores, and $Var\,(X)$ is the total variance. Note that $Var\,[E\,(X|\,Pore)] = 0$ when detecting using a single nanopore.

The following protocols give the practical implementation of the rationale described herein, converting the raw signals into a known concentration with corresponding standard error.

3. Analysis and statistics

3.1 Pre-processing

Data preprocessing includes cleaning up data from irrelevant features such as when the faraday cage of the measurement has been open, or the time before protein has been added. This cleanup step can be carried out automatically, or manually in the recording software due to the variability of the signals. Later the data must be filtered to remove electrical noise and corrected, if necessary, for the current offset. It is important also to extract the baseline current and save it for downstream processing (Listing 1, below).

1. Clean the recorded raw signal based on measurement (Fig. 2A)
   a. *Remove parts of the trace where for example the Faraday cage was opened for sample addition.*
2. Filter the data using a 100 Hz filter (Fig. 2B)
   a. *Y27A-TbpA gets captured into the nanopore for seconds.*
   b. *Conformational transitions are in the order of hundreds of milliseconds.*

```
INPUT
    raw_signal                      # full recording
    start_0, end_0                  # window with no applied potential (offset)
    start, end                      # window containing the nanopore signal
PROCEDURE
    # Gaussian model used for every fit
    G(x) = A * exp(-(x - μ)^2 / (2 * σ^2))
    signal_filtered = GaussianFilter(raw_signal)
    offset          = signal_filtered[start_0 : end_0]
    raw_trace       = signal_filtered[start : end]
    for dataset in {baseline, raw_trace}:
        bin_width      = FreedmanDiaconis(dataset)
        counts, edges  = Histogram(dataset, bin_width)
        centers        = Midpoints(edges)
        (A, μ, σ)      = Fit(G(centers) ≈ counts)
        store μ[dataset], σ[dataset]
    trace   = raw_trace  - μ[offset]
    μ_pore  = μ[raw_trace] - μ[offset]
    σ_pore  = σ[raw_trace]

OUTPUT
    trace                           # offset-corrected nanopore trace
    μ_pore, σ_pore                  # mean and std of the pore current
```

Listing 1 Data preprocessing pseudocode.

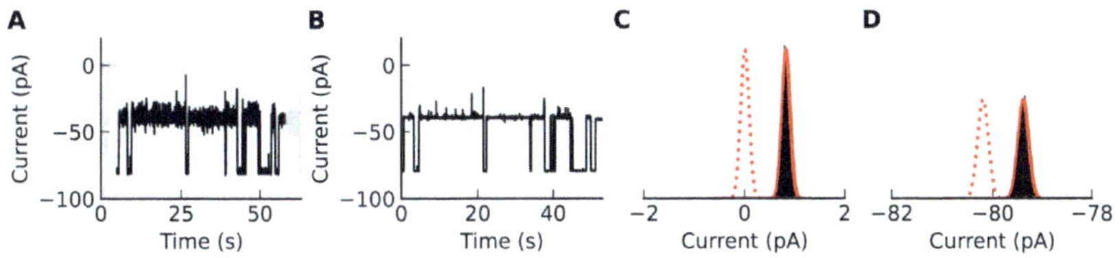

Fig. 2 Signal cleanup and baseline correction. (A) Raw signals are pre-cut into raw traces (black) where non-nanopore related signal is removed (grey). (B) The raw traces filtered at 100 Hz. (C) Histogram of the offset signal (black) is fitted with a normal distribution function (red, solid) and the mean is shifted to 0 (red, dotted). (D) Histogram of the nanopore signal (black) is fitted with a normal distribution function (red, solid) and the mean is shifted by the offset observed in C (red, dotted).

3. Determine the offset current and its standard deviation (σ) (Fig. 2C)
   a. *Create a histogram with bin-width determined by the Freedman-Diaconis rule and fit a normal distribution around the resulting peak*
4. Correct for offset by subtracting the average signal under no applied potential (Fig. 2D)
   a. *Offset causes the signal to become flawed, hence larger corrections may impact your results significantly. Typically, corrections can be 1–2 pA.*

3.2 Segmenting blockades from the baseline

Segmenting the data involves processing it with the RED algorithm to retain only the nanopore traces where the binding protein is inserted. Because a protein inserting into the pore reduces its conductivity, it produces a drop in the absolute value of the current, so events can be identified by finding the points that differ significantly from the baseline (Fig. 3). The procedure begins by taking a reference level (the baseline) and its standard deviation and constructing a list of indexes where the value of the signal sits significantly below the baseline. Scanning along this set of points, the end of each protein insertion is detected where two consecutive points are no longer continuous, with a small overshoot tolerance (2 datapoints at 10 kHz by default), so that brief gaps within an event are not treated as boundaries; the start of the next event then follows immediately after each detected end. Spikes are removed by requiring each candidate event to span at least 2 datapoints, and the surviving start and end positions are finally mapped back onto the original signal to give the event boundaries. This algorithm here is provided below in pseudocode (Listing 2).

3.3 Determination of the (un)bound likelihoods

With our traces segmented between protein blockades and baseline signals, we must next determine whether the protein blockades are active. To do so we first determine the expected distribution of our protein without thiamine added, and with thiamine saturating the sensor. We assume that the majority of blockades in the fully unbound and saturated signals are active. Hence, we can fit a normal distribution around all blockades in these states. However, some contain inactive protein which we reject based on the cumulative distribution function of X^2.

$$X^2 CDF(x, k) = \frac{1}{\Gamma\left(\frac{k}{2}\right)} \gamma\left(\frac{k}{2}, \frac{x}{2}\right) \tag{15}$$

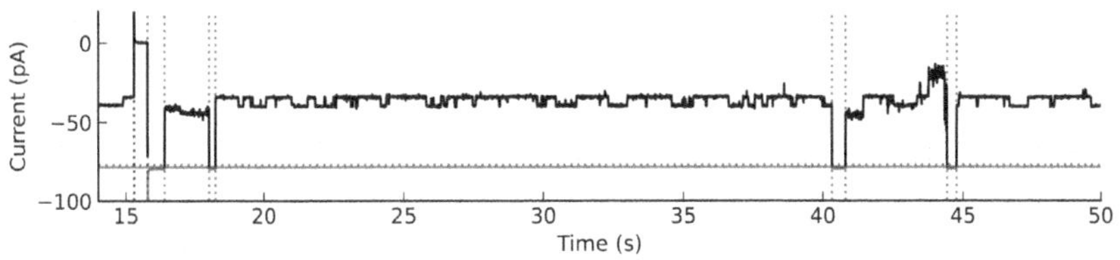

Fig. 3 Segmented nanopore trace. Raw trace is shown in black. The horizontal solid gray line represents the mean baseline current of the nanopore (μ_pore), and the horizontal dotted line represents its positive offset (μ_pore + σ_pore). The vertical dotted gray lines represent the start and the end of each protein blockade.

```
INPUT
    signal
    μ_dataset, σ_dataset          # baseline level and its standard deviation
    sampling_period
    dwell_time                    # minimum dwell time
    skip                          # maximum gap tolerated within an event

PROCEDURE
    candidate_indices = { i : |signal[i]| < |μ_dataset| - |σ_dataset| }

    groups = GroupConsecutive(candidate_indices)
    groups = MergeGroups(groups, gap <= skip)

    minimum_length = max(2, dwell_time / sampling_period)
    groups = Discard(groups, length < minimum_length)

    for group in groups:
        event_start    = first(group)
        event_end      = last(group)
        baseline_start = previous_event_end + 1
        baseline_end   = event_start - 1

OUTPUT
    baseline_start, baseline_end  # baseline segment preceding each event
    event_start, event_end        # blockade boundaries
```

Listing 2 Current blockades segmentation pseudocode.

where x is the chi-squared density, k are the degrees of freedom, and $\Gamma(x)$ is the Gamma function, and $\gamma(a, b)$ is the lower incomplete Gamma Function. The X^2 distribution is implemented in most major programming languages such as Python, Matlab and R. To increase robustness, we ignore lowest 5 % and highest 5 % of the signal for these comparisons (as we may have spikes that don't influence the downstream analysis). We then re-investigate the original blockades to only yield active unbound and saturated blockades to determine their distribution properly.

$$1 - X^2 CDF\left(\frac{\sum \left(\frac{signal - (\mu_{unbound}{}^{*} \text{ or } \mu_{bound}{}^{*})}{3 * (\sigma_{unbound}{}^{*} \text{ or } \sigma_{bound}{}^{*})} \right)^2}{length\,(signal)}, \ length\,(signal) - 3 \right) > 0.95 \quad (16)$$

Where *signal* is the data between the 5 % and 95 % percentile, $\mu_{unbound^*}$ is the mean of all unbound signals, $\sigma_{unbound^*}$ is the standard deviation of all unbound signals, μ_{bound^*} is the mean of all saturated signals, σ_{bound^*} is the standard deviation of all saturated signals. The probability threshold is 95 % and the degrees of freedom at the length of the signal minus the 3 parameters of the normal distribution. The factor of 3 in the denominator widens the tolerance to $\pm 3\sigma$, so that a blockade is rejected only when it deviates from the reference state well beyond normal noise, preventing over-rejection of genuine active protein. It is then used to separate the active from inactive proteins in the unbound and saturated blockades to determine their distribution centers, directly yielding $P(unbound)$ and $P(bound)$ (Listing 3).

1. Pre-process the traces with no thiamine added according to protocol 3.1 and 3.2
2. Join all blockades and fit a normal distribution around the total signal

```
INPUT
    traces_unbound                  # recordings with 0 nM thiamine
    traces_saturated                # recordings with 4096 nM thiamine
    p_low, p_high = 0.05, 0.95      # percentile cut-offs (reject spikes/dips)

PROCEDURE
    # --- reference distribution for one condition ---
    function ReferenceDistribution(traces):
        blockades = Segment(PreProcess(traces))     # protocols 3.1 + 3.2

        # first-pass fit over all blockades (assumes majority active)
        pooled     = Join(blockades)
        (μ*, σ*)   = FitNormal(pooled)

        # keep only blockades consistent with the reference state
        accepted = {}
        for b in blockades:
            core = b[Percentile(b, p_low) : Percentile(b, p_high)]
            k    = length(core) - 3                     # 3 fitted params

            stat = ( Σ_i ((core[i] - μ*) / (3 * σ*))^2 ) / length(core)
            if X²CDF(stat, k) > 0.95:
                accepted += b


        # second-pass fit over accepted blockades only
        counts, edges = Histogram(Join(accepted), FreedmanDiaconis(Join(accepted)))
        (μ, σ)        = FitNormal(counts, edges)
        P             = Normalise((μ, σ), area = 1)      # P(state | ·)
        return P, μ, σ, accepted

    P_unbound, μ_unbound, σ_unbound, _ = ReferenceDistribution(traces_unbound)
    P_bound,   μ_bound,   σ_bound,   _ = ReferenceDistribution(traces_saturated)

OUTPUT
    P_unbound, P_bound              # priors for the bound-fraction calculation
    μ_unbound, σ_unbound            # unbound reference distribution
    μ_bound,   σ_bound              # bound reference distribution
```

Listing 3 Determination of the (un)bound likelihoods.

  a. *This yields a distribution closely matching the unbound distribution*
  b. *We assume that the majority of the blockades are active*
  c. *The mean is denoted as* $\mu_{unbound^*}$
  d. *The standard deviation is denoted as* $\sigma_{unbound^*}$
3. With each blockade from step 1 determine whether the unbound state is actually present, reject all where this is not the case (Fig. 4A)
  a. *Compare the signal between the 5 % and 95 % percentile to reject spikes and dips*
    i. *This means that we don't look at the lowest 5 % and highest 5 % of the signal*
4. Re-fit a normal distribution around the accepted blockades to yield the unbound distribution (Fig. 4B)
  a. *Normalization to a total area of 1 gives* $P(unbound)$
  b. *The mean is* $\mu_{unbound}$ *with standard deviation* $\sigma_{unbound}$
5. Pre-process the traces with saturating 4096 nM thiamine added according to protocol 3.1 and 3.2
6. Join all blockades and fit a normal distribution around the total signal
  a. *This yields a distribution closely matching the ligand bound distribution*
  b. *We assume that the majority of the blockades are active*
7. With each blockade from step 5 determine whether the bound state is actually present, reject all where this is not the case (Fig. 4C)
  a. *Compare the signal between the 5 % and 95 % percentile to reject spikes and dips*
    i. *This means that we don't look at the lowest 5 % and highest 5 % of the signal*

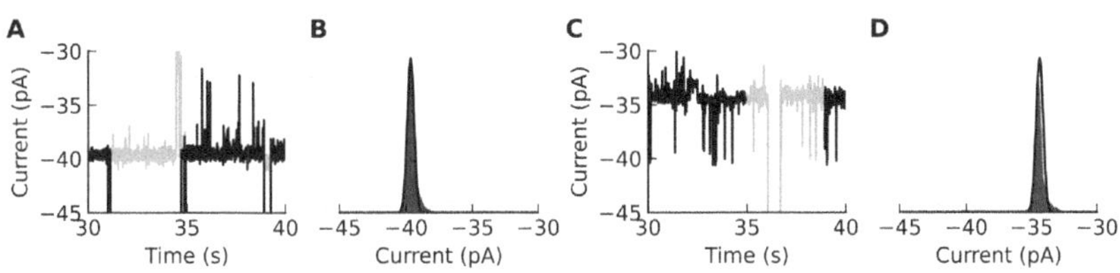

Fig. 4 Trace results to determine $P(unbound)$ and $P(bound)$. (A) Trace of the unbound state of Y27A-TbpA with active protein blockades (black) and inactive blockades (grey). (B) Histogram of the current observed from the active protein of the unbound state of Y27A-TbpA with a fitted normal distribution on top ($P(unbound)$). (C) Trace of the saturated state of Y27A-TbpA with active protein blockades (black) and inactive blockades (grey). (D) Histogram of the current observed from the active protein of the saturated state of Y27A-TbpA with a fitted normal distribution on top ($P(bound)$).

8. Re-fit a normal distribution around the accepted blockades to yield the bound distribution (Fig. 4D)
   a. *Normalization to a total area of 1 gives P(bound)*
   b. *The mean is* μ_{bound} *with standard deviation* σ_{bound}

Listing 3: Determination of the (un)bound likelihoods

3.4 Segmenting active from inactive blockades

With the unbound and bound reference distributions established (protocol 3.3), the individual blockades of a recording can be now classified. Not every blockade contains active protein: a pore may instead trap inactive protein or a contaminant producing a blockade that matches neither reference state. Because only active protein reports on thiamine, these blockades must be removed before the bound fraction is estimated. Here each blockade is tested against the unbound and bound reference distributions using a χ^2 likelihood, retaining only those consistent with the expected two-state (bound/unbound) behaviour and rejecting the rest (Listing 4).

1. Pre-process the traces according to protocol 3.1 and 3.2
2. Run the RED algorithm (protocol 3.2) to segregate the protein blockades
3. For each protein blockade run the RED algorithm (protocol 3.2), but now using $\mu_{unbound}$ and $\sigma_{unbound}$ (protocol 3.3)
   a. *This picks the theoretical levels*
4. For all data in the lower levels (unbound) calculate the χ^2 likelihood relative to $\mu_{unbound}$ and $\sigma_{unbound}$ (protocol 3.3) with total length of the unbound levels – 3 degrees of freedom
5. For all data in higher levels (bound) calculate the χ^2 likelihood relative to μ_{bound} and σ_{bound} (protocol 3.3) with total length of the bound levels – 3 degrees of freedom
6. If any $1 - \chi^2 CDF$ (bound and/or unbound) has a value of more than 0.95, accept the blockade as active, else reject
   a. The result will look similar to Fig. 5

Listing 4: Active blockades segmentation pseudocode

3.5 Calibration

Once the unbound and bound reference distributions are known the sensor must be calibrated so that a measured bound fraction can be

```
INPUT
    signal                           # signal at one concentration
    baseline_model = (μ_0, σ_0)      # unbound reference
    event_model    = (μ_1, σ_1)      # bound reference
    sampling_period

PROCEDURE
    events  = Segment(signal)        # RED segmentation, Listing 2
    results = []

    for event in events:
        if event has no valid baseline:
            continue

        baseline = signal[event.baseline]
        blockade = signal[event.blockade]

        # search for subevents inside the blockade
        subevents = Segment(blockade,
                            threshold = (μ_0, 3 * σ_0))

        baseline_χ2, event_χ2          = 0, 0
        baseline_points, event_points = 0, 0

        for sub in subevents:
            baseline_segment = blockade[sub.baseline]
            event_segment    = blockade[sub.event]

            baseline_χ2 += χ²(baseline_segment, baseline_model)
            event_χ2    += χ²(event_segment,    event_model)

            baseline_points += length(baseline_segment)
            event_points    += length(event_segment)

        # determine which model provides the strongest evidence
        if baseline_points == 0 and event_points == 0:
            p = 0
        else if baseline_points == 0:
            p = 1 - X²CDF(event_χ2    / event_points,
                          event_points - 3)
        else if event_points == 0:
            p = 1 - X²CDF(baseline_χ2 / baseline_points,
                          baseline_points - 3)
        else:
            # compare using the model with the largest amount of data
            if baseline_points >= event_points:
                p = 1 - X²CDF(baseline_χ2 / baseline_points,
                              baseline_points - 3)
            else:
                p = 1 - X²CDF(event_χ2    / event_points,
                              event_points - 3)

        accepted = (p > 0.95)

        results.append({event, baseline, blockade,
                        baseline_χ2, event_χ2,
                        baseline_points, event_points,
                        p, accepted})

OUTPUT
    results                          # per-event χ² evidence and accept/reject flag
```

Listing 4 Active blockades segmentation pseudocode.

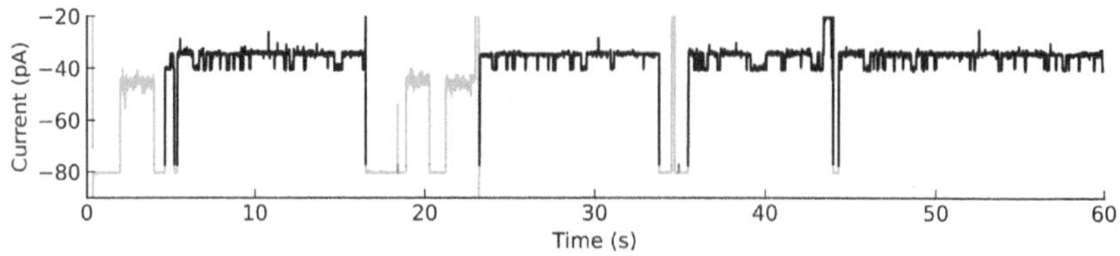

Fig. 5 Segmented nanopore trace. Rejected protein blockades are shown in grey, and accepted active protein is shown in black.

translated into thiamine concentration. The goal of this section is to build the calibration curve from which concentrations will be extrapolated further on (Listing 5).

1. Perform protocol 3.3, which yields the blockades and $P(unbound)$ and $P(bound)$
2. Perform protocol 3.4 to obtain the active protein blockades
3. Multiply each histogram with $P(unbound)$ and separately $P(bound)$
   a. *This yields the marginal likelihoods* $P(unbound|\mathcal{D})$ *and* $P(bound|\mathcal{D})$
4. Calculate the bound fraction (θ) using Eq. 4
5. Perform steps 1–4 for signal with increasing concentration of thiamine
6. Fit the Hill equation to the observed bound fractions for each concentration
   a. Use Eq. 8
   b. *There are many ways to determine fitting parameters. The Levenberg-Marquardt algorithm is implemented by default in most computing languages*
   c. *The x axis is the concentration, the y axis (to be fit) is the bound fraction*
   d. *Starting estimate for the parameters:*
      i. $a = 0.1$
         1. *Y27A-TbpA can also show the bound state without thiamine*
         2. *Ideally this parameter will converge to 0*
      ii. $b = 1$
         1. *If this parameter is larger than 1 it means that there is a scaling effect*
            a. *If it is much larger than 1 (i.e. 1.5) than something went wrong.*
         2. *If this parameter is less than 1 it means that Y27A-TbpA was not saturated in any concentration*
      iii. $K_d = 21\,nM$
         1. *Make sure that the* K_d *is entered with the same units as the concentration*

```
INPUT
    traces_by_conc              # active blockades per thiamine concentration
    concentrations              # e.g. 0, 2, ... 4096 nM (non-linearly spaced)
    P_unbound, P_bound          # reference distributions (protocol 3.3)
    a0, b0, Kd0, n = 0.1, 1, 21 nM, 1   # initial Hill parameters

PROCEDURE
    # --- bound fraction from one set of blockades (shared with 3.6) ---
    function BoundFraction(blockades):
        pooled        = Join(blockades)
        counts, edges = Histogram(pooled, FreedmanDiaconis(pooled))
        m_unbound     = Σ(counts * P_unbound)    # ∝ P(unbound | D)
        m_bound       = Σ(counts * P_bound)      # ∝ P(bound   | D)
        θ             = m_bound / (m_unbound + m_bound)    # Eq. 4
        return θ

    # one bound fraction per concentration
    θ_list = []
    for c in concentrations:
        blockades = ActiveBlockades(traces_by_conc[c])    # protocols 3.3 + 3.4
        θ_list += BoundFraction(blockades)

    # fit the corrected Hill equation θ(L) = a + b · L^n / (Kd + L^n)
    (a, b, Kd) = LevenbergMarquardt(model = θ(concentrations; a, b, Kd, n),
                                    data  = θ_list,
                                    init  = (a0, b0, Kd0))

OUTPUT
    a, b, Kd                    # calibration parameters
    θ_list                      # measured calibration curve
```

Listing 5 Calibration curve construction pseudocode.

3.6 Determination of the concentration

With the unbound and bound reference distributions, and corresponding calibration curve we can determine the concentration in each (new) signal, which we here divide in signals of 1 s long (for real-time detection), however instead a long trace can be loaded to yield the same result. The goal of this section is to determine the concentration and its corresponding error (Listing 6).

1. Segment active blockade signals into snippets of 1 s
   a. Snippet time can be adjusted depending on the event rate
2. Create a histogram of each snippet
3. Multiply each histogram with $P\,(unbound)$ and separately $P\,(bound)$
   a. *This yields the marginal likelihoods* $P\,(unbound\,|\,\mathcal{D})$ *and* $P\,(bound\,|\,\mathcal{D})$
4. Calculate the bound fraction (θ) for each snippet using Eq. 4
5. Calculate the bound concentration for each snippet
   a. Perform protocol 3.5 to yield a calibration curve
6. Average the concentration from the snippets using the weighted average and standard deviation
   a. *Weights are the inverse of the absolute value of the derivative of the Hill equation*

```
INPUT
    signal                          # recording of an unknown sample
    P_unbound, P_bound              # reference distributions (protocol 3.3)
    a, b, Kd, n                     # calibration parameters (protocol 3.5)
    snippet_length = 1 s            # adjust to the event rate
    window = 20..30 s               # moving window, real-time mode only

PROCEDURE
    function SnippetConcentration(snippet):
        θ     = BoundFraction([snippet])               # shared with 3.5, Eq. 4
        L     = ((θ - a) * Kd / (a + b - θ))^(1/n)     # Eq. 8
        dLdθ  = HillDerivative(θ, a, b, Kd, n)         # Eq. 10
        w     = |dLdθ|^(-1)                             # weight = inverse derivative
        return L, w

    blockades = ActiveBlockades(PreProcess(signal))    # protocols 3.3 + 3.4
    snippets  = Split(blockades, snippet_length)

    L_list, w_list = [], []
    for s in snippets:
        L, w = SnippetConcentration(s)
        L_list += L
        w_list += w

    L_mean = Σ(w_i * L_i) / Σ(w_i)
    σ_w    = sqrt( Σ(w_i * (L_i - L_mean)^2) / Σ(w_i) )     # weighted std

    # real-time variant: recompute L_mean, σ_w over the last `window` of snippets

OUTPUT
    L_mean                          # concentration estimate
    σ_w                             # associated error
```

Listing 6 Concentration determination pseudocode with error calculation for a single measurement.

b. *Calculate each concentration using* Eq. 8

c. *Calculate the weight of each concentration using the inverse of* Eq. 10

d. *When performing real-time, update the weights and concentration using a moving window of approximately 20–30 s*

e. *The standard deviation in the obtained concentration can be calculated using* Eq. 14

   i. *Note that* Listing 6 *considers the error of one measurement* (Eq. 13 *is not defined due to division by 0), however,* Eq. 14 *proper should be used when calculating the error across multiple measurements*

3.7 Event rate extraction & determination of K_d

This section describes an alternative, kinetics-based method to obtain the same quantitative information as in Section 3.5. Instead of working from the bound fraction, it extracts on and off rates directly from dwell times, and recovers K_d from their ratio. This can be done with two objectives in mind. First, to provide an independent cross-check on the K_d and bound fractions found by calibration and second to expose the underlying binding kinetics (Fig. 6).

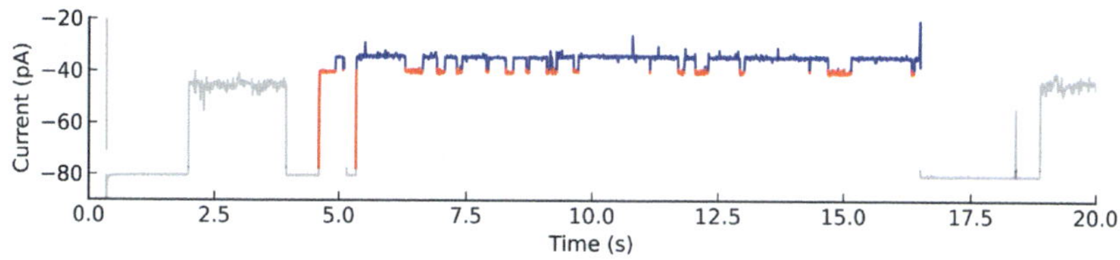

Fig. 6 Detection of bound and unbound parts of active protein. Parts that were rejected to contain active protein are shown in grey, active protein is shown in red and blue, where red represents the unbound level and blue the bound level.

1. Perform protocol 3.3
2. Perform protocol 3.4 to obtain the active protein blockades
3. For each protein blockade run the RED algorithm (protocol 3.2), but now using $\mu_{unbound}$ and $\sigma_{unbound}$ (protocol 3.3)
4. The on rate is the inverse of the average length (in seconds) of the lower (unbound) level
5. The off rate is the inverse of the average length (in seconds) of the high (bound) level
6. Determine the K_d from the k_{on} and k_{off} rate constants (see Section 2). These should correspond to the parameters determined in protocol 3.5

4. Conclusions

We have described a complete pipeline to convert raw nanopore recordings into a thiamine concentration with an associated error. The workflow proceeds from pre-processing (data cleanup, offset correction and filtering), through segmentation of the protein blockades from the baseline by the RED algorithm, to classification of active and inactive blockades, and finally to the extraction of the bound fraction and its conversion into thiamine concentration through a calibration curve.

The central methodological choice is to shift the analysis from a time-dependent to state-dependent problem. Rather than accumulating thousands of transitions to estimate on and off rates we treat each measured current distribution as evidence for bound or unbound state and combine it with reference distributions though empirical Bayesian inference. This allows to extract a bound fraction, and hence a concentration from far shorter recordings, which is needed for any device that was intended to measure concentrations in parallel and in short periods of time. Additionally, a kinetics-based route recovers the same K_d independently and serves as a cross-check. But because dwell times are exponentially distributed it carries a larger error and it is not recommended for routine quantification.

Finally, error estimation is built into the method rather than added afterwards: the derivative of the calibration curve weights each estimate toward the region where the sensor is most informative, and the law of total variance separates the uncertainty expected within a single pore from that accumulated across pores.

Acknowledgements

S.B.R., J.P.D., and C.W. were supported by sectorplan 2.0. G.C.C. was supported by the Research Foundation–Flanders (11Q4E24N). F.L.R.L. was supported by the Dutch Research Council (NWO) file number VI.Veni.242.400 of the research programme VENI under the grant [grant ID https://doi.org/10.61686/GKPYM97413].

Molecular graphics and analyses were performed with UCSF ChimeraX, developed by the Resource for Biocomputing, Visualization, and Informatics at the University of California, San Francisco, with support from National Institutes of Health R01-GM129325 and the Office of Cyber Infrastructure and Computational Biology, National Institute of Allergy and Infectious Diseases.

References

Galenkamp, N. S., Soskine, M., Hermans, J., Wloka, C., & Maglia, G. (2018). Direct electrical quantification of glucose and asparagine from bodily fluids using nanopores. *Nature Communications, 9*(1), 4085. https://doi.org/10.1038/s41467-018-06534-1.

Jo, S., Kim, T., & Im, W. (2007). Automated builder and database of protein/membrane complexes for molecular dynamics simulations. *PLoS ONE, 2*(9), e880. https://doi.org/10.1371/journal.pone.0000880.

Jo, S., Kim, T., Iyer, V. G., & Im, W. (2008). CHARMM-GUI: A web-based graphical user interface for CHARMM. *Journal of Computational Chemistry, 29*(11), 1859–1865. https://doi.org/10.1002/jcc.20945.

Li, P., Liang, D., Yang, E., Zeb, M., Huang, H., Sun, H., ... Ma, W. (2024). Bio-nanopore technology for biomolecules detection. *Advanced Biotechnology, 2*(4), 45. https://doi.org/10.1007/s44307-024-00051-7.

Lucas, F. L. R., Piso, T. R. C., van der Heide, N. J., Galenkamp, N. S., Hermans, J., Wloka, C., & Maglia, G. (2021). Automated electrical quantification of vitamin B1 in a bodily fluid using an engineered nanopore sensor. *Angewandte Chemie International Edition, 60*(42), 22849–22855. https://doi.org/10.1002/anie.202107807.

Montal, M., & Mueller, P. (1972). Formation of bimolecular membranes from lipid monolayers and a study of their electrical properties. *Proceedings of the National Academy of Sciences, 69*(12), 3561–3566. https://doi.org/10.1073/pnas.69.12.3561.

Peng, W., de Souza Santos, M., Li, Y., Tomchick, D. R., & Orth, K. (2019). High-resolution cryo-EM structures of the E. coli hemolysin ClyA oligomers. *PLOS ONE, 14*(5), e0213423. https://doi.org/10.1371/journal.pone.0213423.

Pettersen, E. F., Goddard, T. D., Huang, C. C., Meng, E. C., Couch, G. S., Croll, T. I., ... Ferrin, T. E. (2021). UCSF ChimeraX: Structure visualization for researchers, educators, and developers. *Protein Science, 30*(1), 70–82. https://doi.org/10.1002/pro.3943.

Ramachandran, K. M., & Tsokos, C. P. (2021). *Bayesian estimation and inference. Mathematical statistics with applications in R*. Elsevier, 415–459. https://doi.org/10.1016/B978-0-12-817815-7.00010-5.
Soriano, E. V., Rajashankar, K. R., Hanes, J. W., Bale, S., Begley, T. P., & Ealick, S. E. (2008). Structural similarities between thiamin-binding protein and thiaminase-I suggest a common ancestor. *Biochemistry, 47*(5), 1346–1357. https://doi.org/10.1021/bi7018282.
Soskine, M., Biesemans, A., De Maeyer, M., & Maglia, G. (2013). Tuning the size and properties of ClyA nanopores assisted by directed evolution. *Journal of the American Chemical Society, 135*(36), 13456–13463. https://doi.org/10.1021/ja4053398.

SECTION 2

Riboswitches

Scaffold engineering enables high-resolution cryo-EM analysis of thiamine pyrophosphate riboswitch architecture and dynamics

Madison M. Maille[a], Boris Rudolfs[b], Daniel B. Haack[b], and Navtej Toor[b,*]
[a]Department of Chemistry, University of California, San Diego, La Jolla, CA, United States
[b]Department of Biochemistry and Molecular Biophysics, University of California, San Diego, La Jolla, CA, United States
*Corresponding author. e-mail address: ntoor@ucsd.edu

Contents

Abstract

Cryogenic electron microscopy (cryo-EM) is well suited for studying dynamic macromolecules, but has historically been incompatible with small protein-free RNAs like thiamine pyrophosphate (TPP) riboswitches. This chapter describes a method utilizing an RNA scaffold derived from the *Oceanobacillus iheyensis* group II intron that overcomes these barriers, allowing for high-resolution reconstructions of an embedded target RNA. This approach produced the first reconstruction of the native open conformation of the apo *E. coli* TPP riboswitch aptamer, as well as a high-resolution reconstruction of the bound state. The adaptability of the scaffold design allows for any target RNA sequence to easily be embedded for structure determination by cryo-EM. This chapter outlines protocols for construct design parameters, sample preparation, and the unique data processing steps

Methods in Enzymology, Volume 735
ISSN 0076-6879, https://doi.org/10.1016/bs.mie.2026.06.024

required to use this scaffold to characterize the structures of TPP riboswitch aptamer variants, folding intermediates, and bound to alternate small-molecule aptamer binders.

1. Introduction

The thiamine pyrophosphate (TPP) riboswitch is one of the most abundant classes of riboswitches that spans the genomes of bacteria, plants, fungi, and archaea (McCown et al., 2017; Mukherjee et al., 2018). This RNA element uses the binding of its ligand TPP to regulate the expression of genes essential for thiamine biosynthesis and metabolism (Winkler et al., 2002). Its prevalence in bacteria and absence in the human genome make it a highly studied antibiotic target (Panchal & Brenk, 2021; Hübenthal et al., 2025). The basic riboswitch architecture consists of an aptamer domain and an expression platform (Garst et al., 2011). The aptamer is responsible for binding of the target ligand, folding into a complex 3D structure that shifts between an open (apo) and closed (bound) state (Winkler & Breaker, 2003). This conformational change is an important step for activating a response in the expression platform that controls the expression of the associated gene (Winkler & Breaker, 2003). The sequence and structure of the aptamer core are highly conserved to allow for TPP binding, but variability in other stems across species can contribute to differences in overall activity and tertiary interactions (Antunes et al., 2019; Subki et al., 2020). Understanding how these differences in structure impact ligand binding and overall dynamics can aid in the structure-guided design of antibiotic therapeutics that target this important bacterial regulatory element.

Despite its therapeutic relevance, methods for capturing the 3D structure of TPP riboswitches have been limited to X-ray crystallography. With this method, bound and apo structures of the *Arabidopsis thaliana* TPP riboswitch have been resolved at high-resolution (Thore et al., 2006). However, the apo structures showed only minor conformational changes from the bound state, which is not reflective of small-angle x-ray scattering (SAXS) reconstructions and biochemical probing experiments performed with the RNA in solution (Lee et al., 2023; Olenginski et al., 2024). This is likely due to the constraining conditions that exist within the narrow confines of the crystal lattice, restricting the RNA from fully sampling the native dynamics of this riboswitch class. As a result, no riboswitch was ever captured in its native unbound conformation (Huang et al., 2010; Jenkins et al., 2011; Schroeder et al., 2020; Stagno et al., 2017; Stoddard et al., 2010). Cryogenic electron microscopy (cryo-EM) is an alternative method that is better suited for studying dynamic macromolecules due to there being no requirement for the crystallization of a target (Bonilla & Kieft, 2022). However, it has been historically difficult to work with protein-free RNA samples due to denaturation during the vitrification process, particle aggregation on grids, and/or orientation bias (Bonilla & Jang, 2024). These challenges are further amplified as the molecule size decreases, typically rendering this method unsuitable for small RNAs like TPP riboswitches.

Our lab developed an RNA scaffold that overcomes the typical barriers associated with cryo-EM to enable high-resolution reconstructions of attached target RNAs. The scaffold is derived from a highly structured group II intron from *Oceanobacillus iheyensis* and contains a solvent-exposed stem where target RNAs can be covalently embedded (Toor et al., 2008; Haack et al., 2025). The large size and stability of the *O. iheyensis* intron allows for smaller RNAs to be studied with cryo-EM, with the favorable biophysical qualities of the scaffold translating to the attached RNAs. When developing and testing the scaffold, the *E. coli* TPP riboswitch aptamer domain was used as a control of known structure (Haack et al., 2025; Serganov et al., 2006). Attaching the P1 stem of the aptamer domain to the scaffold enabled a 2.4 Å reconstruction of the bound RNA. Alignment with the solved crystal structure and secondary structure probing of the scaffolded riboswitch with SHAPE-MaP confirmed that attachment to the scaffold didn't interfere with native folding. An additional benefit of using the scaffold was that an open conformation of the apo TPP riboswitch could be solved at a resolution of 6 Å with the scaffolded RNA in the absence of TPP. While this structure did not reach the same resolution as the bound state, the maps provided enough detail to model this open conformation, and resolution could be further improved with additional data collection for a higher particle count. Having the ability to capture these open conformations enables structural studies of ligand recognition steps and determining intermediate folds that support this mechanism of regulating gene expression. Additional testing of the scaffold with other RNAs has confirmed its ability to consistently obtain high-resolution protein-free RNA structures. This provides support that the method can be applied to study cross-species structural variation of other TPP riboswitches, new small-molecule aptamer binders for targeting, or folding intermediates to study riboswitch dynamics. The following

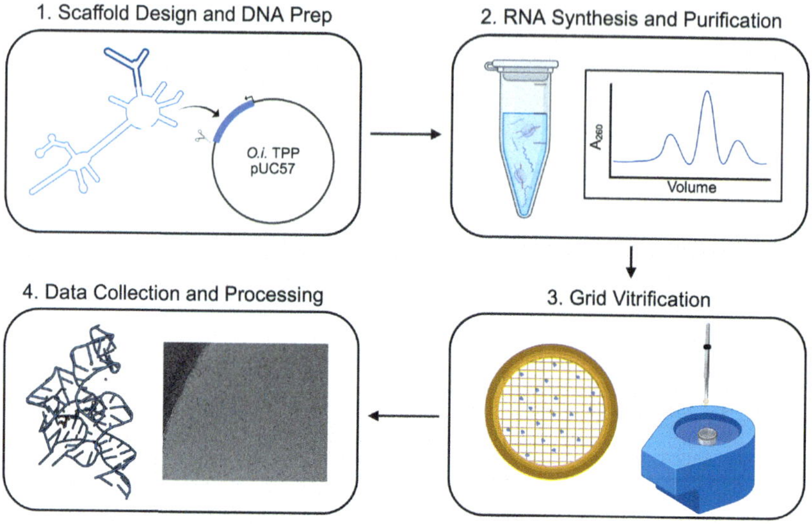

Fig. 1 General scaffold-assisted cryo-EM workflow. The major steps involve designing the scaffolded construct and preparing the DNA template for transcription followed by synthesizing and purifying the RNA. The prepared RNA sample can then be frozen on grids, which will be used for cryo-EM data collection and processing.

sections will outline construct design parameters, sample preparation, and the unique processing protocol required to apply this method to these other applications (Fig. 1).

2. Methods

2.1 Designing constructs with the scaffold and preparing the DNA template for transcription

2.1.1 Rationale

This first part of the protocol will outline the design considerations for creating scaffolded cryo-EM constructs with a TPP riboswitch. The *O. iheyensis* scaffold contains a solvent-exposed stem (domain III) that can be used for attachment of the TPP riboswitch by its P1 stem (Haack et al., 2025). Since this aptamer stem is conserved across TPP riboswitches, our lab's existing scaffold design can be readily adapted for mutant or cross-species variants by simply substituting the riboswitch sequence. The existing design directly embeds the TPP riboswitch sequence into the scaffold sequence, creating a single transcript (Fig. 2). This construct can be ordered as a gene fragment or plasmid, but the protocol below will outline the use of blunt end cloning to generate the DNA template.

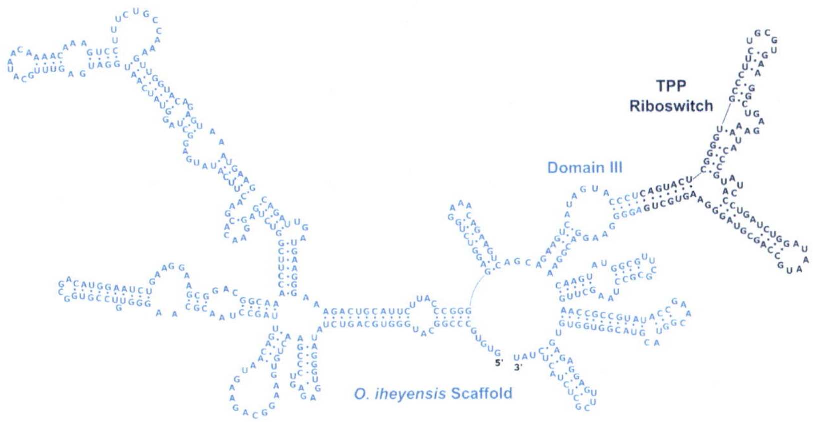

Fig. 2 RNA scaffold design with the TPP riboswitch. The sequence and secondary structure of the *O. iheyensis* scaffold (light blue). The *E. coli* TPP riboswitch (dark blue) sequence is directly embedded within a solvent-exposed stem (domain III) of the scaffold.

2.1.2 Equipment and materials

- Polystyrene petri dishses (VWR; Cat. no. 25384-342)
- Gene Pulser Xcell electroporation system (Biorad; Cat. no. 1652660)
- Gene Pulser/MicroPulser Electroporation Cuvettes, 0.2 cm gap (Biorad; Cat. no. 1652086)
- Incubator
- Round bottom culture tubes with closures (VWR; Cat. no. 60818-725)
- Cryovials (Simport Scientific; Cat. no. T311-2)
- Beveled flask
- Micropipettes and tips
- Nalgene® centrifuge bottle (Sigma-Aldrich; Cat. no. B1033)
- Floor standing centrifuge (Beckman Coulter)
- Water bath
- Fume hood
- NanoDrop 2000c spectrophotometer (Thermo Scientific)

2.1.3 Reagents

- Gene fragment DNA (Genscript)
- pUC57 plasmid (Genscript; Cat. no. SD1176)
- T4 DNA ligase (400,000 units.mL^{-1}; Thermo scientific; Cat. no. EL0011)
- 10X T4 DNA ligase buffer (Thermo scientific; Cat. no. EL0011)
- *Eco*RV-HF (20,000 units.mL^{-1}; New England Biolabs; Cat. no. R3195S)

- Linear polyacrylamide (LPA; Thermo scientific; Cat. no. J67830. XF)
- Sodium acetate trihydrate (Sigma-Aldrich Inc.; Cat. no. 236500)
- Milli-Q water
- DH5α Electrocompetent Cells (Invitrogen™; Cat. no. 11319019)
- LB broth miller (Fisher Scientific; Cat. no. BP1426)
- LB agar miller powder (Fisher Scientific; Cat. no. BP1426-500)
- Carbenicillin disodium salt (Sigma-Aldrich Inc.; Cat. no. C1389)
- ZymoPURE™ II Plasmid Maxiprep Kit (Zymo Research; Cat. no. D4202)
- *Bam*HI-HF (20,000 units.mL^{-1}; New England Biolabs; Cat. no. R3136S)
- 10X rCutSmart™ buffer (New England Biolabs; Cat. no. R3136S)
- UltraPure™phenol:chloroform:isoamyl alcohol (25:24:1, v/v) mixture (Invitrogen; Cat. no. 15593031)
- Ethanol (EtOH) 200 proof (Thermo scientific; Cat. no. T038181000CS)

2.1.4 Protocol

1. Order gene fragment (or plasmid if preferred) for TPP riboswitch inserted into domain III of the *O. iheyensis* scaffold. Sequence of desired TPP riboswitch variant or mutant used for study can be replaced with the bold portion of the sequence in the example construct below.
   a. The gene fragment design includes a T7 promoter (italicized) for transcription initiation upstream of the scaffold sequence and a restriction enzyme recognition sequence (underlined), which will be used to linearize the DNA, placed downstream. We have used *Bam*HI, but the restriction enzyme selection can be changed to a different, preferred enzyme (just ensure it does not have additional cut sites within the scaffold or target sequences). The sequence for the TPP riboswitch is bolded.
   b. *TAATACGACTCACTATAGG*GTTATGTGTGCCCGGCATGGGTGCAG TCTATAGGGTGAGAGTCCCGAACTGTGAAGGCAGAAGTAACA GTTAGCCTAACGCAAGGGTGTCCGTGGCGACATGGAATCTGAAG GAAGCGGACGGCAAACCTTCGGTCTGAGGAACACGAACTTCA TATGAGGCTAGGTATCAATGGATGAGTTTGCATAACAAAACAAA GTCCTTTCTGCCAAAGTTGGTACAGAGTAAATGAAGCAGATTG ATGAAGGGAAAGACTGCATTCTTACCCGGGGAGGTCTGAGCTTT CGAGCTCAGAAGTCAGCAGAAGTCATAGTACCCU**CAGUACUCGG GGUGCCCUUCUGCGUGAAGGCUGAGAAAUACCCGUAUCACCUGAU CUGGAUAAUGCCAGCGUAGGGAAGUGCUG**AGGGGAAGGACGGAA

CAAGTATGGCGTTCGCGCCATGCTTGAACCACCGTATACCGAAC
GGTACGTACGGTGGTGTGGATCC

2. Ligation of the gene fragment into a pUC57 vector.
   a. Dissolve lyophilized gene fragment to a concentration of 25 ng.μL^{-1}.
   b. Set up a one-step digestion-ligation reaction for blunt end cloning of the gene fragment into a pUC57 vector. Combine the following components in a single tube.

Component	Volume (μL)	Final Concentration
T4 DNA ligase buffer (10X)	1	1X
pUC57 plasmid (89 ng.μL^{-1})	0.5	4.4 ng.μL^{-1}
T4 DNA ligase (400,000 units.mL^{-1})	0.5	200 units
*Eco*RV-HF (20,000 units.mL^{-1})	0.5	10 units
Gene fragment (25 ng.μL^{-1})	7.5	18.75 ng.μL^{-1}

   c. Incubate the reaction at room temperature for 24 h.
3. Perform an ethanol precipitation to clean up the digestion-ligation product.
   a. In a new tube, combine the following components.

Component	Volume (μL)	Final Concentration
LPA (5 μg.μL^{-1})	2	0.5 ng.μL^{-1}
Digestion-ligation reaction	10	
Sodium acetate (3 M)	2	0.3 M
Milli-Q water	6	

   b. To the tube, add 50 μL (2.5X volume) of 100 % ethanol.
   c. Place the tube in a −80 °C freezer for 10 min and begin cooling down a temperature-controlled centrifuge to 4 °C.
   d. Spin down the tube at full speed and 4 °C for 10 min.
   e. Gently decant, leaving the pellet undisturbed.
   f. Add 100 μL of 75 % ethanol and let the tube sit at room temperature for 1 min.

   g. Decant completely, leaving the pellet undisturbed, and ensure that no ethanol remains in the tube.
   h. Resuspend pellet with 20 μL of Milli-Q water. The sample can then be used in a transformation or stored in −20 °C for later use.
4. Transform the plasmid into DH5α cells.
   a. Thaw a 25 μL aliquot of DH5α electrocompetent cells and cool an electroporation cuvette on ice.
   b. Add 5 μL of resuspended plasmid (50–200 ng) from step 3 h to the aliquot of cells, gently mix with pipette, and move mixture to the cuvette.
   c. Electroporate cells at a voltage of 2.5 kV, capacitance of 25 μF, and resistance of 200 Ω for the 2 mm cuvette.
   d. Add 1 mL of pure LB to the cells in the cuvette and transfer to a new microcentrifuge tube.
   e. Let the sample recover, shaking for 1 h at 37 °C.
   f. Plate 500 μL of recovered cells on LB carbenicillin plates (100 μg.mL^{-1}).
   g. Incubate plate overnight for 18 h at 37 °C.
5. The next day, take the plate out of the incubator and set up a starter culture. In a culture tube, combine 1 μL of carbenicillin (100 mg.mL^{-1}) with 1 mL of pure LB. Pick a single colony with a P10 pipette tip and inject it into the culture tube. Let culture shake at 37 °C for 8 h or until cloudy. Send off a sample for sequencing (Quintara Biosciences) and use the remainder to create a glycerol stock in a cryovial using a 1:1 ratio of starter culture to 80 % glycerol. Store the glycerol stock in a −80 °C freezer.
6. Once successful sequencing results are obtained, autoclave 250 mL of LB in a beveled flask for a maxiprep. At the end of the day, add 250 μL of carbenicillin (100 mg.mL^{-1}) to the flask with LB and inoculate the culture with a scoop of the glycerol stock. Grow the culture overnight for 18 h in a 37 °C shaker.
7. The following morning, transfer culture to a centrifuge bottle and spin down at 4 °C and 5000 g for 10 min. Pour off the LB and use the cells in the pellet to perform a maxiprep, following the kit's instructions.
8. Linearize the eluted maxiprep DNA with *Bam*HI-HF (or enzyme corresponding to engineered cut site).
   a. Set up the digestion reaction by combining the following components.

Component	Amount	Final concentration
Maxiprep DNA	300 μg	0.67 $\mu g.\mu L^{-1}$
Cutsmart buffer (10X)	45 μL	1X
*Bam*HI-HF (20,000 $units.mL^{-1}$)	30 μL	600 units
Milli-Q water	to 450 μL	

b. Incubate reaction at 37 °C for 2 h.

9. Clean up linear DNA with phenol-chloroform extraction and ethanol precipitation.

a. Add 50 μL of 3 M sodium acetate to the digest reaction and mix. Then add 500 μL of phenol:chloroform:isoamyl alcohol (25:24:1, v/v) to the sample.

i. Handle phenol:chloroform:isoamyl alcohol with caution and perform these steps in a fume hood. Generate proper waste containers for phenol-contaminated tips and tubes.

b. Vortex for 1 min and spin the sample down in a centrifuge for 5 min at full speed and room temperature.

c. Gently remove 300 μL of solution from the top, aqueous layer of the sample and move to a clean microcentrifuge tube.

i. Avoid touching the bottom layer or side of the tube with the pipette tip when performing this step to prevent phenol contamination in your final sample.

d. Add 750 μL of 100 % ethanol to the extracted sample and vortex.

e. Let the tube sit in a −80 °C freezer for 10 min and begin cooling the centrifuge to 4 °C.

f. Spin the sample down in a cooled centrifuge for 10 min at full speed.

g. Decant without disturbing the pellet.

h. Add 400 μL of 75 % ethanol to the pellet and spin down for an additional 2 min at full speed and 4 °C.

i. Decant without disturbing the pellet.

j. Spin down at room temperature and full speed for 1 min. Remove any remaining ethanol from the tube.

k. Resuspend the pellet with 250 μL of Milli-Q water and use a NanoDrop spectrophotometer to measure the concentration at 260 nm of the purified, linear transcription template. Store the linear DNA sample at −20 °C until beginning to set up for an *in vitro* transcription reaction.

i. Check that the A260/280 ratio from the NanoDrop reading is over 1.80 to ensure that the sample has no remaining protein contamination. If the value is lower, re-do the phenol-chloroform extraction and ethanol precipitation steps.

2.2 Synthesizing and purifying RNA

2.2.1 Rationale

The next step involves transcribing the scaffolded construct to make the RNA. To produce enough RNA for the sample and grid preparation steps, a large-scale 1 mL *in vitro* transcription reaction is performed and purified using size exclusion chromatography. This purification method is optimal, as it will remove salts, NTPs, and RNA aggregates while preserving the RNA's native folding from transcription. Other methods, like buffer exchanging and concentrating in spin filters or using denaturing protocols, can be used if a fast protein liquid chromatography (FPLC) system with the appropriate column is not available. However, spin filters can cause some RNA samples to form a hydrogel and denaturing protocols require additional, more complicated refolding steps to be performed after. Once the RNA is purified, it can undergo gentle refolding and concentration.

2.2.2 Equipment and materials

- Micropipettes and tips
- Water bath
- ÄKTA pure™ 25 M chromatography system (Cytiva; Cat. no. 29018226)
- Superdex 200 prep grade SEC resin in HiLoad column, 16/600 (Cytiva; Cat. no. 28989335)
- Amicon Ultra Centrifugal Filter, 100 kDa MWCO (Millipore; Cat. no. UFC9100)
- Eppendorf™ Benchtop Centrifuge 5810R
- NanoDrop 2000c spectrophotometer (Thermo Scientific)

2.2.3 Reagents

- Linear DNA template
- Adenosine 5′-triphosphate disodium salt, ATP (MedChemExpress; Cat. no. HY-B0345A)
- Uridine 5′-triphosphate trisodium salt, UTP (MedChemExpress; Cat. no. HY-W013093)
- Guanosine 5′-triphosphate disodium salt, GTP (MedChemExpress; Cat. no. HY-W010737)

- Cytidine 5′-triphosphate disodium salt, CTP (MedChemExpress; Cat. no. HY-W013100)
- Milli-Q water
- Triton X-100 (Promega; Cat. no. H5142)
- Tris base (Fisher Scientific; Cat. no. BP152)
- Hydrochloric acid (HCl; Fisher Scientific; Cat. no. A144S-500)
- Magnesium chloride ($MgCl_2$) hexahydrate (Macron Fine Chemicals; Cat. no. MK5958-04)
- Spermidine (Sigma-Aldrich Inc.; Cat. no. 85558)
- Dithiothreitol (DTT; Fisher Scientific; Cat. no. BP172)
- Inorganic *E. coli* pyrophosphatase (100 units.mL^{-1}, New England Biolabs; Cat. no. M0361S)
- T7 RNA Polymerase (made in house, but can be purchased from Promega; Cat. no. P4074)
- Calcium chloride ($CaCl_2$) dihydrate (VWR; Cat. no. 97061-904)
- TURBO DNase (Invitrogen™; Cat. no. AM2238)
- Proteinase K (Thermo Scientific; Cat. no. EO0491)
- Sodium cacodylate trihydrate pH 6.5 (Hampton Research; Cat. no. HR2-737)

2.2.4 Protocol

1. Before beginning transcription reactions, prepare a nucleoside triphosphate (NTP) mixture containing 100 mM of each NTP (ATP, UTP, CTP, GTP) with a final concentration of 50 mM Tris and a pH of 7.5. Aliquot and store at −80 °C.
2. Thaw reagents for *in vitro* transcription (linear DNA template, NTPs, Triton X-100, DTT).
3. Set up a 1 mL *in vitro* transcription reaction containing the following, and adding the enzymes last.

Component	Amount	Final concentration
Linearized DNA template	50 μg	0.2 μg.μL^{-1}
NTP mixture (100 mM each)	50 μL	5 mM each
Triton X-100 (1 %)	50 μL	0.05%
Tris-HCl (pH 7.5, 1 M)	50 μL	50 mM
$MgCl_2$ (1 M)	25 μL	25 mM
Spermidine (1 M)	2 μL	2 mM

DTT (1 M)	5 μL	5 mM
E. coli inorganic pyrophosphatase (100 units.mL^{-1})	0.5 μL	0.05 units
T7 RNA Polymerase	5 μL	~400 units
Milli-Q water	to 1000 μL	

4. Incubate at 37 °C in a water bath for 3 h.
5. Add 10 μL of Turbo DNase and 12 μL of 100 mM $CaCl_2$ (final concentration of 1.2 mM).
6. Incubate at 37 °C in a water bath for an additional hour.
7. Add 10 μL of proteinase K.
8. Incubate at 37 °C in a water bath for a final hour.
9. Centrifuge sample to remove any remaining pyrophosphatase precipitation.
10. Load transcription onto FPLC with Superdex 16/600 pg 200 column pre-equilibrated with running buffer (4 mM $MgCl_2$, 5 mM sodium cacodylate).
    a. Sodium cacodylate is toxic and contains arsenic. Handle this reagent with caution and dispose of waste properly.
11. Run FPLC on appropriate method settings for the column.
12. Pool the appropriate fractions containing the scaffolded-RNA. When running scaffolded constructs under these buffer conditions and settings, the desired peak is expected to appear symmetrical with an elution volume around 50–60 mL (Fig. 3A and B).
    a. Since size exclusion chromatography also separates molecules based on folding composition, this step serves as a good indicator for whether a construct can yield a successful high-resolution reconstruction. From our experience, constructs with sharper symmetrical peaks for the scaffolded-RNA had less structural heterogeneity to produce better reconstructions. Despite many efforts at troubleshooting, we have not yet found a robust strategy to correct for an elution profile that does not contain a clear peak at the desired elution volume (Fig. 3C). Based on our previous studies, getting a result like this may indicate that the specific target is not compatible with the scaffold and might be better suited for alternative structural methods, such as x-ray crystallography or SAXS.
13. Gently refold RNA by incubating combined fractions in a 50 °C water bath for 30 min.

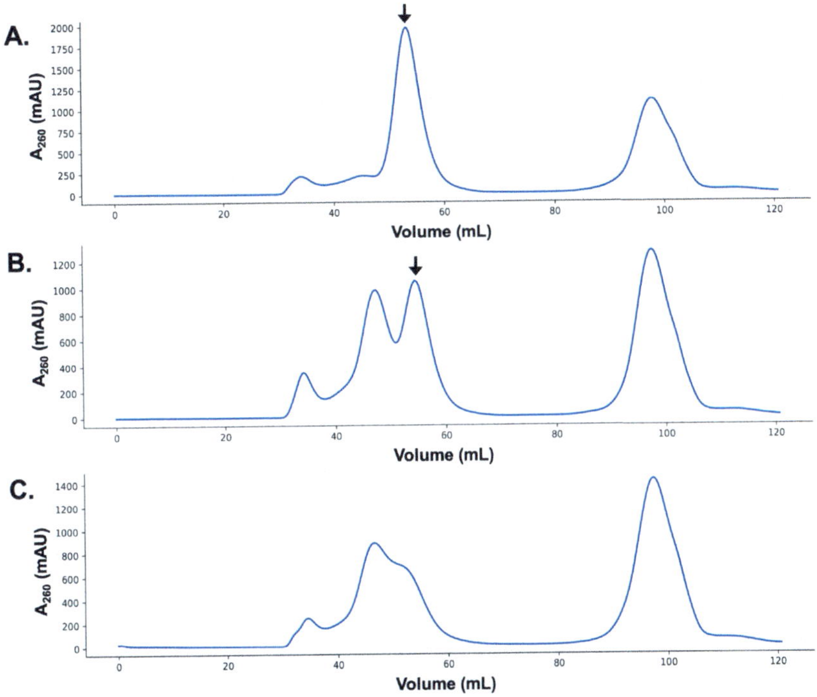

Fig. 3 Size exclusion chromatographs from RNA purification. Examples of elution profiles for purifying *in vitro* transcription reactions. The peak for the properly folded scaffolded construct is expected to elute at a volume around 40–50 mL (arrow pointing to peak). (A) A tall, symmetrical single peak around this elution volume is the ideal profile to look for that has been found to correlate with higher resolution reconstructions. (B) If additional aggregate peaks elute before, the fractions for the distinct desired peak can still be collected and used in the remaining steps of the protocol. (C) If no distinct peak is present and overlaps in with earlier peaks, it is not advised to continue moving forward with this transcription sample.

14. Let the sample gradually cool back to room temperature.
15. Concentrate the sample in 100 kD MWCO spin filters until the concentration reaches a range of 3–5 $\mu g.\mu L^{-1}$.
16. Continue with using the RNA for further sample preparation and vitrification steps or store the sample at 4 °C overnight to proceed with those steps the following day.

2.3 Sample preparation and vitrification

2.3.1 Rationale

After transcription and purification, buffer conditions can be adjusted and the RNA can be incubated with ligand if needed for the desired application. The scaffold can maintain its structured fold over a range of pH values and can tolerate varied buffers. Changing conditions, however, may require additional troubleshooting to ensure micrograph quality is not affected. A vitrification protocol using a manually operated plunging apparatus is described, but this can also be completed using automated methods, like a Vitribot or Leica.

2.3.2 Equipment and materials

- 4 °C cold room
- Manual plunging apparatus (self-constructed [Fig. 4A], but GraVitri Basic by MiTeGen is a similar unit that can be purchased)
- 4 L Liquid Nitrogen Storage Dewar (Scilogex/Taylor Wharton by IC Biomedical; Cat. no. 4LDB)
- PELCO easiGlow™ Glow Discharge Cleaning System (Ted Pella Inc.; Cat. no. 91000S)
- Quantifoil R 2/1 200 mesh, gold grids (Electron Microscopy Sciences; Cat. no. Q2100AR1)
- Hybrid Cryo Grid Box with pin lid (Electron Microscopy Sciences; Cat. no. 71167-56)
- Gripper Tool for Cryo Grid Boxes (Electron Microscopy Sciences: Cat. no. 71166-SP)

A. B.

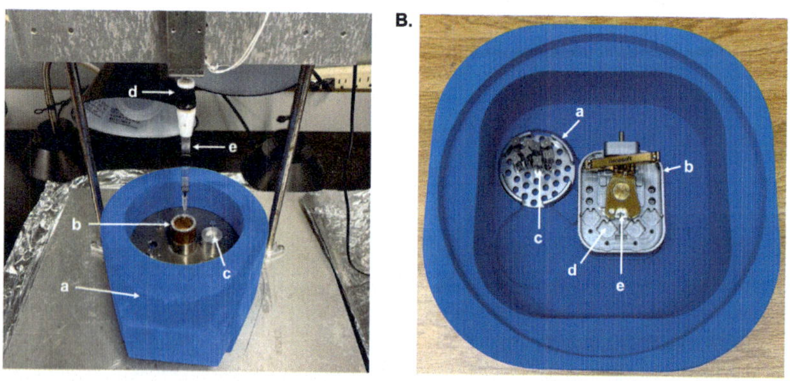

Fig. 4 Grid vitrification and clipping station set-up. (A) Manual plunging apparatus consists of foam dewar (a) with compartments to hold the cryogen cup (b) and grid boxes (c). The plunger (d) on the apparatus contains a magnet for attachment of the L5 tweezers with a clamp ring (e). (B) The clipping station foam dewar has compartments for the clipping tool puck rack (a) and clipping hub (b). The puck rack can hold the pre-loaded clipping tools (c). The clipping hub contains slots for securing grid boxes (d) and an autogrid pocket (e).

- Dumont Style L5 Tweezers (Electron Microscopy Sciences: Cat. no. 72882-D)
- Micropipettes and tips
- Whatman qualitative filter paper, Grade 1 (Cytiva; Cat. no. 1001-070)
- Dewar Flask (Scilogex Dilvac; Cat. no. SS111)
- Clipping station with clipping tools, long grid tweezers, and two clipping tool tweezers (Nanosoft materials; Cat. no. 25011001)
- Autogrid C-clips (Nanosoft materials; Cat. no. 11011002)
- Autogrid ring (Nanosoft materials; Cat no. 11021002)

2.3.3 Reagents

- Thiamine pyrophosphate (Sigma-Aldrich Inc.; Cat. no. C8754)
- Liquid Nitrogen (Airgas; Cat. no. NF230LT22)
- Hydrocarbon gas ethane:propane (37.1/62.9) mixture, liquefied, n.o.s. (Linde Gas & Equipment)

2.3.4 Protocol

1. Adjust buffer concentrations of the RNA sample to match desired application.
   a. The FPLC running buffer is sufficient to maintain native folding of the scaffold and attached riboswitch, but the conditions can be changed if needed. To get a structure of the bound riboswitch, add enough TPP, or chosen ligand, to reach a final concentration of 1 mM. Incubate the sample at room temperature for at least 15 min before moving on to the vitrification steps.
   b. It is important to consider how certain additives may impact micrograph quality. For example, we have seen that with other RNA samples, magnesium concentrations above 50 mM have impacted RNA quality on the grids, with increased aggregation leading to less individual particles that can be used for analysis.
2. Cool manual plunging apparatus with liquid nitrogen in a 4 °C cold room.
   a. Throughout the vitrification process, continue to check the liquid nitrogen level and keep extra liquid nitrogen stored in a 4 L dewar to refill when needed.
3. Fill a pre-chilled conical tube submerged in liquid nitrogen with enough cryogen (liquid propane/ethane mix) to fill the cryogen cup in the apparatus. Transfer liquid into the cup.
4. Secure a grid box in the apparatus to begin cooling down.

5. Glow discharge grids using a plasma current of 15 mA and vacuum pressure of 40 mBar for 30 s with the carbon/foil side facing up. We advise glow discharging enough to prepare at least 4 grids per sample.
6. Use tweezers to gently clamp the grid, grabbing as close to the edge as possible. Lightly shake the tweezers after securing the clamp to ensure that it has been properly secured.
7. Rest the tweezers on the edge of the benchtop with the grid facing up and pipette 3.5 μL of RNA sample onto the grid. Quickly attach the tweezers to the plunging apparatus.
8. Gently blot the grid with a piece of filter paper until observing the termination of wicking. Plunge into the cryogen to freeze the grid.
   a. Slightly modify blotting time for grids of the same sample to create grids with varied ice thickness for screening.
   b. If particles are populated mostly on the carbon rather than in the holes, try preparing grids with a more concentrated RNA sample. A double blotting approach can also be used, where another 3.5 μL of sample is reapplied to the grid and blotted again before plunging the grid into the cryogen.
9. Transfer into a grid box and repeat steps 6–8 with remaining grids.
10. Move full grid boxes to a secondary container (like a 50 mL conical tube) stored in a pre-chilled dewar flask filled with liquid nitrogen.
11. Before screening and data collection, the grids will also need to be clipped. To do this, begin by assembling the clipping station. Place the clipping hub and clipping tool puck rack in the clipping foam dewar (Fig. 4B).
12. Load one C-clip into the barrel of the clipping tool. Prepare one clipping tool per grid and place them in the puck rack of the clipping station.
13. Begin cooling the station down by filling the dewar with liquid nitrogen.
14. Once the station is cooled, transfer the grid boxes to the station and secure them using the grid box lid. At this time, also place the c-clips on the clipping hub to begin cooling down.
15. Place an autogrid ring in the autogrid pocket of the clipping hub and then transfer one of the grids on top and in the center of the ring.
16. Adjust the lever at the top of the clipping hub to shift the brass plate into the clipping position, where the grid and autogrid ring are positioned under the hole in the plate.

17. Insert the loaded clipping tool into the hole and depress the plunger to clip the grid.
18. Remove the clipping tool from the hole and adjust the clipping hub lever to move the brass plate back into the open position. Transfer the clipped grid to an autogrid slot in the grid box.
19. Repeat steps 15–18 for the remaining grids.
20. Store the clipped grids in a pre-chilled dewar filled with liquid nitrogen until ready for data collection.

2.4 Cryo-EM data collection and processing

2.4.1 Rationale

The processing workflow with scaffolded constructs requires extra steps to refine the density of the attached TPP riboswitch. The scaffold expedites particle picking and can serve as an internal control throughout the initial processing steps to validate data quality. The addition of a focused refinement and signal subtraction strategy compensates for the small density of the attached RNA, enabling a high-resolution reconstruction of the riboswitch. The general processing pipeline with these special steps will be outlined in the protocol below.

2.4.2 Equipment and materials

- Titan Krios microscope with K3 Gatan direct electron detector (Thermo Fisher Scientific)
- CryoSPARC software package (Punjani et al., 2017)
- RELION software package (Scheres, 2012)
- Workstation capable of running cryoSPARC (minimum worker node requirements are listed in the cryoSPARC guide, https://guide.cryosparc.com/setup-configuration-and-management/hardware-and-system-requirements)

2.4.3 Protocol

1. Collect movies on a Titan Krios microscope with a K3 Gatan direct electron detector operating at 300 keV and the camera operating in super-resolution mode. All movies can be semi-automatically collected with SerialEM using the following parameters:
   a. Magnification: 105,000x
   b. Number of frames: 50
   c. Total dose: ~50 e-/Å^2
   d. Defocus range: −0.5 μm to −1.5 μm

2. Raw cryo-EM data can be processed using several different software packages, with some of the most popular being cryoSPARC and RELION. Instructions to install this software and basic hardware requirements can be found on their respective websites (cryoSPARC installation guidelines: https://guide.cryosparc.com/setup-configuration-and-management/how-to-download-install-and-configure; RELION installation guidelines: https://relion.readthedocs.io/en/latest/Installation.html). The following protocol will use a walkthrough of a cryoSPARC workflow, but these methods can be replicated across different programs.
   a. The cryoSPARC website also provides access to tutorials and practice datasets that can be used to familiarize yourself with the basic processing workflow (https://guide.cryosparc.com/guides-for-v3/cryo-em-data-processing-in-cryosparc-introductory-tutorial).
3. Cryo-EM movies will often be saved as TIF or MRC files when collected using Gatan detectors or as EER files when collected with Falcon series detectors. These file formats can be imported into cryoSPARC, with the appropriate pixel size, total dose, accelerating voltage, and spherical aberration of the microscope. A gain reference may also be collected with your movies to normalize data against dead or damaged pixels on the electron detector, although your movies may be collected with gain normalization already applied. Be sure to check with the microscopist or microscope settings to know the proper way to import and apply the gain reference.
4. For particle picking, the group II intron scaffold is a strong template for both the blob picker and 2D template picker in cryoSPARC. The density for the group II intron can be imported into cryoSPARC and converted into 2D templates to use as a basis for the template picker job with default settings. If you are unsure of the fidelity of the group II intron scaffold, setting the blob picker to pick elliptical blobs, with diameters of 50 Å and 150 Å, will provide coverage for the size of the scaffold.
   a. If there are difficulties with this step, a smaller micrograph subset can be taken for template picking using a processed *O. iheyensis* map to verify the scaffold has folded properly.
5. Extract particles from micrographs with a box diameter near 240 Å. For the initial classification steps, particles should be typically downsampled to 4 times the pixel size for a final pixel size of 3–4 Å per pixel.
6. Particles should be initially classified using 2D classification (Fig. 5). Given the typical behavior of the *O. iheyensis* scaffold, it is recommended

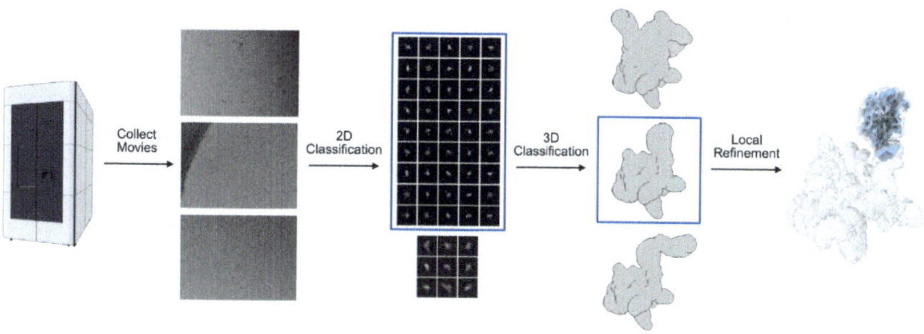

Fig. 5 General overview with example classes from the cryo-EM data processing workflow. After data collection, movies will be imported into software packages, like cryoSPARC, for processing. Picked particles are sorted into 2D classes and those displaying well-defined target volumes undergo further processing. 3D classification further narrows down the particle set used for global and then local refinement to increase the resolution of the attached RNA (TPP riboswitch; EMDB 45248 and 45249). The examples of classes that were selected during classification steps are boxed in blue.

that this step be performed with at least 200 2D classes as an output. Good 2D classes should then be selected for a round of *ab initio* reconstruction with 4 or more classes.

7. With the outputs from *ab initio* reconstruction, perform 3D classification using multiple rounds of heterogeneous refinement. Select classes that fold into the proper *O. iheyensis* scaffold form (Fig. 5). Take these and move on to an initial round of 3D refinement.
8. Re-extract the particles from the 3D refinement job with a box size roughly twice the size of the scaffold, between 250 Å and 300 Å. To speed up computation, particles can be downsampled to between 1−1.3 Å per pixel.
   a. Remove any duplicate particles (ones with 50 Å radius) at this stage.
9. Perform additional rounds of 3D and 2D classification to remove any junk particles that remain.
10. Refine the selected particles and put the particles through a round of refinement and CTF refinement. Continue performing multiple rounds of refinement and CTF refinement until resolution is maximized. The expectation is that the scaffold should be sub-3Å resolution.
    a. At the end of this step, conduct a final 2D classification job and remove particles with poor 2D classes.
    b. Re-extract the final particles with a box size that is double the particle diameter, and include some additional space for the TPP riboswitch. This should be close to 300 Å or more.

11. Perform additional rounds of 3D classification and hetero-refinement focused on the embedded structure, removing particles that lack resolvable densities. Pool together the remaining classes with intact densities.
12. Generate a tight mask around the scaffold and use this for a local refinement of the scaffold. Then generate a second mask for the scaffold, and use that for signal subtraction using the 3D alignment from local refinement.
13. Recenter the signal of subtracted particles on the TPP riboswitch density.
    a. If there is significant signal for the scaffold remaining, a second round of particle subtraction using a mask from step 12 can be performed on the particle stack using the 3D alignment from a global homogenous or non-uniform refinement job (Punjani et al., 2020).
14. Generate a mask for the embedded riboswitch, and then use this mask for local refinement. Local refinement of apo structures can be performed with a larger Gaussian prior for image rotation and shift to account for more uncertainty in particle position, while bound structures can use a tighter Gaussian prior for image rotation and shift. Typical values for more globular structures are 3° rotation and 3 Å shift, while the looser apo structure can be done with 5° rotation and 5 Å. These values and the mask tightness are the optimum way to improve resolution (Fig. 6).
15. Modeling into the local refinement maps can be done using the RCrane plugin in COOT. The 9C6K structure coordinates can be used as a starting point for model building (Emsley & Cowtan, 2004).

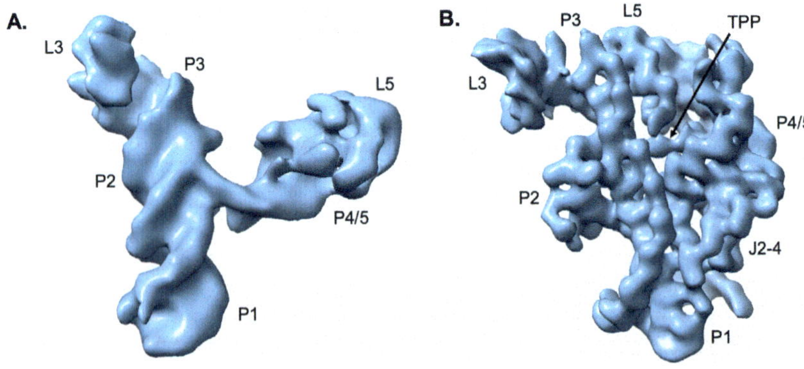

Fig. 6 Final TPP riboswitch aptamer densities. (A) Map of the apo *E. coli* TPP riboswitch aptamer after local refinement shows open conformation (EMDB 45250). (B) Map of the bound *E. coli* TPP riboswitch aptamer after local refinement with an arrow indicating the density of the TPP ligand (EMDB 45249).

3. Conclusion

This chapter outlined the protocol for a scaffold-assisted cryo-EM method that can be used to characterize TPP riboswitch aptamer structures. Embedding the aptamer sequence into the scaffold enables high-resolution reconstructions that capture both ligand-bound and native apo conformations, overcoming limitations of existing crystallographic approaches. After designing the target sequence into the scaffold, RNA is synthesized through a large-scale *in vitro* transcription reaction and purified *via* size exclusion chromatography. The purified RNA can then be incubated with TPP or competing binders prior to grid preparation to capture a ligand-bound conformation. After data collection, additional signal subtraction and local refinement steps are integrated into the processing workflow to specifically improve the resolution for the density of the attached TPP riboswitch aptamer. When using this method with the TPP riboswitch from *E. coli*, the resulting densities captured important structural details for both the apo and ligand-bound states (Haack et al., 2025). These results highlight the potential of this approach for further riboswitch structural characterization studies, as the adaptability of the scaffold design makes it broadly applicable to study TPP aptamer sequences from new species, targeted mutants, folding intermediates, and even other riboswitch classes.

Two results obtained with this protocol illustrate the structural and mechanistic detail that the scaffold-assisted approach can deliver for a target as small as the 86-nucleotide TPP riboswitch aptamer. First, the apo and ligand-bound reconstructions captured a large-scale conformational rearrangement of the aptamer domain that had not previously been visualized (Fig. 6; Haack et al., 2025). In the apo state, the thiamine-sensing (P2/3) and pyrophosphate-sensing (P4/5) helices extend outward into an open Y-shaped conformation, with the J2–4 element and the pyrophosphate stem becoming highly disordered and the thiamine-sensing stem extending through an additional helical groove. Upon TPP binding, these helices collapse into a parallel, compact arrangement that is locked in place by a long-range tertiary contact between the P3 stem and the L5 motif. This is the first direct cryo-EM visualization of a global conformational switch in a riboswitch, and it accounts for why the dynamic apo state had previously eluded crystallographic methods. Second, the 2.5 Å reconstruction of the ligand-binding pocket resolved the functional groups of TPP together with three magnesium ions coordinated to the pyrophosphate moiety, one more than had been observed in earlier crystal structures of this riboswitch. Together these examples demonstrate that a single workflow resolves both the global architecture that governs riboswitch function and the specific metal-ion and ligand contacts that define the binding site, providing a structural foundation for mechanistic interpretation and for structure-guided ligand design. Readers seeking a more detailed analysis of these structures are referred to the full study (Haack et al., 2025).

Acknowledgements

The authors would like to thank Alexa R. Perez for her support and assistance with figures and Simpson Joseph for constructing the manual plunging apparatus for grid vitrification. Figs. 1 and 5 were created in https://BioRender.com. This work was funded by a National Institutes of Health grant R35GM141706 (awarded to N.T.) and M.M.M. was supported by the Chemistry-Biology Interface Training Program, NIH Grant T32 GM112584.

References

Antunes, D., Jorge, N. A. N., Garcia de Souza Costa, M., Passetti, F., & Caffarena, E. R. (2019). Unraveling RNA dynamical behavior of TPP riboswitches: A comparison between Escherichia coli and Arabidopsis thaliana. *Scientific Reports, 9*, 4197. https://doi.org/10.1038/s41598-019-40875-1.

Bonilla, S. L., & Jang, K. (2024). Challenges, advances, and opportunities in RNA structural biology by Cryo-EM. *Current Opinion in Structural Biology, 88*, 102894. https://doi.org/10.1016/j.sbi.2024.102894.

Bonilla, S. L., & Kieft, J. S. (2022). The promise of cryo-EM to explore RNA structural dynamics. *Journal of Molecular Biology, 434*(18), 167802. https://doi.org/10.1016/j.jmb.2022.167802.

Emsley, P., & Cowtan, K. (2004). Coot: Model-building tools for molecular graphics. *Acta Crystallographica. Section D, Biological Crystallography, 60*, 2126–2132. https://doi.org/10.1107/S0907444904019158.

Garst, A. D., Edwards, A. L., & Batey, R. T. (2011). Riboswitches: Structures and mechanisms. *Cold Spring Harbor Perspectives in Biology, 3*(6), a003533. https://doi.org/10.1101/cshperspect.a003533.

Haack, D. B., Rudolfs, B., Jin, S., Khitun, A., Weeks, K. M., & Toor, N. (2025). Scaffold-enabled high-resolution cryo-EM structure determination of RNA. *Nature Communications, 16*(1), 880. https://doi.org/10.1038/s41467-024-55699-5.

Huang, L., Serganov, A., & Patel, D. J. (2010). Structural insights into ligand recognition by a sensing domain of the cooperative glycine riboswitch. *Molecular Cell, 40*, 774–786. https://doi.org/10.1016/j.molcel.2010.11.026.

Hübenthal, A., Panchal, V., Brenk, R., & Mack, M. (2025). Functional analysis of thiamine pyrophosphate-responsive riboswitches in human bacterial pathogens of the ESKAPE group using a dual-luciferase reporter gene assay. *Journal of Bacteriology, 207*(11), e0030825. https://doi.org/10.1128/jb.00308-25.

Jenkins, J. L., Krucinska, J., McCarty, R. M., Bandarian, V., & Wedekind, J. E. (2011). Comparison of a preQ1 riboswitch aptamer in metabolite-bound and free states with implications for gene regulation. *The Journal of Biological Chemistry, 286*, 24626–24637. https://doi.org/10.1074/jbc.m111.230375.

Lee, H. K., Lee, Y. T., Fan, L., Wilt, H. M., Conrad, C. E., Yu, P., ... Stagno, J. R. (2023). Crystal structure of Escherichia coli thiamine pyrophosphate-sensing riboswitch in the apo state. *Structure (London, England: 1993), 31*(7), 848–859. https://doi.org/10.1016/j.str.2023.05.003.

McCown, P. J., Corbino, K. A., Stav, S., Sherlock, M. E., & Breaker, R. R. (2017). Riboswitch diversity and distribution. *RNA (New York, N. Y.), 23*(7), 995–1011. https://doi.org/10.1261/rna.061234.117.

Mukherjee, S., Retwitzer, M. D., Barash, D., & Sengupta, S. (2018). Phylogenomic and comparative analysis of the distribution and regulatory patterns of TPP riboswitches in fungi. *Scientific Reports, 8*(1), 5563. https://doi.org/10.1038/s41598-018-23900-7.

Olenginski, L. T., Spradlin, S. F., & Batey, R. T. (2024). Flipping the script: Understanding riboswitches from an alternative perspective. *The Journal of Biological Chemistry, 300*, 105730. https://doi.org/10.1016/j.jbc.2024.105730.

Panchal, V., & Brenk, R. (2021). Riboswitches as drug targets for antibiotics. *Antibiotics (Basel), 10*(1), 45. https://doi.org/10.3390/antibiotics10010045.

Punjani, A., Rubinstein, J., Fleet, D., & Brubaker, M. A. (2017). cryoSPARC: Algorithms for rapid unsupervised cryo-EM structure determination. *Nature Methods, 14*, 290–296. https://doi.org/10.1038/nmeth.4169.

Punjani, A., Zhang, H., & Fleet, D. J. (2020). Non-uniform refinement: Adaptive regularization improves single-particle cryo-EM reconstruction. *Nature Methods, 17*, 1214–1221. https://doi.org/10.1038/s41592-020-00990-8.

Scheres, S. H. (2012). RELION: Implementation of a Bayesian approach to cryo-EM structure determination. *Journal of Structural Biology, 180*(3), 519–530. https://doi.org/10.1016/j.jsb.2012.09.006.

Schroeder, G. M., Dutta, D., Cavender, C. E., Jenkins, J. L., Pritchett, E. M., Baker, C. D., ... Wedekind, J. E. (2020). Analysis of a preQ1-I riboswitch in effector-free and bound states reveals a metabolite-programmed nucleobase-stacking spine that controls gene regulation. *Nucleic Acids Research, 48*(14), 8146–8164. https://doi.org/10.1093/nar/gkaa546.

Serganov, A., Polonskaia, A., Phan, A. T., Breaker, R. R., & Patel, D. J. (2006). Structural basis for gene regulation by a thiamine pyrophosphate-sensing riboswitch. *Nature, 441*, 1167–1171. https://doi.org/10.1038/nature04740.

Stagno, J. R., Bhandari, Y. R., Conrad, C. E., Liu, Y., & Wang, Y. X. (2017). Real-time crystallographic studies of the adenine riboswitch using an X-ray free-electron laser. *The FEBS Journal, 374284*, 3–3380. https://doi.org/10.1111/febs.14110.

Stoddard, C. D., Montange, R. K., Hennelly, S. P., Rambo, R. P., Sanbonmatsu, K. Y., & Batey, R. T. (2010). Free state conformational sampling of the SAM-I riboswitch aptamer domain. *Structure (London, England: 1993), 18*(7), 787–797. https://doi.org/10.1016/j.str.2010.04.006.

Subki, A., Ho, C. L., Ismail, N. F. N., Zainal Abidin, A. A., & Balia Yusof, Z. N. (2020). Identification and characterisation of thiamine pyrophosphate (TPP) riboswitch in Elaeis guineensis. *PLoS One, 15*(7), e0235431. https://doi.org/10.1371/journal.pone.0235431.

Thore, S., Leibundgut, M., & Ban, N. (2006). Structure of the eukaryotic thiamine pyrophosphate riboswitch with its regulatory ligand. *Science (New York, N. Y.), 312*(5777), 1208–1211. https://doi.org/10.1126/science.1128451.

Toor, N., Keating, K. S., Taylor, S. D., & Pyle, A. M. (2008). Crystal structure of a self-spliced group II intron. *Science (New York, N. Y.), 320*, 77–82. https://doi.org/10.1126/science.1153803.

Winkler, W. C., & Breaker, R. R. (2003). Genetic control by metabolite-binding riboswitches. *Chembiochem: a European Journal of Chemical Biology, 4*, 1024–1032. https://doi.org/10.1002/cbic.200300685.

Winkler, W., Nahvi, A., & Breaker, R. R. (2002). Thiamine derivatives bind messenger RNAs directly to regulate bacterial gene expression. *Nature, 419*(6910), 952–956. https://doi.org/10.1038/nature01145.

CHAPTER THIRTEEN

Targeting bacterial TPP riboswitches using pVEC-conjugated chimeric antisense oligonucleotides

Dimitrios Kaloudas, Nikolet Pavlova, Martina Traykovska, and Robert Penchovsky[*,1]

Laboratory of Synthetic Biology and Bioinformatics, Faculty of Biology, Sofia University "St. Kliment Ohridski", Sofia, Bulgaria

*Corresponding author. e-mail address: robert.penchovsky@hotmail.com

Contents

[1] Website: http://penchovsky.atwebpages.com/

Methods in Enzymology, Volume 735
ISSN 0076-6879, https://doi.org/10.1016/bs.mie.2026.06.015

Abstract

Antisense oligonucleotides (ASOs) provide a sequence-specific approach for targeting bacterial RNAs and suppressing the expression of genes important for bacterial growth and proliferation, making them prominent agents in the design of a novel class of antibacterial agents. This protocol describes the design, preparation, and experimental validation of a chimeric ASO conjugated to the cell-penetrating peptide pVEC and directed against the thiamine pyrophosphate (TPP) riboswitch. The workflow includes bioinformatic identification of conserved and accessible regions within the TPP aptamer, design of a chemically modified ASO, preparation of an *in vitro* transcribed TPP aptamer RNA substrate, and assessment of ASO-dependent RNA cleavage using an RNase H assay. The protocol further describes bacterial growth-inhibition assays in *Listeria monocytogenes* and *Bacillus subtilis*, the use of *Escherichia coli* as a specificity control, and evaluation of pVEC-ASO-1 cytotoxicity in A549 human cells using an MTT-based viability assay. Although presented for the TPP riboswitch, the approach can be adapted to other bacterial mRNA regions. This protocol therefore provides a flexible experimental framework for evaluating riboswitches and other bacterial RNA elements as potential targets for ASO-based antibacterial development.

1. Introduction

The rapid emergence of multidrug-resistant bacterial strains has surpassed the pace of new antibiotic discovery, underscoring the urgent need for alternative therapeutic approaches that are highly selective and less susceptible to resistance development (Elbaiomy et al., 2025). Within this framework, nucleic acid-based therapeutics have attracted attention as a promising class of next-generation antimicrobial agents, capable of precisely regulating gene expression at the molecular level (Martinez, 2014; Ogawara, 2019).

Antisense oligonucleotides (ASOs) have received substantial interest because they can selectively hybridize with complementary mRNA sequences and suppress gene expression. When targeted to essential bacterial genes, ASOs can inhibit key cellular functions, including metabolism, replication, and virulence. In contrast to conventional antibiotics, which typically target a relatively narrow range of proteins, ASOs offer a programmable strategy that can be readily redesigned to target almost any gene of interest. This specificity can improve therapeutic effectiveness while reducing off-target activity and limiting disturbance of the beneficial microbiota (Penchovsky & Traykovska, 2015).

Early-designed ASOs had numerous limitations, including RNase degradation, short half-life, poor stability, and low cellular delivery in bacterial cells. For these reasons, substantial modifications were made in the design of chimeric ASOs. To increase stability and prevent later degradation, ASO designed strategies that used chemical modifications, such as phosphorothioate backbones (PS), locked nucleic acids (LNAs), peptide nucleic acids (PNAs), and 2′-O-modified ribose sugars, within a single molecule (Dhara et al., 2025; Penchovsky et al., 2024).

TPP, a vital cofactor that plays a key role in many bacterial metabolic pathways, such as carbohydrate metabolism and amino acid biosynthesis, makes it a strong target for antisense-based antibacterial strategies. In many bacteria, genes involved in TPP biosynthesis and transport are regulated by highly conserved RNA elements known as TPP riboswitches. These riboswitches are typically located in the untranslated regions (UTRs) of mRNAs and act as metabolite-sensing regulators that modulate gene expression according to intracellular TPP concentrations. Upon TPP binding, the riboswitch adopts an alternative conformation, thereby influencing either transcription or translation of the downstream genes (Pavlova et al., 2023). ASOs can disrupt riboswitch function, leading to destabilization of the mRNA or inhibition of RNA translation (Otcheva et al., 2025; Traykovska et al., 2021). This interference has led to the downregulation of genes vital to the survival of the bacterial cell. Downregulation of these genes can inhibit bacterial growth, leading to growth arrest and, in some cases, cell death. As riboswitches are not found in human cells and are highly conserved across bacterial species, they represent attractive antibacterial targets with a low likelihood of host toxicity (Pavlova & Penchovsky, 2022).

Recent studies have demonstrated the feasibility of this approach, highlighting the effectiveness of chimeric ASOs in targeting TPP-regulated genes across various bacterial pathogens. Delivery technologies, such as conjugation to cell-penetrating peptides and incorporation into nano-carrier systems, have enhanced intracellular uptake and expanded the therapeutic applicability of these molecules (Popova & Penchovsky, 2024; Traykovska et al., 2022). Additionally, the modular nature of ASO design allows for rapid optimization and customization, enabling the development of targeted therapies against emerging resistant strains.

This chapter provides a step-by-step design and experimental evaluation of a chimeric antisense oligonucleotide (pVEC-ASO-1) directed against the thiamine pyrophosphate (TPP) riboswitch aptamer domain. The approach can be applied to target other mRNA regions or mRNAs to downregulate other genes. The described protocol combines bioinformatics identification of conserved riboswitch regions, rational ASO design, *in vitro* RNase H cleavage analysis, and bacterial growth inhibition assays.

The chimeric molecule consists of an antisense oligonucleotide bearing chemically modified oligonucleotides, conjugated to the cell-penetrating peptide pVEC (LLIILRRRIRKQAHAHSK) (Elmquist et al., 2001). The ASO-1 sequence is designed to hybridize with a conserved, accessible region of the TPP riboswitch aptamer. Upon hybridization, the resulting RNA/DNA duplex is recognized by RNase H, which cleaves the target mRNA and inhibits downstream gene expression.

The presented work is based on the original work from our group (Traykovska et al., 2022; Traykovska & Penchovsky, 2022). It focuses on targeting TPP riboswitch aptamer in bacterial mRNAs.

The protocol is organized into five experimental sections and a section designating critical parameters and notes:

a) Bioinformatics identification and design of the TPP-targeting ASO
b) Preparation of the TPP aptamer RNA substrate
c) RNase H cleavage assay

d) Bacterial growth inhibition assay
e) Mammalian-cell toxicity assessment
f) Notes and Critical Parameters

2. Materials

2.1 Bioinformatics resources

- RSwitch (Robert Penchovsky: Bioinformatics) and Rfam (Rfam: The RNA families database) databases for retrieval of bacterial TPP riboswitch aptamer sequences.
- BLASTn for local sequence similarity searches.
- ClustalX or other multiple sequence alignment tools.
- RNAfold (RNAfold web server), and RNAcofold (RNAcofold web server) from ViennaRNA (ViennaRNA Web Services) for RNA secondary-structure prediction (Lorenz et al., 2011).
- DNA/RNA hybrid melting-temperature prediction tool (such as the DNA/RNA TM Calculator) for computing ASO-target stability (Kaloudas et al., 2019).

2.2 Antisense oligonucleotides

The ASO used in the presented study was the pVEC-ASO-1 bearing the sequence of:

LLIILRRRIRKQAHAHSK-$C_1A_1A_1T_2C_2C_2C_2T_2A_2C_1G_1C_1$.
Where:

- **LLIILRRRIRKQAHAHSK** corresponds to the pVEC cell-penetrating peptide
- The ASO-1 is attached to pVEC through the 5′ end of the oligonucleotide
- Subscript 1 denotes 2′-alkyl ribose modifications
- Subscript 2 denotes phosphorothioate modifications (PS) of the phosphodiester linkage.
- The ASO-1 was commercially synthesized bearing the pVEC and the chemically modified nucleotides [pVEC-ASO-1 was commercially synthesized by GeneLink, Inc. (Elmsford, NY, USA)].

2.2.1 Enzymes and kits

- RNase H (Invitrogen/Thermo Fisher Scientific)
- DNA polymerase (Invitrogen/Thermo Fisher Scientific)
- SuperScript II reverse transcriptase (Invitrogen/Thermo Fisher Scientific)

- RiboMAX transcription kit (Promega)
- Synthetic DNA oligonucleotides (Sigma-Aldrich)
- SYBR Green II stain (Invitrogen/Thermo Fisher Scientific)

2.2.2 DNA templates for in vitro TPP aptamer transcription

In this section, oligonucleotides are used to construct the aptamer sequence, which will serve as a template for RNase H digestion assays. Two oligomers are used to extend the full length of the aptamer that bears the ASO-binding sequence (see below). The first oligonucleotide, the non-template strand, carries a T7 promoter sequence followed by two guanines at the 5′ end, and the second oligonucleotide, the template strand, carries an ASO corresponding binding site (underlined sequence in italics).

- Non-template strand:
  5′-TAATACGACTCACTATA**GG**cgttaccacagggggggcttcttagctgagattga *gtccacgtgttttggattctg* -3′
- Template strand
  5′- gaaatgcaatcgcca*caatccctacgc*tcgtattaacgaacaggttcaaaggg*tcagaatccaaaacacgtggac* -3′
- The two DNA strands overlap (nucleotides in *italics*) and are extended to form a double-stranded DNA template for transcription.

Note: The DNA template contains a T7 promoter sequence, TAATACGACTCACTATA, followed by two guanines, **GG**. Therefore, transcription should be performed using a T7-compatible system, such as the RiboMAX™ Large Scale RNA Production System-T7 or the T7 RiboMAX™ Express System.

2.3 Bacterial strains

Bacteria strains used in the original presented research:

- *Listeria monocytogenes* ATCC 8932
- *Bacillus subtilis strain* 168
- *Escherichia coli* strain K12 (negative control)

2.4 Media and reagents

- Luria-Bertani (LB) medium (Thermo Fisher)
- Sterile culture tubes/flasks (ISOLAB)
- RNase-free water (Thermo Fisher)
- Denaturing PAGE reagents (AppliChem)
- Denaturing RNA loading dye (Thermo Fisher)
- Gel staining (gel red- Biotium)

2.5 Equipment

- One Incubator set to 30 °C (Biobase)
- One Incubator set to 37 °C (Faithful)
- UV/visible spectrophotometer (Pharmacia Biotech Ultrospec 100E)
- PAGE electrophoresis apparatus (Biorad)
- UV transilluminator or gel documentation system (Analyticjena)
- Standard RNase-free molecular biology equipment

2.6 Materials for human cell cultures

- A549 human lung cancer cells (from ATCC)
- D-MEM culture medium (Thermo Fisher)
- Fetal calf serum (Apollo Scientific)
- Penicillin (Sigma-Aldrich)
- Streptomycin (Sigma-Aldrich)
- pVEC-ASO-1 (GeneLink)
- Sterile 96-well tissue-culture plates (Biorad)
- MTT reagent (Thermo Fisher)
- Standard mammalian-cell culture equipment (Biobase)
- Plate reader suitable for MTT absorbance measurement (Biobase)

3. Protocol 1: Bioinformatics identification of the TPP riboswitch target region

The objective of this protocol is to identify conserved, structurally accessible regions within the bacterial TPP riboswitch aptamer suitable for pVEC-ASO-1 design. The expected outcome is the selection of a conserved single-stranded target sequence with minimal similarity to human RNAs and suitable predicted hybridization properties for RNase H-mediated cleavage.

3.1 Retrieval of TPP riboswitch sequences

Use bioinformatics to retrieve TPP riboswitch sequences. Target selection begins by collecting TPP aptamer sequences from online databases such as Rfam (accession number: RF00059 for the TPP riboswitch/THI element) and Rswitch (Griffiths-Jones et al., 2003; Penchovsky et al., 2021). These reference sequences can then be used to identify homologous TPP-related RNA sequences in numerous bacterial species. In this context, the Basic Local Alignment Search Tool (BLAST) of the NCBI database (Altschul et al., 1990) can be used to search nucleotide databases for similar sequences. In contrast, multiple-sequence alignment tools such as ClustalX, ClustalW, or Clustal Omega can align the retrieved sequences and identify conserved regions across species (Larkin et al., 2007; Sievers & Higgins, 2014). Conserved regions of the target across several bacterial species of interest represent promising candidates for the design of an ASO with broad-spectrum antibacterial activity.

- Amongst the tested bacterial strains, experimentally tested species can be used for comparative analysis.
- When targeting TPP riboswitches, bacteria lacking a TPP riboswitch can serve as a negative control, as in the original research, which used the *Escherichia coli* strain K12.

3.2 Sequence similarity analysis

Use BLASTn to compare TPP riboswitch-containing regions among bacterial species. Identify conserved regions within the aptamer domain. Retain candidate regions that are conserved amongst the test species. Prioritize target regions with at least 70 % identity and preferably complete conservation across the ASO-binding region in the intended target species. Mismatches within the central RNase H-active region should be avoided. Export the retrieved sequences from the database in FASTA format for sequence alignment.

Note: Exclude regions with poor conservation across the selected bacterial targets if you plan to design an ASO targeting multiple bacterial species.

- Import the selected TPP aptamer sequences into a multiple alignment software such as ClustalX or Clustal Omega.
- Perform multiple sequence alignment using default parameters or parameters optimized for short RNA sequences.
- Based on the alignment, identify conserved regions suitable for antisense targeting (Fig. 1).

Note: Prioritize regions conserved across the target bacteria but absent in non-target organisms.

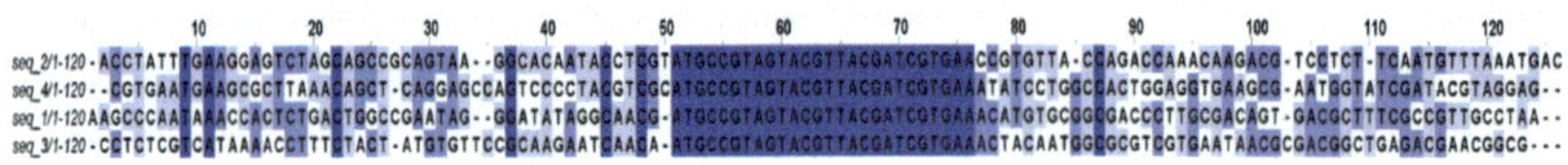

Fig. 1 Example of sequence alignment showing a 26-nt conserved region (central dark blue region) prominent as an ASO target.

3.3 Secondary-structure analysis

- Submit the TPP aptamer sequence to RNAfold and RNAcofold from the ViennaRNA web services.
- Predict the secondary structure of the riboswitch aptamer.

Identify the nucleotide positions of the candidate ASO-binding site in the RNA sequence and highlight these positions on the predicted RNAfold secondary structure or dot-bracket output. Select sites that fall mostly in loops, bulges, junctions, or other unpaired regions.

Note: Select a target region that is predicted to be single-stranded or structurally accessible. Avoid regions buried in stable stems unless disruption of that structure is specifically intended. RNA regions with strong secondary structures will impair ASOs' hybridization to the target.

3.4 ASO design

- Design the ASO as a reverse complement of the selected conserved region of the TPP aptamer sequence.
- Use the RNAfold and RNAcofold Webservers from the ViennaRNA Web Services on the ASO sequence to:
  - Confirm that the ASO does not form stable self-complementary structures.
  - Confirm that the ASO does not strongly self-hybridize.
- Use a program to:
  - Calculate the melting temperature of the DNA/RNA hybrid.
- Select an ASO expected to form a stable hybrid at 37 °C.
- To avoid targeting mRNAs in the host (human) other than the target:
  - Use the BLASTn tool from the NCBI to compare the ASO sequence against the human-expressed RNAs database (Human RefSeqGene sequences).
  - Exclude ASO candidates with high similarity (>70 %) to human RNAs.
  - Select the final ASO candidate for synthesis.

After final selection, the ASO should be ordered from a commercial provider capable of synthesizing the required phosphorothioate and 2′-alkyl modifications and conjugating the oligonucleotide to pVEC; the company that chemically synthesized the pVEC-ASO-1 and its sequence used here are provided in Section 2.2.

4. Protocol 2: Preparation of TPP aptamer RNA Substrate for RNase H test assays

This assay generates a defined TPP aptamer RNA substrate from synthetic DNA oligonucleotides containing a T7 promoter; the expected outcome is a purified full-length RNA transcript containing the pVEC-ASO-1 binding site and suitable for use as the target RNA in the RNase H cleavage assay.

4.1 Purification of synthetic DNA oligonucleotides

- Obtain the synthetic template and non-template DNA strands for generating the aptamer target. In the original work, the Synthetic DNA oligomers were purchased from Sigma-Aldrich.
- Purify each DNA oligonucleotide using a 10% denaturing PAGE.
- Excise the full-length DNA bands.
- Recover the DNA from the gel using the laboratory standard gel-extraction procedure.
  - Excised PAGE bands can be crushed and eluted overnight in nuclease-free water or 0.3 M sodium acetate, then ethanol-precipitated, or purified using an oligonucleotide cleanup kit such as the Zymo Oligo Clean & Concentrator kit or an equivalent product.
- Quantify the purified DNA *via* OD measurements.
  - Quantify purified DNA using a Biodrop, a NanoDrop, or equivalent spectrophotometer. Acceptable purity is approximately 1.8 for the A260/A280 ratio and 2.0–2.2 for the A260/A230 ratio. If purity ratios are outside this range, repeat cleanup before template assembly.
- Store purified DNA at −20 °C until use.

4.2 Preparation of double-stranded transcription template

Combine the purified non-template and template DNA strands.

- Use an equal amount of each oligonucleotide in the reaction mix (20–50 μM) in standard annealing buffer (Tris-HCl 10 mM, pH 7.5−8.0, 50 mM NaCl, 1 mM EDTA) in a nuclease-free PCR tube:

Commercial oligonucleotide annealing buffers from Thermo/Fisher, IDT, or equivalent suppliers may also be used.

- Program the thermocycler to heat the mix for 5 min at 95 °C
- Allow the nucleotide-overlapping region to anneal by gradually cooling down the mixture in the thermocycler (−1 °C per cycle).
- Extend the annealed oligonucleotides using a DNA polymerase.

- Add 2.5 μL of polymerase buffer (x10) to the mixture tube 2.
- dNTPs mix (10 mM): 1 μL
- Taq DNA polymerase: 1 μL (~5 units)
- Bring the total volume of the reaction to 25 μL with Nuclease-free H_2O

- In the thermocycler program, add an extension step at 68 °C for 30 s
- Deactivate the Taq polymerase by heating the reaction at 75 °C for 30 min
- Use a PCR purification kit to purify the reaction.

Note: The protocol may vary depending on the polymerase and the size of the gaps that need to be extended.

4.3 *In vitro* transcription of the TPP aptamer RNA

- Prepare the transcription reaction using the purified double-stranded DNA template containing the T7 promoter.
- Use a transcription system (*i.e.*, RiboMAX T7) according to the manufacturer's instructions to transcribe the aptamer sequence.
  - Brief RiboMAX T7workflow: Combine the purified T7-containing dsDNA template with T7 transcription buffer, rNTP mix (rATP, rCTP, rGTP, and rUTP), T7 enzyme mix, and nuclease-free water; incubate under the manufacturer-recommended conditions, typically at 37 °C; digest template DNA with DNase; purify the RNA by 10 % denaturing PAGE. Follow the Promega RiboMAX T7 kit manual for the exact volumes.
- Incubate the reaction under the recommended conditions for RNA synthesis, as per the manufacturer's instructions.
- Stop the reaction according to the kit protocol.
- Purify the transcribed RNA by 10 % denaturing PAGE.
  - After electrophoresis, the RNA band can be visualized using SYBR Green II or SYBR Gold staining, or by UV shadowing when preservation of RNA integrity is critical. Minimize UV exposure before excising the full-length RNA band. In our work, the gel was stained with SYBR Green II.
- Excise the full-length RNA band.
- Recover the RNA from the gel.
- Quantify the purified RNA.
  - Quantify purified RNA using a Biodrop, NanoDrop, or equivalent spectrophotometer. For a defined RNA transcript, calculate its molar concentration using a sequence-specific extinction coefficient calculator

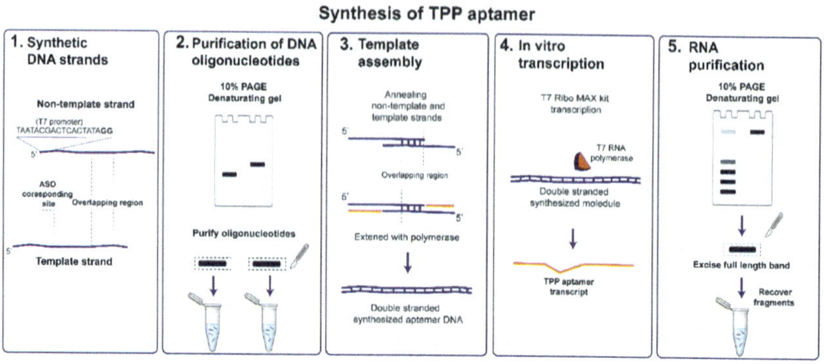

Fig. 2 Steps involved in the generation of the target TPP aptamer region from the synthetic oligonucleotides.

such as the T_m Calculator for Primers from Thermo Fisher (Tm Calculator for Primers | Thermo Fisher Scientific – BG). Concentration can be calculated from A260 using C = A260/(x pathlength), where ε "epsilon" is the molar extinction coefficient in M^{-1} cm^{-1} and the pathlength is normalized to 1 cm.

- Store the RNA at −80 °C under RNase-free conditions.

The overall workflow for preparing the TPP aptamer RNA substrate is summarized in Fig. 2.

5. Protocol 3: RNase H cleavage assay

The objective of this protocol is to determine whether pVEC-ASO-1 hybridizes to the TPP aptamer RNA and promotes RNase H-dependent cleavage. The expected outcome is efficient cleavage of the target RNA in the presence of pVEC-ASO-1 and RNase H, with little or no cleavage in the no-ASO control reaction.

5.1 Experimental design

This assay tests whether the designed pVEC-ASO-1 hybridizes to the TPP aptamer and promotes RNase H-mediated cleavage. The test reaction contains a 3:1 molar ratio of pVEC-ASO-1 to target RNA (Table 1). Prepare triplicates for the following reactions:

- Use Nuclease-free water to adjust the volume of each reaction to 20 μL.
  - The concentrations denoted here are taken from the original research.

Table 1 RNase H cleavage reaction setup for testing pVEC-ASO-1-dependent cleavage of the TPP aptamer RNA.

Target RNA		pVEC-ASO-1	RNase H	Total volume
Test reaction	200 pM	600 pM	10 μg.mL^{-1}	20 μL
Negative control	200 pM	Absent	10 μg.mL^{-1}	20 μL

5.2 RNase H reaction assay

- Thaw purified TPP aptamer and the pVEC-ASO-1 on ice.
- Prepare the RNase H reaction mixture using the manufacturer's recommended RNase H buffer.
  - Typically, add 1 U RNase H in a 20 μL reaction mix.
- Add target RNA to a final concentration of 200 pM.
- Add the synthesized pVEC-ASO-1 to a final concentration of 600 pM for the test reaction.
- Omit pVEC-ASO-1 from the negative-control reaction.
- Add 10 μg.mL^{-1} RNase H to each reaction (1 Unit).
- Mix gently.
- Incubate the reactions for 1 h at 37 °C.
- Stop the reactions by adding denaturing loading dye.

5.3 Analysis by denaturing PAGE

- Prepare a 10 % denaturing polyacrylamide gel.
- Load the RNase H reaction products.
- Run the gel until the cleaved and uncleaved RNA species are resolved.
- Stain the gel with SYBR Green II (Invitrogen).
- Visualize the gel using a UV transilluminator or gel documentation system.
- Compare the test reaction with the negative control.

Quantify the intensity of uncleaved and cleaved RNA bands if required. Image the gel using a gel documentation system and quantify band intensities using programs such as ImageJ/Fiji, GelAnalyzer, or the GelDoc analysis software (Bio-Rad).

The overall workflow for preparing the RNase cleavage assay is summarized in Fig. 3.

- Expected result

  In our study (Traykovska et al., 2022), the target RNA was cleaved by more than 90 % in the presence of pVEC-ASO-1, whereas no cleavage was detected in the reaction lacking pVEC-ASO-1.

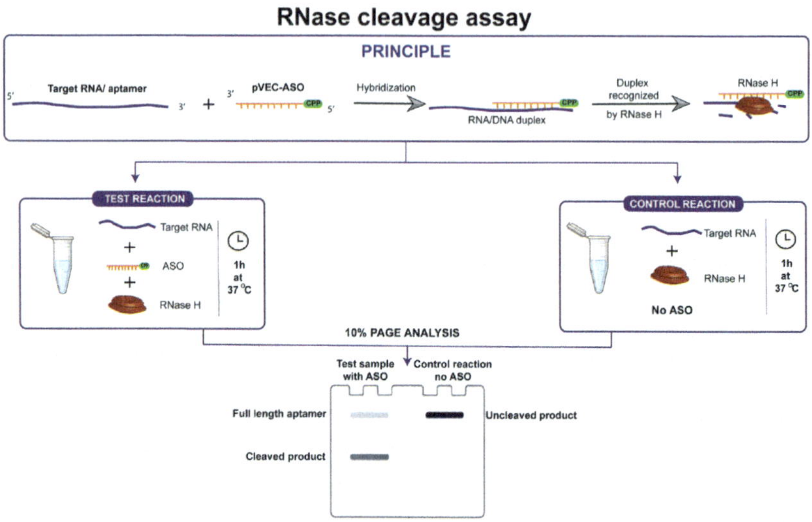

Fig. 3 Schematic representation of the RNase H cleavage assay and obtained results.

- The full-length *in vitro* transcript generated from the template is approximately 112 nt after removal of the T7 promoter sequence. Cleavage products should migrate below the full-length band; their exact sizes depend on the RNase H cut position within the ASO/RNA duplex.

6. Protocol 4: Bacterial growth inhibition assay

The objective of this protocol is to evaluate whether pVEC-ASO-1 inhibits bacterial growth by targeting the TPP riboswitch in living cells. The expected outcome is concentration-dependent growth inhibition of TPP riboswitch-containing bacteria, such as *Listeria* monocytogenes and *Bacillus subtilis*, but no inhibition of the negative control strain *Escherichia coli* K12.

6.1 Experimental design

This assay evaluates whether the designed chimeric ASO suppresses bacterial growth by targeting the TPP riboswitch. The bacterial strains and ASO concentrations shown here are as presented in the original research (Traykovska et al., 2022).

For the presented study, the following bacterial strains were used.

- *L. monocytogenes* ATCC 8932 as the target bacterium

- B. subtilis strain 168 as the target bacterium
- E. coli K12 as a negative-control bacterium lacking the targeted TPP riboswitch sequence

The assayed pVEC-ASO-1 concentrations were 0 nM (untreated control), 150 nM, 350 nM, 700 nM, 1000 nM, and 2000 nM.

Note: Perform three independent experiments for each concentration.

6.2 Overnight bacterial cultures

- Inoculate each bacterial strain into LB medium.
- Incubate overnight.
- Grow cultures to an OD600 of approximately 0.8.
- The bacterial strains used in the original research were grown as follows:
  a) *L. monocytogenes*: 37 °C
  b) *E. coli*: 37 °C
  c) *B. subtilis*: 30 °C

6.3 Dilution of bacterial cultures

- Dilute each overnight culture 200-fold into fresh LB medium.
- Mix gently to obtain a uniform bacterial suspension.
- Dispense diluted cultures into sterile culture vessels.

6.4 ASO treatment

Prepare pVEC-ASO-1 working dilutions.

Add the pVEC-ASO to the diluted cultures to obtain final ASO concentrations between 150 nM and 2000 nM. Prepare a no-ASO control for each bacterial strain.

Note: Presented concentrations are based on the original research.

Include additional controls where appropriate:

a) pVEC alone
b) ASO without pVEC (naked ASO)
c) untreated bacterial culture

- Mix cultures gently after adding the ASO.

6.5 Incubation and OD600 measurement

- Incubate treated and untreated cultures for 12 h.
- Measure OD600 every 30 min

6.6 Data analysis

- Plot OD600 *versus* time for each bacterial strain and ASO concentration.
- Compare treated cultures with untreated controls.

- Calculate percentage growth inhibition using the untreated culture as the reference.
- Determine the ASO concentration that produces approximately 80 % growth inhibition.
- Account for the MIC80 value.

Fig. 4 shows a results plot of the inhibition of bacterial growth at two different ASO-1 concentrations. The negative control sample (Black line) represents untreated cells.

- Expected result

In our study, strong growth inhibition of *L. monocytogenes* and *B. subtilis* was evident at 700, 1000, and 2000 nM pVEC-ASO-1. No meaningful inhibition was observed at 150 or 350 nM, and the negative-control organism, *E. coli* K12, was not inhibited by pVEC-ASO-1, supporting target-specific activity (Traykovska et al., 2022).

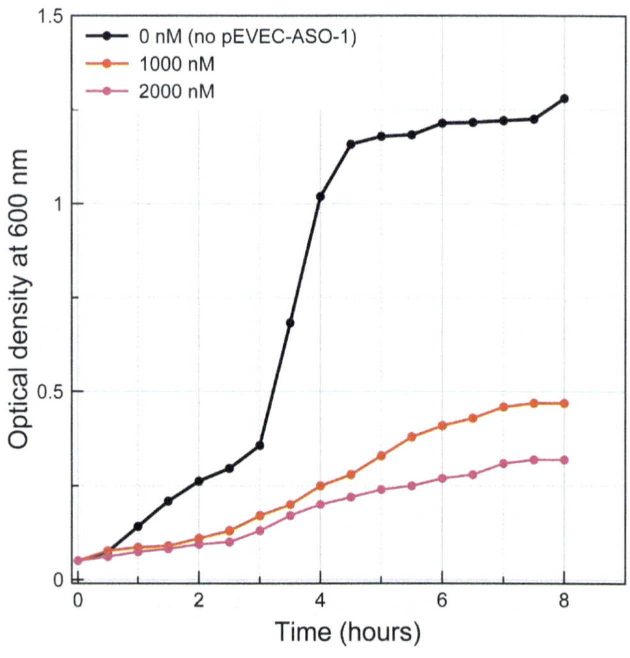

Fig. 4 Schematic representation of L. monocytogenes bacterial growth inhibition assays by the pVEC-ASO-1. The red line shows bacterial growth after treatment with 1000 nM pVEC-ASO-1. The pink line shows bacterial growth after treatment with 2000 nM pVEC-ASO-1. The black line is the untreated control. Measurements are taken every 30 min for an 8 h span.

7. Protocol 5: Mammalian-cell toxicity assessment

The objective of this protocol is to assess the cytotoxicity of pVEC-ASO-1 in A549 human cells using an MTT-based viability assay; the expected outcome is high cell survival at the antibacterial effective concentration, indicating that pVEC-ASO-1 has limited toxicity toward the tested human cell line.

7.1 Principle

This assay evaluates whether the chimeric pVEC-ASO exhibits cytotoxicity toward human cells. The original TPP riboswitch study tested toxicity in the A549 human lung cancer cell line using a previously described protocol. The detailed culture and viability-assay conditions reported here are available and taken from the related FMN riboswitch ASO study (Traykovska & Penchovsky, 2022).

7.2 Cell-culture medium

- Prepare complete D-MEM medium containing:
- D-MEM basal medium
- 10 % fetal calf serum
- Penicillin, 100 U.mL^{-1}
- Streptomycin, 100 μg.mL^{-1}

Note: The A549 cells should be maintained as a monolayer culture in this complete medium.

7.3 Seeding of A549 cells

- Grow A549 cells as a monolayer culture under standard mammalian-cell culture conditions.
- Prepare a single-cell suspension from actively growing A549 cells.
- Count the cells using the laboratory's standard cell-counting method.
- Seed A549 cells into a sterile 96-well tissue-culture plate at 100,000 cells per well
- Add complete D-MEM medium to each well.
- Incubate the cells for 24 h to allow attachment and entry into the exponential growth phase.

7.4 Treatment with pVEC-ASO

Prepare pVEC-ASO working solutions in sterile conditions.

- At 24 h after seeding, add pVEC-ASO directly to the culture medium.

- Use the concentrations tested in your study. For the TPP-related study, these were 700 nM, 1000 nM, and 2000 nM.

Include untreated A549 cells as the 100 % survival control.

Include medium-only wells for background subtraction in the MTT assay.

Incubate the treated cells for 48 h.

Note: Run triplicates for each sample and controls.

The workflow for the mammalian-cell toxicity assessment is summarized in Fig. 3.

7.5 MTT viability assay

After 48 h of pVEC-ASO treatment, measure cell survival using the MTT assay.

Perform the MTT assay according to the Mosmann method (Mosmann, 1983) or the laboratory's validated MTT protocol.

- Add MTT reagent to each well.
- Incubate until formazan crystals develop.
- Solubilize the formazan product with the solvent specified in the MTT protocol. Solvent examples are: DMSO, acidified isopropanol, or SDS/HCl solubilization buffer.
- Measure absorbance at 570 nm using a microplate reader, with optional background/reference subtraction at 630–690 nm depending on the instrument and MTT protocol.
- Subtract the medium-only background absorbance.
- Normalize the absorbance of treated wells to untreated control cells.

Calculate cell survival as:

$$Cell\ survival\ (\%) = \frac{Absorbance\ of\ treated\ cells}{Absorbance\ of\ untreated\ control\ cells}\ x\ 100$$

- Expected result

In our TPP study, pVEC-ASO-1 showed no toxicity at 700 nM, approximately 98.5–100 % A549 survival at 1000 nM, and approximately 82 % survival at 2000 nM.

The overall workflow for the bacterial growth and mammalian-cell toxicity assays is summarized in Fig. 5.

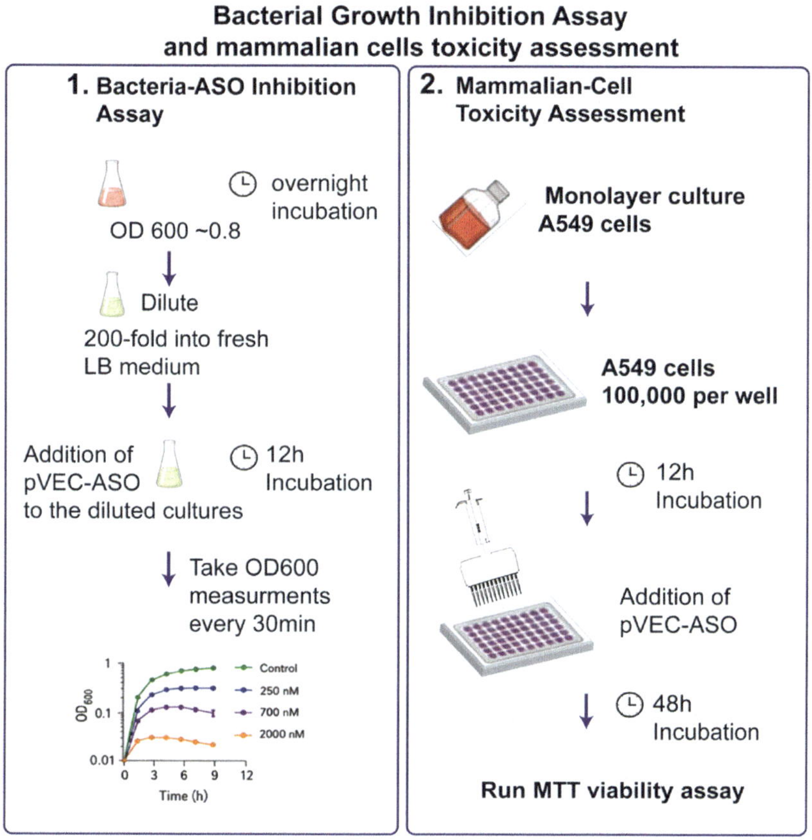

Fig. 5 Workflow graphical representation of the steps involved in the bacterial growth inhibition assay and the ASO toxicity assay on the human A579 cell line.

8. Notes and critical parameters

8.1 Selection of the target region

The ASO should target a conserved and accessible region of the TPP riboswitch aptamer. Regions predicted to be single-stranded are preferred because they are more likely to hybridize efficiently with the ASO.

8.2 ASO chemical design

The ASO combines first- and second-generation modifications. Phosphorothioate modifications support RNase H activity, whereas 2′-alkyl modifications improve nuclease resistance. The balance between these modifications is important because excessive second-generation modification may reduce RNase H-mediated cleavage, induce cytotoxicity, and cause adverse effects in the host.

8.3 pVEC conjugation

Naked ASOs cannot penetrate the cell and cannot inhibit bacterial growth. The pVEC peptide is required for intracellular delivery. Although the pVEC can also pass through eukaryotic cells, no toxicity to pro- or eukaryotic cells has been reported at concentrations up to 2000 nM (Traykovska et al., 2022; Traykovska & Penchovsky, 2022).

8.4 RNase-free handling

RNA preparation and RNase H assays should be performed using RNase-free reagents, tubes, and pipette tips.

8.5 Controls

Essential controls include:

- RNA plus RNase H without ASO
- Untreated bacterial culture
- pVEC alone
- ASO without pVEC (naked ASO)
- Non-target bacterial strain, such as *E. coli* K12 (in the presented protocol)

8.6 Biosafety

Work with pathogenic bacteria should be performed only in an appropriately approved biosafety laboratory using institutional safety procedures.

9. Summary

This protocol describes the workflow for designing and evaluating a pVEC-conjugated chimeric ASO targeting the bacterial TPP riboswitch. The core experimental strategy involves selecting a conserved riboswitch aptamer sequence, synthesizing a chemically modified ASO, validating RNase H-mediated RNA cleavage, and assessing bacterial growth inhibition and mammalian cell toxicity.

10. Conclusion

This protocol describes a step-by-step strategy for designing, preparing, and experimentally validating chimeric antisense oligonucleotides that target the bacterial TPP riboswitch. The workflow combines bioinformatic selection of a conserved and accessible RNA target site, preparation of a defined RNA substrate, *in vitro* confirmation of RNase H-mediated cleavage, bacterial growth inhibition assays, and mammalian cell toxicity testing. Although the protocol is presented for the TPP riboswitch aptamer, the same approach can be adapted to other bacterial mRNA regions by redesigning the ASO and generating the corresponding RNA substrate from a gene-specific cDNA-derived PCR template containing a T7 promoter. In this broader application, total bacterial RNA can be reverse-transcribed using a 3′ gene-specific primer, amplified across the ASO-binding region, transcribed *in vitro*, and tested for ASO-dependent RNase H cleavage. Thus, the method provides a flexible platform for evaluating riboswitches, untranslated regions, coding sequences, or other accessible bacterial RNA elements as potential antisense antibacterial targets.

Using this protocol, the TPP aptamer RNA substrate should be efficiently generated from the T7-containing DNA template and purified by denaturing PAGE. In the RNase H assay, incubation of target RNA with pVEC-ASO-1 and RNase H for 1 h at 37 °C is expected to produce strong ASO-dependent cleavage, with the original study reporting more than 90 % cleavage in the presence of pVEC-ASO-1 and no detectable cleavage in the no-ASO control. In bacterial growth assays, pVEC-ASO-1 should inhibit the growth of tested strains in a concentration-dependent manner. If a negative control bacterial strain is used, it should remain unaffected. At the effective antibacterial concentration, pVEC-ASO is expected to show low toxicity toward A549 cells, with the original TPP study reporting no toxicity at 700 nM, approximately 98.5−100 % survival at 1000 nM, and 82 % survival at 2000 nM.

11. Troubleshooting

Step	Problem	Possible cause	Suggested solution
ASO design	No RNase H cleavage is observed	ASO does not form a stable RNA/DNA hybrid with the target RNA	Recheck complementarity between the ASO and target RNA. Confirm that the ASO-binding region is single-stranded or accessible in the predicted RNA structure.

ASO design	Weak cleavage despite correct complementarity	The target site is structurally buried in the RNA	Redesign the ASO to target a more accessible single-stranded region.
ASO design	ASO appears inactive in cells but active *in vitro*	Poor cellular uptake or instability	Confirm that the ASO is conjugated to pVEC.
Synthetic DNA template preparation	Low yield of double-stranded DNA template	Poor annealing between overlapping oligonucleotides	Confirm the 10-nt overlap, oligonucleotide quality, and equimolar mixing. Increase annealing time.
Synthetic DNA template preparation	No RNA transcription product	Missing or incorrect T7 promoter	Confirm that the non-template strand or forward PCR primer contains the T7 promoter sequence TAATACGACTCAC TATA followed by GG.
Synthetic DNA template preparation	Multiple transcription products	Impure DNA oligonucleotides or nonspecific template extension	Purify synthetic DNA oligonucleotides by 10% denaturing PAGE before template assembly. Gel-purify the final dsDNA template if needed.
T7 *in vitro* transcription	Low RNA yield	Poor-quality dsDNA template or suboptimal transcription reaction	Confirm template concentration and integrity. Use a fresh T7-compatible RiboMAX reaction. Ensure the template contains a functional T7 promoter.
T7 *in vitro* transcription	RNA is degraded	RNase contamination	Use RNase-free water, tubes, tips, and gloves. Clean surfaces with RNase decontamination solution. Keep RNA on ice and avoid repeated freeze-thaw cycles.
RNA purification	Weak or smeared RNA band on denaturing PAGE	RNA degradation or overloading	Load less RNA, denature the sample properly before loading, and prepare fresh denaturing gel and running buffer.
RNA purification	Poor recovery from gel	Inefficient gel extraction	Crush gel slices thoroughly and elute RNA for sufficient time using RNase-free buffer.
RNase H assay	No cleavage in the test reaction	RNase H inactive, wrong buffer, or ASO/RNA hybrid not forming	Include a positive-control RNA/DNA hybrid if available. Use fresh RNase H and the manufacturer's buffer. Pre-anneal ASO and RNA before adding RNase H.

RNase H assay	Cleavage occurs in the no-ASO control	RNA contamination with nucleases	Prepare fresh RNA, repeat with RNase-free reagents, and include an RNA-only control without RNase H.
RNase H assay	Partial cleavage only	Insufficient ASO concentration, short incubation, or inaccessible target region	Increase the ASO: RNA ratio and extend the incubation time.
RNase H assay	Smearing on denaturing PAGE	RNA degradation after reaction or incomplete denaturation	Add denaturing loading dye immediately after incubation. Heat-denature before loading.
RNase H assay	Cleavage bands are hard to distinguish	The target RNA fragment is too short, or the cleavage products are similar in size.	Use higher-percentage denaturing PAGE for increased resolution.
Bacterial growth assay	No growth in the untreated control	Poor culture quality, wrong incubation temperature, or medium problem	Start from fresh colonies. Confirm LB composition and incubation temperature.
Bacterial growth assay	High variation between replicates	Unequal starting OD or poor mixing	Standardize overnight cultures to an OD600 ~0.8 before diluting 1:200. Mix cultures thoroughly after adding ASO. Use at least three biological replicates.
Bacterial growth assay	No inhibition	ASO degradation, incorrect ASO sequence, poor pVEC conjugation, or the target not expressed	Confirm ASO identity and conjugation. Check the target sequence in the tested strain. Verify that the target mRNA is expressed under the culture conditions.
Endogenous mRNA RT-PCR validation	PCR band still present after ASO treatment	Incomplete mRNA degradation, primers outside the cleavage region, or insufficient ASO exposure	Confirm primers flank the ASO-binding site. Increase ASO concentration or treatment time.
Endogenous mRNA RT-PCR validation	No PCR band in the untreated sample	Poor RNA quality or failed reverse transcription	Confirm RNA integrity and cDNA synthesis using a housekeeping transcript control.

A549 toxicity assay	Low viability in untreated cells	Poor cell health, overcrowding, contamination, or medium issue	Use actively growing A549 cells. Confirm complete D-MEM medium with 10 % fetal calf serum, penicillin, and streptomycin.
A549 toxicity assay	High variability in MTT readings	Uneven seeding, edge effects, or incomplete formazan solubilization	Mix the cell suspension well before seeding. Avoid using the plate's edge wells or fill them with sterile PBS/medium. Ensure complete solubilization before reading absorbance.
Data analysis	MIC80 is inconsistent between experiments	Different growth phase, inoculum size, or ASO batch	Normalize starting cultures carefully. Use the same ASO stock and passage number. Calculate inhibition relative to the matched untreated control for each experiment.

References

Altschul, S. F., Gish, W., Miller, W., Myers, E. W., & Lipman, D. J. (1990). Basic local alignment search tool. *Journal of Molecular Biology, 215*(3), 403–410. https://doi.org/10.1016/s0022-2836(05)80360-2.

Dhara, D., Hill, A. C., Ramesh, A., Traore, D., Radzikowska-Cieciura, E., Wood, M. J. A., & Brown, T. (2025). Biophysical and biological properties of splice-switching oligonucleotides and click conjugates containing LNA-phosphothiotriester linkages. *Nucleic Acids Research, 53*(21), gkaf1263. https://doi.org/10.1093/nar/gkaf1263.

Elbaiomy, R. G., El-Sappah, A. H., Guo, R., Luo, X., Deng, S., Du, M., ... Zhang, Z. (2025). Antibiotic resistance: A genetic and physiological perspective. *MedComm (2020), 6*(11), e70447. https://doi.org/10.1002/mco2.70447.

Elmquist, A., Lindgren, M., Bartfai, T., & Langel, Ü. (2001). VE-Cadherin-derived cell-penetrating peptide, pVEC, with carrier functions. *Experimental Cell Research, 269*(2), 237–244. https://doi.org/10.1006/excr.2001.5316.

Griffiths-Jones, S., Bateman, A., Marshall, M., Khanna, A., & Eddy, S. R. (2003). Rfam: An RNA family database. *Nucleic Acids Research, 31*(1), 439–441. https://doi.org/10.1093/nar/gkg006.

Kaloudas, D., Pavlova, N., & Penchovsky, R. (2019). EBWS: Essential bioinformatics web services for sequence analyses. *IEEE/ACM Transactions on Computational Biology and Bioinformatics, 16*(3), 942–953. https://doi.org/10.1109/tcbb.2018.2816645.

Larkin, M. A., Blackshields, G., Brown, N. P., Chenna, R., McGettigan, P. A., McWilliam, H., ... Higgins, D. G. (2007). Clustal W and clustal X version 2.0. *Bioinformatics (Oxford, England), 23*(21), 2947–2948. https://doi.org/10.1093/bioinformatics/btm404.

Lorenz, R., Bernhart, S. H., Höner zu Siederdissen, C., Tafer, H., Flamm, C., Stadler, P. F., & Hofacker, I. L. (2011). ViennaRNA package 2.0. *Algorithms for Molecular Biology, 6*(1), 26. https://doi.org/10.1186/1748-7188-6-26.

Martinez, J. L. (2014). General principles of antibiotic resistance in bacteria. *Drug Discovery Today: Technologies, 11*, 33–39. https://doi.org/10.1016/j.ddtec.2014.02.001.

Mosmann, T. (1983). Rapid colorimetric assay for cellular growth and survival: Application to proliferation and cytotoxicity assays. *Journal of Immunological Methods, 65*(1), 55–63. https://doi.org/10.1016/0022-1759(83)90303-4.

Ogawara, H. (2019). Comparison of antibiotic resistance mechanisms in antibiotic-producing and pathogenic bacteria. *Molecules (Basel, Switzerland), 24*(19), https://doi.org/10.3390/molecules24193430.

Otcheva, L. A., Traykovska, M., & Penchovsky, R. (2025). Targeting bacterial adenylate kinase mRNA with a chimeric antisense oligonucleotide for rational antibacterial drug development. *Molecules (Basel, Switzerland), 30*(16), https://doi.org/10.3390/molecules30163425.

Pavlova, N., & Penchovsky, R. (2022). Bioinformatics and genomic analyses of the suitability of eight riboswitches for antibacterial drug targets. *Antibiotics, 11*(9), https://doi.org/10.3390/antibiotics11091177.

Pavlova, N., Traykovska, M., & Penchovsky, R. (2023). Targeting FMN, TPP, SAM-I, and glmS riboswitches with chimeric antisense oligonucleotides for completely rational antibacterial drug development. *Antibiotics, 12*(11), https://doi.org/10.3390/antibiotics12111607.

Penchovsky, R., Georgieva, A. V., Dyakova, V., Traykovska, M., & Pavlova, N. (2024). Antisense and functional nucleic acids in rational drug development. *Antibiotics, 13*(3), 221. https://www.mdpi.com/2079-6382/13/3/221.

Penchovsky, R., Pavlova, N., & Kaloudas, D. (2021). RSwitch: A novel bioinformatics database on riboswitches as antibacterial drug targets. *IEEE/ACM Transactions on Computational Biology and Bioinformatics, 18*(2), 804–808. https://doi.org/10.1109/tcbb.2020.2983922.

Penchovsky, R., & Traykovska, M. (2015). Designing drugs that overcome antibacterial resistance: Where do we stand and what should we do? *Expert Opinion on Drug Discovery, 10*(6), 631–650. https://doi.org/10.1517/17460441.2015.1048219.

Popova, K. B., & Penchovsky, R. (2024). General and specific cytotoxicity of chimeric antisense oligonucleotides in bacterial cells and human cell lines. *Antibiotics, 13*(2), https://doi.org/10.3390/antibiotics13020122.

Sievers, F., & Higgins, D. G. (2014). Clustal omega. *Current Protocols in Bioinformatics, 48*, 3.13.11–13.13.16. https://doi.org/10.1002/0471250953.bi0313s48.

Traykovska, M., Otcheva, L. A., & Penchovsky, R. (2022). Targeting TPP riboswitches using chimeric antisense oligonucleotide technology for antibacterial drug development. *ACS Applied Bio. Materials, 5*(10), 4896–4902. https://doi.org/10.1021/acsabm.2c00628.

Traykovska, M., & Penchovsky, R. (2022). Engineering antisense oligonucleotides as antibacterial agents that target FMN riboswitches and inhibit the growth of staphylococcus aureus, listeria monocytogenes, and Escherichia coli. *ACS Synthetic Biology, 11*(5), 1845–1855. https://doi.org/10.1021/acssynbio.2c00013.

Traykovska, M., Popova, K. B., & Penchovsky, R. (2021). Targeting glmS ribozyme with chimeric antisense oligonucleotides for antibacterial drug development. *ACS Synthetic Biology, 10*(11), 3167–3176. https://doi.org/10.1021/acssynbio.1c00443.

## CHAPTER FOURTEEN

Characterization of metabolite-sensing riboswitches using *in vivo* reporter gene assays

Anna Hübenthal and Matthias Mack*
Department of Biotechnology, Institute for Technical Microbiology, Technische Hochschule Mannheim, Mannheim, Germany
*Corresponding author. e-mail address: m.mack@hs-mannheim.de

Contents

Methods in Enzymology, Volume 735
ISSN 0076-6879, https://doi.org/10.1016/bs.mie.2026.06.027

Abstract

Riboswitches are genetic regulatory elements based on RNA molecules that control the expression of genes. Most known metabolite-sensing riboswitches control expression of genes involved in anabolic reactions but a few also control catabolic reactions. In addition, riboswitches control the expression of genes encoding metabolite uptake systems. Binding of a specific small molecule ligand (< 1500 Da) to the aptamer ("binder") part of the riboswitch typically results in repression of the downstream genes, in many cases by either transcriptional termination or inhibition of translation. The downstream expression control is supported by the so called "expression platform" which is part of the riboswitch. In this chapter the analysis of thiamine pyrophosphate (TPP)-responsive riboswitches using a specialized dual-luciferase reporter gene assay is exemplarily described and may serve as a blueprint to study other bacterial riboswitches. Also, the mode of action of the riboswitches can be determined by testing transcriptional as well as translational gene fusions using different test plasmids. The effect of candidate ligands which negatively interfere with riboswitch functions can be investigated using this assay, provided the chemicals are taken up by the test organism. The assay can be thus to identify and characterize novel compounds that have the capacity to act as antibiotics. As riboswitches constitute widespread genetic control elements, they represent novel targets for antimicrobial compounds.

Abbreviations

TPP thiamine pyrophosphate.

1. Introduction

Metabolite-sensing bacterial riboswitches are RNAs typically located in the 5′-untranslated parts of mRNAs and control the synthesis of metabolism-related proteins (Mandal & Breaker, 2004; Kavita & Breaker, 2023). Riboswitches consist of an aptamer domain and a downstream expression platform. While the aptamer ("binder"; from the Latin *aptus*, meaning "to fit," and the Greek *meros*, meaning "part") is responsible for binding a specific small molecule ligand, the expression platform translates the ligand-induced conformational change into a response of the gene expression machinery or platform. Various types of expression platforms are exploited by riboswitches, but the two most common mechanisms involve regulation of premature transcription termination ("transcription control") and control of ribosome binding to mRNAs ("translation control") (Fig. 1). Across these known 55 classes and based on the type of ligand, riboswitches can be divided into several groups the biggest of which comprises riboswitches that bind and respond to the enzyme cofactors flavin mononucleotide (FMN), adenosylcobalamin or thiamine pyrophosphate (TPP) (Kavita & Breaker, 2023). All these metabolites are derived from RNA nucleotides or their precursors. Riboswitch aptamers typically bind their ligand with high affinities and dissociation constants appear to be in the nanomolar or even picomolar range (Blount & Breaker, 2006). While most reported riboswitches consist of one aptamer domain and the accompanying expression platform, aptamer regions have also been reported to exist in tandem, which allows an even more fine-tuned regulation, observed for example for glycine riboswitches that contain two adjacent aptamers, which function cooperatively. In contrast to the highly conserved aptamer regions, the expression platforms exhibit greater diversity in sequence and structure, corresponding with the variety of regulatory mechanisms identified.

For some riboswitches like the FMN riboswitch of *Escherichia coli*, also a combined mechanism of transcriptional and translational control was observed (Pedrolli et al., 2015a). Additionally, transcriptional pause sites within the riboswitch sequence were discussed to play an important role in

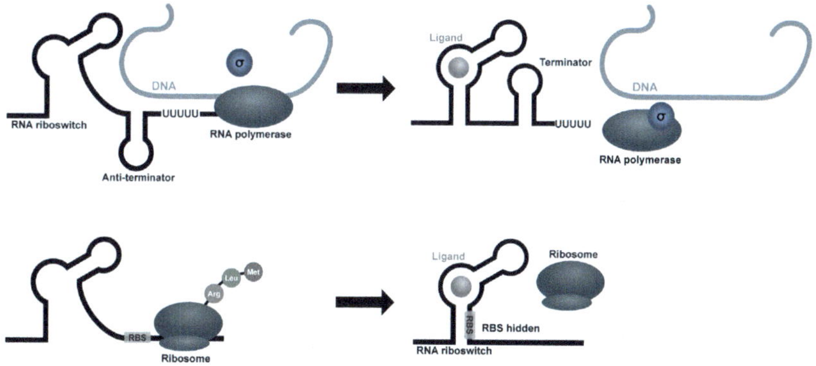

Fig. 1 Riboswitch regulation of gene expression upon ligand binding. *Via* formation of a Rho-independent transcriptional terminator (top) and (bottom) *via* sequestration of the ribosomal binding site (RBS). Sigma factors are components of the RNA polymerase holoenzyme that direct bacterial core RNA polymerase to specific promoter elements.

gene regulation by riboswitches. Other mechanisms of riboswitch regulation have been reported. These include Rho-dependent transcription termination, where binding of a ligand to the aptamer provides access to a Rho-binding sequence located in the riboswitch region, self-cleavage of RNA and RNA destabilization (Bastet et al., 2017). In addition a regulator protein (RibR) was discovered which modulates riboswitch activity (Pedrolli et al., 2015b; Hübenthal & Mack, 2025). In principle co-production of such a factor in an *in vivo* system may be used to help to characterize its function.

One of the well-studied riboswitches is the TPP riboswitch, which binds thiamine pyrophosphate, the active form of so called vitamin B1 (Winkler et al., 2002). Thiamine is an essential nutrient for animals and humans. The vitamin is synthesized not only by microorganisms but also by fungi and plants. TPP is an important cofactor for various enzymes involved in the carbohydrate and amino acid metabolism of a cell, including transketolase, pyruvate dehydrogenase, and α-ketoglutarate dehydrogenase. It is an important cofactor and has a molecular mass of 424 Da and consists of an aminopyrimidine ring linked by a methylene bridge to a thiazole ring with two phosphate groups attached to it. In all organisms producing thiamine, first thiamine monophosphate is synthesized from two precursors, the thiazole moiety 4-methyl-5-(ß-hydroxyethyl) thiazole monophosphate and the pyrimidine moiety 4-amino-5-hydroxymethyl-2-methylpyrimidine pyrophosphate. The monophosphate is then further phosphorylated to thiamine pyrophosphate, either by direct phosphorylation or by dephosphorylation to thiamine and subsequent pyrophosphorylation.

Soon after the first riboswitches had been experimentally validated, it was suggested that they may represent novel antibiotic targets. This is plausible, as riboswitches are important regulatory elements, involved in main metabolic pathways of bacteria. Targeting of these RNA elements should massively interfere with crucial biochemical processes within the cell. Even though TPP riboswitches have been found in some eukaryotic genomes, so far, no riboswitches have been identified in animals or humans. This should allow a specific targeting of bacterial cells and minimize the risk of side effects of a riboswitch targeting compound. Among the first thiamine analogs studied for their binding affinity to the TPP riboswitch was oxythiamine (Grabowska et al., 2021), a molecule very similar to thiamine, but with a hydroxyl group in place of the exocyclic amino group of the pyrimidine ring of thiamine. Another studied thiamine analog is pyrithiamine (Sudarsan et al., 2005). Here, a pyridine ring replaces the central thiazole ring of thiamine. Pyrithiamine has initially been synthesized to study thiamine metabolism, but turned out to be an effective antimicrobial negatively affecting bacterial species and fungi. Pyrithiamine and oxythiamine were used to validate the described *in vivo* test systems.

1.1 Definition

The following protocol is useful to characterize predicted or putative metabolite-sensing riboswitches, to determine their mode of action and to identify small molecule compounds which may negatively interfere with riboswitch function and thus lead to growth reduction or growth inhibition of the corresponding bacterium (Blount & Breaker, 2006; Panchal & Brenk, 2021; Hübenthal et al., 2025).

1.2 Rationale

RNA polymerases locate to promoters which in some bacterial species are well studied, while in other species they are less well characterized. Different sigma factors combine with RNA polymerases and only through the binding of a sigma factor to the RNA polymerase core enzyme is the functional holoenzyme formed, which is ready for the initiation of transcription (Fig. 1). This event is probably slightly different in all bacteria since different, species specific polymerases and sigma factors are used and it is therefore difficult to reliably test riboswitches in an environment such as *E. coli* which is used here. Riboswitches are dynamic structures which in bacteria build up at a speed of about 40–80 nucleotides per second and thus best are studied using a dynamic system involving an actively transcribing RNA polymerase. Riboswitches are composed of 34 to >300 nucleotides and it thus takes a few seconds for them to fully form and it is difficult to predict when exactly the metabolite ligands start to interact with the aptamer. The binding event may start as soon as the first secondary structures form, and the speed of interaction may be more important for folding the expression platform than the final dissociation constant, which *e.g.* can be determined by testing the fully folded riboswitch. Metabolite-sensing bacterial riboswitches are located in the 5′ untranslated regions of mRNA. The strength of the promoter, the nature of the sigma factor, and the speed of RNA polymerase in a test system may affect riboswitch function. Some riboswitches depend on additional factors that may not be present in an *E. coli* test system. Additionally, the temperature and ionic environment may not be ideal for testing a given riboswitch. However, from our experience of testing a wide variety of riboswitches from different bacterial species, we were surprised to find that almost all of the putative metabolite-sensing bacterial riboswitches predicted by bioinformatic tools were indeed functional in our test system. We are therefore confident that our test system is useful, at least for an initial functional characterisation of the riboswitches. The system is technically simple. The DNA sequences of the predicted riboswitches are synthesised using either a polymerase chain reaction (PCR), with an upstream and downstream primer in addition to a template, or by overlapping oligonucleotides (ordered as "synthetic genes"), if the riboswitches are derived from pathogenic organisms. The candidate riboswitch DNA is ligated to a test plasmid that contains the promoter. This promoter ensures the production of riboswitch RNA. Depending on the availability of a certain metabolite, the RNA structure of the riboswitch changes, which triggers the expression or non-expression of the downstream reporter gene. If the riboswitch contains a ribosomal binding site, the start codon and the first six codons of the original gene controlled by this riboswitch are fused in frame to the reporter gene. If the riboswitch contains a transcriptional terminator, binding of a certain metabolite will lead to termination of transcription, meaning that the downstream reporter gene will also not be expressed, even if a functional ribosomal binding site is present. The plasmids are constructed and used to transform the *E. coli* test system. The cells are grown to the exponential growth phase where *E. coli* RNA polymerase is present in large amounts in addition to one specific sigma factor, the sigma 70 factor (RpoD). The test promoter is sigma 70 dependent and thus the downstream gene is comparably strongly expressed. Again, it is important to keep in mind that the riboswitch to be tested may be transcribed by a another, species specific RNA polymerase in combination with other, species specific sigma factors and thus our test system may produce results which do not reflect the respective physiological conditions. The cells can be challenged with different metabolites which are added to the cultures of the test system. Subsequently, the cells are lysed and the reporter enzyme can be measured.

2. Materials, equipment and reagents

2.1 Chemicals

For all buffers and reagents, standard laboratory-grade chemicals from the following suppliers were used, if not stated otherwise: AppliChem GmbH, Darmstadt, Germany, Carl Roth GmbH & Co. KG, Karlsruhe, Germany, Caymen Chemical, Ann Arbor, MI, USA, Merck KGaA, Darmstadt, Germany, Thermo Fisher Scientific, Darmstadt, Germany and VWR

International GmbH, Darmstadt, Germany. All media and buffers listed below were prepared with ultrapure water (Milli-Q® water purification system), except for the Lysogeny broth (LB) medium, which was prepared with double distilled water. All heat-stable media and components were sterilized by autoclaving at 121 °C for 20 min (for 1 L). Heat-sensitive components like sugar solutions were filter sterilized using a 0.22 μm cellulose-acetate filter. LB routinely was prepared from a premix (Carl Roth GmbH + Co. KG, Karlsruhe, Germany), using 25 g of the compound mixture per liter of medium. All bacterial strains used were either grown in LB medium or in a minimal medium. For the β-galactosidase and luciferase assays, *E. coli* strains were grown in M9 medium with 0.4 % (w/v) glucose as a carbon source and 0.1 % (w/v) casamino acids. Cultures of *E. coli* DH5α were additionally supplemented with 20 nM of thiamine. Pre-cultures were grown in test tubes in 6–10 mL of medium or in a 24-well plate with a culture volume of 1 mL per well, while cultures with a larger volume were grown in baffled shake flasks.

3. Manipulation of DNA and construction of test plasmids

Plasmids were isolated using the GeneJET Plasmid Miniprep Kit (Thermo Fisher Scientific, Waltham, USA) according to the instructions of the manufacturer. Pre-cultures of *E. coli* strains harbouring the respective plasmid were grown overnight in LB medium supplemented with the appropriate antibiotics. 1–2 mL of the culture volume was transferred to a microcentrifuge tube and centrifuged at 6000 × *g* for 3 min at room temperature (Microcentrifuge 5415 R, Eppendorf, Hamburg, Germany). The supernatant was discarded, and plasmid extraction was performed from the cell pellets. For preparation of plasmids, the GeneJET Plasmid Maxiprep Kit was used with adjusted cell culture volumes, according to the recommendations of the manufacturer. Plasmids were always eluted with 30–40 μL of ultrapure water in case of the Miniprep kit and 700 μL of water for the Maxiprep kit.

4. Polymerase chain reactions

For amplification of specific DNA sequences polymerase chain reactions (PCR) were performed using Phusion Hot Start II DNA Polymerase (Thermo Fisher Scientific, Waltham, USA).

Reaction mixtures for PCR amplifications with the Phusion HS II DNA Polymerase.

Component	Volume/μL
Template DNA	20–50 ng
5x High GC buffer	10
Reverse primer (10 μM)	2.5
Forward primer (10 μM)	2.5
dNTP mix (10 mM)	2
Phusion HS II DNA polymerase (2 U/μL)	0.5
ddH_2O	*ad* 50 μL

The annealing temperatures for all primer pairs used were calculated using the New England Biolabs T_m calculator. For Phusion polymerase reactions, the actual annealing temperature was set two degrees lower than the calculated annealing temperature.

5. Restriction and ligation of DNA fragments and plasmids

DNA fragments and plasmids were digested for cloning with Fast Digest Restriction Enzymes (Thermo Fisher Scientific, Waltham, MA, USA). For a 10 μL reaction, 1 μL of total restriction enzyme were used. To prevent religation, the FastAP Thermosensitive Alkaline Phosphatase (Thermo Fisher Scientific, Waltham, MA, USA) was added to the restriction reaction of plasmids, according to the instructions of the manufacturer. For DNA fragments, 200–300 ng per 10 μL reaction were digested, for plasmids 1 μg per 10 μL were used. Either the 10x FastDigest buffer or the 10x FastDigest Green buffer (Thermo Fisher Scientific, Waltham, MA, USA) were used in the restriction reactions. Reaction mixtures were incubated at 37 °C for a specified time, depending on the required duration indicated for each restriction enzyme. Reactions were stopped with an incubation at 80 °C for 10–20 min.

6. DNA purification

Linear DNA from polymerase chain reactions or restriction reactions was purified with the GeneJET PCR Purification Kit (Thermo Fisher Scientific, Waltham, MA, USA), according to the instructions of the manufacturer. The DNA was eluted from the column with 20–40 µL of ultrapure H_2O. *Hin*dIII-linearized plasmid pSP64 was purified from the 5 mL restriction reaction *via* phenol-chloroform extraction: 5 mL of a phenol/chloroform/isoamyl alcohol mix (25:24:1, v/v) (Carl Roth GmbH + Co. KG, Karlsruhe, Germany) were added to the restriction reaction in a 50 mL conical tube. The tube was centrifuged at 4 °C and maximum speed for 20 min. The aqueous phase was then carefully transferred to a fresh tube and mixed with one volume of a chloroform/isoamyl alcohol mixture (24:1, v/v). Subsequently, the contents of the tube were mixed again and centrifuged at 4 °C and maximum speed for 20 min. The aqueous phase was transferred to a fresh tube and mixed with 1.2 volumes of ice-cold isopropanol (100 %). After the mixture was incubated at −20 °C for 20 min, a third centrifugation was performed under the same conditions as before. The supernatant was carefully discarded, and the DNA pellet was washed with 2 mL of 70 % ethanol. The tube was centrifuged again at 4 °C and maximum speed for 15 min. Ethanol was carefully removed, and the DNA pellet was dissolved in 200 µL of ultrapure H_2O. DNA fragments or linearized plasmids used for cloning were usually purified from an agarose gel (1 % w/v) to ensure that there were as little contaminations with unwanted DNA or unrestricted vector as possible. Gel purification was performed with the Monarch DNA Gel Extraction Kit (New England Biolabs GmbH, Frankfurt a. M., Germany), following the instructions of the manufacturer. Elution was performed in 10–15 µL of prewarmed (room temperature), ultrapure water.

7. Measuring DNA and RNA concentrations

Concentrations of DNA and RNA solutions were determined by absorption measurement using a microtiter plate reader (with a built in spectrophotometer). For this purpose, 2 µL of the respective nucleic acid solution were applied to a NanoQuant plate (Tecan, Männedorf, Switzerland) and the absorption at 260 nm was measured in a plate reader (Tecan Spark). Additionally, the ratios of the absorptions at 260/230 nm and 260/280 nm were determined to check for DNA/RNA purity, with expected 260/280 ratios of about 1.8 for pure DNA and about 2.0 for pure RNA samples, as well as an expected 260/230 ratio of 2.0–2.2.

8. DNA sequencing

Sequencing of PCR products and plasmids was performed by LGC Genomics GmbH (Berlin, Germany). In accordance with the suggestions of the service company, plasmids were diluted to a concentration of 100 μg/μL, while linear DNA fragments were diluted to a concentration of 10–40 μg/μL, depending on the fragment size. 2 μL of the respective sequencing primer (10 μM) were added to 10 μL of the DNA solution and the total volume of the sequencing mixture was adjusted to 14 μL with ddH_2O. Sequencing was performed according to the Sanger sequencing method by LGC Genomics GmbH (Berlin, Germany). The obtained sequencing results were analyzed using the software CLC Main Workbench 20.

9. Construction of the β-galactosidase reporter plasmid pHAt191l

For construction of the β-galactosidase reporter gene plasmid pHAt191l, pUC19 was used as a backbone. To delete the *lac* promoter, the whole plasmid was amplified by PCR with the overlapping oligonucleotides (GCG CAA CGC AAT CAC ACA GGA AAC AGC T) and (GTT TCC TGT GTG ATT GCG TTG CGC TCA C) and the resulting sequence with complementary overhangs was ligated to create pUC19del. The *lacZ* reporter gene for a translational fusion was amplified from the expression plasmid pDG268, using the primer pair (TAC ATT CTA GAC GGG TCG ACC GGG AAA AC and TAC ATG AAT TCG AAA TAC GGC AGA CAT GG) and inserted into *Xba*I/*Eco*RI restricted pUC19del plasmid, creating pHA191. The terminator sequences for the *lac* terminator (GCC TGA TAC GCT GCG CTT ATC AGG CCT ACA AGT TCA GCG ATC TAC ATT AGC CGC ATC CGG CAT GAA CAA AGC GCA GGA ACA AGC GTC GCA TCA TGC CTC TT) and the *rrnB* T1 terminator (GAA GAG CTA GGG AAC TGC CAG GCA TCA AAT AAA ACG AAA GGC TCA GTC GAA AGA CTG GGC CTT TCG

TTT TAT CTG TTG TTT GTC GGT GAA CGC TCT CCT GAG TA) were both amplified from genomic DNA of *E. coli* MG1655 with the modifying primer pairs (TAC TAG AAT TCGC CTG ATA CGC TGC GCT TA and TAC TAG AAT TCA AGA GGC ATG ATG CGA CG) and (TAT AAA GCG GAA GAG CTA GGG AAC TGC CAG GCA TTA CAT AAG CTT TAC TCA GGA GAG CGT TCA). The *rrnB* T1 terminator was inserted between the *Sap*I and *Hin*dIII restriction sites upstream of *lacZ*, while the *lac* terminator was inserted downstream of *lacZ* at the *Eco*RI restriction site, creating the plasmid pHAt191l (Fig. 2). The TPP riboswitch region Ec01 from *E. coli*, including the respective promoter, ribosomal binding site and the first few codons of the downstream gene *thiC* was amplified from genomic DNA with the oligonucleotide pair (TCA CTA AGC TTG CTG GAA GAT AAG CTG ATT and TAT ATT CTA GAG CGT TGT TCG CGG CGG GT) and inserted between the *Hin*dIII and *Xba*I restriction sites of the plasmid to generate the reporter plasmid pHAt191l::Ec01 (see Table 1 for the corresponding nucleotide sequences). To create the promoter control plasmid pHAt191l::pEc01, the *E. coli thiC* promoter region was individually amplified with the oligonucleotides (P11 TCA CTA AGC TTG CTG GAA GAT AAG CTG ATT)

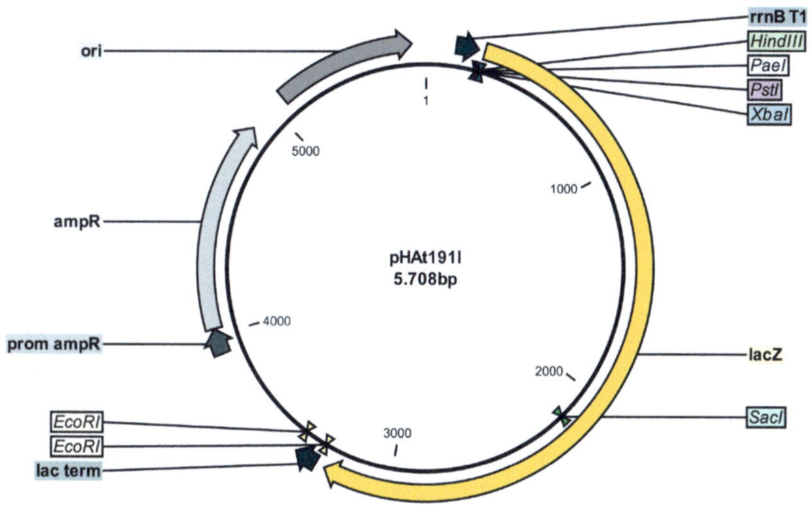

Fig. 2 Plasmid map of pHAt191l. Depicted are the β-galactosidase coding sequence (*lacZ*), the origin of replication (ori), the ampicillin resistance gene (*ampR*) and its promoter (prom *ampR*), the *rrnB* T1 terminator (*rrnB* T1), the *lac* terminator (*lac* term) and the relevant restriction sites.

Table 1 Sequences of TPP riboswitch regions (including promoters, riboswitches, ribosomal binding sites and the first few codons of the downstream genes) in the test plasmids (*e.g.* Ec01) and the respective promoter control regions (promoter + ribosomal binding sites and start of coding sequence; *e.g.* pEc01). Relevant sequences are highlighted in different colors according to the following color code: Start codon, (putative) promoter region, predicted riboswitch RS region, (putative) RBS, *thi* gene. Restriction sites are displayed in bold.

Ec01 (pHAt191l)	**AAGCTT**GCTGGAAGATAAGCTGATTCTGCTGGTGCTT GACGCCGCCCGCGTCAAACATCCTGCTTGAGTTCTGCG CTGTTAACGCGTAATTTACATTCAATGCCCCATTTGCG GGGCTAATTTCTTGTCGGAGTGCCTTAACTGGCTGAGA CCGTTTATTCGGGATCCGCGGAACCTGATCAGGCTAAT ACCTGCGAAGGGAACAAGAGTTAATCTGCTATCGCATC GCCCCTGCGGCGATCGTCTCTTGCTTCATCCGTCGTCT GACAAGCCACGTCCTTAACTTTTTGGAATGAGCTATGT CTGCAACAAAACTGACCCGCCGCGAACAACGC**TCTAGA**
pEc01 (pHAt191l)	**AAGCTT**GCTGGAAGATAAGCTGATTCTGCTGGTGCTTG ACGCCGCCCGCGTCAAACATCCTGCTTGAGTTCTGCGC TGTTAACGCGTAATTTACATTCAATGCCCC**CTGCAG**TT TTGGAATGAGCTATGTCTGCAACAAAACTGACCCGCCG CGAACAACGC**TCTAGA**
Kp04 (pHAt191l)	**AAGCTT**GATAATATTGCGCCGACCGCTTGAGTTGTGAA CCGTTAACGCGTAATTTACGTACATTATCCCTCTGCGG AGGGATTTCATCTTGTCGGAGTGCCTATTTTCCATGCG CAGGCTAACGTGGAAAAGGCTGAGACCGTTAATTCGGG ATCCGCGGAACCTGATCAGGCTAATACCTGCGAAGGGA ACAAGAGTAAACTGCTGTTCGCGTCAGTCCGTCAGGGC CGATCGCATCTGTTACTCCATCCGTCGTCTGACAAGCC ATGTCCTTTTTTAACTGGAATGCGCTATGTCTACTACC AAACTAACCCGT**TCTAGA**
pKp04 (pHAt191l)	**AAGCTT**GATAATATTGCGCCGACCGCTTGAGTTGTGAA CCGTTAACGCGTAATTTACGTACATTATCCCTCTGCGG AGGGATTTCAT**CTGCAG**GCCATGTCCTTTTTTAACTGG AATGCGCTATGTCTACTACCAAACTAACCCGT**TCTAGA**
pEc01 (pDluc)	**CAGCTG**GCTGGAAGATAAGCTGATTCTGCTGGTGCTTG ACGCCGCCCGCGTCAAACATCCTGCTTGAGTTCTGCGC TGTTAACGCGTAATTTACATTCAATGCCCC**CCATGG**GG GAGGGAAACAAATGT**GGATCC**

Ec01 (pDluc)	**CAGCTG**GCTGGAAGATAAGCTGATTCTGCTGGTGCTT GACGCCGCCCGCGTCAAACATCCTGCTTGAGTTCTGC GCTGTTAACGCGTAATTTACATTCAATGCCCCATTTG CGGGGCTAATTTCTTGTCGGAGTGCCTTAACTGGCTG AGACCGTTTATTCGGGATCCGCGGAACCTGATCAGGC TAATACCTGCGAAGGGAACAAGAGTTAATCTGCTATC GCATCGCCCCTGCGGCGATCGTCTCTTGCTTCATCCG TCGTCTGACAAGCCACGTCCTTAACTTTTTGGAATGA GCTATGTCTGCAACAAAACTGACCCGCCGCGAACAAC GCT**GGATCC**
Kp04 (pDluc)	**CAGCTG**GATAATATTGCGCCGACCGCTTGAGTTGTGA ACCGTTAACGCGTAATTTACGTACATTATCCCTCTGC GGAGGGATTTCATCTTGTCGGAGTGCCTATTTTCCAT GCGCAGGCTAACGTGGAAAAGGCTGAGACCGTTAATT CGGGATCCGCGGAACCTGATCAGGCTAATACCTGCGA AGGGAACAAGAGTAAACTGCTGTTCGCGTCAGTCCGT CAGGGCCGATCGCATCTGTTACTCCATCCGTCGTCTG ACAAGCCATGTCCTTTTTTAACTGGAATGCGCTATGT CTACTACCAAACTAACCCGTT**GGATCC**
pKp04 (pDluc)	**CAGCTG**GATAATATTGCGCCGACCGCTTGAGTTGTGA ACCGTTAACGCGTAATTTACGTACATTATCCCTCTGC GGAGGGATTTCAT**CTGCAG**GCCATGTCCTTTTTTAAC TGGAATGCGCTATGTCTACTACCAAACTAACCCGTT **GGATCC**

and inserted between the *Hin*dIII and *Pst*I restriction sites of pHAt191l. Subsequently, the ribosomal binding site together with the start of the coding sequence of *thiC* was amplified with the primer pair (P12 TAC ATC TGC AGT TTT GGA ATG AGC TAT G) and inserted downstream of the promoter region between the *Pst*I and the *Xba*I restriction site. The reporter plasmid pHAt191l::Kp04 and control vector pHAt191l::pKp04 for the *thiC* riboswitch from *K. pneumoniae* were generated in a similar way. Both, the riboswitch region and the control insert were designed and obtained as synthetic DNA constructs ligated to the pMA plasmid from GeneArt (now Thermo Fisher Scientific, Waltham, MA, USA). The two inserts were amplified from the pMA plasmid using PCR and ligated to *Hin*dIII/*Xba*I restricted pHAt191l. All the generated plasmids were propagated in *E. coli* DH5α and proof sequenced.

10. Dual-luciferase reporter plasmid pDluc

The dual-luciferase plasmid was constructed based on the plasmid pPrib-RibDG-RFN-luc which originated from the single-luciferase reporter plasmid pT7luc (Promega, Mannheim, Germany). To generate a dual-luciferase plasmid, a second luciferase gene with a codon-adapted sequence coding for the *Renilla reniformis* luciferase (luc^R) was inserted downstream of the firefly luciferase gene (luc^F) on the pPrib-RibDG-RFN-luc plasmid. The luc^R gene was adapted for efficient expression in *E. coli*. The luc^R gene was placed under control of the strong constitutive promoter OXB15 (derived from the *recA* promoter by random mutagenesis, Oxford Genetics Limited, Oxford, UK), and was used as an internal control reporter. The *rrnB* T1 terminator sequence was produced by PCR from *E. coli* chromosomal DNA with the primer pair (TAC ATG AGC TCT AGG GAA CTG CCA GGC ATC A and TAC ATG AGC TCC AGA ACT CGA GTA CTC AGG AGA GCG TT), introducing *Sac*I restriction sites at each end of the fragment and an additional *Xho*I restriction site at the 3'-end of the terminator sequence. The *rrnB* T1 fragment was ligated to *Sac*I restricted pPrib-RibDG-RFN-luc, downstream of luc^F to generate pPrib-RFN-luc-t. A *Kpn*I restriction site was introduced into the plasmid 12 base pairs downstream of the newly introduced *Xho*I site by amplifying the whole plasmid using the overlapping primers (ATA GGT CGG TAC CGA GCT CCC AAA AAA AAA A and GGT ACC GAC CTA TCA GAA CTC GAG TAC TCA G) and ligation of the resulting sequence. The DNA sequence comprising the OXB15 promoter and the codon-adapted luc^R gene (see above) was obtained as a synthetic DNA (Thermo Fisher Scientific, Waltham, MA, USA), was amplified by PCR using the modifying oligonucleotide pair (TCA GAC TCG AGT ACT TGT and TCG TAG GTA CCT GTC ACT GCA GTT ACT GTT) and ligated to *Xho*I/*Kpn*I restricted pPrib-luc.t, creating the plasmid pPrib-RFN-Dluc. Since the luc^R sequence contains a *Hin*dIII restriction site, the *Hin*dIII site upstream of luc^R in pPrib-RFN-Dluc was replaced with an *Nco*I site by a site-directed mutagenesis reaction using the overlapping oligonucleotides (GCT AAT CAT TAG CGT TAT AGT GAA TCC GCT CCA TGG TTT AAG GAC AAA TGA ATA AAG ATT GTA TCC GGA TAC AAT CTT TAT TCA TTT GTC CTT AAA CCA TGG AGC GGA TTC ACT ATA ACG CTA ATG ATT AGC). To delete the *Bacillus subtilis* FMN riboswitch sequence from this plasmid, the complementary oligonucleotides (CAT GGG GGA GGG AAA CAA ATG TG and GAT CCA CAT TTG TTT CCC TCC CC), containing an RBS and an ATG start codon, were used. The two oligonucleotides (10 μM each) were mixed in 1x NEBuffer™ 2 (New England Biolabs GmbH, Frankfurt am Main, Germany) and annealed by heating the mixture to 98 °C for five minutes, followed by slow cooling to room temperature (leave on the bench). A linker with sticky ends was therefore generated and ligated into the *Nco*I/*Bam*HI-restricted pPrib-RFN-Dluc, creating pPrib-Dluc. To remove the *rib* promoter sequence from this plasmid, the complementary oligonucleotides (CTG TGC TGC CCA AGG TAT ATC TCC CAT GGG AGA TAT ACC TTG GGC AGCA CAG) were annealed to create a second linker with a blunt end and a sticky end to be inserted into *Pvu*II/*Nco*I restricted pPrib-Dluc. The generated double-strand linker was ligated to pPrib-Dluc to create the translational fusion plasmid pDluc (Fig. 3). Another site-directed mutagenesis reaction with the primer pair (CAT GGG AGA TAT ACC TTG GGC AGC ACA G and AAA CAA ATG TGG ATC CAA TAG AGG GCC CGC ATC CAA ATG GAA GAC GCC AAA AAC) was used to insert a stop codon and a ribosomal binding site upstream of luc^F, as well as a start codon for the luc^F gene to generate the transcriptional fusion plasmid pDlucTC (Fig. 4). For testing the

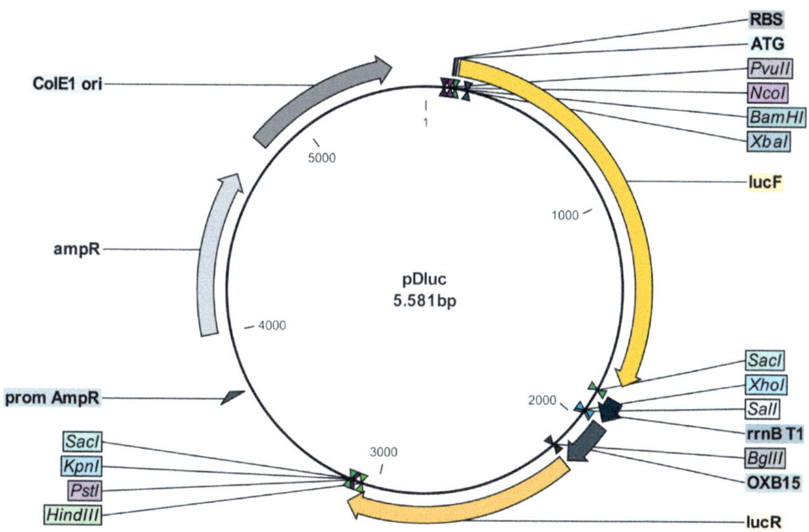

Fig. 3 Plasmid map of pDluc. Depicted are the firefly luciferase coding sequence (luc^F) with the upstream ribosomal binding site (RBS) and additional start codon (ATG), the Renilla luciferase coding sequence (luc^R) the origin of replication (ColE1 ori), the ampicillin resistance gene (*ampR*) and its promoter (prom *ampR*), the *rrnB* T1 terminator (*rrnB* T1), the OXB15 promoter (OXB15) and the relevant restriction sites.

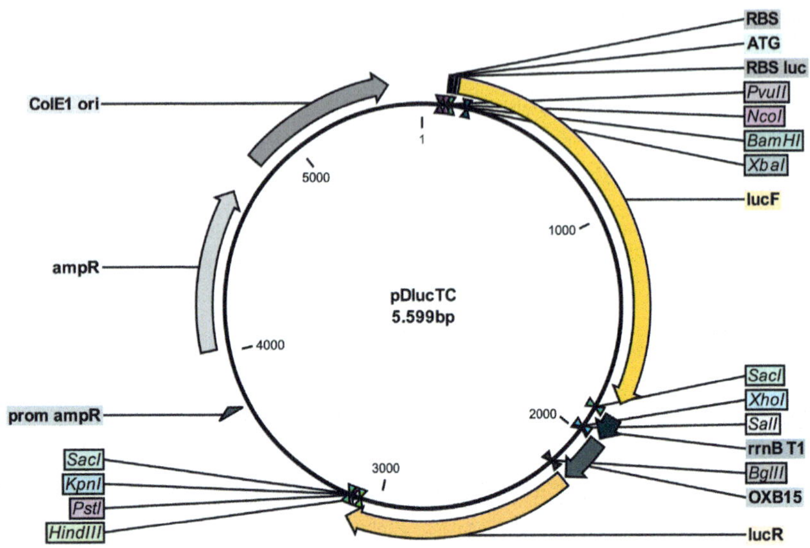

Fig. 4 Plasmid map of pDlucTC. Depicted are the firefly luciferase coding sequence (*luc*[F]) with the upstream ribosomal binding site (RBS), the *luc*[F] RBS (RBS *luc*) and additional start codon (ATG), the Renilla luciferase coding sequence (*luc*[R]) the origin of replication (ColE1 ori), the ampicillin resistance gene (*ampR*) and its promoter (prom *ampR*), the *rrnB* T1 terminator (*rrnB* T1), the OXB15 promoter (OXB15) and the relevant restriction sites.

TPP riboswitch function, the *thiC* promoter of *E. coli* was amplified with the primers (TCA CTC AGC TGG CTG GAA GAT AAG CTG ATT) and TCA CTC CAT GGG GGG CAT TGA ATG TAA AT and the promoter region was ligated to *Pvu*II/*Nco*I restricted pDluc and pDlucTC, respectively, creating plasmids pDluc::pEc01 and pDlucTC::pEc01 (see Table 1 for the corresponding nucleotide sequences). The TPP riboswitch regions from different bacterial species including the promoter, riboswitch, RBS together with the first few codons of the downstream gene were obtained as synthetic DNA constructs cloned into the GeneArt pMA plasmid (Thermo Fisher Scientific, Waltham, MA, USA). The riboswitches Ab01, Kp04, Eb01, Eb03, Ms01, Ms02, Pa01 and Sa02 were amplified from these templates using the respective primer pairs creating inserts with sticky ends for ligation with the *Nco*I/*Bam*HI restricted pDluc::pEc01 and pDlucTC::pEc01 For the TPP riboswitches Ec01, Ab01, Ef02, Kp04, Ms01, Ms02, Sp06, Kp01, Kp10 and Kp11 the whole regions including the native riboswitch promoter were amplified from the custom templates using the respective primer pairs. The amplified riboswitch inserts were cloned between the *Pvu*II/*Bam*HI restriction sites of pDluc and pDlucTC. The respective promoter control regions consist of the native riboswitch promoter and the RBS together with the first codons of the downstream gene but lack the riboswitch sequence itself. These sequences were also obtained as synthetic DNA constructs. All cloning steps described were individually verified by DNA sequencing. The plasmids were propagated using *E. coli* DH5α and then used to transform *E. coli* DH5α and *E. coli* MG1655.

11. Step-by-step procedure for using the β-galactosidase reporter plasmid pHAt191l

1. To analyze the activity of TPP riboswitches and their response to thiamine-analogs, at first a β-galactosidase reporter gene assay was employed. For the assay, *E. coli* DH5α strains carrying the plasmid pHAt191l with a translational fusion of the promoter and TPP riboswitch region with the *lacZ* reporter gene were used.
2. Pre-cultures of the strains were grown in 1 mL of M9 medium with 100 μg/mL ampicillin, supplemented with 20 nM of thiamine.
3. The cultures were incubated overnight in a 24-well plate at 37 °C and shaking at 160 rpm.
4. The OD_{600} of each overnight culture was determined and the culture volume was transferred to a 1.5 mL microcentrifuge tube.
5. The cells were pelleted by centrifugation at 6000 × *g* for 5 min and washed once with fresh M9 medium without supplements.
6. The cell pellets were then resuspended in fresh M9 medium and 10 μL of the cell suspension were used to inoculate 200 μL of fresh M9 medium in a transparent, flat-bottom 96-well plate (Greiner Bio-One GmbH, Frickenhausen, Germany) to an initial OD_{600} of 0.1.
7. Thiamine and different thiamine analogs dissolved in DMSO were added to the cultures to a concentration of 1 mM and the plate was incubated for 4 h at 37 °C with shaking at 180 rpm.
8. After the incubation time, 100 μL of each culture were transferred to a new 96-well plate and the OD_{600} was determined in a plate reader.
9. 10 μL of a 1:1 mixture of PopCulture reagent (Merck KGaA, Darmstadt, Germany) and lysozyme solution (50 mg in 1 mL of Z buffer; see Table below) were added to each well.
10. The plate was incubated at room temperature for 15 min to enable cell lysis.

11. Afterwards, 15 μL of each cell lysate were mixed with 155 μL of Z buffer containing 0.135 % of dithiothreitol (DTT).
12. 30 μL of the LacZ substrate *ortho*-nitrophenyl-β-galactoside (ONPG) (4 mg in 1 mL of Z buffer) were added to each well.
13. After addition of the substrate, the β-galactosidase activity was determined by following the conversion of ONPG to *ortho*-nitrophenol (ONP).
14. Absorption at 420 nm was measured at 30 °C for up to 1 h. The activity of the β-galactosidase in Miller units was calculated with the following formula:

$$Activity = \frac{OD_{420}}{\Delta t \min \times V\ mL \times OD_{600}} \times 1000$$

Z buffer for the β-galactosidase reporter assay.

Component	Mass conc.	Final conc.
$Na_2HPO_4{\cdot}7H_2O$	16.08 g/L	60 mM
NaH_2PO_4	5.52 g/L	40 mM
KCl	0.746 g/L	10 mM
$MgSO_4$	0.247 g/L	1 mM

12. Application of the described classical β-galactosidase reporter gene assay: Validation of riboswitch functions

The following part describes the successful application of a classical β-galactosidase reporter gene assay to study riboswitches. One of the goals regarding the establishment of *in vivo* reporter gene assays to test the functions of riboswitches was to validate these RNA molecules as unique targets for antibacterial compounds. As a positive control the natural known or suspected ligands were used. In addition, our test systems were used to validate the functions of riboswitches predicted by bioinformatic tools. A variety of TPP riboswitches from different bacteria were used to test, fine-tune and validate the system. The results are presented in the subsequent sections and could be useful for studying riboswitches that have not yet been explored. The six leading pathogens are *E. coli*, *Staphylococcus aureus*, *Klebsiella pneumoniae*, *Streptococcus pneumoniae*, *Acinetobacter baumannii* and *Pseudomonas*

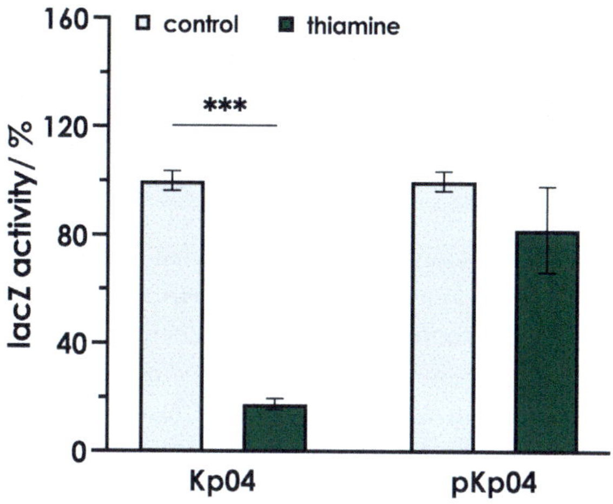

Fig. 5 Thiamine decreases β-galactosidase (LacZ) activity in an *Escherichia coli* strain designated as Kp04 with a translational fusion of the *Klebsiella pneumoniae thiC* riboswitch and the reporter gene *lacZ*. *E. coli* strains transformed with the reporter plasmids were grown in M9 minimal medium in the absence (control) or presence of 10 μM thiamine. LacZ activity was measured using *o*-nitrophenyl-β-D-galactopyranoside as a substrate. In strain Kp04 the native *K. pneumoniae* promoter *thiC* is coupled to the riboswitch *thiC* and thus expression of the reporter gene is controlled by this riboswitch. Accordingly, the reduced gene expression upon addition of thiamine indicates that thiamine turns the riboswitch off. In strain pKp04 the reporter gene is constitutively expressed from the *K. pneumoniae thiC* promoter as no riboswitch is present. LacZ activity is given as relative activity compared to the control. Cultures were grown in triplicates in a plate reader in a 12-well plate. Depicted are the mean values ± standard deviations of the data obtained from the triplicates. Asterisks indicate statistically significant differences determined by a two-tailed Students *t*-test (*** $p \leq 0.001$).

aeruginosa. Additional pathogens that are responsible for a considerable number of deaths around the world are *Enterobacter* spp. and *Enterococcus faecium*. A special focus has been placed on six of these species with the potential to become multidrug-resistant: the ESKAPE pathogens. These are *E. faecium*, *S. aureus*, *K. pneumoniae*, *A. baumannii*, *P. aeruginosa* and *Enterobacter spp*. Prior to testing compounds which potentially affect riboswitch activity, the putative riboswitches had to be validated in these bacterial strains. *E. coli* was used as a host to test riboswitch activity *in vivo* and, accordingly, a plasmid was constructed which allowed functional coupling of a range of riboswitches under control of different promoters to a reporter gene. Initially, the *E. coli* gene encoding β-galactosidase (LacZ) was employed as a reporter (*E. coli* K-12 MG1655, Gene ID: 945006). The first riboswitch sequence that was exemplarily examined was the *thiC* riboswitch from the pathogen *K. pneumoniae* (Kp04), which was predicted by bioinformatic tools and which is listed in Rfam TPP (RF00059). To validate its riboswitch activity, a translational fusion plasmid (pHAt191l) (Fig. 2) with the *lacZ* reporter gene fused to the first few codons of the native downstream gene *thiC* was used. In Table 1 the cloning procedures and the elements of the test plasmids exemplarily for a few constructs can be followed at the nucleotide level. For the assay, thiamine was added to the growth medium, which is expected to be taken up by the *E. coli* cells and phosphorylated to its active form thiamine pyrophosphate (TPP). When the reporter gene was fused to the complete riboswitch sequence including the native promoter a decrease in *lacZ* expression was observed when thiamine was present in the cultures (Fig. 5) This effect was absent when only the Kp04 promoter (pKp04) was regulating reporter gene expression.

13. Application of the described classical β-galactosidase reporter gene assay: Testing the effect of different small molecule compounds on the *thiC* riboswitch of *K. pneumoniae*

Following the validation of the *K. pneumoniae thiC* riboswitch (see preceding section), 18 small compounds (Table 2) were exemplarily tested for their effect on this riboswitch. The method should in principle also be

Table 2 List of small molecule compounds 1–18, tested for their effect on riboswitch activity in the β-galactosidase reporter gene assay.

Substance 1	**1-[4-(1H-imidazol-2-yl)phenyl] methanamine**
Substance 2	1-[4-(1H-pyrazol-5-yl)phenyl] methanamine
Substance 3	1-[4-(1,3-Thiazol-2-yl)phenyl] methanamine hydrochloride
Substance 4	2-(1H-imidazol−2-yl) pyridine
Substance 5	1-[4-(1H-pyrazol-1-yl)phenyl] methanamine
Substance 6	4-(2H-tetrazol-5-yl) pyridine
Substance 7	1-[4-(1H-1,2,4-triazol-1-yl)phenyl] methanamine dihydrochloride
Substance 8	1-[4-(5-Methyl-1H-1,2,4-triazol-3-yl)phenyl] methanamine hydrochloride

Substance 9	4-(1H-1,2,4-triazol-3-yl)phenyl] methanamine hydrochloride
Substance 10	4-(2H-tetrazol-5-yl)-benzenemethanamine hydrochloride
Substance 11	1-[4-(1H-imidazol-4-yl)phenyl] methanamine dihydrochloride
Substance 12	4-(1,2,3-Thiadizol-4-yl) benzylamine hydrochloride
Substance 13	Thieno [2,3-b]pyrazin-7-amine
Substance 14	1-[4-(1H-Pyrazol-4-yl)-2-thienyl] methanamine
Substance 15	1-[4-(1H-Pyrazol-3-yl)-2-thienyl] methanamine
Substance 16	1-[5-fluoro-2-methyl-4-(1H-1,2,3-triazol-5-yl)phenyl] methanamine
Substance 17	1-[5-(4H-1,2,4-Triazol-3-yl)-2-pyrimidinyl] methanamine
Substance 18	6-(pyrazin-2-yl)-1,2,3,4-tetrahydroisoquinoline

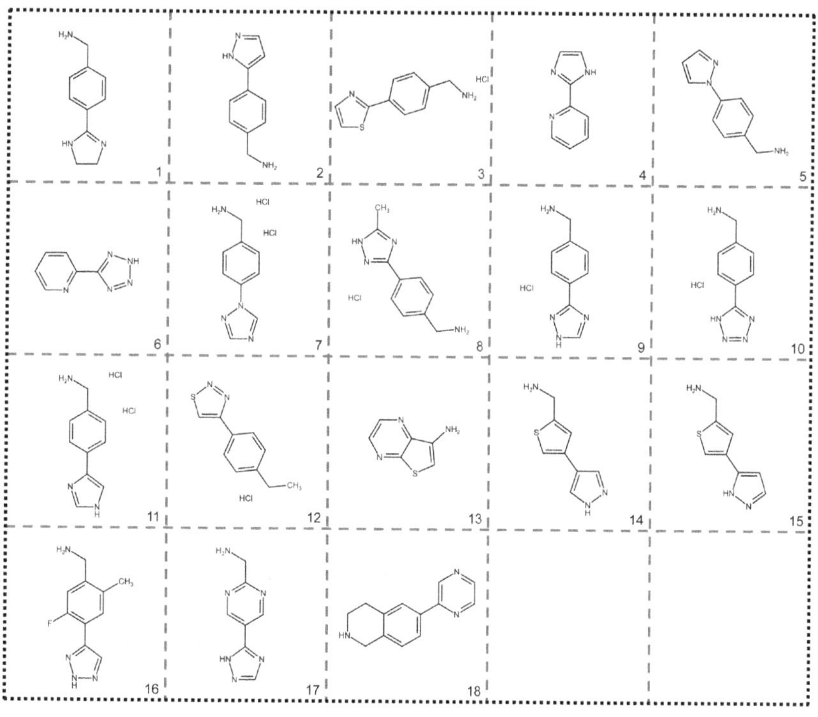

Fig. 6 Chemical structures of the compounds listed in Table 2.

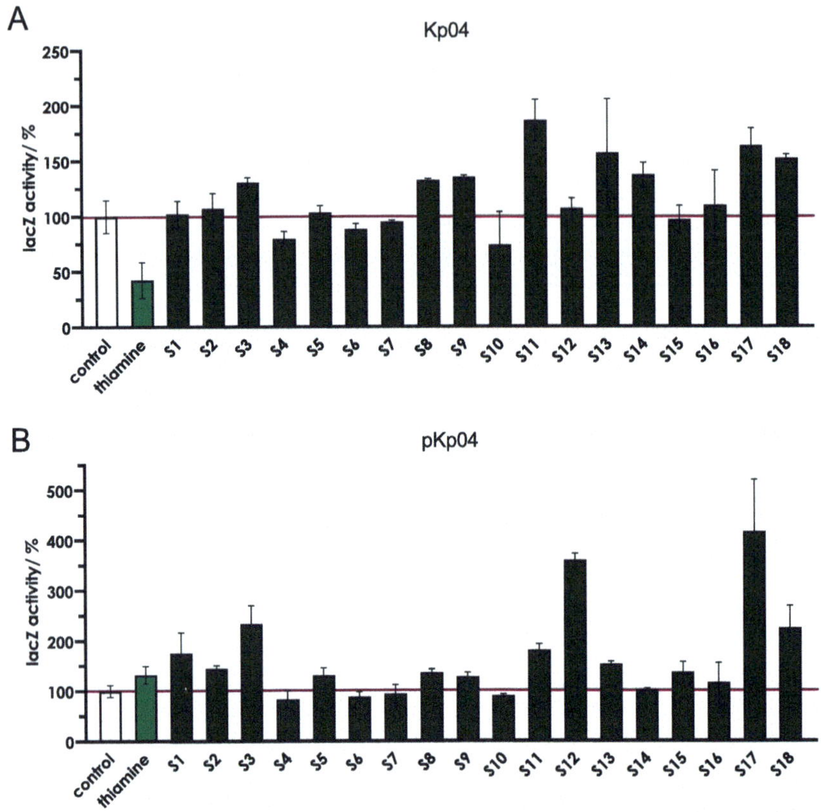

Fig. 7 β-galactosidase reporter gene assays for (A) the *K. pneumoniae thiC* riboswitch Kp04 and (B) the promoter control strain pKp04. *E. coli* DH5α strains transformed with the pHAt191I reporter plasmids were grown in M9 minimal medium in the absence (control) or presence of 1 mM of thiamine or one of the 18 small compounds with structural similarities. LacZ activity was measured using *o*-nitrophenyl-β-D-galactopyranoside as a substrate. In strain Kp04 the native *K. pneumoniae* promoter *thiC* is coupled to the riboswitch *thiC* and thus expression of the *lacZ* reporter gene is controlled by the riboswitch. Accordingly, the reduced gene expression upon addition of thiamine indicates that thiamine turns the riboswitch off. In strain pKp04 the reporter gene is constitutively expressed from the *K. pneumoniae thiC* promoter as no riboswitch is present. LacZ activity is given as relative activity compared to the control. Cultures were grown in triplicates in a plate reader in a 96-well plate. Depicted are the mean values ± standard deviations of the data obtained from the triplicates.

useful to test potential ligands from *e.g.*, a novel substance library. Table 2 lists the selected compounds that were chosen to challenge the TPP riboswitch. The respective molecular structures are depicted in Fig. 6. All substances used were soluble in dimethyl sulfoxide (DMSO). All chemicals, including thiamine, were first tested in a comparably high final concentration of 1 mM. Since no previous *in vitro* tests could be run with these substances, the high concentration was used to ensure that the binding affinity was met. Addition of none of the 18 small compounds to the medium resulted in a significant decrease in β-galactosidase expression, as would be expected if a substance regulated riboswitch activity comparable to thiamine pyrophosphate (Fig. 7A). On the contrary, some substances even led to an increase of β-galactosidase activity, thus even had an opposite effect on the expression of the downstream gene (see *e.g.*, substance 11 and 17). However, the effect of these compounds on reporter gene activity in the promoter control was comparable to the values measured for the riboswitch test plasmid (Fig. 7B).

14. Step-by-step procedure for the dual-luciferase assay based on the reporter plasmids pDluc or pDlucTC

The dual luciferase assay (Fig. 8) was used as a more sensitive reporter gene assay to test putative TPP riboswitches and their response to thiamine and thiamine analogs. The system (Tecan Spark) allows to acquire luminescence of both firefly and *Renilla* luciferases and simplifies normalization.

1. Pre-cultures of *E. coli* DH5α or MG1655 containing the pDluc or pDlucTC reporter plasmids with the different TPP riboswitches were inoculated in M9 minimal medium with 100 μg/mL ampicillin.
2. Pre-cultures of *E. coli* DH5α were additionally supplemented with 20 nM of thiamine.
3. The overnight cultures were incubated in a 24-well plate at 37 °C with shaking at 160 rpm.
4. The OD_{600} of each overnight culture was measured, and the corresponding culture volume was transferred to a 1.5 mL microcentrifuge tube.
5. The cells were then pelleted by centrifugation at 6000 g for 5 min and washed once with fresh M9 medium without supplements.
6. Finally, the cell pellets were resuspended in fresh M9 medium to the same OD_{600}.
7. 10 μL of the resulting cell suspension were used to inoculate the 2 mL main cultures in triplicates in a non-treated 12-well plate (VWR International GmbH, Darmstadt, Germany) to an initial OD_{600} of 0.1.

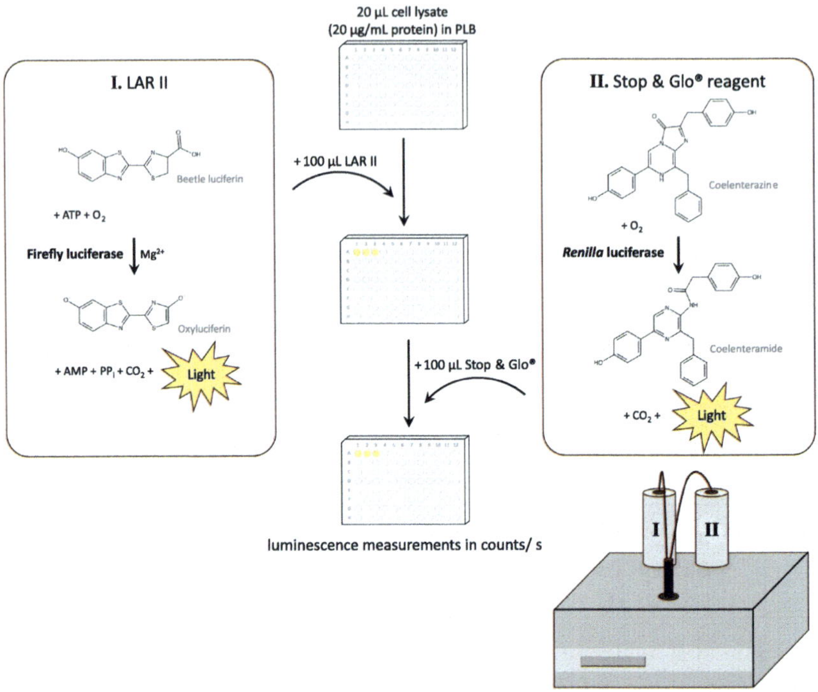

Fig. 8 **Schematic representation of the dual-luciferase assay principle and the corresponding measurement setup.** The figure illustrates the sequential measurement of firefly and *Renilla* luciferase activities, enabling sensitive and quantitative analysis of promoter activity and normalization to a control reporter. The figure was adapted from the manufacturer's manual provided by Promega, which supplies the Dual-Luciferase® Reporter Assay System and the Dual-Luciferase® Reporter 1000 Assay System.

The M9 minimal medium for the main cultures was supplemented with 10 µM of either thiamine, pyrithiamine or bacimethrin, or the equivalent volume of water as a control.

8. The 12-well plates were incubated at 37 °C and 180 rpm for four hours.
9. The whole culture volume was then transferred to a 2 mL microcentrifuge tube and cells were harvested by centrifugation at 4000 × *g* and 10 °C for 10 min.
10. After discarding the supernatant, the cell pellets were washed once with 500 µL of Phosphate Buffered Saline (PBS).
11. The centrifugation step was repeated and the supernatant discarded again.
12. The cell pellets were resuspended in 100–200 µL of Passive Lysis Buffer (PLB, Dual Luciferase Reporter Assay System, Promega,

Mannheim, Germany), depending on the pellet size. This was followed by an incubation on ice for 15 min.

13. To determine the amount of total protein in the samples, a standard Bradford assay was performed.
14. The cell extracts were diluted to a total protein amount of 20 μg/mL with PLB buffer.
15. 20 μL of each diluted sample were transferred to one well of a white, flat-bottom 96-well microplate (Greiner).
16. The plate was incubated inside a microplate reader (Tecan Spark) for 10 min at room temperature to decrease self-luminescence during the measurement.
17. Meanwhile, the connected injectors were flushed three times with dH_2O.
18. Injector A was primed with 1 mL of LARII and injector B with 1 mL of Stop & Glo® reagent (Dual Luciferase Reporter Assay System, Promega). The dual-luciferase measurement was performed using the well-wise luminescence measurement. Per well, 100 μL of LARII were injected, followed by a three-second waiting period, before the luminescence measurement was performed with automatic attenuation, an integration time of 10,000 ms and a settling time of 0 ms.
19. Subsequently, 100 μL of Stop & Glo® reagent were injected per well and luminescence was measured using the same parameters as before. The output was given in counts per second.

15. Application of the described dual-luciferase reporter assay for testing riboswitch activity in *E. coli* and comparison to the *lacZ* based assay

With the β-galactosidase reporter gene assay, the TPP riboswitch from *E. coli* was successfully analyzed. In addition, the *thiC* riboswitch from *K. pneumoniae* was experimentally validated for the first time. However, when other putative TPP riboswitches from other pathogenic bacteria were examined, some of them showed only weak responses and for some the results seemed to be not reproducible (data not shown). Therefore, another reporter gene assay was developed, employing a specific, newly designed plasmid harboring two different luciferase reporter genes. The introduction of a second luciferase gene, constitutively expressed from the same reporter plasmid, allowed to normalize the data with regard to differences in cell lysis efficiency and plasmid copy numbers thereby making the reporter assay even more reliable. Accordingly, a new dual-luciferase plasmid was constructed based on the single luciferase plasmid pPrib-luc (Pedrolli et al., 2015b) by inserting the *Renilla* luciferase gene luc^R under control of the constitutive promoter OXB15. The corresponding test plasmid was named pDluc. In this dual-luciferase assay, which so far has been mostly used in mammalian cells, one reporter gene originating from the firefly (luc^F) was used to probe the riboswitch activity, whereas the second luciferase gene from *Renilla reniformis* (luc^R) was placed under control of a constitutive promoter and served for normalization. The thiamine-auxotrophic strain *E. coli* DH5α was used again as a host for this reporter assay. To validate the newly developed reporter gene assay based on the dual-luciferase reactions, the assay was first tested with the *E. coli thiC* riboswitch (Ec01) that is known to negatively affect expression of *thiC* at the transcriptional as well as at the translational level. Notably, for *E. coli* a dual-luciferase reporter assay has not been described before. To differentiate between regulation at the transcriptional as well as at the translational level two different constructs were generated. In case of a "translation control" mechanism, TPP binding to the aptamer domain of the forming TPP riboswitch mRNA causes an alternative folding of this RNA, prevents ribosomes to access the ribosomal binding site of the reporter gene mRNAs and thus prevents formation of the reporter protein (Fig. 9A).

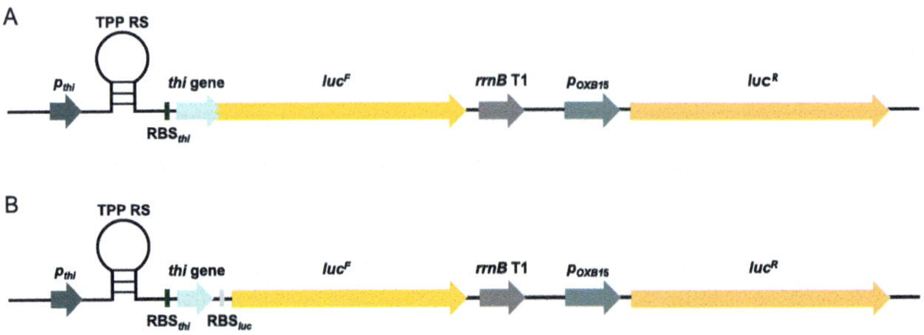

Fig. 9 Schematic structure of (A) the translational reporter gene fusion plasmid pDluc and (B) the transcriptional fusion plasmids pDlucTC. On both plasmids expression of the gene encoding *Renilla* luciferase luc^R is controlled by the artificial, constitutive *E. coli* promoter p_{OXB15}. An *rrnB* T1 terminator sequence upstream of the p_{OXB15} promoter prevents interference of upstream transcription with luc^R expression. The firefly luciferase reporter gene luc^F is located downstream of the first few codons of the respective thiamine biosynthesis or transport gene (*thi* gene), the corresponding ribosomal binding site (RBS_{thi}), the TPP riboswitch (TPP-RS) and the employed promoter (p_{thi}). In the translational fusion plasmid pDluc (A) the firefly luciferase gene luc^F is directly fused to the *thiC* gene codons, while in the transcriptional fusion plasmid pDlucTC (B) an additional ribosomal binding site (RBS_{luc}) is placed between the *thi* gene and the luc^F coding sequence.

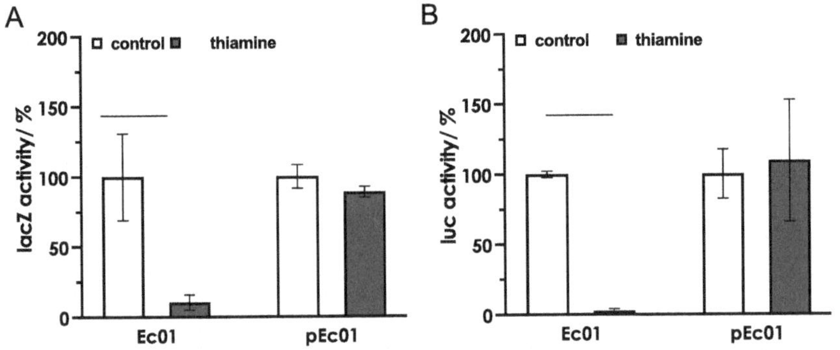

Fig. 10 Comparison of the (A) β-galactosidase and (B) luciferase reporter gene assays for testing the *E. coli thiC* riboswitch activity in *E. coli* DH5α. Shown are the results from the β-galactosidase (LacZ)-based assay using pHAt191I (A) and compared to the results of the novel dual-luciferase reporter gene assay based on pDluc for a translational fusion (B) for the activity of the *E. coli thiC* riboswitch. *E. coli* DH5α strains transformed with the reporter plasmids were grown in M9 minimal medium in the absence (control) or presence of 10 μM thiamine. Firefly luciferase (Luc^F) activity was followed by measuring the light emission caused by the conversion of luciferin to oxyluciferin, whereas LacZ activity was measured using *o*-nitrophenyl-β-D-galactopyranoside as a substrate. In strain Ec01 the native *E. coli* promoter *thiC* is coupled to the riboswitch *thiC* and thus expression of the reporter gene is controlled by the riboswitch. In strain pEc01 the reporter gene is constitutively expressed from the *E. coli thiC* promoter as no riboswitch is present. Luciferase activity was normalized by the respective *Renilla* luciferase activity and Luc^F and LacZ activities are given as relative activities compared to the respective controls. Cultures were grown in triplicates in a plate reader in a 12-well plate. Depicted are the mean values ± standard deviations of the data obtained from the triplicates. Asterisks indicate statistically significant differences determined by a two-tailed Students *t*-test (** $p \leq 0.01$, *** $p \leq 0.001$).

In case of a "transcription control" mediated by a TPP riboswitch, binding of TPP to the nascent transcript leads to formation of a transcriptional terminator, which in turn leads to dissociation of the RNA polymerase from the DNA template and thus termination of reporter gene expression (Fig. 9B). Transcriptional terminators and RNA secondary structures which sequester ribosomal binding sites can be predicted using bioinformatic tools and, accordingly, it can tentatively be predicted whether riboswitches exert a "transcription control" or a "translation control". For an initial characterization of the putative TPP riboswitches, it was important to ensure that no active TPP riboswitch sequence was overlooked and thus two different constructs were generated for each TPP riboswitch DNA using the two different reporter plasmids. In pDluc (Fig. 3) the (predicted) ribosomal binding sites of the TPP riboswitches to be tested replaced the ribosomal

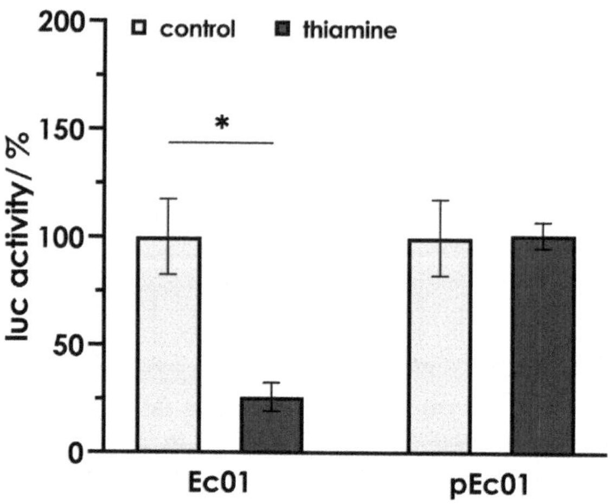

Fig. 11 Thiamine decreases luciferase (Luc^F) activity in a transcriptional fusion with the *E. coli thiC* riboswitch Ec01, but not in the respective promoter control pEc01. *E. coli* strains transformed with the reporter plasmids based on pDlucTC were grown in M9 minimal medium in the absence (control) or presence of 10 μM thiamine. Firefly luciferase (Luc^F) activity was followed by measuring the light emission caused by the conversion of luciferin to oxyluciferin. Luc^F activity was normalized by the respective *Renilla* luciferase activity and is given as relative activity compared to the control. Cultures were grown in triplicates in a plate reader in a 12-well plate. Depicted are the mean values ± standard deviations of the data obtained from the triplicates. Asterisks indicate statistically significant differences determined by a two-tailed Students *t*-test (* $p \leq 0.05$).

binding site of *lucF*, while in pDlucTC the ribosomal binding site of *lucF* was conserved. If the putative riboswitch reduced gene expression upon ligand treatment employing pDluc, control may have occurred at both the transcriptional and translational levels. If the putative riboswitch reduced gene expression upon ligand treatment employing pDlucTC (Fig. 4), control occurred mainly at the transcription level.

First, the translational riboswitch fusion using pDluc was tested to validate the new reporter gene assay. To this end, it was compared with the previously performed reporter gene assay, which was based on the β-galactosidase reaction using pHAt1911 (Fig. 2). While thiamine addition decreased β-galactosidase activity to around 20 % of the activity of the control reaction (Fig. 10A), the luciferase activity was consistently decreased to approximately 1 % of the control culture without thiamine. No decrease in activity was observed for the respective promoter control plasmid (Fig. 10B). Therefore, it was concluded that the dual-luciferase assay represents a more sensitive reporter assay to measure

TPP riboswitch activity than the colorimetric β-galactosidase assay. To be able to also gain information about the mode of action of the riboswitches, a translational as well as a transcriptional fusion plasmid was constructed, allowing to determine the main way of function of the different riboswitches. This plasmid is similar to the translational fusion plasmid, but additionally contains a second Shine-Dalgarno sequence (ribosomal binding site) upstream of the firefly luciferase gene. This allows reporter gene translation, even when the riboswitch sequesters the first ribosomal binding site of its downstream gene. For the transcriptional fusion of the *E. coli thiC* riboswitch, as well as the promoter control, the reporter gene expression pattern is similar to the one of the translational fusion indicating that the riboswitch is active at both levels (Pedrolli et al., 2015a). Addition of thiamine decreased firefly luciferase activity significantly only in presence of the riboswitch (Fig. 11).

16. Application of the described dual-luciferase reporter assay and the importance of using native promoters for *in vivo* testing

After the assay was successfully established with the *E. coli thiC* riboswitch further putative TPP riboswitches from various pathogens were tested. Table 3 lists the potential riboswitch regions within the ESKAPE pathogens, as well as *M. sciuri* and *S. pneumoniae* that were investigated. Riboswitch candidates were identified by searching the Rfam database for TPP riboswitches, and their genomic localization was determined by matching the corresponding genome data in the NCBI database. The identified regions showed sequence similarity to known TPP riboswitch sequences and were additionally located upstream of genes involved in thiamine synthesis or transport. The last column of Table 3 shows the PASIFIC web server (https://www.weizmann.ac.il/molgen/Sorek/PASIFIC/) prediction of the putative way of action of the riboswitch, based on the predicted presence of a transcriptional terminator-anti-terminator structure within the riboswitch sequence. A value above 0.5 was considered reliable for the prediction of such a structure. Riboswitches with a score below 0.5 were inversely assigned to the translational mode of action. The first luciferase reporter gene assays with putative TPP riboswitches were performed employing the translational fusion plasmid pDluc in combination with the *E. coli thiC* promotor. This was to ensure that the tested promoter was also functional in the *E. coli* host. For the riboswitch sequences Pa01 from *P. aeruginosa* and the two sequences Eb01 and Eb03 from *Enterobacter* spp.,

Table 3 Identified thiamine pyrophosphate riboswitches in ESKAPE pathogens, including *M. sciuri* and *S. pneumoniae*. Listed are all the riboswitches that showed a decrease in reporter gene activity upon addition of thiamine. The *in silico* predicted mode of action (PASIFIC) is compared to the experimental outcome (exp.). The mode of action, translational or transcriptional, was predicted with the PASIFIC web server (https://www.weizmann.ac.il/molgen/ Sorek/PASIFIC/). A prediction score of over 0.5 for formation of a terminator structure was considered reliable.

Organism	RS name	Gene	Predicted	Mode of action (exp)
Acinetobacter baumannii	Ab01	*thiC*	Transcriptional	Transcriptional/ (translational)
Enterobacter spp.	Eb01	*thiC*	Transcriptional	Transcriptional
Enterobacter spp.	Eb03	*thiBPQ* operon	Translational	Translational
Enterococcus faecium	Ef02	*thiT*	Translational	Translational/ transcriptional
Klebsiella pneumoniae	Kp04	*thiC*	Transcriptional	Translational/ transcriptional
Pseudomonas aeruginosa	Pa01	*thiC*	Translational	Transcriptional
Mammaliicoccus sciuri	Ms02	*thiE* operon	Transcriptional	Translational
Streptococcus pneumoniae	Sp06	*thiD*	Translational	-

as well as Ms02 from *M. sciuri*, a significant reduction in luciferase activity was observed upon addition of 10 μM thiamine to the bacterial cultures, compared to the control without thiamine (data not shown). For the riboswitch sequences Kp04 from *K. pneumoniae*, Ab01 from *A. baumannii* and Ms01 from *M. sciuri*, no effect of thiamine on the reporter activity was measured (data not shown). For the reporter plasmid with the putative *thiBPQ* riboswitch from *S. aureus* Sa02, no reporter gene activity was detected and therefore the response of this region to thiamine could not be determined with this assay (data not shown). Since the putative *thiC* riboswitch sequence Kp04 from *K. pneumoniae* showed a high sequence similarity to the *thiC* riboswitch sequence from *E. coli* and, moreover, was found to respond to thiamine in a previously performed *lacZ* reporter gene assay, using the native *K. pneumoniae* promoter it was suspected that the employed *E. coli thiC* promotor was too strong and thus prevented

Kp04 from functioning properly. Therefore, translational fusions of the riboswitches Kp04, Ab01, Ms01 and Ms02 were tested with their respective native promoters. In addition, sequences from *Enterococcus faecium* (Ef02) and *Streptococcus pneumoniae* were included (Sp06), which did not respond to thiamine in combination with the *E. coli thiC* promoter. A specific control plasmid was constructed for each of these putative riboswitches, only containing the respective promotor and ribosomal binding site (RBS) but not the riboswitch sequence itself. For none of the strains with a promoter-control plasmid, a reduction in luciferase activity was measured when thiamine was present (data not shown). In all strains with a riboswitch plasmid reporter activity was significantly decreased, except for the *M. sciuri* Ms01 riboswitch. While the Ms02 riboswitch in conjunction with its native promoter did not respond to thiamine to a greater extent than with the *E. coli thiC* promoter, the effect of thiamine was much more pronounced for the natural combination of promotor and riboswitch for the *A. baumannii* Ab01 and the *K. pneumoniae* Kp04 regions. In total, the employment of the novel dual-luciferase assay identified eight new TPP riboswitches, at least one in each of the ESKAPE pathogen strains, excluding *S. aureus*, but including *M. sciuri* and *S. pneumoniae* (Table 3).

17. Application of the described dual-luciferase reporter assay and investigation of the mode of action for the newly identified TPP riboswitches

In the translational fusion plasmids based on pDluc (Fig. 3), the putative riboswitch, the respective RBS and the first few codons of the downstream gene are directly fused to the firefly luciferase gene. This construct allows the detection of a regulation of downstream gene expression that can be achieved either by the sequestering of the RBS, as well as by the formation of a transcriptional terminator. This makes it a suitable construct to identify putative riboswitches. However, with this plasmid it is not test possible to distinguish between the two modes of action. Therefore, a transcriptional fusion plasmid was assembled for all the riboswitches that were tested employing pDlucTC (Fig. 4). This plasmid contains an additional RBS for the luc^F gene, enabling a translation of mRNA even in case the first RBS has become inaccessible due to RNA folding. Hence, a reduction in luciferase activity only occurs, when a terminator structure is formed that already aborts transcription. For the transcriptional reporter gene fusions of the riboswitches from *K. pneumoniae*, *A. baumannii*, *P. aeruginosa* and the Eb01 riboswitch from *Enterobacter* spp., a significant reduction in luciferase activity was detected, when 10 μM of thiamine were added (data not shown). For the riboswitch regions of *M. sciuri* and the Eb03 region of *Enterobacter*, on the other hand, no reduction in reporter activity was measured. The riboswitches that have been identified before in a translational fusion in combination with their native promoter were examined again for their transcriptional function with the respective promoter region. This is to ensure that possible responses are not masked by some influence of the *E. coli* promoter. The riboswitch regions Kp04 and Ab01 were included as well, even though a significant effect of thiamine was already measured for these two when employing the Ec01 promoter. In contrast to the translational fusions for these regions, no stronger effect of thiamine is observed here on the reporter gene activity, when the native promoter is used. For the riboswitch region Ef02 from *E. faecium*, a decrease in luciferase activity to 73 % of the control was measured, when thiamine was added. The riboswitch region Sp06 from *S. pneumoniae* and the promoter control region pMs02 from *M. sciuri* could not be successfully tested in the transcriptional reporter gene assay. No firefly luciferase activity was detected in the cell-free extracts of the strains carrying the respective plasmids, although the reaction catalyzed by the control enzyme, the *Renilla* luciferase, was measurable and for the same riboswitch regions Luc^F activity could be measured for the translational fusion plasmid. In total, eight different putative TPP riboswitch regions showed a response to thiamine in the reporter gene assay performed. Translational and transcriptional fusions were successfully tested for six of them.

18. Precursor techniques

In the literature *lacZ* based assays are described to test riboswitches *in vivo*. Luciferase based test systems were also described. In addition, *in vitro* test systems employing fluorescent dyes which bind to riboswitch RNAs were described (Bains et al., 2023).

19. Safety considerations and standards

We are not aware of any safety concerns that go beyond standard microbiological and genetic engineering procedures. Some of the chemicals described are harmful to health and must be handled in accordance with standard safety protocols.

20. Related techniques

In vitro transcription and translation protocols use e.g. bacterial cell-free extracts and in most cases these systems are based on *E. coli* as this organism is easy to grow using inexpensive growth media. Moreover, due to the Gram-negative cell envelope cell-free extracts are comparably easily generated. In principle the same components as in an *in vivo* system are present. However, in a life cell uptake of test chemicals depends on membrane solubility or a membrane transporter and thus is not guaranteed. This is a disadvantage of an *in vivo* system.

21. Pros and cons

Pros	Cons
Reliable and highly reproducible	Many cloning procedures are necessary and some riboswitch DNA sequences are difficult to clone and propagate *in vivo* due to secondary structures also at the DNA level

22. Troubleshooting & optimization

Problem	Solution
Cloning of riboswitch sequences with extensive secondary structures	Use low copy plasmids for cloning in case toxicity is observed
Promoter strength affects riboswitch activity	Use different promoters for initial characterization

23. Summary and conclusion

To test the functionality of a metabolite-sensing riboswitch, a reporter assay was developed in which the riboswitch controls the expression of a reporter gene located downstream of the riboswitch sequence and which also should be useful to test other riboswitches in a similar manner. At first, a reporter plasmid with the β-galactosidase gene *lacZ* was constructed and tested. Compared to cells grown without thiamine supplementation, a significant reduction in reporter activity was observed for the cultures grown in the presence of thiamine, while reporter gene activity of the promoter control strain (not containing a riboswitch) remained unaffected by the addition of thiamine. This was a replication of the results obtained by Winkler et al. (2002), who first experimentally validated the response of the *E. coli thiC* riboswitch to thiamine using an *in vivo* system. In our hands the results of the LacZ assay were reproducible for the *E. coli* and *K. pneumoniae thiC* riboswitch regions, the measured reporter activity, however, showed high standard deviations for some of the constructs. Therefore, we constructed another reporter gene plasmid, based on the firefly luciferase gene luc^F as a reporter. The newly designed reporter plasmid was tested for its functionality, again using the *E. coli thiC* promoter and the corresponding riboswitch. While thiamine addition decreased β-galactosidase activity to around 20% of the activity of the control, the luciferase activity was consistently decreased to approximately 1% of the control culture without thiamine. No decrease in activity was observed for the respective promoter control plasmid. Therefore, it was concluded that the dual-luciferase assay represents a more sensitive reporter assay to measure TPP riboswitch activity than the colorimetric β-galactosidase assay. To be able to also gain information about the mode of action of the riboswitches, a translational as well as a transcriptional fusion plasmid was constructed, allowing to determine the main way of function of the different riboswitches. Since metabolite-dependent riboswitches from various bacteria function in combination with a specific promoter and possibly also with a specific terminator or even a specific ribosomal binding site, it would be preferable to test these genetic elements in the respective host. *E. coli* is still the organism that is easiest to genetically engineer, and is therefore likely still the first choice for such experiments.

Acknowledgments

This study was supported by Bundesministerium für Bildung und Forschung (FKZ 01KI1826) and Deutsche Forschungsgemeinschaft (MA 2510/9–3).

References

Bains, J. K., Qureshi, N. S., Ceylan, B., Wacker, A., & Schwalbe, H. (2023). Cell-free transcription-translation system: A dual read-out assay to characterize riboswitch function. *Nucleic Acids Research, 51*, e82.

Bastet, L., Chauvier, A., Singh, N., Lussier, A., Lamontagne, A.-M., Prévost, K., et al. (2017). Translational control and Rho-dependent transcription termination are intimately linked in riboswitch regulation. *Nucleic Acids Research, 45*, 7474–7486.

Blount, K. F., & Breaker, R. R. (2006). Riboswitches as antibacterial drug targets. *Nature Biotechnology, 24*, 1558–1564.
Grabowska, E., Czerniecka, M., Czyżewska, U., Zambrzycka, A., Łotowski, Z., & Tylicki, A. (2021). Differences in the efficiency of 3-deazathiamine and oxythiamine pyrophosphates as inhibitors of pyruvate dehydrogenase complex and growth of HeLa cells *in vitro*. *Journal of Enzyme Inhibition and Medicinal Chemistry, 36*, 122–129.
Hübenthal, A., & Mack, M. (2025). The monofunctional FMN Riboswitch-Binding regulator RIBR from *Bacillus amyloliquefaciens* promotes riboflavin biosynthesis. *Molecular Microbiology, 124*, 373–385.
Hübenthal, A., Panchal, V., Brenk, R., & Mack, M. (2025). Functional analysis of thiamine pyrophosphate-responsive riboswitches in human bacterial pathogens of the ESKAPE group using a dual-luciferase reporter gene assay. *Journal of Bacteriology, 207*, e00308-25.
Kavita, K., & Breaker, R. R. (2023). Discovering riboswitches: The past and the future. *Trends in Biochemical Sciences, 48*, 119–141.
Mandal, M., & Breaker, R. R. (2004). Gene regulation by riboswitches. *Nature Reviews. Molecular Cell Biology, 5*, 451–463.
Panchal, V., & Brenk, R. (2021). Riboswitches as drug targets for antibiotics. *Antibiotics, 10*, 45.
Pedrolli, D., Langer, S., Hobl, B., Schwarz, J., Hashimoto, M., & Mack, M. (2015a). The *ribB* FMN riboswitch from *Escherichia coli* operates at the transcriptional and translational level and regulates riboflavin biosynthesis. *The FEBS Journal, 282*, 3230–3242.
Pedrolli, D. B., Kühm, C., Sévin, D. C., Vockenhuber, M. P., Sauer, U., Suess, B., & Mack, M. (2015b). A dual control mechanism synchronizes riboflavin and sulphur metabolism in *Bacillus subtilis*. *Proceedings of the National Academy of Sciences, 112*, 14054–14059.
Sudarsan, N., Cohen-Chalamish, S., Nakamura, S., Emilsson, G. M., & Breaker, R. R. (2005). Thiamine pyrophosphate riboswitches are targets for the antimicrobial compound pyrithiamine. *Chemistry & Biology, 12*, 1325–1335.
Winkler, W., Nahvi, A., & Breaker, R. R. (2002). Thiamine derivatives bind messenger RNAs directly to regulate bacterial gene expression. *Nature, 419*, 952–956.

CHAPTER FIFTEEN

Mapping ligand-RNA interactions of thiamine analogues using transcriptome-wide sequencing

Daphne A.L. van den Homberg, Stefan Crielaard, and Willem A. Velema*
Institute for Molecules and Materials, Radboud University, Nijmegen, The Netherlands
*Corresponding author. e-mail address: Willem.Velema@ru.nl

Contents

Abstract

This chapter describes the use of transcriptome-wide sequencing to identify ligand-RNA interactions. Photoaffinity probes, based on small molecules of interest, can directly report on binding of ligands to their target RNA. Combined with pulldown enrichment experiments and next-generation sequencing (NGS), this chemical tool allows for transcriptome-wide screening to identify novel RNA targets. Here, we describe protocols to identify and validate ligand-RNA interactions using pulldown enrichments followed by transcriptome-wide sequencing or Reverse Transcription-quantitative PCR (RT-qPCR), and to investigate these interactions using Reverse-

Methods in Enzymology, Volume 735
ISSN 0076-6879, https://doi.org/10.1016/bs.mie.2026.06.022

Transcription (RT) stop, Selective 2′-hydroxyl acylation analyzed by primer extension (SHAPE) and *in vitro* transcription and translation (IVTT) assays.

1. Introduction

RNA has emerged as an important regulator in cellular processes, carrying genetic information, regulating gene expression, catalyzing enzymatic reactions and more (Cech & Steitz, 2014; Lieberman, 2018). Important to their molecular functions, RNA molecules are able to form stable and complex tertiary RNA structures (Spitale & Incarnato, 2023). Within bacteria, these intricate RNA structures can be recognized and selectively bound by small molecules to influence their function, as demonstrated by riboswitches (Breaker, 2012).

Riboswitches are RNA structures that can act as biosensors regulating gene expression based on the presence of their cognate ligand (McCown et al., 2017). Riboswitches are comprised of two distinct domains in the 5′ untranslated region (UTR) of bacterial mRNAs, a ligand-binding aptamer domain and an expression platform. In response to binding of a ligand to the aptamer domain, the downstream expression platform changes its structural conformation to regulate gene expression on a transcriptional or translational level (Winkler, Nahvi, et al., 2002). In general, these ligands include cellular small molecules including a wide variety of metabolites, enzyme cofactors, amino acids or ions (Breaker, 2022; Scull et al., 2021). Across bacteria, more than 55 riboswitch classes have been experimentally validated (Salvail & Breaker, 2023), of which the Thiamine Pyrophosphate (TPP) riboswitch (Winkler, Nahvi, et al., 2002), Cobalamin riboswitch (Vitreschak et al., 2003), S-Adenosylmethionine (SAM) riboswitch (Winkler et al., 2003), Flavin Mononucleotide (FMN) Riboswitch (Winkler, Cohen-Chalamish, et al., 2002) and prequeosine1 ($PreQ_1$) riboswitch (Roth et al., 2007) are among the most common. Importantly, as riboswitches regulate key functions within pathogenic bacteria, they have emerged as interesting antibiotic targets (Giarimoglou et al., 2022; Panchal & Brenk, 2021).

By studying interactions between small molecules and RNA, potential new RNA ligands and targets can be identified. When applied to bacteria, novel antibiotic compounds targeting riboswitches can be discovered. Recent studies have demonstrated the potential of chemical probes that can directly report on target binding of small molecules. These chemical probes are characterized by a small molecule of interest combined with a cross-linking moiety and a purification tag to allow for photoaffinity labeling and purification of target RNAs (Chen et al., 2020; Mortison et al., 2018). Early examples of photoaffinity probes applied towards RNA include pioneering work by the groups of Schultz and Johnson (Wang et al., 2018), Myers (Mortison et al., 2018) and Disney (Chen et al., 2020). When combined with pulldown enrichment experiments and next-generation sequencing (NGS), this tool allows for transcriptome-wide screening of targets of small molecules of interest (Gibaut et al., 2023; Mukherjee et al., 2020; Tong et al., 2022). It has been demonstrated that chemical probing is an effective method to investigate riboswitches (Balaratnam et al., 2021; Crielaard et al., 2022). Chemical probes based on the molecular structure of $PreQ_1$ and FMN were shown to be useful in unraveling the riboswitch-ligand interaction and identifying new riboswitch sequences in different species (Balaratnam et al., 2021; Crielaard et al., 2022; Hanisch et al., 2025).

Recently, our group synthesized a photoaffinity probe of Thiamine Monophosphate (TMP) (Crielaard et al., 2025). TMP is a small molecule metabolite without known biological function that is an intermediate in the TPP biosynthesis. We employed a diazirine based photoaffinity probe to identify potential binding pockets within the *E. coli* transcriptome (Fig. 1). The photoaffinity probe is incorporated into an RNA molecule upon irradiation, typically at ~330–370 nm. (Step 1) The diazirine moiety undergoes photolysis and creates a highly reactive carbene intermediate under release of molecular nitrogen. The short lifetime of the reactive intermediate facilitates spatial selective cross-linking of nearby nucleotides. (Step 2) The probe can be further functionalized using click-chemistry to attach a biotin enrichment handle to pulldown bound RNA for downstream analysis. Here, we describe protocols for the identification and validation of small molecule-RNA interactions with the use of photoaffinity probes containing a diazirine cross-linking moiety and alkyne purification handle using pulldown enrichments followed by transcriptome-wide sequencing and RT-qPCR. Moreover, we describe techniques to investigate small molecule-RNA interactions using competition, RT-stop, SHAPE and IVTT assays.

2. Equipment

Refrigerated centrifuge.
Shaking heat block.
UV Stratalinker™ 1800 (365 nm) (Stratagene).

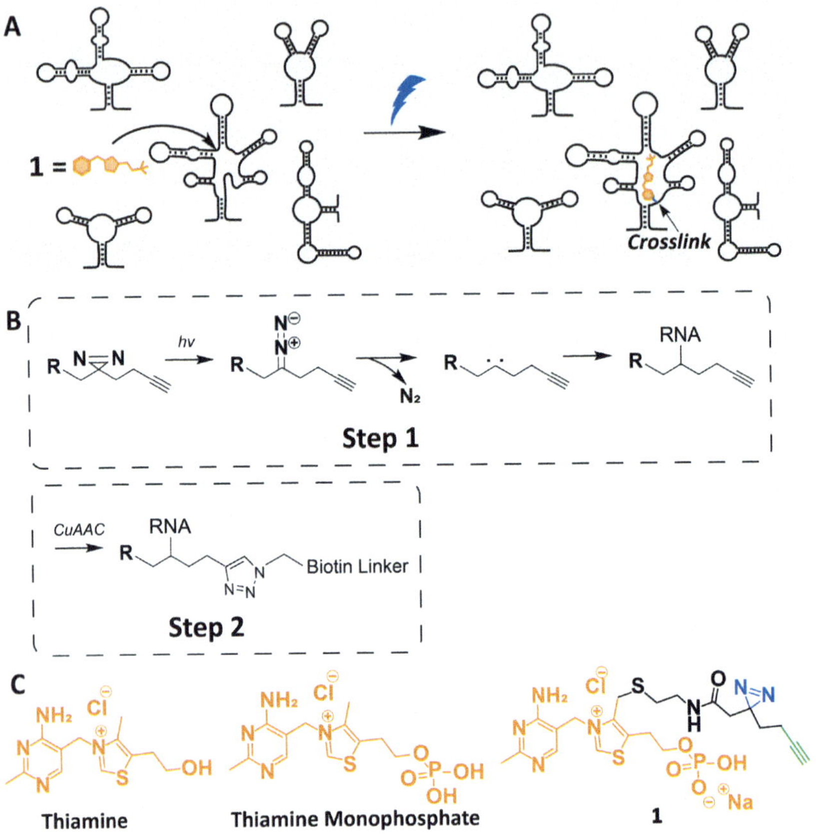

Fig. 1 (A) Schematic illustration of chemotranscriptomic profiling using photocrosslinking. (B) Reaction of diazirine crosslinking of probe to RNA and subsequent biotin functionalization. (C) Molecular structures of thiamine, thiamine monophosphate (TMP) and TMP probe **1**.

Magnetic rack that fits 1.5 mL Eppendorf tubes.
Overhead rotator.
Vortex.
Qubit™ 4 Fluorometer (Invitrogen).
QuantStudio RT-qPCR instrument (Thermo Fisher Scientific).
20 cm × 45 cm x 1 mm Owl S4S Vertical Gel Electrophoresis apparatus (Thermo Fisher Scientific).
20 cm × 20 cm x 1 mm V1517 Vertical Gel Electrophoresis apparatus (Life Technologies).
Amersham Typhoon Imager.
BioTek Synergy H1 Microplate reader (BioSPX).

3. Materials

Nuclease-free water.
Ultrapure milliQ water.
Sodium Chloride (NaCl).
Magnesium Chloride ($MgCl_2$).
4-(2-hydroxyethyl)-1-piperazineethanesulfonic acid (HEPES).
Copper(II) Sulfate ($CuSO_4$).
Sodium Ascorbate (NaAsc).
Tris-(3-hydroxypropyltriazolylmethyl)amine (THPTA).
Disulfide biotin azide (Tocris Bioscience).
6-FAM-azide (Jena Bioscience).
Tris base, adjusted pH using HCl (Tris-HCl).
Ethylenediaminetetraacetic acid (EDTA).
Tween-20.
Dithiothreitol (DTT).
Glycogen.
Sodium Acetate (NaOAc).
Absolute ethanol.
Boric acid.
Urea.
40 % Polyacrylamide (19:1).
Ammonium persulfate (APS).
TEMED.
1 μm Magnetic Streptavidin Beads (New England Biolabs).
RNA Clean & Concentrator kit – 5 (Zymo Research).
Qubit™ High Sensitivity RNA assay (Invitrogen).
Luna® one-step RT-qPCR kit (New England Biolabs).
Target specific primers.
SuperScript™ III (Invitrogen).
2x RNA Loading Dye.
Sybr™ Gold (Invitrogen).
1-Methyl-7-nitroisatoic anhydride (1M7) (Sigma Aldrich).
E. coli T7 S30 Extract System for Circular DNA (Promega).
D-luciferin (Sigma Aldrich).

3.1 Solutions & buffers

3.3x RNA folding buffer.

Component	Final concentration	10 x Folding buffer	Amount
NaCl	100 mM	333 mM	584.0 mg
$MgCl_2$	10 mM	33 mM	95.1 mg
HEPES	100 mM	333 mM	2.4 g

Mix using 30 mL of nuclease-free water. Adjust to pH 8.

10x CuAAC click mixture.

Component	Final concentration	10 x Click mixture	Stock	Amount
$CuSO_4$	0.5 mM	5 mM	25 mM	20 µL
NaAsc	5.0 mM	50 mM	250 mM	20 µL
THPTA	2.5 mM	25 mM	200 mM	12.5 µL
Disulfide biotin azide/ Fluorescein azide	0.2 mM 0.4 mM	2 mM 4 mM	10 mM 10 mM	20 µL 40 µL

Mix with nuclease-free water up to 100 µL.

Binding Buffer.

Component	Final concentration	Stock	Amount
NaCl	0.5 M	5 M	5 mL
Tris-HCl	20 mM	1 M (pH 7.5)	1 mL
EDTA	1 mM	0.5 M (pH 8)	100 µL

Mix with nuclease-free water up to 50 mL.

High-salt Washing Buffer.

Component	Final concentration	Stock	Amount
NaCl	0.5 M	5 M	5 mL
Tris-HCl	20 mM	1 M (pH 7.5)	1 mL
EDTA	1 mM	0.5 M (pH 8)	100 µL
Tween-20	0.2 % (v/v)		100 µL

Mix with nuclease-free water up to 50 mL.

Low-salt Washing Buffer.

Component	Final concentration	Stock	Amount
NaCl	0.15 M	5 M	1.5 mL
Tris-HCl	20 mM	1 M (pH 7.5)	1 mL
EDTA	1 mM	0.5 M (pH 8)	100 μL

Mix with nuclease-free water up to 50 mL.

Elution Buffer.

Dissolve 7.7 mg DTT in 1 mL nuclease-free water. Prepare the elution buffer fresh on the day of experiment.

3 M NaOAc, pH 5.2.

Dissolve 20.4 g NaOAc in 40 mL water; adjust the pH to 5.2 with glacial acetic acid. Add water to 50 mL and filter-sterilize the solution.

Glycogen (20 mg/mL).

Mix 20 mg of glycogen with 1 mL of nuclease-free water. Store in small aliquots of 100 μL in −20 °C.

Ethanol 70 %.

Mix 35 mL absolute ethanol and 15 mL nuclease-free water

1x TBE.

Component	Final concentration	Stock	Amount
Tris	89 mM		108 g
Boric acid	89 mM		55 g
EDTA	2.2 mM		7.5 g

Adjust with milliQ water to 10 L.

Denaturing 8 % PAGE buffer.

Component	Final concentration	Stock	Amount
Tris	89 mM		2.69 g
Boric acid	89 mM		1.35 g
EDTA	2.7 mM		0.23 g

Dissolve in 25 mL of milliQ water.

40 % Polyacrylamide (19:1)	8 %		50 mL
Urea	8 M		120 g

Add milliQ water up to 250 mL and stir buffer until dissolved.

Denaturing 8 % PAGE gel.

Component	Final concentration	Stock	Amount
8 % PAGE buffer			30 mL
TEMED			40 µL
APS		10 % (w/v)	100 µL

TEMED is a corrosive chemical and APS can cause skin and respiratory irritation. Wear gloves and handle these chemicals in a fume hood.

IVTT master mix – 1 reaction.

Component	Final concentration	Stock	Amount
DNA template	7.5 nM		
Amino Acid mixture			1.11 µL
S30 Premix			4.44 µL
T7 S30 Extract			0.22 µL
D-luciferin	2 mM	88.25 mM	0.25 µL

Add nuclease-free water up to a final volume of 11.1 µL per reaction. Prepare one IVTT master mix for the number of samples plus excess.

4. Protocols

4.1 Protocol A: Photoaffinity labeling and enrichment followed by transcriptome-wide sequencing

4.1.1 Overview

Photoaffinity labelling and Streptavidin enrichment using a chemical probe based on a small molecule of interest followed by transcriptome-wide sequencing allows for the identification of RNA targets interacting with the small molecule Fig. 2.

4.1.2 Duration

2 days.

Fig. 2 Flowchart of protocols A-C.

1. In 1.5 mL Eppendorf tubes, prepare the amount of RNA (5−50 μg) in nuclease-free water for each sample. Denature the samples at 75 °C for 5 min and store on ice to keep the RNA in a denatured state.
2. Add the folding buffer to each sample to ensure proper folding of the RNA and incubate RNA at 37 °C for 10 min with gentle agitation.
3. Then add the respective probe (final concentration 10−100 μM) or equal amount of DMSO and mix well, before incubating the samples at 37 °C for 30 min with gentle agitation in the dark. The final volume will be 100 μL and the final concentration of DMSO should be below 5%.
4. To covalently attach the probe to RNA, irradiate the samples at a wavelength of 365 nm for 30 min in *Time* mode of the UV Stratalinker™. Ensure that the Eppendorf tubes are open and directed to the light source.
5. Wash away excess unbound probe by performing ethanol precipitation. Add 0.1 vol of 3 M NaOAc, 2.5−3 volumes of absolute ethanol and 1 μL of 20 mg/mL glycogen to precipitate the RNA. Mix well and let the RNA precipitate for at least 1 h at −20 °C. The precipitated RNA is pelleted by centrifuging the samples at 17,000 × *g* for 1 h in a cooled centrifuge. The pellets are washed twice using 70 % ethanol and dissolved in 9 μL nuclease-free water.
6. Modify the probe-bound RNA with disulfide-biotin azide using copper-click chemistry. Denature the RNA samples at 65 °C for 5 min and store on ice. Prepare a 10x CuAAC click mixture and add to the denatured samples. Incubate samples at 37 °C for 60 min with gentle agitation. Wash away the excess click mixture using ethanol precipitation as described above.
7. Denature samples in 10 μL nuclease-free water at 65 °C for 5 min and then add binding buffer up to 100 μL. In the meantime, prepare new 1.5 mL Eppendorf tubes and add 25 μL magnetic streptavidin beads. Remove the supernatant using a magnetic rack to separate the magnetic beads. Wash the magnetic beads by mixing with binding buffer and removing the supernatant again. Then, mix the RNA samples with the magnetic beads by pipetting up and down until a suspension is reached. Incubate samples at room temperature with overhead rotation at 18 rpm for 1 h.
8. Remove unbound RNA by washing the beads thrice with a high-salt washing buffer and thrice with a low-salt washing buffer. Each washing

step consists of briefly spinning down the samples, collecting the beads at the side of the tube using the magnetic rack, removing the supernatant, mixing the beads with 400 μL of respective washing buffer and rotating the samples at room temperature for 3–5 min. When the washing steps are completed, remove the supernatant as much as possible.

9. To elute the RNA, add an equal volume elution buffer to beads and mix well by pipetting up and down. Incubate samples at 37 °C for 30 min with gentle agitation. The disulfide bonds are cleaved by DTT and RNA is released from the beads. Place the samples once again on the magnetic rack and transfer the supernatant into new 1.5 mL Eppendorf tubes.
10. Clean up the RNA using Zymo Clean & Concentrator columns according to manufacturer's protocol and elute the RNA in 15 μL nuclease-free water.
11. Quantify the samples using the Qubit High Sensitivity assay. RNA samples can now be used for library prep and sequencing.

4.1.3 Tip

A control probe without the RNA-targeting moiety is advisable to account for background binding.

4.1.4 Tip

Keep 1 % of reaction mixture of step 2 aside as input control. The 1 % input control represents RNA levels before enrichment.

4.1.5 Tip

If too little RNA is left after the enrichment protocol, we recommend combining multiple replicates before cleaning the samples using the Clean & Concentrator columns to increase RNA yield. Make sure to use the same amount of RNA total input for all conditions. More input RNA or increased photoaffinity probe concentration can also increase enriched RNA yield, however these methods can result in higher background labeling.

4.2 Protocol B: Validation of identified targets

4.2.1 Overview

It is necessary to validate the sequencing results before continuing to ensure reproducibility and reliability of the identified targets by excluding bias caused by lowly expressed genes or targets that show relatively small enrichments. Photoaffinity labeling pulldown experiments followed by

RT-qPCR allow for the measurement of specific RNA targets. See 'Explanatory chapter: Quantitative PCR' for more background information on RT-qPCR.

4.2.2 Duration

2 days.

1. Design target specific and housekeeping gene primers for RT-qPCR.
2. Follow step 1–10 of protocol A as described above.
3. Make the primer specific one-step RT-qPCR master mixes according to manufacturer's protocol.
4. Pipette 1 μL of enriched RNA in a 96-well plate as input for the RT-qPCR.
5. Pipette 9 μL of master mix in each respective well making sure not to touch the RNA.
6. Run the RT-qPCR protocol as instructed for your RT-qPCR machine model.

4.3 Protocol C: Competition using cognate ligand

4.3.1 Overview

To further investigate if the identified RNA targets truly interact with the small molecule of interest, competition experiments using the unmodified ligand are useful to distinguish between on-target effects and off-target binding, especially at high concentrations of probe. If it is a true on-target effect, the ligand will displace the photoaffinity probe and the enrichment will decrease.

4.3.2 Duration

2 days.

1. In 1.5 mL Eppendorf tubes, prepare RNA (5–50 μg) in nuclease-free water for each sample. Denature the samples at 75 °C for 5 min and store on ice to keep the RNA in a denatured state.
2. Add folding buffer to each sample to ensure proper folding of the RNA and incubate at 37 °C for 10 min with gentle agitation.
3. Then add an equimolar or excess of competing ligand compared to probe to the RNA and incubate at 37 °C for 20 min with gentle agitation. The final volume will be 100 μL and the final concentration of DMSO should be below 5 %.
4. After initial incubation with the ligand, add the respective probe and mix well, before incubating the samples at 37 °C for 30 min with gentle agitation in the dark.

5. Irradiate the samples at a wavelength of 365 nm for 30 min in *Time* mode of the UV Stratalinker™. Ensure that the Eppendorf tubes are open and directed to the light source.
6. Perform step 5−17 as described in Protocol A and B.

4.4 Protocol D: Selective photoaffinity labeling of in vitro transcribed RNA

4.4.1 Overview

Photoaffinity labelling of *in vitro* transcribed RNA directly measures the interaction between the photoaffinity probe and the target RNA. Combined with a fluorophore, the interaction can be visualized on a gel. Competition using the unmodified ligand will displace the photoaffinity probe and will decrease the fluorescent signal. *In vitro* transcription of RNA using T7 polymerase is described elsewhere (See '*In Vitro* Transcription from Plasmid or PCR-amplified DNA') Fig. 3.

4.4.2 Duration

1.5 days.

1. In 1.5 mL Eppendorf tubes, prepare the amount of in vitro transcribed RNA (10 pmol, 1 μL of 10 μM stock) in nuclease-free water for each sample. Denature the samples at 95 °C for 5 min and store on ice to keep the RNA in a denatured state.
2. Add the folding buffer to each sample to ensure proper folding of the RNA and incubate the RNA at 37 °C for 10 min with gentle agitation.
3. Add an equimolar or excess of competing ligand compared to probe to the RNA and incubate at 37 °C for 20 min with gentle agitation.
4. Then add the respective probe (final concentration 10−100 μM) or DMSO and mix well, before incubating the samples at 37 °C for 30 min with gentle agitation in the dark. The final volume will be 10 μL and the final concentration of DMSO should be below 5 %.
5. Irradiate the samples at a wavelength of 365 nm for 30 min. Ensure that the Eppendorf tubes are open and directed to the light source.
6. Wash away the excess unbound probe by performing ethanol precipitation. Add 0.1 vol of 3 M NaOAc, 2.5−3 volumes of absolute ethanol and 1 μL of 20 mg/mL glycogen to precipitate the RNA. Mix well and let the RNA precipitate for at least 1 h at −20 °C. The precipitated RNA is pelleted by centrifuging the samples at 17,000 × *g*

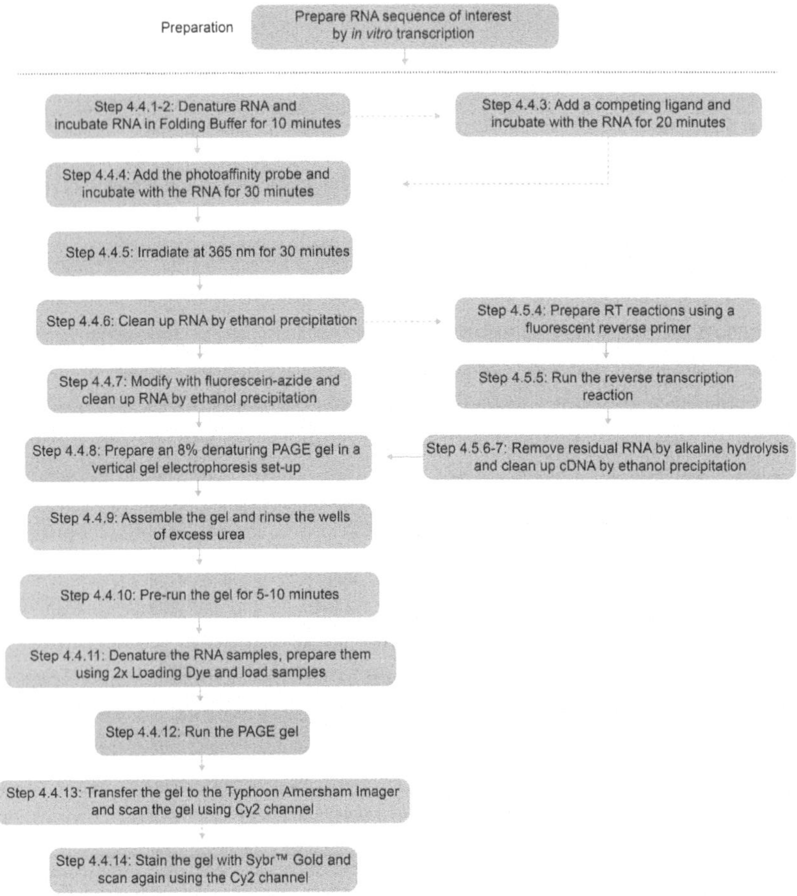

Fig. 3 Flowchart of protocols D-E.

for 1 h in a cooled centrifuge. The pellets are washed twice using 70 % ethanol and dissolved in 9 μL nuclease-free water.

7. Click 6-FAM-azide on the probe-bound RNA using standard copper-click chemistry. Denature the RNA samples at 65 °C for 5 min and store on ice. Prepare a 10x CuAAC click mixture and add to the denatured samples. Incubate samples at 37 °C for 60 min with gentle agitation. Wash away the excess click mixture using ethanol precipitation as described above.
8. Assemble a 20 cm × 20 cm × 1 mm vertical gel electrophoresis casting sandwich according to manufacturer's instructions. Prepare a

denaturing 8 % polyacrylamide gel and pour the gel immediately. Take care to avoid any air bubbles. Insert the comb and let the gel polymerize for 45 min.

9. After polymerization, assemble the gel in the gel electrophoresis tank per instructions. Fill the upper and lower chamber with 1 x TBE running buffer. Remove the comb carefully and rinse the wells with running buffer of excess urea using a syringe and needle.
10. Pre-run the polyacrylamide gel for 5–10 min at 300–500 V.
11. In the meantime, denature the RNA samples at 75 °C for 5 min and add 2x Loading Dye to the RNA samples. When the pre-run is finished, rinse the wells again to remove excess urea and load the samples into the wells.
12. Run the gel at 300–500 V for 30–120 min depending on the size of RNA.
13. Remove the gel from between the glass plates and transfer the gel onto the scanning plate of the Amersham Typhoon Imager. Scan the fluorescein signal (Abs/Em = 496/516 nm) using the CY2 channel.
14. Next, transfer the gel into a dish to stain the gel for nucleic acid content using Sybr™ Gold or a similar nucleic acid dye. Scan the gel for Sybr™ Gold signal (Abs/Em = 495/537 nm) using the CY2 channel on the Amersham Typhoon Imager.

4.4.3 Tip

Include controls without UV irradiation or photoaffinity probe to visualize fluorescent background signal.

4.5 Protocol E: Crosslinking induced reverse transcription (RT) stop assay

4.5.1 Overview

The RT-stop assay gives information about the binding site of the photoaffinity probe. The crosslinked probe creates an adduct that causes a stop during reverse transcription resulting in a truncated cDNA product. The RNA of interest is embedded between structure (SHAPE) cassettes described by Wilkinson et al. allowing all RNA positions to be evaluated (Wilkinson et al., 2006). A fluorescent reverse primer binds to the 3′ SHAPE cassette, and the RT stop sites can be visualized on a sequencing gel. Because of the universal SHAPE cassette and reverse primer, any RNA of interest (<500 nt) can be used without the need for optimization.

4.5.2 Duration

2.5 days.

1. In 1.5 mL Eppendorf tubes, prepare the amount of in vitro transcribed RNA containing SHAPE cassettes flanking the RNA of interest (5 pmol, 1 μL of 5 μM stock) for each sample. Denature the samples at 95 °C for 5 min and store on ice to keep the RNA in a denatured state.
2. Perform photoaffinity labeling as described in step 25–28 as described in Protocol D.
3. Wash away the excess unbound probe by performing ethanol precipitation. Add 0.1 vol of 3 M NaOAc, 2.5–3 volumes of absolute ethanol and 1 μL of 20 mg/mL glycogen to precipitate the RNA. Mix well and let the RNA precipitate for at least 1 h at −20 °C. The precipitated RNA is pelleted by centrifuging the samples at 17,000 × *g* for 1 h in a cooled centrifuge. The pellets are washed twice using 70% ethanol and dissolved in 10 μL nuclease-free water.
4. For the reverse transcription reaction, mix RNA with 6 pmol FAM-Primer (5′-/56-FAM/GAA CCG GAC CGA AGC CCG −3′) and incubated at 65 °C for 5 min then stored on ice. After incubation, add 2 μL of 10 mM dNTPs, 4 μL 5x first-strand buffer 1 μL 0.1 M DTT and 1 μL SuperScript™ III enzyme. Final reaction volume will be 20 μL.
5. The reactions are incubated at 45 °C for 2 min, 55 °C for 60 min and 70 °C for 15 min.
6. RNA is removed from the resulting cDNA samples by alkaline hydrolysis using 10 μL of 1 M NaOH and incubating the mixtures at 95 °C for 10 min. The reaction is quenched with an equal volume of 1 M HCl.
7. Samples are cleaned using ethanol precipitation as described above. cDNA pellets are dissolved in 5 μL of nuclease-free water.
8. Assemble a 20 cm × 45 cm × 1 mm vertical gel electrophoresis casting sandwich according to manufacturer's instructions. Prepare a denaturing 8 % polyacrylamide gel and pour the gel immediately. Take care to avoid any air bubbles. Insert the comb and let the gel polymerize for 45 min.
9. After polymerization, assemble the gel in the gel electrophoresis tank per instructions. Fill the upper and lower chamber with 1 x TBE running buffer. Remove the comb carefully and rinse the wells with running buffer to remove excess urea using a syringe and needle.
10. Pre-run the polyacrylamide gel for 5–10 min at 1000–1500 V.

11. In the meantime, denature the RNA samples at 75 °C for 5 min and add 2x Loading Dye to the RNA samples. When the pre-run is finished, rinse the wells again to remove excess urea and load the samples into the wells.
12. Run the gel at 1000−1500 V for ~6 h, or until the bromophenol blue dye reaches the end of the gel.
13. Remove the gel from between the glass plates and transfer onto the scanning plate of the Amersham Typhoon Imager. Scan the fluorescein signal using the CY2 channel.

4.5.3 Tip

Include sequencing lanes by preparing reverse transcription reactions containing 1 mM of each dideoxynucleotide to inform on the specific nucleotide location in the sequence. Include a no UV irradiation or photoaffinity probe control to visualize structure-induced stops by the reverse transcriptase enzyme. Include a competition control with the cognate ligand to check if it out competes the photoaffinity probe. cDNA products can be stored at −20 °C for several months to over a year.

4.6 Protocol F: Selective 2′-hydroxyl acylation analyzed by primer extension

4.6.1 Overview

In the research field of riboswitches, SHAPE is a convenient method to investigate RNA conformations in the presence or absence of a certain ligand. It is based on a chemical reagent that reacts preferentially with the 2′-hydroxyl group in flexible nucleotides resulting in a stable 2′-O-ester adduct. Several SHAPE reagents are available, in this protocol we will use 1M7 Fig. 4.

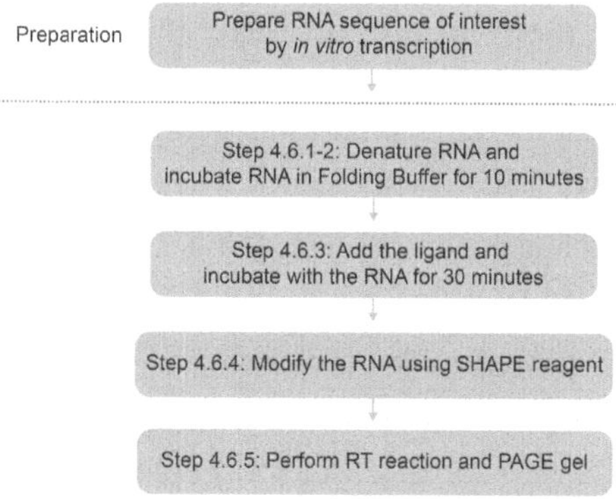

Fig. 4 Flowchart of protocol F.

4.6.2 Duration

2.5 days.

1. In 1.5 mL Eppendorf tubes, prepare the amount of *in vitro* transcribed RNA containing SHAPE cassettes flanking the RNA of interest (5 pmol, 1 µL of 5 µM stock) for each sample. Denature the samples at 95 °C for 5 min and store on ice to keep the RNA in a denatured state.
2. Add the folding buffer to each sample to ensure proper folding of the RNA and incubate the RNA at 37 °C for 10 min with gentle agitation.
3. Add different concentrations of ligand and incubate at 37 °C for 30 min with gentle agitation. The final volume will be 10 µL and the final concentration of DMSO should be below 5 %.
4. Mix 1.1 µL of 160 mM of 1M7 SHAPE reagent dissolved in DMSO with the RNA and incubate for 2 min on ice.
5. Perform reverse transcription reaction and gel visualization as described in Section 4.5.

4.6.3 Tip

Include a no 1M7 control to check the efficiency of the acylation reaction. No full-length product in the + 1M7 lanes can be a sign of extensive acylation. Lower the concentration of SHAPE reagent or reaction time. Inversely, an intense full-length product band and a low amount of RT stop bands can be a sign of insufficient modification by 1M7. Increase the concentration of SHAPE reagent or reaction time. Besides optimization of reaction conditions, a different SHAPE reagent can be used.

4.7 Protocol G: *In vitro* transcription/translation assay

4.7.1 Overview

The regulatory function of a small molecule-RNA interaction can be investigated using an IVTT assay. A template containing a luciferase reporter protein under the regulation of the RNA of interest is incubated in an *in vitro* transcription and translation system. Ligands can be tested to determine their effect on the regulation of luciferase on a transcriptional or translational level Fig. 5.

4.7.2 Duration

3 h.

1. Prepare the IVTT master mix by combining T7 S30 extract, S30 pre-mix, amino acids, luciferin and 7.5 nM template plasmid and keep on ice. The final volume per reaction will be 11.1 µL.
2. Add different concentrations of ligand to a 200 µL tube-strip.

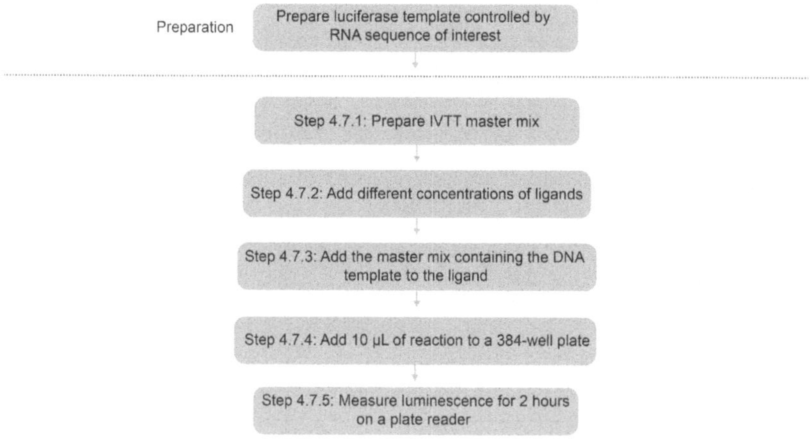

Fig. 5 Flowchart of protocol G.

3. Pipette the DNA template containing IVTT master mix on the wall of the tubes. Quickly spin down the tube-strips to start the IVTT reactions at the same time.
4. Add 10 μL of each reaction into a 384-wells flat bottomed plate. Be careful not to pipette any air bubbles into the well.
5. Measure luminescence on a plate reader every 2 min after a shaking step, while incubating at 30 °C–37 °C.

4.7.3 Tip

Include a control with only a luciferase template to check if there is an interfering effect of the ligand with the IVTT system. High concentrations of small molecules have been described to have an interfering effect independent of gene regulation.

4.7.4 Tip

Instead of measuring luminescence continuously, it is also possible to measure end-point luminescence, by only adding luciferin after quenching the translation reaction.

5. Analysis and statistics

Here we describe different methods to analyze the techniques reported in this chapter.

5.1 RNA sequencing analysis

RNA sequencing results can be analyzed using the following software: Bowtie2 (https://bowtie-bio.sourceforge.net/bowtie2, Langmead & Salzberg, 2012), RNA-Framework (https://rnaframework-docs.readthedocs.io, Incarnato et al., 2018), FeatureCounts (Liao et al., 2013) and the EdgeR package (https://bioconductor.org/, Robinson et al., 2010).

The following parameters are used in an example using paired-end sequencing reads of *E. coli* RNA:

1. Download the genome file from the FTP database and decompress the downloaded file

   ```
   > wget https://ftp.ncbi.nlm.nih.gov/genomes/all/GCF/000/005/845/GCF_000005845.2_ASM584v2/GCF_000005845.2_ASM584v2_genomic.fna.gz
   > gzip -d GCF_000005845.2_ASM584v2_genomic.fna.gz
   ```

2. Build a reference genome. Bowtie2 is recommended for prokaryotic genomes.

   ```
   > bowtie2-build GCF_000005845.2_ASM584v2_genomic.fna ecoli_ref
   For each genome, this step needs to be performed only once. The index file can be reused for new analyses.
   ```

3. Pre-process the reads and map them to the reference genome. We use rf-map from the RNA Framework, that combines these two steps into one step using Cutadapt, Bowtie2 and Samtools.

   ```
   > rf-map -b2 -bi ecoli_ref -ca5 (sequence of 5' adapter) -ca3 (sequence of 3' adapter) -kl Sample_1_R1.fastq,Sample_1_R2.fastq,Sample_N_R1.fastq,Sample_N_R2.fastq
   ```

4. Summarize the read counts using FeatureCounts.

   ```
   > featureCounts -t gene -g gene -F GTF -a GCF_000005845.2_ASM584v2_genomic.gtf -o counts.txt Sample_1_S1.bam Sample_N_SN.bam
   ```

5. Analyze differential expression using R and the EdgeR package to analyze enrichment. Here, an example is used with two conditions consisting of three replicates.

   ```
   > counts < - read.delim("counts_raw.csv", header = TRUE, row.names = 1, sep = ",")
   ```

```
> counts_lengths < - read.delim("counts_withlengths.
csv", header = TRUE, row.names = 1, sep = ",")
> length_genes < - counts_lengths[,1,drop=F]

# Set conditions for each sample in the dataset
> condition < - factor(c("Probe", "Probe", "Probe",
"Ctrl", "Ctrl", "Ctrl"))

# Create DGE object
> dgList < - DGEList(counts = counts, group = factor
(condition), genes = length_genes)
> rpkm < - rpkm(dgList)

# Filter low expressed genes
> countsPerMillion < - cpm(dgList)
> countCheck < - countsPerMillion > 1
> keep < - which(rowSums(countCheck) > = 2)
> dgList < - dgList[keep,]
> dgList$samples$lib.size < - colSums(dgList$counts)

# Normalize data using TMM method
> dgList < - calcNormFactors(dgList, method="TMM")

# Estimate dispersion
> design.mat < - model.matrix(~ 0 + dgList$samples
$group)
> colnames(design.mat) < -  levels(dgList$samples
$group)
> dgList < - estimateGLMCommonDisp(dgList,design.
mat)
> dgList < - estimateGLMTrendedDisp(dgList,design.
mat, method="power")
> dgList < - estimateGLMTagwiseDisp(dgList,design.
mat)

# Differential expression analysis
> fit < - glmFit(dgList, design.mat)

# Conduct likelihood ratio tests
> lrt < - glmLRT(fit, contrast=c(-1,1))

# Create a new object containing the differential
expression analysis results
```

```
> edgeR_results< - topTags(lrt, n=Inf)

# Merge with rpkm dataset
> resdata< -merge(as.data.frame(edgeR_results),
as.data.frame(rpkm), by= "row.names")

> names(resdata)[1]< - "Gene"
> resdata=na.omit(resdata)
> write.csv(resdata,"edgeR-diffexpr-results.csv")
```

The differential expression analysis results can be visualized using a volcano plot, MA plot or heatmap. Enriched genes can be identified using the following parameters, padj > 0.05 and log2FC > 0.58 (Fold change of 1.5) or > 1 (Fold change of 2).

RT-qPCR analysis.

There are different methods to analyze RT-qPCR results from pull-down experiments. Here, we describe the 'Relative fold enrichment' and the 'Percentage of Input' method (Solomon et al., 2021). The Relative fold enrichment is a method to quantify RNA of interest relative to a control condition. The Percentage of Input represents the amount of enriched RNA relative to the starting RNA (input). Define statistical significance by performing unpaired student's *t*-tests between conditions.

Relative fold enrichment method:

1. Calculate ΔCt

$$\Delta Ct = Ct\,(gene\ \ of\ \ interest) - Ct\,(housekeeping\ gene)$$

2. Calculate ΔΔCt

$$\Delta\Delta Ct = \Delta Ct\,(probe) - \Delta Ct\,(control\ probe)$$

3. Calculate Relative fold enrichment

$$Relative\ fold\ enrichment = 2^{-\Delta\Delta Ct}$$

Percentage of Input method:

1. Calculate adjusted input

$$Adjusted\ Input = Ct\,(1\%\ of\ \ Input) - \log 2\,(Dilution\ factor)$$

If 1% of Input is used, the dilution factor is 100.

2. Calculate Percentage of Input

$$Percentage\ \ of\ \ Input = 2^{(Adjusted\ Input - Ct\,(sample))}$$

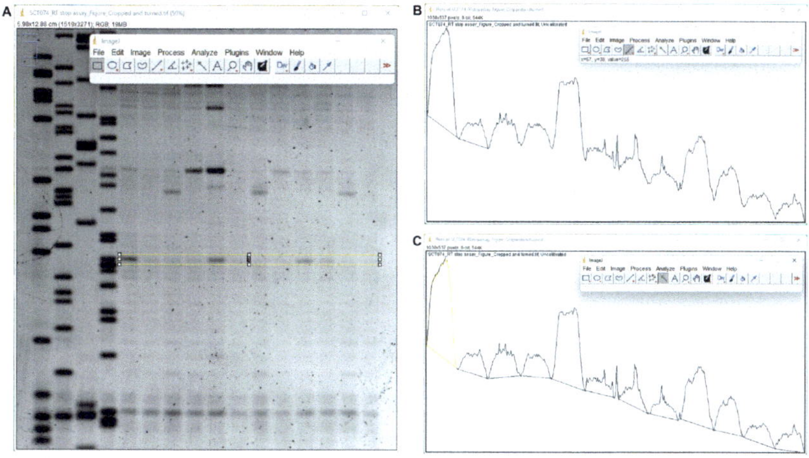

Fig. 6 Example of gel analysis using ImageJ.

5.2 Gel analysis

Fluorescent photoaffinity labeling or RT-stop and SHAPE assays can be conveniently quantified using ImageJ (Schneider et al., 2012).

1. Import the TIFF file into ImageJ.
2. Start by adjusting the overall brightness and contrast of the image until signal is adequately visible: Image > Adjust > Brightness/Contrast.
3. If necessary, adjust the rotation of the image until rows are nicely vertically oriented: Image > Transform > Rotate.
4. Using the rectangle tool, select the first row of bands. See Fig. 6A.
5. Mark this selection as the first lane: Analyze > Gels > Select First Lane (Ctrl + 1).

   If the gel contains only one row of bands, continue with step 9.
6. Move the rectangular selection to the second row.
7. Mark this selection as the next lane: Analyze > Gels > Select Next Lane (Ctrl + 2).
8. Repeat steps 6 and 7 as many times as needed.
9. Generate lane profile plots: Analyze > Gels > Plot Lanes (Ctrl + 3).

   Each peak represents the intensity of one band. Measuring the area under the curve of each peak allows for quantification.
10. Draw lines at the base of each peak using the line drawing tool, so that a closed area is created. Holding down the shift key, constrains the lines to be vertical. See Fig. 6B.

11. Use the wand tool and click inside each peak to measure the area under the curve. See Fig. 6C.
12. The resulting areas can be copied to excel for further analysis.

For the photoaffinity labeling coupled with a fluorescein azide, quantify both the fluorescent band intensity and the Sybr™ Gold band intensity. Calculate normalized intensities by normalizing the fluorescent intensities with the respective Sybr™ Gold intensities.

For the RT-stop and SHAPE assays, normalize the band intensities using a band of the SHAPE cassette as a representation of the amount of loaded cDNA.

5.3 IVTT analysis

1. Luminescent signals are measured over time on a plate reader.

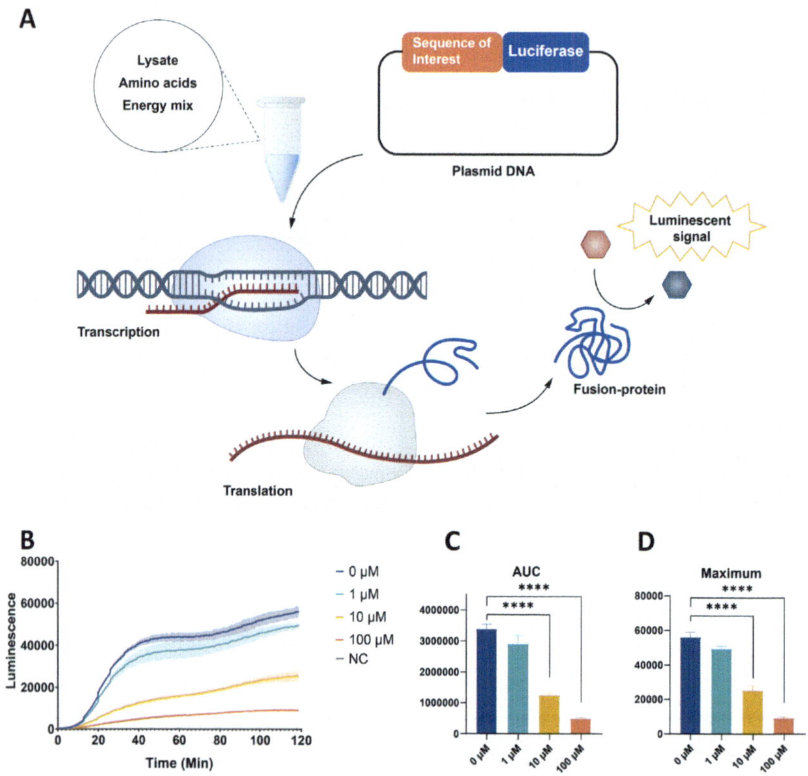

Fig. 7 (A) Schematic illustration of IVTT assay. (B) Example of IVTT data visualized overtime, (C) visualized as Area Under the Curve (AUC) or (D) Maximum value of each curve with 0, 1, 10 or 100 μM ligand concentration or a negative control (NC) without DNA template.

2. Calculate the peak height or area under the curve for each condition.
3. Calculate relative luminescence by normalizing the ligand condition versus the no ligand condition Fig. 7.

6. Summary

In this chapter, we described protocols to identify and validate ligand-RNA interactions using pulldown enrichments followed by transcriptome-wide sequencing or RT-qPCR, and to investigate these interactions using RT-stop, SHAPE and IVTT assays.

References

Balaratnam, S., Rhodes, C., Bume, D. D., Connelly, C., Lai, C. C., Kelley, J. A., ... Schneekloth Jr, J. S. (2021). A chemical probe based on the PreQ1 metabolite enables transcriptome-wide mapping of binding sites. *Nature Communications, 12*(1), 5856. https://doi.org/10.1038/s41467-021-25973-x.

Breaker, R. R. (2012). Riboswitches and the RNA world. *Cold Spring Harbor Perspectives in Biology, 4*(2), https://doi.org/10.1101/cshperspect.a003566.

Breaker, R. R. (2022). The biochemical landscape of riboswitch ligands. *Biochemistry, 61*(3), 137–149. https://doi.org/10.1021/acs.biochem.1c00765.

Cech, T. R., & Steitz, J. A. (2014). The noncoding RNA revolution-trashing old rules to forge new ones. *Cell, 157*(1), 77–94. https://doi.org/10.1016/j.cell.2014.03.008.

Chen, J. L., Zhang, P., Abe, M., Aikawa, H., Zhang, L., Frank, A. J., ... Disney, M. D. (2020). Design, optimization, and study of small molecules that target Tau Pre-mRNA and affect splicing. *Journal of the American Chemical Society, 142*(19), 8706–8727. https://doi.org/10.1021/jacs.0c00768.

Crielaard, S., Maassen, R., Vosman, T., Rempkens, I., & Velema, W. A. (2022). Affinity-based profiling of the flavin mononucleotide riboswitch. *Journal of the American Chemical Society, 144*(23), 10462–10470. https://doi.org/10.1021/jacs.2c02685.

Crielaard, S., Peters, C. F. M., Slivkov, A., van den Homberg, D. A. L., & Velema, W. A. (2025). Chemotranscriptomic profiling with a thiamine monophosphate photoaffinity probe. *Chemical Science, 16*(11), 4725–4731. https://doi.org/10.1039/D4SC06189F.

Giarimoglou, N., Kouvela, A., Maniatis, A., Papakyriakou, A., Zhang, J., Stamatopoulou, V., & Stathopoulos, C. (2022). A riboswitch-driven era of new antibacterials. *Antibiotics, 11*(9), 1243. https://www.mdpi.com/2079-6382/11/9/1243.

Gibaut, Q. M. R., Bush, J. A., Tong, Y., Baisden, J. T., Taghavi, A., Olafson, H., ... Disney, M. D. (2023). Transcriptome-wide studies of RNA-targeted small molecules provide a simple and selective r(CUG)(exp) degrader in myotonic dystrophy. *ACS Central Science, 9*(7), 1342–1353. https://doi.org/10.1021/acscentsci.2c01223.

Hanisch, M., Flemmich, L., Mitteregger, C., Bauer, I., Velandia-Huerto, C. A., Hofacker, I., ... Lusser, A. (2025). Experimental identification of preQ1-binding RNAs in the pathogenic bacterium Listeria monocytogenes. *RSC Chemical Biology, 6*(12), 1867–1878. https://doi.org/10.1039/D5CB00102A.

Incarnato, D., Morandi, E., Simon, L. M., & Oliviero, S. (2018). RNA framework: An all-in-one toolkit for the analysis of RNA structures and post-transcriptional modifications. *Nucleic Acids Research, 46*(16), e97. https://doi.org/10.1093/nar/gky486.

Langmead, B., & Salzberg, S. L. (2012). Fast gapped-read alignment with Bowtie 2. *Nature Methods, 9*(4), 357–359. https://doi.org/10.1038/nmeth.1923.

Liao, Y., Smyth, G. K., & Shi, W. (2013). featureCounts: An efficient general purpose program for assigning sequence reads to genomic features. *Bioinformatics (Oxford, England), 30*(7), 923–930. https://doi.org/10.1093/bioinformatics/btt656.

Lieberman, J. (2018). Tapping the RNA world for therapeutics. *Nature Structural & Molecular Biology, 25*(5), 357–364. https://doi.org/10.1038/s41594-018-0054-4.

McCown, P. J., Corbino, K. A., Stav, S., Sherlock, M. E., & Breaker, R. R. (2017). Riboswitch diversity and distribution. *RNA (New York, N. Y.), 23*(7), 995–1011. https://doi.org/10.1261/rna.061234.117.

Mortison, J. D., Schenone, M., Myers, J. A., Zhang, Z., Chen, L., Ciarlo, C., ... Myers, A. G. (2018). Tetracyclines modify translation by targeting key human rRNA substructures. *Cell Chemical Biology, 25*(12), 1506–1518. https://doi.org/10.1016/j.chembiol.2018.09.010 e1513.

Mukherjee, H., Blain, J. C., Vandivier, L. E., Chin, D. N., Friedman, J. E., Liu, F., ... Petter, J. C. (2020). PEARL-seq: A photoaffinity platform for the analysis of small molecule-RNA interactions. *ACS Chemical Biology, 15*(9), 2374–2381. https://doi.org/10.1021/acschembio.0c00357.

Panchal, V., & Brenk, R. (2021). Riboswitches as drug targets for antibiotics. *Antibiotics (Basel), 10*(1), https://doi.org/10.3390/antibiotics10010045.

Robinson, M. D., McCarthy, D. J., & Smyth, G. K. (2010). edgeR: A bioconductor package for differential expression analysis of digital gene expression data. *Bioinformatics (Oxford, England), 26*(1), 139–140. https://doi.org/10.1093/bioinformatics/btp616.

Roth, A., Winkler, W. C., Regulski, E. E., Lee, B. W. K., Lim, J., Jona, I., ... Breaker, R. R. (2007). A riboswitch selective for the queuosine precursor preQ1 contains an unusually small aptamer domain. *Nature Structural & Molecular Biology, 14*(4), 308–317. https://doi.org/10.1038/nsmb1224.

Salvail, H., & Breaker, R. R. (2023). Riboswitches. *Current Biology: CB, 33*(9), R343–r348. https://doi.org/10.1016/j.cub.2023.03.069.

Schneider, C. A., Rasband, W. S., & Eliceiri, K. W. (2012). NIH image to ImageJ: 25 years of image analysis. *Nature Methods, 9*(7), 671–675. https://doi.org/10.1038/nmeth.2089.

Scull, C. E., Dandpat, S. S., Romero, R. A., & Walter, N. G. (2021). Transcriptional riboswitches integrate timescales for bacterial gene expression control [mini review]. *Frontiers in Molecular Biosciences, 7 - 2020*. https://doi.org/10.3389/fmolb.2020.607158.

Solomon, E. R., Caldwell, K. K., & Allan, A. M. (2021). A novel method for the normalization of ChIP-qPCR data. *MethodsX, 8*, 101504. https://doi.org/10.1016/j.mex.2021.101504.

Spitale, R. C., & Incarnato, D. (2023). Probing the dynamic RNA structurome and its functions. *Nature Reviews. Genetics, 24*(3), 178–196. https://doi.org/10.1038/s41576-022-00546-w.

Tong, Y., Gibaut, Q. M. R., Rouse, W., Childs-Disney, J. L., Suresh, B. M., Abegg, D., ... Disney, M. D. (2022). Transcriptome-wide mapping of small-molecule RNA-binding sites in cells informs an isoform-specific degrader of QSOX1 mRNA. *Journal of the American Chemical Society, 144*(26), 11620–11625. https://doi.org/10.1021/jacs.2c01929.

Vitreschak, A. G., Rodionov, D. A., Mironov, A. A., & Gelfand, M. S. (2003). Regulation of the vitamin B12 metabolism and transport in bacteria by a conserved RNA structural element. *RNA (New York, N. Y.), 9*(9), 1084–1097. https://doi.org/10.1261/rna.5710303.

Wang, J., Schultz, P. G., & Johnson, K. A. (2018). Mechanistic studies of a small-molecule modulator of SMN2 splicing. *Proceedings of the National Academy of Sciences, 115*(20), E4604–E4612. https://doi.org/10.1073/pnas.1800260115.

Wilkinson, K. A., Merino, E. J., & Weeks, K. M. (2006). Selective 2′-hydroxyl acylation analyzed by primer extension (SHAPE): Quantitative RNA structure analysis at single nucleotide resolution. *Nature Protocols, 1*(3), 1610–1616. https://doi.org/10.1038/nprot.2006.249.

Winkler, W. C., Cohen-Chalamish, S., & Breaker, R. R. (2002). An mRNA structure that controls gene expression by binding FMN. *Proceedings of the National Academy of Sciences of the United States of America, 99*(25), 15908–15913. https://doi.org/10.1073/pnas.212628899.

Winkler, W. C., Nahvi, A., & Breaker, R. R. (2002). Thiamine derivatives bind messenger RNAs directly to regulate bacterial gene expression. *Nature, 419*(6910), 952–956. https://doi.org/10.1038/nature01145.

Winkler, W. C., Nahvi, A., Sudarsan, N., Barrick, J. E., & Breaker, R. R. (2003). An mRNA structure that controls gene expression by binding S-adenosylmethionine. *Nature Structural & Molecular Biology, 10*(9), 701–707. https://doi.org/10.1038/nsb967.

CHAPTER SIXTEEN

Monitoring structural dynamics of nascent bacterial riboswitches responding to TPP within transcriptional complexes

Mirette Hashem[a], Patrick St-Pierre[a], Adrien Chauvier[a,1], J. Carlos Penedo[b,c,*], and Daniel A. Lafontaine[a,*]

[a]Department of Biology, Faculty of Science, RNA Group, Université de Sherbrooke, Sherbrooke, QC, Canada

[b]Laboratory for Biophysics and Biomolecular Dynamics, SUPA School of Physics and Astronomy, Centre of Biophotonics, University of St. Andrews, St Andrews, United Kingdom

[c]Laboratory for Biophysics and Biomolecular Dynamics, Biomedical Sciences Research Complex, School of Biology, Centre of Biophotonics, University of St. Andrews, St Andrews, United Kingdom

*Corresponding author. e-mail address: jcp10@st-andrews.ac.uk; Daniel.Lafontaine@USherbrooke.ca

Contents

[1] Present address: Single Molecule Analysis Group, Department of Chemistry, University of Michigan, Ann Arbor, Michigan 48109, United States.

Methods in Enzymology, Volume 735
ISSN 0076-6879, https://doi.org/10.1016/bs.mie.2026.05.043

Abstract

This chapter describes a single-molecule fluorescence resonance energy transfer (smFRET) approach to monitor cotranscriptional RNA folding within native transcription complexes. The approach combines stepwise transcription by the *Escherichia coli* RNA polymerase (RNAP) with site-specific RNA labeling with a Cy3-Cy5 FRET pair via copper-free click chemistry. The chapter also details single molecule approaches to monitor the structural aspects of nascent transcripts and the analysis of observed RNA populations. Using the thiamin pyrophosphate (TPP)-sensing *E. coli tbpA* riboswitch as a model system, we detail procedures for DNA template preparation, RNA labeling and smFRET acquisition. Key advantages of the described smFRET approach include the use of native transcripts and the ability to monitor RNA structural changes within transcription complexes that are either stalled at naturally occurring pause sites or that are actively elongating along the DNA template. This method permits the study of cotranscriptional regulatory mechanisms taking place either in bacterial or eukaryotic elongation complexes and could be extrapolated to the use of any biochemical probes compatible with the click chemistry approach.

1. Introduction

Riboswitches are structured RNA motifs that are present in the 5′-untranslated regions (5′-UTR) of mRNA and control various genetic mechanisms such as transcription, translation, mRNA decay and splicing (Bédard et al., 2020; Breaker, 2022; Salvail & Breaker, 2023; Serganov & Nudler, 2013). These RNA switches are composed of a highly conserved metabolite-sensing aptamer domain and an expression platform that controls genetic expression. It was previously shown that the adenine-sensing *Bacillus subtilis pbuE* riboswitch performs ligand binding exclusively cotranscriptionally (Lemay et al., 2006; Lemay et al., 2011; Rieder et al., 2007; Wickiser et al., 2005), therefore showing the importance of riboswitch cotranscriptional regulation. The cotranscriptional sensing of adenine is crucial as it regulates the premature transcription termination of the *pbuE* gene. Furthermore, it has been shown that cotranscriptional sensing is important for several translationally-controlling riboswitches for which the metabolite binding affinity is dictated by the position of the elongating RNAP along the transcriptional landscape (Bédard et al., 2020). Collectively, these data show that riboswitches primarily rely on various cotranscriptional mechanisms in which nascent RNA structures play crucial roles in both ligand sensing and genetic regulation (Bédard et al., 2020; Chiaruttini & Guillier, 2020, p. 202; Jayasinghe & Babitzke, 2025; Schärfen & Neugebauer, 2021).

The folding of RNA molecules typically occurs on a millisecond to minute timescale, which can be studied using single-molecule approaches (Roy et al., 2008; Shaw et al., 2014). The smFRET technique allows monitoring conformational changes using Cy3 and Cy5 dyes that are incorporated at specific nucleotides of RNA molecules (Perez-Gonzalez et al., 2016; Roy et al., 2008; Shaw et al., 2014). These assays provide information that may be difficult to extract from ensemble studies due to the dynamic heterogeneity of the population (Roy et al., 2008; St-Pierre et al., 2014; Weiss, 2000; Zhuang et al., 2000). While chemical RNA synthesis has been invaluable in producing fluorescent molecules to study RNA dynamics at the single-molecule level, novel enzymatic approaches have been developed to reconstitute fluorescent elongation complexes that allow to directly probe crucial aspects of transcriptional regulatory mechanisms (Chatterjee et al., 2021; Chauvier, Ajmera, et al., 2021; Chauvier, St-Pierre, et al., 2021; Hien, St-Pierre, et al., 2024). For instance, due to its ability to incorporate cyanine-modified nucleotides, T7 RNAP has been instrumental to site-specifically introduce reactive group or dyes within RNA molecules (Liu et al., 2015; Uhm et al., 2018). However, because T7 RNAP exhibits a fast elongation rate (Chamberlin & Ring, 1973) and does not efficiently respond to bacterial transcriptional pause sites (Artsimovitch et al., 2000), its use to study the folding and activity of bacterial transcripts has remained unsuccessful (Chao et al., 1995; Lewicki et al., 1993; Pan & Sosnick, 2006; Perdrizet 2nd et al., 2012; Wong et al., 2007). Together, these papers suggest that the cotranscriptional folding of bacterial transcripts must be studied using the native RNAP, i.e., from the same organism as the investigated RNA molecule.

A new smFRET approach has recently been developed to characterize the folding of *E. coli* nascent transcripts in native transcription conditions using the *E. coli* RNAP (Chauvier et al., 2021; Hien, St-Pierre, et al., 2024). Using this approach, a Cy3-Cy5 pair was introduced in the thiamin pyrophosphate (TPP)-sensing *thpA* riboswitch to monitor RNA structural dynamics at the single-molecule level. The incorporation of Cy3 is performed by initiating transcription reactions with a Cy3-labeled trinucleotide, therefore restricting its presence to the RNA 5′ extremity. However, in contrast to the T7 RNAP, the introduction of Cy5 cannot be achieved using cyanine-modified nucleotides since the latter cannot be accommodated in the bacterial RNAP active site (Vassylyev et al., 2002). This limitation is alleviated by incorporating in the RNA sequence an azido-uridine analog that can be coupled with DBCO-Cy5 through a copper-free click reaction (Paredes & Das, 2011; Winz et al., 2012) (Fig. 1A). The site-specific introduction of the azido group in nascent RNAs is achieved using stepwise transcription reactions, which enables to stall transcription elongation complexes (EC) at any desired positions along the DNA template by omitting specific nucleotides in the transcription reaction (Chauvier et al., 2021). Because smFRET assays can record the folding of several single molecules over time, it allows reconstitution of the folding pathway of RNA molecules

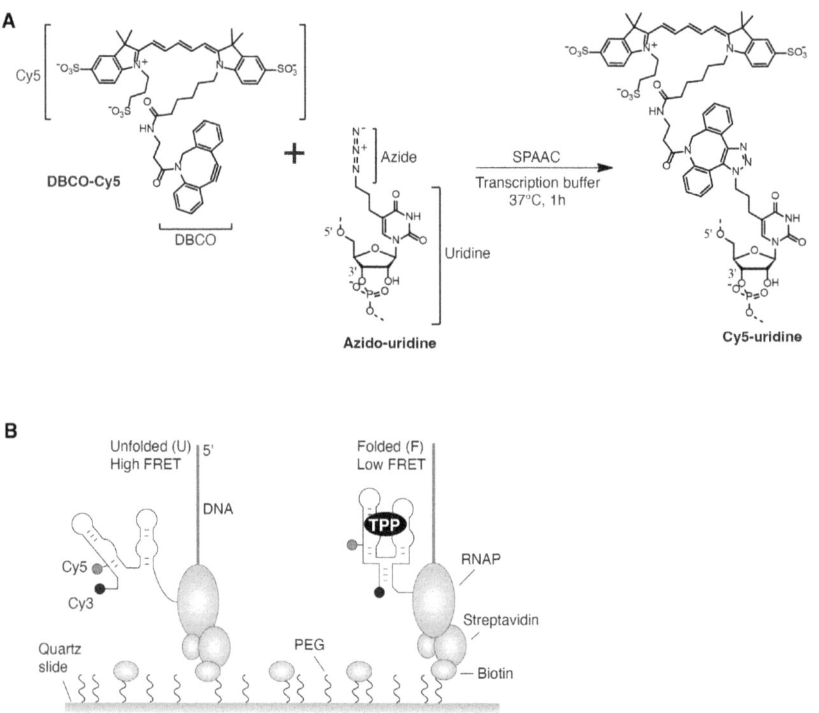

Fig. 1 (A) Chemical structure of the labels used for the click chemistry. The structure of the DBCO-Cy5 and azido-uridine are shown. The resulting Cy5-uridine obtained through the strain-promoted azide-alkyne cycloaddition (SPAAC) is also shown. (B) Schematic representation of the *tbpA* riboswitch in the absence (left, U state) and presence (right, F state) of TPP. The transcripts are attached to a quartz slide using the biotin-streptavidin roadblock. The slide was functionalized using PEG and biotin-PEG to allow streptavidin attachment.

and to reveal the kinetics of the associated structural states. Monitoring the folding of many single RNA molecules simultaneously requires the use of a microscope setup equipped with total-internal reflection (TIR) illumination methods and an EMCCD camera (iXon Life) to detect the FRET signal (Roy et al., 2008; Shaw et al., 2014). To prevent diffusion and enable the visualization of single RNA molecules for long periods of time, transcription complexes are attached to a microscope slide via a biotin-streptavidin interaction (Fig. 1B) (Hien, St-Pierre, et al., 2024; McCluskey et al., 2019; Roy et al., 2008; Shaw et al., 2014). In the case of the *tbpA* TPP-sensing riboswitch, it was determined that the unbound and TPP-bound RNA molecules exhibit an unfolded (U) and folded (F) structures (Fig. 1B), respectively, that can be studied using smFRET analysis (Chauvier et al., 2021; Hien, St-Pierre, et al., 2024). This approach can be used to study nascent transcripts in any prokaryotic or eukaryotic transcription systems and can be expanded to use any click chemistry compatible reagents.

In this chapter, we describe the design, fluorescent labeling and characterization of the TPP-sensing *tbpA* riboswitch using smFRET measurements.

2. DNA template design and construction

2.1 Rationale

Given that transcription elongation is influenced by key parameters such as RNAP pausing, elongation rate and RNA-protein interactions (Zhang & Landick, 2016), *E. coli* riboswitches must be transcribed using the *E. coli* RNAP to adopt native structures. Consequently, the design of DNA templates used for transcription reactions should include a bacterial promoter to ensure functional assays. In the present case, the lacUV5 promoter is used as it yields robust in vitro transcription reactions (Chauvier et al., 2017; Chauvier et al., 2021; Hien, Chauvier, et al., 2024; Hien, St-Pierre, et al., 2024).

Previous studies have revealed that nascent riboswitches show varying binding affinities depending on the position of elongating RNAP along the DNA template (Chauvier et al., 2017; Chauvier et al., 2021). These findings can be rationalized by the fact that growing nascent transcripts exhibit different structures during transcription elongation, which can influence the RNA binding affinity. In the case of the *E. coli tbpA* riboswitch, it was determined that transcription complexes located at position 117 (Fig. 2A), which is a natural pause site, exhibit a strong affinity for TPP (~360 nM) (Chauvier et al., 2021). Importantly, the biotin-streptavidin complex can be used as a transcriptional roadblock to stall transcription complexes at the pause site 117 (EC-117) by attaching it at the 5′ end of DNA template strand (Fig. 2B) (Chauvier et al., 2017; Chauvier et al., 2021; Hien, Chauvier, et al., 2024; Hien, St-Pierre, et al., 2024). It should be noted that the biotin moiety must be positioned 9 nt downstream of the RNAP stalled position (Chauvier et al., 2017), therefore meaning that the biotin is attached at position 126 of the DNA strand to obtain the EC-117 complex (Chauvier et al., 2021).

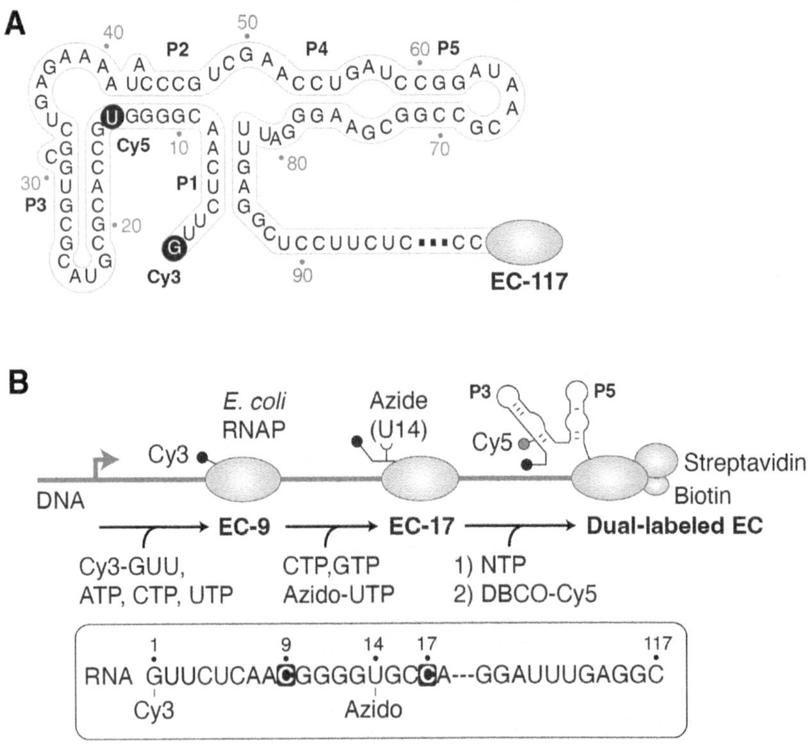

Fig. 2 (A) Predicted secondary structure of the thiamin pyrophosphate (TPP)-sensing *tbpA* riboswitch in the ligand-bound state. Cy3 and Cy5 are indicated at their respective positions. RNA is shown in the context of the EC-117 elongation complex. The helical domains are denoted P1 to P5. (B) Schematic approach to obtain Cy3-Cy5 dual-labeled elongation complexes. Each step allowing the formation of EC-9, EC-17 and full-length (EC-117) are shown. The corresponding riboswitch sequence is shown where the EC-9 and EC-17 are indicated in black.

The following protocol is directed toward the making of DNA templates for the transcription of the *tbpA* riboswitch that encompasses the region contained within positions 1–117 (Fig. 1A), thus allowing efficient TPP sensing.

2.2 Materials, equipment and reagents

- Bio Rad T100 TM Thermal Cycler
- *E. coli* genomic DNA 50 μg/mL
- Primer forward 10 μM (with lacUV5 promoter sequence)
- Primer reverse 10 μM (with biotin at 5′ extremity)
- dNTP 10 mM
- Taq buffer 10X (Bioshop)
- Taq DNA polymerase (Bioshop, #TAQ001.400)
- Milli-Q water
- QIAquick PCR purification Kit (Cat. No. 28106, Qiagen)

2.3 Protocols

2.3.1 Design of the DNA primers used for the PCR amplification

The final sequence required for transcription reactions is:

GGGCACCCCAGGCTTTACACTTTATGCTTCCGGCTCG TATAATGTGTGGGTTCTCAACGGGGTGCCACGCGTACGCG TGCGCTGAGAAAATACCCGTCGAACCTGATCCGGATAACGC CGGCGAAGGGATTTGAGGCTCCTTCTCAAGTCCTTTGCCAC TCTTTTT**TGAGGTGCA**.

The promoter sequence and the last nine residues required to obtain the biotin-streptavidin roadblock are in bold. The complete 117 nt sequence of the transcribed *tbpA* riboswitch is underlined.

The DNA primers required for the PCR amplification are:

Forward Primer:

5′-**GGGCACCCCAGGCTTTACACTTTATGCTTCCGGCTC GTATAATGTGTGG**GTTCTCAACGGGGTGCCAC.

Reverse Primer:

5′biotin-**TGCACCTCA**AAAAAGAGTGGCAAAGG.

For both primers, the promoter (Forward primer) and the biotin-streptavidin roadblock sequence (Reverse primer) are shown in bold.

2.3.2 PCR reaction and purification steps

In a 100 μL reaction volume:

- 2 μL *E. coli* genomic DNA 50 μg/mL
- 5 μL Forward primer 10 μM

- 5 μL Reverse primer 10 μM
- 2 μL dNTP 10 mM
- 10 μL Taq buffer 10X
- 1 μL Taq DNA polymerase
- 75 μL Milli-Q water

Using the Bio Rad T100 TM Thermal Cycler, follow the PCR cycle shown below:

- 95 °C, 3 min (denaturation step)
- 95 °C, 30 s (start of PCR cycle)
- 60 °C, 30 s
- 72 °C, 30 s (repeat 30X from the start of PCR cycle)
- 72 °C, 5 min
- 4 °C, 1 min

It is suggested that the PCR products are purified with the QIAquick PCR purification kit following the manufacturer's protocol.

Notes:

- If required, an agarose gel can be done to visually inspect the quality of the obtained PCR products. Using the above protocol, it is expected that a single PCR product of ~175 bp is produced.
- In cases where non-desired PCR products are obtained during the amplification, the addition of the Taq DNA polymerase could be done right after the 95 °C/3 min step of the PCR reaction. This procedure ensures that no partially paired DNA duplex is amplified by the Taq DNA polymerase during the first PCR cycle. In addition, the temperature of the annealing step (60 °C) could be raised to increase the stringency of the DNA hybridization, which should reduce the presence of non-specific products.
- An HPLC purification of the biotinylated primer could improve the efficiency of smFRET imaging.

3. Stepwise transcription reactions and Cy5 labeling of nascent transcripts

3.1 Rationale

The smFRET technique provides quantitative measurements when the distance occurring between the Cy3 and Cy5 dyes ranges from 2 to 8 nm, which is relevant for various biological processes such as RNA folding

(Clegg, 1992, p. 199; Joo et al., 2008; Shaw et al., 2014). To ensure that the introduced Cy3 and Cy5 dyes are within a meaningful FRET range within the characterized RNA molecule, a crystal structure should be used whenever possible to estimate the distance between both dyes. For example, to study the TPP-sensing *E. coli tbpA* riboswitch, the crystal structure of the TPP-bound *E. coli thiM* riboswitch (Edwards & Ferre-D′Amare, 2006; Serganov et al., 2006; Thore et al., 2006) may be used to estimate fluorophore placements as it is highly similar to the *tbpA* variant (Bastet et al., 2017; Chauvier et al., 2021). Another point that needs to be considered for riboswitches is whether the locations of both dyes may detect metabolite-dependent structural changes. Taking these criteria into account, the smFRET study of the *tbpA* riboswitch was done by introducing the Cy3 and Cy5 dyes at positions 1 and 14, respectively, as they are expected to report on the formation of the TPP-dependent P1 stem (Fig. 2) (Chauvier et al., 2021; Hien, St-Pierre, et al., 2024).

In contrast to the Cy3 dye that is incorporated at the 5′ RNA extremity, the azido-UTP analog needs to be introduced internally within the RNA sequence at position 14 (Fig. 2A). The site-specific introduction of the reactive nucleotide requires a stepwise transcription approach in which a subset of nucleotides is used to elongate specific regions of the RNA. In particular, when using a mixture containing the trinucleotide Cy3-GUU, ATP, CTP and UTP, the transcription elongation complex polymerizes RNA up to position 9 (EC-9) (Fig. 2B). Next, after removing unincorporated residues with G-50 columns, the use of azido-UTP, CTP and GTP leads to the formation of EC-17 (Fig. 2B). Because U14 is the only position corresponding to an uracil residue during this step, the azido-UTP analog is only introduced at this position. Transcription elongation can then be resumed by adding NTP to the reaction, thus yielding the full-length RNA (EC-117). Finally, the Cy5 labeling at position U14 is achieved through copper-free click chemistry by incubating DBCO-Cy5 with the EC-117 complex that contains an azido group at U14.

3.2 Materials, equipment and reagents

- Thermo SCIENTIFIC Sorvall Legend Micro 21 centrifuge
- Isotemp Fisher Scientific dry bath
- EDTA (Disodium Salt, Dihydrate) Reagent Grade (Bioshop)
- Streptavidin 15 μM
- Transcription buffer 5X and 1X

- RNAP E. coli 6 μM (Protein Purification Platform of the Université de Sherbrooke)
- Sigma70 12 μM (Protein Purification Platform of the Université de Sherbrooke)
- Cy3-GUU 50 μM (Eurogentech, #39236)
- High purity ATP, GTP, CTP, UTP 25 μM (Cytiva, #27202501)
- azido-UTP 25 μM (Chromatographic Specialties Inc, #JBNU157S)
- DBCO-Cy5 1 mM (Lumiprobe Corp, #A30F0)
- NTPs 20 mM
- Sephadex G-50 Fine DNA Grade (Cytiva)
- Micro Bio-Spin Chromatography Columns (Bio-Rad)
- Milli-Q water

3.3 Protocols

3.3.1 Solutions preparation

All solutions are prepared with Milli-Q water.

- EDTA 0.5 M, pH 8.0: Dissolve 186 g of EDTA in water, adjust the pH to 8.0 and bring the final volume to 1 L with water.
- Transcription buffer 5X: Mix 1.5 mL Tris-HCl 1 pH 8.0, 1.5 mL MgCl2 1 M, 1.5 mL NaCl 1 M, 15 μL EDTA 0.5 M and 10.5 mL water.
- Transcription buffer 1X: Mix 3 mL Transcription buffer 5X and 12 mL water.
- GTP 25 μM, CTP 25 μM, azido-UTP 50 μM: Mix 1 μL ATP 1 mM, 1 μL CTP 1 mM, 2 μL azido-UTP 1 mM and 36 μL water.
- ATP 25 μM, CTP 25 μM, UTP 25 μM: Mix 1 μL ATP 1 mM, 1 μL CTP 1 mM, 1 μL UTP 1 mM and 37 μL water.
- NTP 20 mM: Mix 20 μL GTP 100 mM, 20 μL, CTP 100 mM, 20 μL UTP 100 mM, 20 μL ATP 100 mM and 20 μL water.
- Solution of Sephadex G-50 resin: 2.13 g of Sephadex G-50 Fine DNA Grade resin, 400 μL Tris-HCl 1 pH 8.0, 80 μL EDTA 0.5 M and water up to 40 mL.

3.3.2 Stepwise transcription reactions and Cy5 labeling

All solutions are prepared with Milli-Q water.

In a 1.5 mL Eppendorf tube:

1. Add 1 μL (~6 pmol) of DNA template obtained by PCR (Sections 2.3.2), 1 μL of streptavidin 15 μM, 4 μL transcription buffer 5X and 6 μL water.
2. Incubate 5 min at 37 °C.

3. Add 2 μL RNA polymerase 6 μM and 2 μL sigma70 12 μM.
4. Incubate 5 min at 37 °C.
5. Add 2 μL Cy3-GUU 50 μM and 2 μL ATP 25 μM, CTP 25 μM, UTP 25 μM. Incubate 10 min at 37 °C → **EC-9 (Cy3-labeled)**. Arrows denote transcription products formed at this stage.
6. Wash the elongation complexes using G-50 columns (see section below for details).
7. Add 2 μL GTP 25 μM, CTP 25 μM, azido-UTP 50 μM. Incubate 10 min at 37 °C → **EC-17 (Cy3-labeled)**.
8. Add 2 μL NTP 20 mM. Incubate 5 min at 37 °C → **EC-117 (Cy3-labeled)**.
9. Add 2 μL DBCO-Cy5 1 mM. Incubate 1 h at 37 °C → **EC-117 (Cy3-Cy5-labeled)**.
10. Wash the elongation complexes using G-50 columns (see section below for details).

3.3.3 Preparation of G-50 columns and washing steps

Column preparation:

For a ~20 μL reaction, prepare two G-50 columns:

1. Place a Micro Bio-Spin column in a 1.5 mL Eppendorf tube.
2. Add 500 μL of solution of resin Sephadex G-50 Fine DNA Grade.
3. Centrifuge 700 rcf for 1 min
4. Discard flow-through and place the column back in the same Eppendorf tube.
5. Add 100 μL transcription buffer 1X to the G-50 column.
6. Centrifuge 700 rcf for 1 min
7. Place the ready to be used G-50 column in a fresh 1.5 mL Eppendorf tube in which the elongation complexes will be collected.

G-50 column washing steps:

1. Add ~20 μL transcription reaction to a freshly made G-50 column that is in a clean 1.5 mL Eppendorf tube.
2. Centrifuge at 700 rcf for 1 min
3. Pipet the flow-through and add it to a second freshly prepared G-50 column that is in a clean 1.5 mL Eppendorf tube.
4. Centrifuge at 700 rcf for 1 min
5. Collect the flow through for further use.

Notes:

- Prepare G-50 columns 5 min before using them to prevent the columns from drying out and losing their efficiency.
- During the washing steps, add the samples to the center of the resin without disturbing the surface.

4. Quality control for stepwise transcription and RNA Cy5 labeling

4.1 Rationale

During our studies of the *tbpA* riboswitch, we have observed that one of the most critical steps in obtaining meaningful smFRET measurements is the efficient Cy5 labeling of nascent transcripts (Chauvier et al., 2021; Hien, St-Pierre, et al., 2024). Indeed, in the event that RNA molecules are inefficiently labeled with Cy5, it will result in the majority of molecules being Cy3-labeled only, thus leading to most molecules being non-informative about the studied structural changes. To our knowledge, it is difficult to predict the efficiency of RNA labeling with Cy5 using copper-free click chemistry. We expect that the Cy5 labeling efficiency should depend on accessibility to the RNA labeling site and the structural environment that could affect the labeling chemical reaction. For these reasons, we have adopted an approach in which the Cy5 labeling efficiency is monitored using polyacrylamide gel electrophoresis (PAGE) (Chauvier et al., 2021; Hien, St-Pierre, et al., 2024). In this approach, the successful labeling of nascent transcripts with Cy5 leads to a gel retardation, which can be used to estimate the efficiency of labeling.

4.2 Materials, equipment and reagents

- Tris (Tris (Hydroxymethyl) Aminomethane) BioUltraPure (Bioshop)
- EDTA (Disodium Salt, Dihydrate) Reagent Grade (Bioshop)
- Formamide (Biotechnology Grade) (Bioshop)
- Boric acid ACS GRADE (Bioshop)
- TEMED (N, N, N′, N′-Tetramethylethylenediamine) (Bioshop)
- APS (Ammonium persulfate) (Sigma)
- Urea Reagent Grade (Bioshop)
- Acrylamide Ultra Pure Grade (Bioshop)
- Bis-acrylamide BioUltraPure Grade (N, N-Methylene-bis-acrylamide) (Bioshop)

- Filter paper 150 mm Ø (Whatman Schleicher & Schuell)
- Aluminium foil
- VWR N95 Disposable Respirator
- **Refer to "Stepwise transcription reactions and Cy5 labeling of nascent transcripts"**, sections "3.2 Materials, equipment and reagents" and "3.3.1 Solutions preparation".

4.3 Protocols

4.3.1 Solutions preparation

All solutions are prepared with Milli-Q water.

- EDTA 0.5 M, pH 8: Dissolve 186.12 g of EDTA in water, adjust the pH to 8, and bring the final volume to 1 L with water.
- TBE 5X: Dissolve 54.4 g TRIS and 27.5 g boric acid, add 20 mL of 0.5 M EDTA pH8, then bring the final volume up to 1 L with water.
- TBE 1X: Add 100 mL TBE 5X, then bring the volume up to 500 mL with water.
- TBE 1X Urea 8 M: In a beaker, add 100 mL TBE 5X and 240.24 g of urea, then bring the volume up to 500 mL with water. Stir until fully dissolved and then filter the solution using filter paper. Store in room temperature.
- TBE 1X Urea 8 M Acrylamide 20 %: In a beaker, add 100 mL TBE 5X, 240.24 g of urea, 95 g of acrylamide, 5 g of bis acrylamide, then bring the volume up to 500 mL with water. Stir until fully dissolved and then filter the solution using filter paper. Store in room temperature.
- APS 1 %: In a Falcon tube, add 1 g of APS and then bring the volume up to 10 mL with water. Vortex until fully dissolved. Cover the Falcon tube with aluminum foil and store it at 4 °C.
- Denaturing buffer: In a Falcon tube, add 340 μL of 10 % SDS, 1.44 mL EDTA 0.5 M (pH 8), and 38 mL of formamide.

Notes:

- When preparing the solutions TBE 1X Urea 8 M Acrylamide 20 %, place the beaker on a magnetic stirrer hot plate (~60 °C) to prepare the solution.
- When handling acrylamide, wear a VWR N95 disposable respirator mask because the acrylamide and bis-acrylamide are neurotoxic. Also, cover the beaker with aluminium foil while heating to prevent the release of acrylamide and bis-acrylamide vapors into the air.

- The solution TBE 1X Urea 8 M Acrylamide 20 % should be stored in a brown glass bottle as it is sensitive to light.
- When preparing the denaturing buffer, handle the formamide in the fume hood due to its toxicity.

4.3.2 Samples preparation

All solutions are prepared with Milli-Q water.

This part follows the protocol in the section "Stepwise transcription reaction and Cy5 labeling" with a few modifications. Please refer to Fig. 3A to see a scheme showing the different fractions that are resolved on PAGE.

1. Add 18 pmol TPP riboswitch DNA template (obtained by PCR), 3 μL of streptavidin 15 μM, 12 μL Transcription buffer 5x and water (aim for final volume of 60 μL).
2. Incubate 5 min at 37 °C.
3. Add 6 μL RNA polymerase *E. coli* 6 μM and 6 μL sigma70 12 μM.
4. Incubate 5 min at 37 °C.
5. Add 6 μL Cy3-GUU 50 μM and 6 μL ATP 25 μM, CTP 25 μM, UTP 25 μM. Incubate 10 min at 37 °C → **EC-9 (Cy3-labeled)**.
6. Washing step using G-50 columns (see Section 3.3.3 for details).
7. Split the reaction obtained after the washing steps (60 μL) into three Eppendorf tubes: 10 μL in tube **A**, 40 μL in tube **B** and 10 μL in tube **Extra**.
8. Add 10 μL of denaturing buffer to tube **A** → **EC-9 labeled with Cy3**.
9. In tube **B**, add 4 μL of GTP 25 μM, CTP 25 μM, and azido-UTP 50 μM. Incubate 10 min at 37 °C → **EC-17 (Cy3-labeled)**.
10. Split the reaction of tube **B** into two Eppendorf tubes: 10 μL in the tube **C** and 30 μL in the tube **D**.
11. Add 10 μL of denaturing buffer to tube **C** → **EC-17 labeled with Cy3**.
12. Add 3 μL NTP 20 mM to tube **D**. Incubate 5 min at 37 °C.
13. Split the reaction of tube **D** into two Eppendorf tubes: 10 μL in tube **E** and 20 μL in tube **F**.
14. Add 10 μL of denaturing buffer to tube **E** → **EC-117 labeled with Cy3**.
15. Add 2 μL of DBCO-Cy5 1 mM to tube **F**. Incubate 1 h at 37 °C.
16. Washing step using G-50 columns (see Section 3.3.3 for details).
17. Add 20 μL of denaturing buffer to tube **F** → **EC-117 labeled with Cy3 and Cy5**.
18. Load tubes **A**, **C**, **E** and **F** on a denaturing PAGE gel.

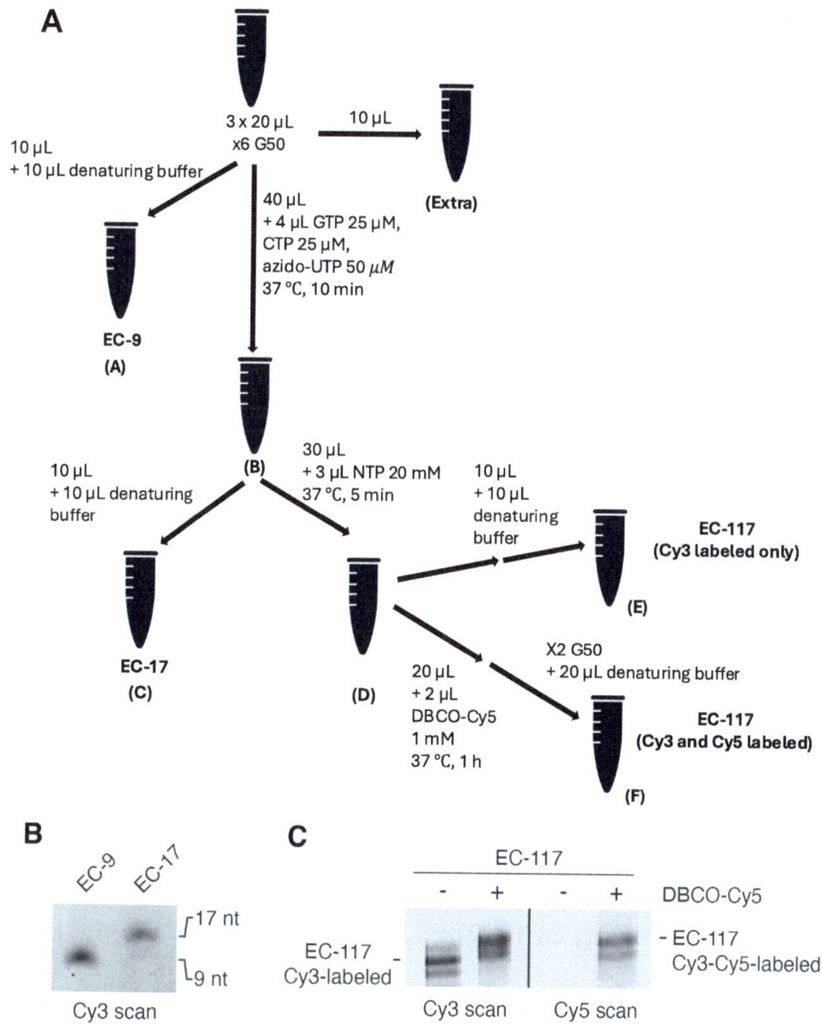

Fig. 3 (A) Scheme representing the different fractions obtained during stepwise transcription and Cy5 labeling. The respective manipulations are shown for each tube that is obtained (A to F) to monitor the efficiency of transcription and fluorophore labeling. (B) Stepwise transcription reactions to obtain EC-9 and EC-17, which correspond to transcripts of 9 and 17 nt, respectively. The transcripts were detected using Cy3 fluorescence. Sizes of transcripts are indicated to the right of the gel. (C) Analysis of the Cy5 labeling reaction. EC−117 transcripts are allowed to react with DBCO-Cy5, which leads to Cy5 labeling. The successful labeling of transcripts results in a slower migrating RNA species that is due to the presence of Cy5. While the left part of the gel was detected using Cy3 fluorescence, the right part was detected using Cy5 fluorescence.

Notes:

- If required, other controls could be done such as replacing the azido-UTP analog with UTP to determine whether the Cy5 labeling is specific to the presence of the azido reactive group.

4.4 Results

Stepwise transcription reactions allow the efficient polymerization of defined portions of nascent transcripts (Nudler et al., 2003). By performing transcription in a stepwise manner, the EC-9 complex was successfully obtained (Fig. 3B). Furthermore, upon washing the unincorporated NTP and adding the required mix of nucleotides, the complex EC-17 was generated (Fig. 3B). Importantly, nascent RNA in EC-17 contains a unique azido-UTP analog at position U14, which is due to the presence of a single uracil in between position 9 and 17. By adding a complete mix of nucleotides, the full-length RNA is transcribed at EC-117 due to the presence of the biotin-streptavidin transcriptional roadblock (Fig. 3C, Cy3 scan, – DBCO-Cy5). The incorporation of Cy5 on nascent transcripts is performed by incubating with DBCO-Cy5, which leads to an apparent retardation in gel migration (Fig. 3C, Cy3 scan, + DBCO-Cy5). The presence of Cy5 can also be detected by directly scanning the Cy5 dye (Fig. 3C, Cy5 scan). Together, these results show that our approach allows to efficiently obtain EC-117 transcription complexes that are labeled both with Cy3 and Cy5 dyes.

5. Monitoring of nascent RNA and smFRET analysis

5.1 Rationale

Due to the lower stability of Cy5, the addition of an oxygen scavenger system (PCA/PCD) and a triplet-state quencher (Trolox) are required to improve the efficiency of analysis (Roy et al., 2008; Shaw et al., 2014). As a result, the recording of time traces of selected molecules can be obtained to observe the folding dynamics of nascent RNA within transcription complexes (Roy et al., 2008). The following section does not cover how to prepare microscope slides or how to operate an smFRET optical setup, which was previously described in an excellent review (Roy et al., 2008). We rather provide information about how to prepare fluorescent samples for smFRET measurements and show data that was collected for the *tbpA* nascent riboswitch transcripts with and without TPP.

5.2 Materials, equipment and reagents

- PCA (3,4-protocatechuic acid) (Fisher Scientific, #10162810)
- PCD (protocatechuate dioxygenase) (Sigma-Aldrich, #P8279-25UN)
- Trolox (6-hydroxy-2,5,7,8-tetramethylchroman-2-carboxylic acid) (Sigma-Aldrich, #238813)

5.3 Protocols

5.3.1 Solutions preparation

All solutions are prepared with Milli-Q water.

- Transcription buffer 1X: 3 mL of Transcription buffer 5X (section 3.3.1) and 12 mL water.
- Imaging buffer 1X: 95 μL Transcription buffer 1X, 2 μL Trolox, 2 μL PCA and 1 μL PCD.
- PCD: Resuspend PCD bottle with 450 μL of PCD buffer 1X. Aliquot and keep at −20 °C.
- PCD buffer 1X: 250 μL glycerol and 250 μL Tris-HCl 100 pH 8.0
- Trolox: Resuspend 125 mg Trolox in 1 mL of methanol and add 9 mL water. Adjust to pH 9.5. Aliquot and keep at −20 °C.
- PCA: Resuspend 77 mg PCA in 8 mL water. Adjust to pH 9.0 and complete to 10 mL with water. Aliquot and keep at −20 °C.

5.3.2 Data processing and analysis workflow

The data analysis workflow outlined below assumes a single-molecule wide-field FRET configuration using total-internal reflection excitation and a split-view of donor and acceptor into a single EMCCD camera. The flowchart below includes each step from raw data acquisition to the analysis of FRET histograms and dwell times of each FRET state (Fig. 4).

1. Dual channel acquisition: for a Cy3/Cy5 FRET pair or equivalent, a 532 nm laser operating typically at 0.5–5 mW in TIR fluorescence mode is adjusted to give a signal of approximately 2000 counts above background per frame. For TIR fluorescence specifically, the beam must hit the prism or objective at the critical angle (~63–66° for glass/water interface), producing an evanescent field that decays exponentially with ~100–200 nm depth. This restricts excitation to surface-immobilized molecules and is what gives smFRET its low background. Since both donor (D) and acceptor (A) emission must be captured simultaneously, a split-view mode using one camera is the usual choice. A dichroic mirror inside a commercial image splitter (e.g. Cairn Optosplit, Photometrics

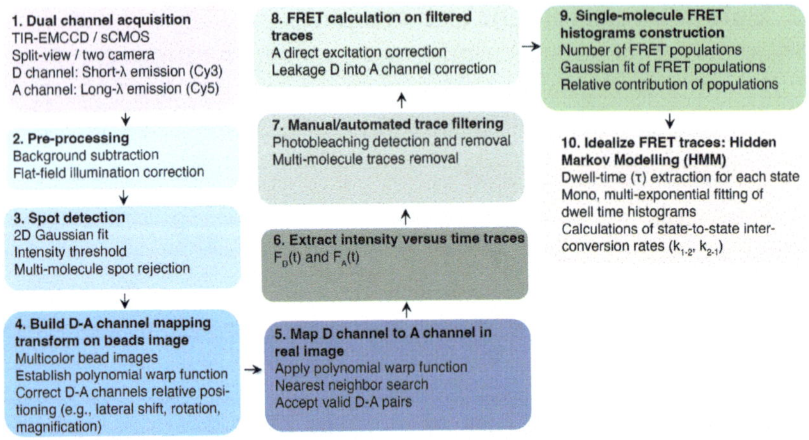

Fig. 4 Data analysis workflow of the smFRET approach. Each step of the flowchart shows key aspects of the analysis that are described in more details in the text.

DualView) splits the emission onto two halves of the same camera detector. The short wavelength emission corresponding to the donor is imaged in the left half of the camera and the acceptor intensity in the right half.

2. Pre-processing: before any analysis, subtracts a camera background image from each channel and apply flat-field correction to normalize the non-uniform laser illumination profile.
3. Spot detection: candidate single molecules are detected in one channel (usually D, or whichever has better signal-to-noise ratio). Next, fitting a difference-of-Gaussians (DoG) bandpass filter is applied to suppress noise followed by thresholding on intensity. Each detected spot provides a (x,y) coordinate. Next, spots that are too dim, too bright (aggregates), or too close together are rejected.
4. Build D-A channel mapping: because the D and A channels are never perfectly aligned, a geometric transform relating D-pixel coordinates to A-pixel coordinates is required prior to extracting FRET traces. This is done using multicolor beads (e.g. TetraSpeck beads) that emit in both channels so that the same bead appears in both halves of the camera. A set of corresponding point pairs is detected and their positions fitted using a polynomial warp that provides a geometrical transform.
5. The transform is then applied to the experimental D spots to predict where each molecule should appear in the A channel. Co-localization (typically within 1–2 pixels after mapping) confirms a valid D-A pair.

6. and 7. Intensity extraction and filtering: for each accepted pair, the pixel intensities are integrated in both channels for each correlated D and A pair across every frame of the movie, giving time traces $F_D(t)$ and $F_A(t)$. Traces with multiple bleach steps suggesting multi-molecule spots, acceptor-only signals, or excessive noise are rejected at this stage either manually or using semi-automated software (see notes).

8. FRET calculation on filtered traces: for most applications that require only to determine relative changes in FRET efficiency, this can be calculated using the following expression:

$$E_{app} = \frac{I_A}{I_A + I_D}$$

which corresponds to an apparent FRET efficiency (E_{app}) value. In this expression, I_A indicates the intensity of the acceptor and I_D denotes the intensity of the donor. However, to correlate changes in FRET efficiency with distance changes, corrections due to leakage of the donor into the acceptor channel, direct excitation of the acceptor and different detection efficiency of detectors at the donor and acceptor emission wavelengths are required. This is beyond the scope of this review, and we refer to a recently published protocol regarding accuracy and precision in single-molecule FRET methods (Hellenkamp et al., 2018).

9. Generation of single-molecule FRET histograms: calculate the FRET average of the first ten frames for each molecule and plot them as a histogram where the y-axis represents number of molecules in a particular FRET bin (see notes). Next, single-molecule FRET histograms are fitted to a distribution of gaussians where the number of gaussians depends on the number of FRET states observed. The area under each gaussian can be used to extract the percentual contribution of the corresponding FRET population. As a rule of thumb, the number of molecules added to the single-molecule histograms is enough when the number of gaussians and the relative contribution is independent on the number of FRET bins used and does not change with the addition of more molecules.

10. Dwell time analysis: to extract kinetic information including the dwell time of each state and the interconversion rates between them, the FRET traces are idealized to recover from the noisy trace a step function of discrete FRET states. The standard is to implement variational Bayes Hidden Markov Modelling (HMM) using tools such as vbFRET, ebFRET or HaMMy (see reference (Hellenkamp et al., 2018)). These tools simultaneously infer the number of states and their transition rates. From the idealized trace, the dwell times in each state are extracted and pooled across hundreds of molecules and the resulting survival functions (or histograms) are fitted to mono- or multi-exponential decays using maximum likelihood to provide the state-to-state interconversion rate constants.

Notes:

- Too high laser power will photobleach the sample within seconds. The use of a too low laser power may result in the signal-to-noise ratio being insufficient for HMM analysis. In practice, the optimal configuration is the one that allows to collect traces that last at least 30–60 s with clear anti-correlation between donor and acceptor fluorescence efficiencies.
- Correction for non-uniform illumination is critical because uneven illumination creates FRET gradients across the field.
- In our lab, we use custom built routines coded in IDL language (NV5 Geospatial) and Matlab (Mathworks) but there are multiple software packages freely available to carry out all steps from spot detection and extraction of intensities to calculating single-molecule FRET histograms and dwell times. Some of these include iSMS (Hatzakis group, Aarhus University, https://inano.au.dk/about/research-groups/single-molecule-biophysics-and-chemistry-group/software/download), HaMMy (Ha lab, Harvard University, https://github.com/Ha-SingleMoleculeLab/HaMMy), TwoTone and MASH-FRET (Kapanidis lab, Oxford University, https://kapanidis.web.ox.ac.uk/software).
- Single-molecule FRET histograms can also be built cumulative from entire traces. In these conditions, traces should have the same integration time and accumulated over the same temporal window. The advantage of using the initial 10-frame average snapshot is that it removes the contribution from photobleaching as most traces will last longer than this 10-frame window.

5.4 Results

The approach described above allows the production of transcription complexes that are stalled at the naturally occurring pause site 117 of the *E. coli thpA* riboswitch. Since elongating complexes are stalled at the pause site using a biotin-streptavidin roadblock, the latter may be used to anchor complexes on a microscope slide that is covered with polyethylene glycol (PEG) and biotin-PEG (Fig. 1B). By performing single-molecule imaging of EC-117 nascent transcripts using wide-field TIRF microscopy, time traces of the fluorescence intensities of Cy3 and Cy5 were obtained. In the absence of TPP, the intensities of Cy3 and Cy5 are relatively stable and the intensity of Cy5 is slightly higher than Cy3 (Fig. 5, No TPP). The analysis of the trace

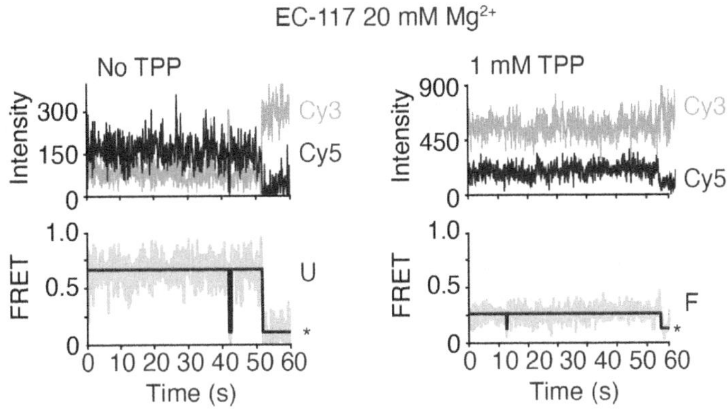

Fig. 5 Single-molecule time traces obtained for Cy3-Cy5 dual-labeled EC-117 nascent transcripts in a background of 20 mM Mg^{2+} ions. Traces were obtained in the absence (left) and presence (right) of TPP. For both traces, the fluorescence intensities of Cy3 and Cy5 and the corresponding FRET traces are shown. An asterisk denotes a photobleaching event.

reveals an smFRET value of ~0.6, which is attributed to the unfolded structure (Fig. 5). However, in the presence of 1 mM TPP, the Cy5 fluorescence signal is lower than the Cy3 signal, thus resulting in a FRET value of ~0.3 (Fig. 5, right panel). In this case, the different FRET value indicates that the TPP-bound nascent transcripts adopt the folded state where Cy3 and Cy5 are located further apart from each other (Fig. 1B). Time traces obtained both without and with TPP remain stable during the experiments, indicating that the structure of nascent transcripts does not fluctuate between both states but remains committed to a structure once it is formed.

6. Conclusion

The folding of nascent transcripts into biologically relevant structures is a highly regulated process that is crucial for cellular functions. However, due to the short-lived nature of some nascent RNA structures, it is difficult to characterize the process by which nascent structures are adopted during transcription elongation. Furthermore, because our approach relies on the use of bacterial RNAP, it ensures that the studied RNA molecules fold into their native structures and that meaningful data are collected. Lastly, since azido-UTP nucleotide analogs may be used by eukaryotic RNAP (Jao & Salic, 2008; Sawant et al., 2016), our approach can be used to monitor at the single-molecule level the structural dynamics of nascent transcripts in the context of eukaryotic systems.

Related techniques

Single-molecule FRET techniques using chemically synthesized and fluorescently labeled RNAs have been extensively employed and provided insights into equilibrium folding pathways and ligand-induced conformational changes. However, these techniques examine RNA function and structure outside of transcription complexes. Cotranscriptional folding has been studied using T7 RNAP transcripts combined with smFRET. Although this approach enables the incorporation of modified nucleotides, the intrinsic features of T7 RNAP might not accurately mimic native bacterial transcription complexes.

Ensemble probing methods such as SHAPE, DMS mapping and in-line probing are powerful tools to investigate RNA structure in the absence and presence of folding cofactors. When applied in a cotranscriptional context, these methods provide nucleotide-resolution but averaged over molecule populations.

Single-molecule force spectroscopy techniques, including optical and magnetic tweezers allow mechanical interrogation of RNA structural changes during transcription and provide complementary information to smFRET. However, they require specialized instrumentation, and real-time folding can only be proven for a statistically limited number of molecules.

Pros and cons

Pros	Cons
Enables direct observation of RNA folding during transcription using native RNA polymerase.	Cy5 labeling efficiency can be variable and requires optimization.
Allows site-specific fluorescent labeling of nascent transcripts within transcription elongation complexes.	Stepwise transcription and purification are labor-intensive.
Preserves transcriptional pausing and regulatory features essential for riboswitch function.	Requires access to purified RNA polymerase and single-molecule instrumentation.
Compatible with single-molecule analysis and detection of heterogeneous RNA populations	Limited throughput when compared to ensemble biochemical assays.
Can be adapted to other transcription systems and click chemistry-compatible probes.	Optimization may be required for new RNA targets.
Allows to obtain very large fluorescently labeled RNA molecules (>0.5 kb).	

Alternative methods/procedures

Recent alternative approaches have been developed to produce fluorescently-labeled transcripts. For example, the method PLOR (Position-specific Labeling of RNA) has been modified to allow the use of the *E. coli* RNAP (Xue et al., 2023). Furthermore, nucleic scaffold assembly systems have been employed to produce nascent transcripts (Chauvier et al., 2024; Widom et al., 2018; Yadav et al., 2022). Together, these methods contribute to a versatile toolbox to ensure the making of fluorescent nascent transcripts.

Troubleshooting & optimization

Problem	Solution
Low yields of transcription elongation complexes (EC) might result from inefficient promoter recognition or sub-optimal RNAP concentration.	Verify DNA template integrity and promoter sequence. Increase RNAP and σ^{70} concentrations slightly ($\leq$1.5 ×). Ensure incubation steps at 37 °C are respected.
Incomplete stalling at desired EC position could result from inefficient roadblock or incorrect biotin placement.	Confirm that the biotin moiety is positioned 9 nt downstream of the intended stall site. Use freshly prepared streptavidin and avoid excess free biotin.
Poor incorporation of azido-UTP due to competition with natural UTP or incorrect nucleotide.	Ensure that azido-UTP is the only uridine source during the EC-17 step. Verify nucleotide concentrations and remove unincorporated NTP thoroughly using G-50 columns.
Low Cy5 labeling efficiency might result from limited accessibility of azido groups or suboptimal click reaction.	Increase incubation time with DBCO-Cy5 or slightly increase dye concentration. Perform labeling at 37 °C and avoid excessive Mg^{2+} concentrations that may stabilize inaccessible RNA conformations.
High background fluorescence in smFRET imaging due to excess of free dye or high density of immobilized complexes.	Perform two sequential G-50 purification steps after click labeling. Dilute samples to obtain ~300 molecules per field of view.
Rapid Cy5 photobleaching.	Ensure fresh preparation of the PCA/PCD oxygen scavenging system and Trolox. Minimize laser power.
Heterogeneous or unstable FRET signals possibly due to RNA misfolding or non-specific interaction with the surface.	Confirm PEGylation quality of microscope slides using a non-biotinylated sample or by removing streptavidin. Reduce surface density or immobilized complexes.
No observable FRET changes upon ligand addition due to low ligand concentration or RNA in non-native fold.	Pre-equilibrate samples before imaging to ensure ligand binding has occurred.

Summary

Stepwise transcription by native *E. coli* RNAP allows the generation of fluorescent labeled transcription elongation complexes stalled at different positions along the DNA template or engaged in active elongation.

Site-specific fluorescent labeling of nascent RNA can be achieved using a combination of 5′ Cy3 labeled trinucleotide and internal azido-UTP labeled with Cy5 via copper-free click chemistry.

Dual-labeled nascent transcripts can be studied either within transcription elongation complexes or following transcript release.

The approach enables detection of heterogeneous RNA populations and stable folding states that are not accessible by ensemble measurements.

Although demonstrated here using a bacterial riboswitch, the strategy can be adapted to other RNA systems and transcription machineries compatible with azido-modified nucleotides and click chemistry.

References

Artsimovitch, I., Svetlov, V., Anthony, L., Burgess, R. R., & Landick, R. (2000). RNA polymerases from Bacillus subtilis and Escherichia coli differ in recognition of regulatory signals in vitro. *Journal of Bacteriology, 182*(21), 6027–6035.

Bastet, L., Chauvier, A., Singh, N., Lussier, A., Lamontagne, A.-M., Prévost, K., ... Lafontaine, D. A. (2017). Translational control and Rho-dependent transcription termination are intimately linked in riboswitch regulation. *Nucleic Acids Research, 45*(12), 7474–7486. https://doi.org/10.1093/nar/gkx434.

Bédard, A.-S. V., Hien, E. D. M., & Lafontaine, D. A. (2020). Riboswitch regulation mechanisms: RNA, metabolites and regulatory proteins. *Biochimica et Biophysica Acta. Gene Regulatory Mechanisms, 1863*(3), 194501. https://doi.org/10.1016/j.bbagrm.2020.194501.

Breaker, R. R. (2022). The biochemical landscape of riboswitch ligands. *Biochemistry, 61*(3), 137–149. https://doi.org/10.1021/acs.biochem.1c00765.

Chamberlin, M., & Ring, J. (1973). Characterization of T7-specific ribonucleic acid polymerase. 1. General properties of the enzymatic reaction and the template specificity of the enzyme. *The Journal of Biological Chemistry, 248*(6), 2235–2244.

Chao, M. Y., Kan, M. C., & Lin-Chao, S. (1995). RNAII transcribed by IPTG-induced T7 RNA polymerase is non-functional as a replication primer for ColE1-type plasmids in Escherichia coli. *Nucleic Acids Research, 23*(10), 1691–1695.

Chatterjee, S., Chauvier, A., Dandpat, S. S., Artsimovitch, I., & Walter, N. G. (2021). A translational riboswitch coordinates nascent transcription-translation coupling. *Proceedings of the National Academy of Sciences of the United States of America, 118*(16), e2023426118. https://doi.org/10.1073/pnas.2023426118.

Chauvier, A., Ajmera, P., Yadav, R., & Walter, N. G. (2021). Dynamic competition between a ligand and transcription factor NusA governs riboswitch-mediated transcription regulation. *Proceedings of the National Academy of Sciences of the United States of America, 118*(47), e2109026118. https://doi.org/10.1073/pnas.2109026118.

Chauvier, A., Dandpat, S. S., Romero, R., & Walter, N. G. (2024). A nascent riboswitch helix orchestrates robust transcriptional regulation through signal integration. *Nature Communications, 15*(1), 3955. https://doi.org/10.1038/s41467-024-48409-8.

Chauvier, A., Picard-Jean, F., Berger-Dancause, J.-C., Bastet, L., Naghdi, M. R., Dubé, A., ... Lafontaine, D. A. (2017). Transcriptional pausing at the translation start site operates as a critical checkpoint for riboswitch regulation. *Nature Communications, 8*, 13892. https://doi.org/10.1038/ncomms13892.

Chauvier, A., St-Pierre, P., Nadon, J.-F., Hien, E. D. M., Pérez-González, C., Eschbach, S. H., ... Lafontaine, D. A. (2021). Monitoring RNA dynamics in native transcriptional complexes. *Proceedings of the National Academy of Sciences of the United States of America, 118*(45), e2106564118. https://doi.org/10.1073/pnas.2106564118.

Chiaruttini, C., & Guillier, M. (2020). On the role of mRNA secondary structure in bacterial translation. *Wiley Interdisciplinary Reviews. RNA, 11*(3), e1579. https://doi.org/10.1002/wrna.1579.

Clegg, R. M. (1992). Fluorescence resonance energy transfer and nucleic acids. *Methods in Enzymology, 211*, 353–388.

Edwards, T. E., & Ferre-D'Amare, A. R. (2006). Crystal structures of the Thi-Box riboswitch bound to thiamine pyrophosphate analogs reveal adaptive RNA-small molecule recognition. *Structure (London, England: 1993), 14*(9), 1459–1468. https://doi.org/10.1016/j.str.2006.07.008.

Hellenkamp, B., Schmid, S., Doroshenko, O., Opanasyuk, O., Kühnemuth, R., Rezaei Adariani, S., ... Hugel, T. (2018). Precision and accuracy of single-molecule FRET measurements-a multi-laboratory benchmark study. *Nature Methods, 15*(9), 669–676. https://doi.org/10.1038/s41592-018-0085-0.

Hien, E. D. M., Chauvier, A., St-Pierre, P., & Lafontaine, D. A. (2024). Structural characterization of the cotranscriptional folding of the thiamin pyrophosphate sensing *thiC* riboswitch in *Escherichia coli*. *Biochemistry, 63*(13), 1608–1620. https://doi.org/10.1021/acs.biochem.3c00665.

Hien, E. D. M., St-Pierre, P., Penedo, J. C., & Lafontaine, D. A. (2024). Cotranscriptional folding of a 5′ stem-loop in the Escherichia coli tbpA riboswitch at single-nucleotide resolution. *Journal of Molecular Biology, 436*(22), 168771. https://doi.org/10.1016/j.jmb.2024.168771.

Jao, C. Y., & Salic, A. (2008). Exploring RNA transcription and turnover in vivo by using click chemistry. *Proceedings of the National Academy of Sciences of the United States of America, 105*(41), 15779–15784. https://doi.org/10.1073/pnas.0808480105.

Jayasinghe, O. T., & Babitzke, P. (2025). Applying the brakes to transcription: Regulation of gene expression by RNA polymerase pausing. *Journal of Bacteriology, 207*(7), e0008425. https://doi.org/10.1128/jb.00084-25.

Joo, C., Balci, H., Ishitsuka, Y., Buranachai, C., & Ha, T. (2008). Advances in single-molecule fluorescence methods for molecular biology. *Annual Review of Biochemistry, 77*, 51–76. https://doi.org/10.1146/annurev.biochem.77.070606.101543.

Lemay, J. F., Penedo, J. C., Tremblay, R., Lilley, D. M. J., & Lafontaine, D. A. (2006). Folding of the adenine riboswitch. *Chemistry and Biology, 13*(8), 857–868. https://doi.org/10.1016/j.chembiol.2006.06.010.

Lemay, J.-F., Desnoyers, G., Blouin, S., Heppell, B., Bastet, L., St-Pierre, P., ... Lafontaine, D. A. (2011). Comparative study between transcriptionally- and translationally-acting adenine riboswitches reveals key differences in riboswitch regulatory mechanisms. *PLoS Genetics*, 7(1), e1001278. https://doi.org/10.1371/journal.pgen.1001278.

Lewicki, B. T., Margus, T., Remme, J., & Nierhaus, K. H. (1993). Coupling of rRNA transcription and ribosomal assembly in vivo. Formation of active ribosomal subunits in Escherichia coli requires transcription of rRNA genes by host RNA polymerase which cannot be replaced by bacteriophage T7 RNA polymerase. *Journal of Molecular Biology, 231*(3), 581–593. https://doi.org/10.1006/jmbi.1993.1311.

Liu, Y., Holmstrom, E., Zhang, J., Yu, P., Wang, J., Dyba, M. A., ... Wang, Y.-X. (2015). Synthesis and applications of RNAs with position-selective labelling and mosaic composition. *Nature, 522*(7556), 368–372. https://doi.org/10.1038/nature14352.

McCluskey, K., Boudreault, J., St-Pierre, P., Perez-Gonzalez, C., Chauvier, A., Rizzi, A., ... Penedo, J. C. (2019). Unprecedented tunability of riboswitch structure and regulatory function by sub-millimolar variations in physiological Mg^{2+}. *Nucleic Acids Research, 47*(12), 6478–6487. https://doi.org/10.1093/nar/gkz316.

Nudler, E., Gusarov, I., & Bar-Nahum, G. (2003). Methods of walking with the RNA polymerase. *Methods in Enzymology, 371*, 160–169. https://doi.org/10.1016/S0076-6879(03)71011-8.

Pan, T., & Sosnick, T. (2006). RNA folding during transcription. *Annual Review of Biophysics and Biomolecular Structure, 35*, 161–175.

Paredes, E., & Das, S. R. (2011). Click chemistry for rapid labeling and ligation of RNA. *Chembiochem: a European Journal of Chemical Biology, 12*(1), 125–131. https://doi.org/10.1002/cbic.201000466.

Perdrizet 2nd, G. A., Artsimovitch, I., Furman, R., Sosnick, T. R., & Pan, T. (2012). Transcriptional pausing coordinates folding of the aptamer domain and the expression platform of a riboswitch. *Proceedings of the National Academy of Sciences of the United States of America, 109*(9), 3323–3328. https://doi.org/10.1073/pnas.1113086109.

Perez-Gonzalez, C., Lafontaine, D. A., & Penedo, J. C. (2016). Fluorescence-based strategies to investigate the structure and dynamics of aptamer-ligand complexes. *Frontiers in Chemistry, 4*(AUG), 33. https://doi.org/10.3389/fchem.2016.00033.

Rieder, R., Lang, K., Graber, D., & Micura, R. (2007). Ligand-Induced folding of the adenosine deaminase a-riboswitch and implications on riboswitch translational control. *Chembiochem: a European Journal of Chemical Biology, 8*, 896–902.

Roy, R., Hohng, S., & Ha, T. (2008). A practical guide to single-molecule FRET. *Nature Methods, 5*(6), 507–516. https://doi.org/10.1038/nmeth.1208.

Salvail, H., & Breaker, R. R. (2023). Riboswitches. *Current Biology: CB, 33*(9), R343–R348. https://doi.org/10.1016/j.cub.2023.03.069.

Sawant, A. A., Tanpure, A. A., Mukherjee, P. P., Athavale, S., Kelkar, A., Galande, S., & Srivatsan, S. G. (2016). A versatile toolbox for posttranscriptional chemical labeling and imaging of RNA. *Nucleic Acids Research, 44*(2), e16. https://doi.org/10.1093/nar/gkv903.

Schärfen, L., & Neugebauer, K. M. (2021). Transcription regulation through nascent RNA folding. *Journal of Molecular Biology, 433*(14), 166975. https://doi.org/10.1016/j.jmb.2021.166975.

Serganov, A., & Nudler, E. (2013). A decade of riboswitches. *Cell, 152*(1–2), 17–24. https://doi.org/10.1016/j.cell.2012.12.024.

Serganov, A., Polonskaia, A., Phan, A. T., Breaker, R. R., & Patel, D. J. (2006). Structural basis for gene regulation by a thiamine pyrophosphate-sensing riboswitch. *Nature, 441*(7097), 1167–1171. https://doi.org/10.1038/nature04740.

Shaw, E., St-Pierre, P., McCluskey, K., Lafontaine, D. A., & Penedo, J. C. C. (2014). Using sm-FRET and denaturants to reveal folding landscapes. *Methods in Enzymology, 549*(C), 313–341. https://doi.org/10.1016/B978-0-12-801122-5.00014-3.

St-Pierre, P., McCluskey, K., Shaw, E., Penedo, J. C. C., & Lafontaine, D. A. A. (2014). Fluorescence tools to investigate riboswitch structural dynamics. *Biochimica et Biophysica Acta (BBA) - Gene Regulatory Mechanisms, 1839*(10), 1005–1019. https://doi.org/10.1016/j.bbagrm.2014.05.015.

Thore, S., Leibundgut, M., & Ban, N. (2006). Structure of the eukaryotic thiamine pyrophosphate riboswitch with its regulatory ligand. *Science (New York, N. Y.), 312*(5777), 1208–1211. https://doi.org/10.1126/science.1128451.

Uhm, H., Kang, W., Ha, K. S., Kang, C., & Hohng, S. (2018). Single-molecule FRET studies on the cotranscriptional folding of a thiamine pyrophosphate riboswitch. *Proceedings of the National Academy of Sciences of the United States of America, 115*(2), 331–336. https://doi.org/10.1073/pnas.1712983115.

Vassylyev, D. G., Sekine, S., Laptenko, O., Lee, J., Vassylyeva, M. N., Borukhov, S., & Yokoyama, S. (2002). Crystal structure of a bacterial RNA polymerase holoenzyme at 2.6 A resolution. *Nature, 417*(6890), 712–719. https://doi.org/10.1038/nature752.

Weiss, S. (2000). Measuring conformational dynamics of biomolecules by single molecule fluorescence spectroscopy. *Nature Structural Biology, 7*(9), 724–729.

Wickiser, J. K., Cheah, M. T., Breaker, R. R., & Crothers, D. M. (2005). The kinetics of ligand binding by an adenine-sensing riboswitch. *Biochemistry, 44*(40), 13404–13414. https://doi.org/10.1021/bi051008u.

Widom, J. R., Nedialkov, Y. A., Rai, V., Hayes, R. L., Brooks, C. L., Artsimovitch, I., & Walter, N. G. (2018). Ligand modulates cross-coupling between riboswitch folding and transcriptional pausing. *Molecular Cell, 72*(3), 541–552.e6. https://doi.org/10.1016/j.molcel.2018.08.046.

Winz, M.-L., Samanta, A., Benzinger, D., & Jäschke, A. (2012). Site-specific terminal and internal labeling of RNA by poly(A) polymerase tailing and copper-catalyzed or copper-free strain-promoted click chemistry. *Nucleic Acids Research, 40*(10), e78. https://doi.org/10.1093/nar/gks062.

Wong, T. N., Sosnick, T. R., & Pan, T. (2007). Folding of noncoding RNAs during transcription facilitated by pausing-induced nonnative structures. *Proceedings of the National Academy of Sciences of the United States of America, 104*(46), 17995–18000. https://doi.org/10.1073/pnas.0705038104.

Xue, Y., Li, J., Chen, D., Zhao, X., Hong, L., & Liu, Y. (2023). Observation of structural switch in nascent SAM-VI riboswitch during transcription at single-nucleotide and single-molecule resolution. *Nature Communications, 14*(1), 2320. https://doi.org/10.1038/s41467-023-38042-2.

Yadav, R., Widom, J. R., Chauvier, A., & Walter, N. G. (2022). An anionic ligand snap-locks a long-range interaction in a magnesium-folded riboswitch. *Nature Communications, 13*(1), 207. https://doi.org/10.1038/s41467-021-27827-y.

Zhang, J., & Landick, R. (2016). A two-way street: Regulatory interplay between RNA polymerase and nascent RNA structure. *Trends in Biochemical Sciences, 41*(4), 293–310. https://doi.org/10.1016/j.tibs.2015.12.009.

Zhuang, X., Bartley, L. E., Babcock, H. P., Russell, R., Ha, T., Herschlag, D., & Chu, S. (2000). A single-molecule study of RNA catalysis and folding. *Science (New York, N. Y.), 288*(5473), 2048–2051.

Printed and bound by CPI Group (UK) Ltd, Croydon, CR0 4YY
25/09/2026
02222751-0001